William Adolph

The Simplicity of the Creation

New theory of the solar system, thunderstorms, waterspouts, aurora borealis

William Adolph

The Simplicity of the Creation
New theory of the solar system, thunderstorms, waterspouts, aurora borealis

ISBN/EAN: 9783741112348

Manufactured in Europe, USA, Canada, Australia, Japa

Cover: Foto ©berggeist007 / pixelio.de

Manufactured and distributed by brebook publishing software
(www.brebook.com)

William Adolph

The Simplicity of the Creation

THE

SIMPLICITY OF THE CREATION;

OR, THE

Astronomical Monument of the Blessed Virgin,

A

NEW THEORY OF THE SOLAR SYSTEM,

THUNDERSTORMS, WATERSPOUTS,

AURORA BOREALIS, ETC., AND THE TIDES.

DEDICATED TO THE MOTHER OF GOD.

BY

WILLIAM ADOLPH.

SECOND ENLARGED EDITION.

LONDON:

BURNS & LAMBERT, 17, PORTMAN STREET, PORTMAN SQUARE, AND
63, PATERNOSTER ROW.

T. RICHARDSON & SON, DERBY ; 9, CAPEL STREET, DUBLIN ; AND
26, PATERNOSTER ROW.

1864.

DEDICATION.

WHEN works are dedicated to other parties, it is done because they have shown some act of kindness, of friendship, to the author, have taken him by the hand, and nursed him under their protection, or are by him loved and respected.

I cannot boast, that, in the progress and development of my new theory of the solar system, I have had help from friends, or the great, or learned ones of mankind ; but it pleased God so to direct my thoughts and change the proposed end of a business speculation, as to lead to the result of my discovering and propounding "the Simplicity of the Creation."

In the subsequent pages I have shown how God created matter, to which He gave laws and properties ; that He created no forces, because Heal one is THE self-acting force Himself, and that this self-acting force is exercised by Him then only, when He supersedes, abrogates, or suspends those laws which He laid down at the creation ; and HERE I have the gratification of stating, that, with regard to the latter, besides many other favours, the singular proof and privilege has been bestowed upon me and mine, through the intercession of the Blessed Virgin, and by a special act of divine goodness, of seeing one of the laws of nature superseded in my own house. And hence my dedication.

It is now more than six years since my dear wife has been ill. During that time up to the 8th of December, 1855, she had been lame for two years and could neither stand, walk, nor lift up her legs by herself when hanging down from the chair or sofa on which she might be reclining; and it was my daily office during these two years to carry her in my arms from the bed to the sofa and from the sofa to the bed. Often she said, that if God would but give her the use of her legs again, she would willingly suffer.

Thus the grand festival of the proclamation of the Immaculate Conception was approaching, when a novena, or nine days' prayer° for her intention was to be made to the Blessed Virgin. Her confessor, the Rev. Daniel Gilbert, had promised to offer up the Holy Sacrifice for her on that memorable day, and the evening before asked all his penitents to offer up their prayers and holy communions for his intention, which he gave to my dear wife. The 8th of December, 1855, arrived; but for her no signs of improvement. "The festival has done you no good as yet," I said to her in the morning, upon which, full of suffering, she replied: "God is very good; if He chooses to do anything, He can do it still, the day is not yet over. About 6 o'clock in the evening, sitting on the sofa, she observed to me: "Had we not better say some prayers?" to which I replied, that we might as well finish our novena. Upon this I knelt down at the little table by her side and finished the prayers, to which, from pain, she was hardly able to pay any attention. After getting up I saw her pray very fervently by herself, intently looking upon a picture of the Assumption on the wall opposite to her, when all at once she said: "Thank God, I can walk again!" and with these words she got up, walked to the picture (and myself following trem-

---

* This custom, to obtain a particular blessing, dates from the time of the Apostles who, after the ascension of our Lord, were assembled in prayer for nine days, and on the tenth received the Holy Ghost.

blingly behind lest she might fall), gave thanks to God and the Blessed Virgin for the favour bestowed upon her, and went back again to the sofa. Since that time, though continuing to be afflicted with illness more or less trying, she notwithstanding, never lost the use of her legs again.

Whether, in this instance, God restored to her the gift of walking by a natural law, or by a special exercise of His Omnipotence, by a supersession of the natural order of things : it is always HE who accomplished it, He who allowed and allows the law of grace to supersede the law of nature ; and in this instance He evidently 'granted it through the intercession of His Blessed Mother.

May then this token of gratitude to our heavenly Mother be acceptable to her and to Him who created her for us ; may this monument travel far and wide to spread the beauty of her name, and wherever the starry sky is beheld and explored by the human race, may (not overlooking the Star of the Sea,) " generations and generations still call her Blessed," and experience the efficacy of her powerful and never-failing intercession like

THE AUTHOR.

August, 1859.

PREFACE TO FIRST EDITION.

MANY years ago I became acquainted with a foreign architect who possessed a perfect knowledge of ventilation, although personal peculiarities prevented his success in life. By my intercourse with him, his science of ventilation became my own, and I am only sorry that it lies beyond my sphere to carry it into practice.

The science of ventilation requires an intimate knowledge of the nature of the air and its movements; and stepping beyond ventilation I was first led to my new theory of the tides, of sound, thunder, &c., several articles from me on these subjects being kindly allowed space in the columns of the *Builder* of 1851 and 1852. Since then each one of these subjects began more and more to develop itself in my mind; and though I had scarcely any time to devote to scientific pursuits, save now and then, the fruit began to ripen by degrees, until it dropped upon paper in the shape of the " Simplicty of the Creation." It has been a work of love, the brightest ideas often coming across my mind when I ought and thought to have been buried in devotion ; and thus I lay it before the world as the product of an amateur in science, as one who cannot claim it as the result of his own learning or capacious intellect ; and I thefore trust it will be received with indulgence, and judged and criticised with kindness. It is true, I have taken a dogmatic stand, inseparable from

the conviction of a good cause; yet I am far from presuming to say, that my New Theory of the Solar System is absolutely true, because, though simpler than the old one, God may have created all in a still more simple manner, may have arranged all in a more simple way, by more simple laws, than I have assumed to exist. But, if not true even, I still hope that my work will lead others to more successful inquiries, and open a new field to science.

I am aware that many points are wanting, many points but feebly or defectively carried out in my present work; but want of time from many causes, want of intercourse with men of science, their seeming reluctance to give an opinion of my new theory, either *pro* or *con*, whenever I have applied to them in writing, and my consequent inability of rectifying defects or objections, must plead an excuse. Thus not only left to myself, not only not assisted in the progress and development of my New Theory of the Solar System by the great, or the learned ones, of mankind, I have not even been encouraged by those few to whom I proposed to read, or who were condescending enough to read my MS. in some of its stages. The one did not think it of a character likely to be received by the Royal Society, but gave no opinion of the system; the other thought they (scientific men) were satisfied with the old system, as they could calculate everything in astronomy; a third evaded my request to read the MS. though exceedingly kind to me; judging only from the little precursor which I published in 1856, a fourth, in one of the weekly papers, declared it one of the most remarkable productions that could well be imagined, and absolutely dangerous to those possessing but an imperfect acquaintance with science, whilst but few who had attended lectures on mathematics at any college would coincide with me; the fifth had read a great part of my work, admired the vastity of physical knowledge embodied therein, but considered it too difficult to treat of the subjects agitated in my MS. in the present state of science ; thought a work of this kind would not be well received by scientific people, and (yet anticipating its publication) advised me to leave out the religious part of it.

Thus, not only without help, but left in the dark respecting the actual merits or demerits of my theory, I have nevertheless had a few tokens of approbation from others, though non-professors of astronomy, to cheer me on.

Judging merely by the above-mentioned little precursor of my work, the one kindly wrote " that he saw many things in it with which he could agree, and much vigour and originality of thinking ;" the other was satisfied " that my theory was ingenious and truthful, that it would astonish the world indeed, if I succeeded, AS HE THOUGHT I SHOULD, to account for the phenomena of the universe on my hypothesis." Providing for the proofs of some of the facts, a third, most eminent, though not professed, authority, considered my theory SIMPLE AND COMPLETE ; a fourth would be happy to see my theory acknowledged, though it threatened to give the go-by to a mass of painfully-acquired learning ; the fifth, the head of one of the largest colleges in Europe, thought very highly of the MS. (though still imperfect), and so did the party who sent it to him.

All this, the absence of censure, which I consider the highest encomium, my own conviction of the untenability of the old, and the utmost probability, if not certainty, of the truth of my new system, had fixed my mind upon its publication ; but a singular incident confirmed me in my resolution, particularly IN REGARD TO MY INTENDED DEDICATION.

Whilst my MS. was in the hands of the astronomer, who, about three months later, without the slightest indication respecting the merits of my theory, gave me the advise to leave out the dedication and other semi-religious expressions, though on these points entirely unasked for, I was travelling in the Tyrol in the month of March 1858, and through the kind introduction of his Eminence Cardinal Wiseman to his Lordship, John Nepom. Tschiderer, Bishop of Trent, obtained permission to see Maria Mörl, the estatica at Caldern, with whom England was first made acquainted by the pious Earl of Shrewsbury about twenty-five years ago.

When with the Pater Provincial of the Capuchins, her confessor, and three women of the country, I entered her apartment, she was in ecstacy kneeling on her bed. Being recalled into the consciousness of this world, I recommended myself and all mine to her prayers, and her face lit up with a peculiar delight when I told her of the miraculous event which happened to my dear wife. Whilst standing before her conversing with her confessor, I heard the rustling of clothes, and looking up—she was already on her knees—in her previous angelic posture of ecstatic adoration. I had one more request to make of her, and so the good father recalled her again from her communion with God. "Will you, Miss Maria,"—I said to her, "pray for me, that I may succeed in the monument I wish to erect to the Blessed Virgin Mary?" to which (as she is dumb to all except her confessor) she repeatedly nodded a most cheerful consent. Her confessor, however, asked, "What monument?"—to which I replied : "Not one as usual;" and turning to her almost at the same moment she smiled at me so significantly, making at the same time with her right hand the sign of writing, that I understood her at once to know that the monument I meant was a WRITTEN one.

Now, THOUGH THIS IS NO TEST OF THE WORK ITSELF, yet I took and take it for granted, that I am doing right and something pleasing to God, who gave her this supernatural knowledge of a written monument in honour of that sublime and exalted model of womankind, the B.V., whose " Magnificat," although not with

our affection, now even resounds still in the cathedrals of the
Anglican Church, if I publish the whole work as it is, without
any shrinking as to what may be said of it, fondly loving to think
and acknowledge myself one of *Her* seed, one of *Her*—I hope,
accepted—children. The dedication, then, is a thing personal to
myself, an act of gratitude, which will be respected by generous
minds, though they may differ ; but the work itself, with all its
faults or shortcomings, is launched forth into a world full of contention, to be perfected or torn to pieces, and must stand or fall
on its own deserts. I do trust however, that, though crude, the
foundation and outlines of the monument are good, and that by
the truly learned and well-disposed it will by degrees be more
clearly and distinctly developed, and its form and features assume
all that comeliness and beauty of aspect with which master-hands
are able to endow it ; that it will expand over the whole globe,
and, in return, still more unfold the power, wisdom and goodness of
God in His creation, as well as the favours and blessings He has
bestowed upon His children.

Before passing over to my New Solar System I beg to observe,
that, in order to facilitate the comprehension of my new theory,
the reader must picture to himself the whole universe turned over
from its present representation, in such a manner, that the
ecliptic, or zodiac, appears to him vertical, like a hoop, like a ring
of brilliant stars, standing upright and encircling our solar
system, and the planets rolling and revolving underneath between
it and the sun, in progressive order and elevation, instead of
ecliptic and planets lying horizontally in a plane to our solar
centre, according to the theory of Newton.

THE AUTHOR.

August, 1859.

PREFACE TO SECOND EDITION.

SUBSEQUENTLY to the first edition of my work, I published a concise view of the same, adding a most important simplification of the theory of the tides.

I therein also begged " for any remarks, suggestions, or objections," which, however, I never received, save from my friend, J. Graeme, Esq., to whom I owe many thanks for his hearty encouragement in my labours and many valuable suggestions.

Him alone excepted, and he no PROFESSOR, the New Theory—though competent gleaners have been in the rear—was not taken up by the astronomical world, and I am not sorry for it; for, had it been otherwise, my task would have been over, and the monument left unfinished, if not become disfigured. As it is, however, the hand that began was also to complete it, and thus the unity of design has been preserved, the original outlines have grown, diffused, and been filled up, in natural order and form, and carried to their legitimate conclusion.

And thus I trust that my monument will not be the least among its rivals of wood, stone, marble, and canvass, and that one day, by a higher hand, I may see it crowned with approbation.

Religion being so intimately interwoven with the Simplicity of the Creation, it here becomes my duty to make a profession of faith, negative and positive.

In the negative I am not one of those who, in the 19th century, so inconsistently pretend to understand the holy scriptures, or know their primitive explanation, or, professing their ignorance, and then pay another layman like themselves and similarly situated, in order to instruct them in their meaning and preach to them a sermon. In the positive I am one of those who, to the utmost, value and follow THEIR OWN genuine private judgment, as will be seen in the succeeding pages, and THEREFORE truly believe in ONE HOLY CATHOLIC CHURCH FROM THE BEGINNING, and to her DIVINE AUTHORITY submit, freely and unasked for, whatever I have written.

THE AUTHOR.

London, August, 1864.

CONTENTS.

NON NOBIS DOMINE, NON NOBIS,

SED NOMINI TUO DA GLORIAM.

NEW. THEORY OF THE SOLAR SYSTEM.

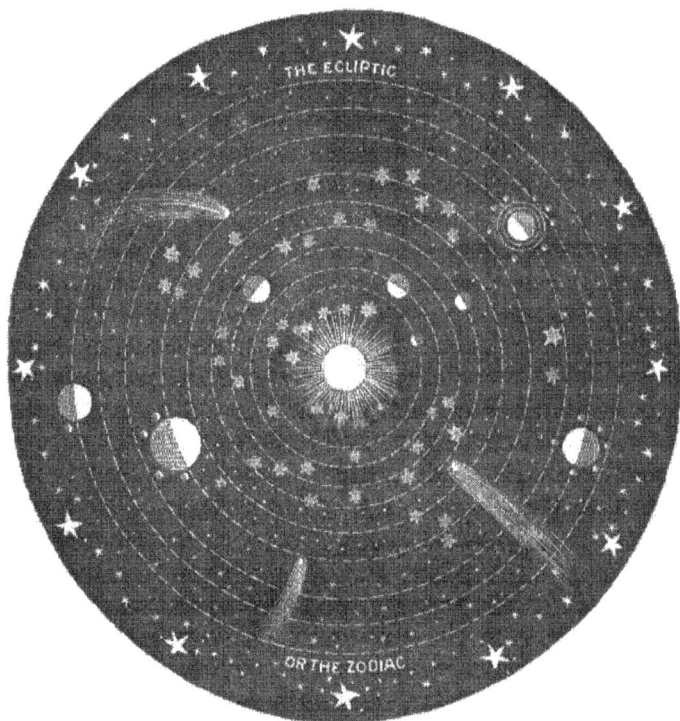

I.

CREATION OF THE WORLD.—NON-EXISTENCE OF SELF-ACTING FORCES.
—SUBSTANCES PRODUCING EFFECTS.—GROWTH OF THE HEAVENLY
BODIES.—ELECTRICITY THE CHIEF ACTING AGENT.—THE SIX DAYS
OF CREATION.—HARMONY WITH THE RELATION OF MOSES.—CREA-
TION OF PLANTS AND LIVING CREATURES.—DARWIN'S ORIGIN OF
SPECIES.—PARADISE.—COMMENCEMENT OF TIME.—LENGTH OF THE
DAYS OF CREATION.

EVERYTHING had a beginning; God alone had no beginning, and
has no end; He exists from eternity to eternity, and He alone
fills all eternity. What He created, He created out of nothing,

B

because nothing existed but Himself. Out of this nothing He created spirits, the spirits of angels and of man ; and of all things created, these spirits are next in degree to God.

He also, in the beginning, created heaven and earth, that same heaven and earth which, in the words of Christ, shall pass away; He laid the foundation of all the heavenly bodies as well as of our own terrestrial abode, that is : He created things visible and invisible, "visibilium omnium et invisibilium," substances ponderable and imponderable, which He moulded according to, and into the instruments of, His divine will, and which He endowed with certain properties to carry out His wonderful designs; and these instruments, thus endowed, produce the effects which we witness in nature. God created material causes, instruments, as I have said, to produce effects ; but He did not create effects, or abstract, self-acting forces, whether these forces be centripetal or centrifugal, attractive or repulsive, moving or reposing. For, there is no ABSTRACT, no SELF-ACTING FORCE, even in animated creation, and though endowed with a soul, there is no self-acting force, properly so called, even in man himself, unless he is CONSCIOUS OF HIMSELF, and IN POSSESSION OF FREE WILL ; a SELF-ACTING FORCE, therefore, cannot be imagined to exist, without pre-supposing these two conditions, and necessarily something like a soul and body.

Thus, without, at THE VERY LEAST, creating substances, instruments, by which forces are exercised, and effects produced, forces themselves are nothing but so many immediate and direct operations of the divine omnipotent will, without the instrumentality of matter; their exercise, independent of matter, unconnected with the means which He created for the production of effects, is a PROPERTY OF HIS OWN, an attribute of His divinity.

Matter and its properties God subjected to certain laws ; matter, properties, and laws, He continues to preserve, sustain, and direct according to the objects of creation; through these, and by their means, He operates; and hence we may say, that, in the ordinary course of nature, God does not act by mere abstract forces, that He does not act in a direct and special manner, but that the operations of nature proceed from material causes which He created; and which He continues to wield and to sustain.

God rested on the seventh day of creation; hence we may conclude that He operated but once materially, that forces went out from Him when He created and endowed everything with certain properties. When thus the clock-work of creation was finished, He only, by a further direct exercise of force, though without. trouble or labour, set the pendulum in motion; and it is the pendulum of preservation and compensation alone which, by His persevering will, He continues to maintain and to guide; but He does not in a direct manner, by an abstract centrifugal power, turn the wheel of the earth on its axis, or by direct impulse waft it round the sun. If the earth rotates on its own axis and revolves round the sun, He causes it by some material means, but not independently of these means, the instruments designed for this end; if flowers bloom, the fields become green, and forests put on the garb of spring: it is not by the exercise of God's power in the abstract, in a direct manner, by which all this is brought about, but by the instruments of electricity, oxygen, carbon, and other substances which He devised and organized for the purpose ; and · thus it will be seen, that the exercise of force or forces in the abstract emanates from His omnipotent arm in a direct manner, unconnected with matter, then only, when He creates, or when He suspends, abrogates, or supersedes, any of those laws or properties, which He impressed upon all matter and bodies at their creation; or, in a few words, when He performs, what is called, a miracle. Hence also, all calculations made upon the assumption of mere forces, must necessarily be false, unless based upon the substances, the instruments from which these forces proceed ; for, upon the attributes of God, upon the exercise of His omnipotent power or will, the force, for instance, with which He is said to have projected the earth into space, no astronomical calculations can be made. The fallacy of calculating the centre of gravity of the planets as at a distance from the centre of the sun ; of calculating the centre of gravity between the earth and the moon to lie at a considerable distance from the earth's centre, and both the earth and the moon revolving round this common centre of gravity ; and of calculating the amount of abstract attraction of the sun and the moon upon the waters of the earth to produce the tides, will hereafter be made apparent.*

When God, in the beginning, had created heaven and earth, substances visible and invisible, ponderable and imponderable, electricity, in its twofold nature, as the binding and repelling agent of creation, had also been called into life, and with its first quivering and quickening motion chaos began to tremble, to sunder, and matter to conglomerate according to the design with which He had cast and distributed the seed of worlds within it. Silently and, as it were, unconsciously, negative electricity principally allied itself to matter in order to form these worlds† of ponderable bodies, which, at the same time, it repelled from each other like so many drops of water falling from the cloud ; positive electricity expanded the more, chiefly to keep company with imponderable substances, and to embrace the nascent systems and constellations, until magnetic worlds—inclosing within their womb the kernel, the germ of their own foundation, the

---

* Dealing with abstract, or self-acting, forces, is less than beating the air or fighting with real shadows ; it is, in fact, fighting with shadows without a substance to come from.

† " Rocks and mountains are nothing but immense masses of crystallised matter . . . . and negative voltaic electricity instantly determines the crystalline arrangements, whilst positive voltaic electricity counteracts it."—*Mackintosh*, p. 101.

FIAT of the Eternal Word—were moulded into shape, the Spirit of God moving over the dark, formless, and silent deep, and infusing fire and life into their veins to be called forth into activity with the dawn of light.

When, in their circular course, the currents of air and water flow from the equator to the poles, and from the poles to amalgamate again at the equator : so the electric elements. Thus, when the separation of negative from positive electricity had in part been accomplished by their respective incorporation and alliance with solid, liquid, and gaseous matter ; when by the electric precipitation of oxygen and hydrogen the medium between gas and solids covered the whole earth in the form of water ; when by this means the nucleus of growing worlds had been defined, and that they began to manifest the polarity of their bodies ; when thus ponderables had been separated from imponderables, so as to unveil the face of creation, and leave untroubled and unimpeded the mutual inclination of positive and negative electricity in a free state, for the entire dissipation of its darkness, God said : " LET THERE BE LIGHT !" and it was so. Accumulated to overflowing at the poles of creation, electricity returned again towards the centre of the universe ; positive and negative electricity coursing back from their respective poles, for the first time met for mutual embrace in chemical recombination ; all became active with life ; a spark, and, in the twinkling of an eye, all was light ; light, the fruit, the product of reunion, suddenly flashed from the centre, embraced the unbounded equator of the universe and shone over and into the depths of the whole creation ; darkness gave way to the surprise of the first grand morning, and in obedience, fear and humility retired behind the enlightened portions of the heavenly bodies ; towards the vast girdle of the universe all was bright, away from it all became dark ; and as the systems began to separate, and each one to move round its own magnetic centre within the electric globe of the world, light and darkness in succession passed over the surface of their bodies during their systemal revolution.

When the immeasurable space of creation was illumed, and light first fell upon the faces of the electric infant worlds, in anxious expectation, as it were, of approaching animation, inclining to their common centre, to the primary, which from its own interior heat also threw a radiant warmth upon them, they first began to pulsate and quicken into life; light had given motion to the element within, and it broke forth in copious exhalation. The separated electricities within and without also endeavoured to unite; they strove, in a manner, for mastery over the as yet unapportioned and undisposed of vapours, gases, and solutions that still hung over the abyss of the deep ; and as their contending activity both contracted and precipitated, decomposed and distended the substances coming under the influence of their recombination, their first, and most likely unresisted effort at amalgamation, augmented to a greater degree still the light shed upon created worlds from the equator of the universe ; and their shaded sides even, during that long systemal revolution, shared in that glimmer of light which is never absent from us in a clear, though starless, night, and which always, more or less, is attendant upon the union of the opposite electricities when free and unaccumulated.

Thus it was that the first evening of the world had been ushered in ; it lasted until systems had performed a revolution round their own respective axes, or the globe of the universe itself having accomplished, perhaps, a revolution round its own centre; and, as the moon during one of her revolutions round the earth has but one night and one day, so this first long evening, with the morning that preceded the second day of creation, were but one day, during which, light and darkness passing once over

every part of the orbs moving with their systems within the revolving creation, the electricity within them continued to attract
the chaotic particles of superincumbent elements, until, as unfolded to us by geology, each stratum in succession was formed
according to the attraction exercised upon the various kinds of
matter;* and thus these orbs were moulded and consolidated
within the deep for the ulterior occupation and benefit of those
beings who were to inhabit them.

No suns having as yet been enkindled in the vault of heaven
to divide the day from the night, but light merely called into
existence to separate the light from the darkness, it was the light
of evening rather, the first AURORA CENTRALIS UNIVERSI, that
first shone upon the airless creation. Hence, day and night are
not named by the sacred penman, but evening and morning only ;
and the DAY of the world having commenced with the evening,
increasing in light with every successive day of creation, he continues to introduce every subsequent day of creation as beginning
with the evening ; and from this, no doubt, it also arises that the
Jews, the once chosen people, still count their days from sunset
to sunset.

On the second day God separated, for good, the contending,
yet to union-inclining electric elements, by the creation of the
firmament, the non-conducting atmosphere of the earth, and

* As regards this deposition of matter, the respective order of which is
left to the learned to find out, Sir David Brewster says, in his " More Worlds
than One,"—" If the Almighty, then, since the creation of man, broke up
the foundations of the deep, and opened the windows of heaven, and thus,
by apparently natural causes, covered the whole earth with an ocean that
rose above the Himalaya and the Andes, why may he not at different periods, or during the whole course of the earth's formation, have deposited
its strata by a rapid precipitation of their elements from the waters in
which they were dissolved or suspended ? Chemical and physical forces of
high activity may have been the agents in such summary operations ; and in
the beautiful process of the electrotype, we have a striking example of the
influence of electricity in effecting a rapid precipitation of metals from
their solutions."

probably also the atmospheres of the other planets, suns and stars of the universe. Thenceforward, heated by the gradually increased electric action of the interior, and more and more warmed by a stronger radiation from the primary, mists arose from, and floated over, the liquid waters of the earth ; and by thus moistening the atmosphere, and neutralizing to a great degree the non-conducting nature of the air, the two electricities continued to amalgamate, and though with increased obstruction, yet at a greater distance from the earth, and within a clearer and more rarefied space ; and in this manner the process of precipitation and decomposition went on as before, the same as it does now under similar conditions.

When on the third day God separated the dry land from the water, the mists rising from the earth took a higher direction, they became less with the diminished extent of the surface of the waters ; with them, positive electricity retired still further from the solid body of our globe, whilst the negative element became more strictly confined within the ponderable spheres with which it had associated ; and with this closer confinement the heat of the exterior crust of the earth gradually decreased in proportion, whilst interior heat and activity augmented.

By the partial removal of superabundant moisture, as well as by the wider separation by the firmament between the waters below, and the vapours, mineral solutions, oxygen and hydrogen, and perhaps other gases above which as yet had not been, or were not to be, incorporated with the solid or liquid masses of the earth, the air became clarified, and began more and more to approach the state in which we enjoy it at present. Deprived, therefore, to a great degree of the moist conductor by which it had been saturated before, still more resistance had to be overcome in the union of the two electricities. This, however, was counteracted by the warmth imparted by the earth to the atmos-

phere; and recombining at a still more increased distance from
the earth, and within a still more rarefied space than before, we
may even suppose that the third day of creation was lighter than
the first and the second. Land, still soft after its emersion from
the deep, and fertilized by depositions from the atmosphere, the
absence of sunshine made good by an increase of light, the want
of solar fire made good by the highest degree of lightless radiation of heat from the opaque body of the sun, and further compensated by interior warmth and activity, and the absence of rain
amply and more universally supplied by heavy dews, the earth
became an electricity-exhaling and inhaling hothouse from pole to
pole, to receive. the Word of God, and to bring forth in gigantic
forms the wonders of the vegetable kingdom.

On the fourth day of creation the mists of the earth, of the
planets, and suns had still further retired above their firmaments;
positive electricity likewise withdrew into still higher regions, and
began more and more to accumulate at its increased distance from
the attracting twin confined within the solid masses below; at the
bidding of God, their tendency to unite arriving at the point of
culmination, at the highest and purest pitch of concentration; and
the orbs of creation having arrived at manhood, as it were, and
the planets with their envelopes ripened, peeled off from, to move
freely within, the now insurpassably dilated and transparent
atmosphere of their system, the electric envelopes of the suns
broke through the charm that had bound them in slumbering
action, they burst into fire, and from SIMPLE HEAVENLY BODIES
became LIGHTS and STARS, to shine and sparkle in the expanse of
the heavens, and to behold and hold fast the inferior bodies that
had nestled and grown up within their invisible bosoms. Negative electricity having drawn around itself the firm nucleus and
the still accumulating crusts of the heavenly bodies; having robed
its habitation with the festive verdure of youthful existence, the

virgin soil uncursed as yet through the fall of man : its heart expanded with accelerated vigour, and, like the newly-born infant beginning with LIVING LIFE, so to say, its mysteriously-nourished pulsation, it bounded in the light and warmth of its glorified kindred; and it was then, in the exuberance of joy, that, having from the commencement partaken of the rotation of the ball of the universe, the matured worlds of creation first began to rock to and fro in the cradle of heaven ; that, prepared for it by the separation of the dry land from the water, they began their individual motions and courses, that the sun commenced to rotate round his own axis once in twenty-five days, and that our days and nights began to reckon twenty-four hours. For, until then there had been but evening and morning, one night and one day for each period, and each succeeding period shorter than the preceding one, and on earth a uniform climate, with the most luxuriant vegetation, because no lights were made to divide the day and the night, to be " for signs, and for seasons, and for days and years," such as we count them, and the light and darkness of the first (to us long) days of creation gave way to, and were absorbed by, the concentrated light and darkness of solar systems without number, and comparative repose of the heavenly bodies gave way to revolving and wandering motion.

With God there is neither space nor time, and as the present is at this moment present to us, so the whole universe as it is, was, and will be, always is, was, and will be present to Him.

With Him also there is nothing either great or small ; and as with the increasing power of the microscope, the photographic point, scarcely perceptible to the unassisted human eye,—unfolds to us, step by step, more and more, the minutest lineaments of a perfect picture, really present in that little dot : so in the seed of plants first cast by the Almighty upon the earth, He saw all their numerous and variegated offspring covering our globe ; so in each respective individual little seed, His Omnipresent eye beheld the blooming flower in all its beauty and fragrance, the ear filled with corn, the lofty cedar and the towering mammoth tree, all in their minutest detail and organisation ; the earth, the sun, the whole universe, each and all and all and each were and are to Him but a small photographic point.

Hence it is clear, that God, in one instant, might have done all that was done in the first four days of creation, might have called into existence, in the twinkling of an eye, everything as it is at this moment ; but, as if to confound the future unbeliever, to adapt the work of creation to our understanding, and leave no room for cavilling at the divinely-inspired account given by Moses, and as such perhaps transmitted to him from Adam himself : He was pleased to give time to the development, the realisation, of His designs, the same as those of man, however conceived, require time to carry out, whether flowing from the mind through pen, pencil or brush, upon paper and canvas, or, based upon massive stones, to rise up into the magnificent temple of the living God, with its pillars and arches and the vaulted roof, its ornaments and decorations in most wonderful variety—all copied from the dome of the universe, from the adorned temple of our own globe,—and pervaded by the Spirit of Him by whom all things were made, " per quem omnia facta sunt," and who in return is reflected and glorified in the productions which He inspired, but not in any others.　God's temple of nature is not naked and cold, nor the temples, raised to Him by man in the spirit of truth ; for, the one must necessarily be the reflex of the other.　Nakedness has no share in the praise and glorification of infinite Majesty, and naked ideas are not inspired or nourished by the Spirit of Truth ; there is no poetry therein, and God and truth are the essence of poesie in its elevated fact, form and expression.

Thus, God willed it, that His idea, His eternal conception of the universe as it is, should take form and receive expression by degrees; and as a thousand years with Him are but as one day, so He chose but one day for each of the six divisions in which He created heaven and earth and all that exists.

Sun, planets, moons, all the heavenly bodies and all things created enclose within their essence the thought of God; beyond this, man alone was designated for the dignity of being made according to His own likeness and to receive the impression of His image.

When the thoughts of the Creator were being clothed in the forms of heavenly bodies, the whole chaotic ball of the universe was as yet silently floating through eternity, until by degrees, with the dawn of light, it began its rotating course. Suns, almost stationary within the centres of their systems, partook of this motion, until, more and more repelled by each other as they became more consistent, they slowly, with their systems, began their own revolving motions. Infantlike, unborn, as it were, their first complete revolution round their own axis may have taken ages of our own time, whilst planets and satellites, floating merely within their sphere, and carried round in their then most likely narrow orbits, may have taken more of our time still until the first journey round their primaries was accomplished.

Thus, from the creation of light, but one single day elapsed, long enough, however, to our poor comprehension, for the most protracted development of creation during and for the first day of creation, as related by Moses, and sufficient for all the requirements of scientific doubters and unbelievers; and still, in all truth, it was but one day, one single day, though such as no sinner would like to live and to suffer.

With the progress of growing worlds and their increasing life and activity, their motions and courses became accelerated; and when the second day of creation, the second revolution had been accomplished, a shorter space of our time had elapsed than was required for the first day, but sufficiently long to satisfy the most scrupulous of the scrupulous, for the production of all the phenomena which may be ascribed to have taken place on the second day, the second literal day of creation.

On the third day the motions of the heavenly bodies became swifter, the time of their revolution shorter ; yet it was but one day, literally one day, long enough to prepare the earth for the reception of the seeds of vegetation, for the thoughts of God to germ, to root and bring forth the grass and green herbs and bushes and the most colossal trees, each one bearing within itself seed of its own kind for insuring its propagation.

When these, to us, long days were over, and everything was ready for receiving the living creation, but a short moment was required,—having been wound up as it were,—to set the heavenly bodies going in full motion and speed, and on the fourth day of creation to give them their present rotation and progression.

The whole universe having been lighted up and illuminated in one moment by countless hosts of suns and stars, and shining planets and satellites rising into higher altitudes in the now more dilated spheres of their primaries,—all being brought into visible communion from one end of creation to the other, and individually and collectively proclaiming the majesty of Him who clothed them in light, who made THEM and the whole universe in radiant splendour the befitting garment of His Divinity, the palpable evidence of His existence,—God created all living creatures, and last of all man, in the prime of life introducing him into the fullness of this overwhelmingly glorious world, and the especially beautiful habitation of the earth.

God was pleased to let the earth and the heavenly bodies grow and be organised by degrees ; to let the vegetable world, as we see it now, spring up from the earth and thrive, and, arriving at maturity, every plant to bear within itself seed of its own kind to provide for its continued reproduction aud diffusion (Gen. i. 12, ii. 9); of all this we have the evidence daily before our eyes; but we have no proof that mustard seed brought forth an oak, and the acorn a mammoth tree, we have not seen the violet change into a rose, nor the willow into a cedar.*

In the animal world we see that kind begets kind, though kind may multiply and vary into species of the same kind, according to climate and other conditions; we also behold the caterpillar transformed into a butterfly; but we have it neither from reason, tradition, scripture, or any record whatever among any of the tribes of the earth, nor from any observed fact or analogy in nature, that a goose was brought forth from the egg of a duck, or that a swan slipped out from the egg of a goose, or that one of these birds was improved or changed into the other, that the caterpillar was changed into a snail, or the butterfly into a bird; that asses were turned into horses, or horses into asses, or that mules, or other hybrid productions, are capable of generation; these hybrids, neither horse nor ass, neither hot nor cold,—as if constantly to vindicate the narration of Moses,—God WILL NOT have them fructify, because He did not create them, could not therefore and did not say to them: "increase and multiply!" †

---

* "As in hybridisation, so with grafting, the capacity is limited by systematic affinity, for no one has been able to graft trees together, belonging to quite distinct families; and on the other hand, closely allied species can usually, but not invariably, be grafted with ease."—*Darwin on the origin of Species*, p. 283.

† It is distressing to see the wisdom of God's will questioned by the learned aberration of Darwin in his work on the "*Origin of Species*." Page 282 he says: "Why, it may even be asked, has the production of hybrids been permitted? To grant to specie the special power of producing hybrids, and then to stop their future propagation by different degrees of sterility, not strictly related to the family of the first union between their parents, seems to be a strange arrangement."

And yet this arrangement of God, found fault with by the writer as "strange," is exactly one of the most convincing proofs of the inspiration of Moses, a test of the fact, that one plant or animal did not proceed from the other, but that "God created plants and animals of different kinds, each one bearing seed and offspring of its own kind," and not bringing forth seed or offspring of another kind.

There is no trace of remembrance or tradition anywhere, even amongst the lowest tribes of mankind, that their founders or fore-fathers, that the parents of the human race, had from monkeys progressed into the form and mind of man, had been monkeys themselves, or been nursed by monkeys, though, alas! we have but too much evidence, how man, so nobly created and endowed by a good God, degenerates below the monkey, even below the beasts of the field. And yet, for all that, we cannot come to the conclusion, that the animal creation, beginning with the monkey tribe, has descended from man, notwithstanding the fact of his awful degradation in body and mind, in some instances even living like a brute among the brutes; for, in his most abandoned state, man still retains some vestige of the type of special creation, a remnant of the stamp and impression of a higher order of beings, of a creation separate and independent of the animal world (Gen. i. 25, 27); and as for the reasons stated already, and from the very helplessness of infancy, man neither proceeded from animal, no more than one kind of animal changes into, or is brought forth by, the other, either by crossing, or hybridisation, and as neither the one nor the other grows or grew out of the earth: so we are irresistibly forced to the conclusion, agreeably to the account of holy writ, that the seed of man was to proceed from man,—and of animal from animal,—in the perfect state and development of his manhood, and that, consequently, God was obliged to create him, and all other living creatures, in the full spring tide of their existence; and for this purpose, humanly speaking, He did not want days like the first ones of creation, not even a day of twenty-four hours was required,—a single moment was sufficient for His omnipotence.

Hence, the creation of the animal world, and lastly of man, on the fifth and sixth days of creation, in days like our own, is con-sistent with reason, with the order of divine providence, and with the revealed account of Moses.

Nevertheless, a still more probable account may perhaps be given of the last three days of creation.

When, on the fourth day, the sun for the first time threw his golden rays upon the earth, his own rotation began to quicken ; and that of the earth, and its progression round the sun, increased in proportion to the stimulus given to the electric blood within its bowels, and the body began slightly to be polarized, and to incline and tend towards its present oscillation for the ultimate production of the seasons. Plants, at the same time, which hitherto, not without light but without sunshine, had grown and flourished as in a hothouse, began to change their succulent and pulpy character, trees to harden into greater consistency, and little by little show in their growth the gradual commencement of seasons, grass and herbs to become more solid and substantial, fit as food for a living creation, and flowers to assume the magnificent display of colors and designs, to delight and fill with admiration the future lord of the earth, and to remind him of the Lord of All who bade them come forth from the ground.

The earth once more had made a revolution round its own axis, and finished another course round the sun whilst all this was being accomplished ; but its course and rotation was much shorter than the previous one; the great day of the creation of man drew nearer.

On the fifth day God created the great whales, and every living and moving creature which the waters brought forth, and every fowl flying over the earth ; and whilst again with still more accelerated speed, rotating but once on it own axis, the earth ran its course round the sun, becoming more and more polarized and inclined to oscillation, the fifth day of creation, shorter still than the preceding one, was still long enough for the creeping creature and the winged fowl to increase and multiply, and to present a wonderful and animated spectacle to the intended living image of Him who made them.

On the sixth day of creation, the earth, once void and empty had become a paradise full of joyous life, and nothing was wanted but the tenant of this terrestrial heaven, prepared for him by God Himself.

But, still more to afford happiness to the future ruler of the earth, his Creator wished to surround him with another living creation, more noble, as it were, to have dominion over, and with creatures likewise becoming attached and familiarized to him, to be incorporated, so to say, into the society of man.

Accordingly, God made the beasts of the earth and the cattle, and after having done so, He made man to his own image and likeness, brought to him all the beasts of the earth, and the cattle, and all the fowls of the air, * that he would call them, and Adam called them all by their names.

To Adam, however, the happiness of paradise, of the loveliest spot of the whole earth, was still incomplete; therefore, God

---

* "I believe that animals have descended from at most only four or five progenitors, and plants from an equal or lesser number." *Darwin on the Origin of Species, p. 518.*

That so few KINDS of plants and animals should have been created does not seem likely, neither from the words of Scripture I have quoted, from the different kinds of plants and animals which we see, nor from the words of the writer himself when at page 27 he says: "Now, it is difficult, perhaps IMPOSSIBLE, to bring forward ONE case of the hybrid offspring of two animals CLEARLY DISTINCT, being themselves perfectly fertile."

Now, every one of my readers will clearly distinguish, looking about at home merely, RATHER MORE than four or five CLEARLY DISTINCT animals, and of plants the same, neither of them—to use the the the writer's own language—the offspring of TWO ANIMALS CLEARLY DISTINCT, or of trees, the product of "quite distinct families." (See page 14.)

But this fact shows all the more the greatness, power, and wisdom of God, that, as from the seed of the Church, all the various orders and institutions sprang forth to become, in the most variegated form and manner, the ornament of the kingdom of God upon earth: so, perhaps, from the created suns, from the seed of worlds, there sprang forth the magnificent planets, and the twinkling heavenly bodies, ornamenting the universal temple of God, and harmoniously circling around the centre by whom they are lighted and nourished; so from the plants and animals originally created in the unity of their kind or genus, there sprang up the most surprising and wonderful variety of species, an ornament of God's earth which they enliven, each one adapting itself to the state and condition of man, and the place and climate which he inhabits.

c

threw him into a profound sleep, the ecstatic sleep of primitive
innocence, and from one of his ribs made and gave to him Eve,
his wife, the transformed and lovely object of his own substance.

What, but vice, the spirit of evil, and the doctrine of the
serpent, of Satan in disguise, could ever mar the bliss, or sever the
bond of such a union of husband and wife?

As yet their happiness was complete; for them, time did not
as yet exist, nor had the sun as yet set upon Eden and their
paradisial existence; there was no night weighing upon their
minds, nor had darkness overtaken and closed their eyes; they
still saw God, communed with Him, and beheld creation in all its
beauty and harmony; and if they slept, they slept or reposed in
the cool of the shade, whilst their souls were awake. But when
the serpent had done its work, when by the first act of disobedience
to the will of God the first drop of bitterness had been put into
the chalice of Gethsemani, their minds became overcast, THE
SUN OF PARADISE WAS SETTING TO THEM FOR THE FIRST TIME;
and when cast out of their special abode of happiness, it was then
that the law of nature became the handmaid of the law of grace
and justice, that their life became day and night, that their existence
became one of joy and sorrow, that night set in upon them to give
them rest after a day of labour; that then only Adam knew his
wife, (Gen. iv. 1.), that the generation of man began with the
beginning of time, with the present rotation of the sun and solar
system, and the present rotation and course of the earth; that
with the beginning of time the return of the seasons was established;
that the whole harmony of the heavens was organized as we see
it now, so that "all the days of the earth," (Gen. viii, 22), until
its final consummation by fire, " seed time and harvest, cold and hot,
summer and winter, night and day, " were not to cease.

In treating of the slow, but gradual, development of the earth,
the transition from a primitive state of comparative rest of the

heavenly bodies into that of a floating progression, and ultimately rotating progressive motion, I do not mean to insist that one rotation of the earth round its own axis coincided with one of its revolutions round the sun, that the length of one day coincided with the length of one year, whatever that length respectively may have been. For, it is possible, that on the first day of creation the earth floated twice, or oftener, round the sun, before it had once turned on its own axis and presented every portion of its surface to the then unenkindled, but still gradually warming, central body; or it may have turned once round its axis, before its first course around him was accomplished. And so on, on the sixth day of creation, the earth may have gone beyond one complete course round the sun before it had once turned on its own axis; or it may have made one rotation on its own axis before it had gone the twelfth part of its orbit. The DAYS of creation, according to my theory, depending on the time of rotation of the earth round its own axis, and not on its course round the sun, the former case would give the length of more than one of our present years, and the latter but one month of our time to the length of the sixth day of creation, to the work done on that day, to the beginning and end of a paradise upon earth.

When the seemingly insignificant little match first enkindles the fire under the boiler, by degrees changes the fuel into one burning mass, slowly warms and finally heats and expands the water into steam, and step by step sets the locomotive in motion until it acquires that degree of velocity and speed which the constructor had intended it to attain : so in the same manner, from the dawn of creation, the germ of the celestial bodies and of planetary life began to develop, until the whole of our solar system, with all the constellations of the universe, attained that regularity and harmony of motion which the Divine Constructor had ordained.

If, then, the revolution of our earth on its own axis, and its course round the sun, was arrived at in unerring geometrical progression, it becomes an interesting speculation to find out the steady advance of each day of creation, the length of our time which it took for each day, until our earth acquired its present velocity of rotation, until it made one complete rotation on its own axis in twenty-four hours.

According to my theory, as lastly propounded, our present rapidity of rotation was attained on the seventh day of creation. If then we multiply 24 hours by 24 hours, we obtain 576 hours, or 24 of our days for the rotation of the earth round its axis on the sixth day of creation, being the duration of the paradisial state of existence. If again we multiply 576 hours by 576 hours, we find that the time of the previous rotation of the earth, the length of the fifth day of creation, lasted 331,775 hours, or 13,800 days, or about 38 years of our present time, and so on.

The progressive calculation back to the first day of creation, and then only to the " beginning," when " God created heaven and earth," and the harmonizing calculations concerning other planets, systems and spheres, if not of the whole universe, I must leave to the reader, who, however, may chose, to calculate and speculate upon what interpretation and extent of time or period he pleases; but I trust I have shown, almost to superfluity, that the days of creation spoken of by the inspired pontif of the Jews, may, and no doubt have been, literally, as well as figuratively, days, on each one of which the light, first separated from the darkness, and later the light of the sun, made to separate the day from the night, rose and set upon the earth but once during the length of their respective duration.

Thus we have seen, how by degrees everything grew into existence, how everything was prepared for man, the most exalted object of creation, and how the organization of the body of the earth itself, and the solar system, and the whole universe, received the finishing touch, when the light, previously diffused, was absorbed, when the transparent veil, spread over and mildly shining upon awakening infant orbs, was drawn away and superceded by the radiating light of millions of suns bursting out upon the whole of creation; but all would fade, wandering worlds would stand still again, if the resplendent bodies were extinguished with which the unfathomable firmament of heaven was thus ornamented and sprinkled all over; with a dissolution of the electric element, CREATION ITSELF WOULD AGAIN RETURN INTO CHAOS.

II.

ELECTRICITY, THE FIRST ELEMENT AND BASIS OF CREATION—PRIVILEGE OF MAN TO EXPLORE THE DEPTHS AND MYSTERIES OF CREATION. —SIMPLICITY OF THEORY ABOVE CALCULATION.—THE UNIVERSE GOVERNED BY SYSTEMAL REPULSION AND BY SOLAR AND PLANETARY ATTRACTION.—CENTRES OF SYSTEMS.—ENVELOPES OF THE HEAVENLY BODIES; THE OUTER ONE OF THE EARTH, ONE OF PURE HYDROGEN.—COMETS AFFECTED BY THE SOLAR ATMOSPHERE.— LEXELL'S AND OTHER COMETS.—FALLACIES CONNECTED WITH THE OLD THEORY OF COMETS.—HYDROGEN THE REFLECTOR OF LIGHT.

THE universe of creation, of which God Himself is the centre, whilst, at the same time, it reposes within Him, this universe of created things, limited within His infinite Godhead, He filled, as we have assumed, and as we shall see more particularly hereafter, with that subtle, imponderable, and elastic matter, which, as confirmed by all human researches and observations, of all substances occupies the first rank in the gradation of created elements— electricity.

As all throughout nature everything seems to consist of two kindred elements, so electricity exists in a twofold kind; it forms two genders, and these are known and distinguished as positive and negative, the latter no doubt proceeding from the former, as woman was taken from man; and these two mysterious twin elements are, and still more will be found to be, the very life and soul, the ever-active agents, the essence, of creation; they are the most expansive and the most contractible, the most subtle, and most powerful, and they, more or less, pervade or combine, with every substance; IT IS FROM THEM THAT PROCEED LIGHT AND HEAT, EXPANSION AND CONTRACTION, ATTRACTION AND REPULSION, MOTION.

When suns had been lighted, positive electricity alone appears to have taken possession of, and filled the entire immeasurable expanse of the universe, as with a globe of cohesive and elastic fluid, distended by its own expansive nature, and by all the heavenly systems poised within it, and yet tending to collapse towards the centres of attraction distributed within its unfathomable space.

Beyond this universal globe, this universal sphere of positive electricity, beyond which there is eternal darkness, the darkness separated from the light, there God alone is reigning in endless being and presence, holding creation upon His hand like a precious pearl, and turning it round without effort; and within this moving creation travelling through eternity, all the heavenly systems are harmoniously revolving, most likely round a centre of their own, perhaps round our own solar system.

To each of these systems, repelling each other like so many electrified pith-balls, the omnipotent Creator gave a centre of its own, a sun, as a centre of light and warmth, and as the point of attraction.   Thus, our own sun is the centre, the electro-magnetic point of concentration of our solar system, as the earth is the centre of gravitation of everything that moves within its sphere; and in the same manner as He gave atmospheres of their own to our earth and to all the planets, so He gave an atmosphere to the sun, within which all the planets of our systems were to REVOLVE in regular order and orbits, the same as the moon FLOATS within our own sphere.

True, God can do anything; He can, according to the old theory, fill creation with vacant spaces, and move through them, in arbitrary positions, the heavenly bodies, by the most arbitrary laws, in any manner He pleases; the creation, however, alone was arbitrary, but the laws which govern it are simple, and only appear arbitrary as long as we do not know them. To find out these laws, to wield power over unconscious matter, and make it subject to his wants and will, is one of the noble tasks left to man; and as his success therein evidences the high position in which God has placed him upon earth, evidences that brilliant spark of conscious intelligence which is not of his own making, but which the Supreme and Eternal Intelligence infused into him from His own nature; so it glorifies and evinces that wisdom and intelligence of the Creator and Master of all matter, in man himself, and in every thing created. And, as if to mock the intellect, naturally limited by vice and pride, of the professed descendants of monkeys in human form; as if to confound, shame, if possible, and increase still more the culpable responsibility of the infidel, materialist, or pantheistical philosopher, who ignores and disowns the name of God, of an Almighty Being, who worships his own imaginary wisdom, the dust he is made from, or some abstract absurdity, as the origin of things of which he can give no account:—the more man penetrates into the mysteries and the mysterious simplicity of nature and of the heavens, the more enlarged becomes the sphere of observation; it grows with, and ever outgrows, the progress of science; for, " He has de-

livered the world to their consideration, so that man cannot find
out the work which God hath made from the beginning to the
end;" Ecclesiastes iii. 2.; and thus with every new discovery,
with every deeper plunge into the abyss of the ocean of the stars,
the goal of human researches and investigations is enlarged and
further removed, in order that the CHRISTIAN philosopher may
never cease more and more to magnify God in His, to us, infinite
creation, called into existence for His own glorification, and the
consequent happiness here, and eternal bliss hereafter, of those
who DO glorify Him in the simplicity and yet endless variety of
creation, and in the production of works and effects so stupendous
by means so simple.

When these means, these arrangements of God are truly
discovered, all becomes plain and intelligible to ordinary comprehension ; they are sure not to be true laws if they present
to us a theory, intricate and anomalous, and indigestible to the
mass of mankind.

When Thomas Young, and Fresnel, first shook the foundation
of Newton's now discarded theory of light, Fresnel, in one of his
eminent works said : "In choosing a system, regard must be had
only to the SIMPLICITY OF AN HYPOTHESIS ; that of CALCULATION
can have no weight in testing its admissibility. Nature did not
trouble itself about the difficulties of analysis ; it only avoided
complicated means for the attainment of an object. It seems to
have made up its mind to accomplish much by little ; and this is a
principle, which the physical sciences in their progress incessantly
confirm by new proofs."

If then my theory be true as it is simple, I shall feel happy to
have been the humble instrument in promoting the knowledge of
the simplicity of the laws which govern creation, to have propounded, and perhaps satisfactorily proved, a "UNIVERSAL SYSTEMAL
REPULSION ;" and investing the sun with an atmosphere of his own,

to having reduced the law of " universal gravitation " into that of " solar and planetary gravitation," by which the planets gravitate to the CENTRE OF THE SUN, and the moons gravitate to the CENTRES of their respective planets, that is: all the planets, however numerous, are attracted by the centre of the sun, in whose sphere they progressively REVOLVE, and the moons are attracted by the centres of the planets, in whose sphere they FLOAT onward.

I am borne out in this new theory by Mr. Montagu Lyon Philipps, who says, in his " *Worlds beyond the Earth*:" "It is difficult to conceive a system, like that of the universe, to be stable without some agent to counteract gravitation. The fact of the fixed stars keeping in their relative positions so unchanged, would be inexplicable upon the hypothesis of gravitation alone." In another place he says: " Gravitation alone will not account for the stability of such (planetary) systems. The one sun must be repulsive of the other, as we have explained. Each would maintain its own SYSTEM OF PLANETS unaffected by the other."

Nearly half the ponderable matter of our globe consists of oxygen, the vital part of the air ; and this gas, on being freed from union with solid bodies, expands into a space two thousand times greater than that which it previously occupied : what space, then, would it not fill if it were released from the combination with the greater part of the crust of the earth !

By some, the sun is considered as an ever-burning body; by some, a vast volcano in action ; by some, as a body, black and cold, like our own earth, but enveloped by a luminous cloud. If we consider him as a globe of fire, what may not have been the liberation of matter when God created him, and bade him give light and life to a world! We may suppose a crust to be dissolved, whose liberated gases rose up from the surface, until they filled the space assigned to them, and in which they are kept expanded by the same fiery agency, gravitating still to their centre, and, like our own atmosphere, becoming dense as they approach, and more and more rarefied as they recede from it. If we consider the sun and fixed stars as encircled by a luminous cloud, formed, as I believe, by a concentration of that positive electricity which I assume to pervade the whole universe, to transgress the limits of every solar system, to know of no barrier save the embrace of the omnipotent God: that concentration would, on my theory, manifest itself as a permanent state of sheet lightning, a constant overspreading aurora-borealis, produced by laws similar to those which operate on our own planet, as we shall have occasion to see more clearly hereafter; and this cloud of electric fire, in the same manner as if the body of the sun itself were burning, would tend to keep his atmosphere in that expanded and attenuated state in which, beyond doubt, it exists, the same as the heat of the sun himself contributes to the expansion and extension of our own atmosphere, particularly round the equator.

Of the pressure of this inconceivably rarefied atmosphere of the sun, of an atmosphere of gases gradating in wonderful affinity, in increased attenuation and elasticity from the surface of his body to the confines of his sphere, and, no doubt, capable also of transmitting light with an infinitely greater velocity than our own atmosphere, we may conceive but a faint idea, by bearing in mind the extraordinary pressure of a little air in a bladder expanding within an exhausted receiver, so that the atmosphere of the sun forms a resisting medium, modified to the highest degree by its extreme elastic condition.

Within this atmosphere it pleased God to station the planets, invested with correspondingly elastic and resisting envelopes, to facilitate their passage through seemingly vacant spaces. And for this purpose He not only endowed them with one atmosphere of matter, but, like the sun, with a succession of envelopes of various descriptions, with ocean above ocean, as I shall prove later on, each envelope, ring, or ocean, becoming attenuated and elastic in proportion to its approaching the extremity of our sphere, and forming, as it were, an elastic web or network that keeps the whole together.

Thus, our own atmosphere of air, ending at a calculated distance of about forty-five miles above us, would not have been sufficient, and on that account cannot terminate the extent of our sphere ; for it was not only neccessary to swell the bulk of the earth, as well as that of all the planets, by a more fluid, rarefied, and elastic body than the air, to keep it afloat in the atmosphere of the sun, and to lessen the effect on the solid part of it in which we live, which a passing planet or comet might otherwise exercise, but also to encircle them by a substance corresponding to the rapidity of their revolution and progression, as well as to the substance they were to move in, as it is not, and cannot be true, that the heavenly bodies should move through empty regions, without any resisting medium, without anything to support them.

That they move within a resisting medium is proved by comets, whose transparent and magnifying envelopes or atmospheres begin to form into tails, and to elongate or shorten, according to the velocity of their motion and the density of the fluid in which they move ; and this effect could not take place if they were moving in a vacuum, whatever their envelopes or atmospheres may be composed of, whatever the nature of their cohesion.

A comet approaching the sun, when seen at any considerable distance from that body, has little or no tail. The tail begins to appear when the body draws near the sun, moving with prodigious velocity ; with the proximity and velocity of the body the length of the tail increases, but does not acquire its greatest extent till after the perihelion passage ; it is then by degrees drawn in again, until, in the remote regions of the system, moving slower, the comet reassumes its spherical form, and even expands, like that of Encke and the one of 1835, as it ascends into the more and more dilated atmosphere of the sun.

The velocity of the comet's motion through a resisting medium changes its hair, or nebulous envelope, into a tail, which becomes expanded or contracted in bulk according to the distance from the sun and the consequent immersion into a more dense or more attenuated atmosphere; and it is this expansion and contraction, with the elongation and shortening of the tail, that causes the eccentricity of the comet's orbit. Or rather, destined, perhaps, to carry light and warmth, in a latent state, into the remotest parts of our solar system, to planets, but sparingly supplied by the rays of the sun, they probably pass in cycles round the sun, similar to the motion of the moon round the earth; or, ascending balloon-like from perihelion until they find their level, they then only commence and pursue their real journey within the solar sphere, at distances and in curves unknown. Having given off its superabundance of light and warmth, the reduced or exhausted body of the comet is attracted by one of the polar regions of the sun, or by some place or other laid bare by the immense openings of his veil of fire; and gravitating with ever increasing velocity to the centre of its attraction, it is replenished with a fresh store of vitality, and its exposed nucleus endowed with renewed polarization; and thus charged to overflowing with the bounty of the Almighty, it is repelled back with prodigious power on its distant errand, to reappear at some uncalculated future period, a wonder of creation, in another part of the heavens, attracted by the other pole, or some other spot of the sun, or its electric helix.

One of the greatest visible proofs of a resisting solar atmosphere was established by the appearance of Donati's comet in 1858. On the third of September the discoverer observed already that the nucleus of the comet had assumed an elliptic form, as shown in the drawing; and this could not but be owing to the resisting medium through which it was passing.

The comet, moreover, contrary to what might have been expected from the expanding influence of the sun, did not only not increase, but actually decreased in diameter whilst getting four times nearer to the sun than it was on the second of June, on the day of its discovery; and more remarkable still, a manifest densation and shrinking of the nucleus took place. By a careful measurement in 1838 of Encke's comet, having no tail, it was found that

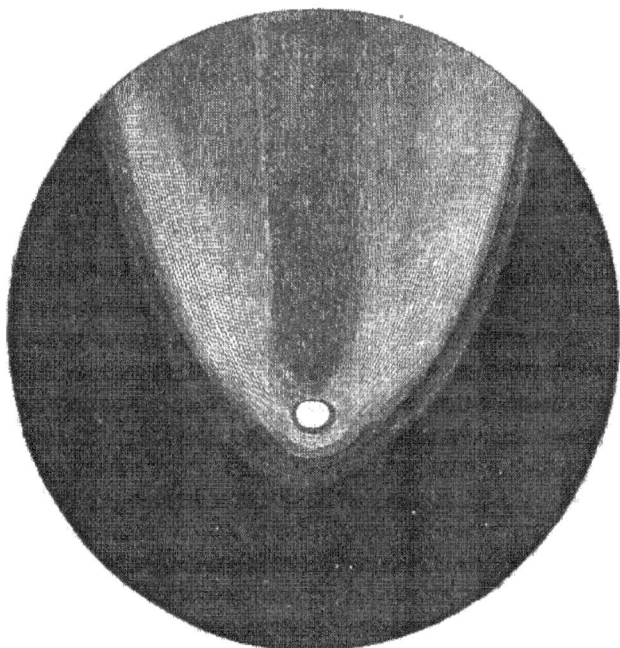

its nebulous envelope contracted more and more the nearer it approached the sun in its elliptical course.

This telescopic comet of Encke, moreover, seen about the end of November, 1861, is said to describe a spiral round the sun, which brings it nearer to him with every revolution, and must in the end fall into the same, unless, like the other comets, it has its ultimate perihelion, though distant in time, at which it is repelled again into the higher regions of the solar atmosphere.

In the mean-while,—to quote a writer in *Gaglinani's Messenger*, taken from the *Times* of the 19th December, 1861.—"Encke's comet is perhaps the most important of celestial bodies in a scientific point of view, since it deals the theory of universal gravitation a heavy blow . . . . . . it is attracted according to a law which is not recognised or followed by any other celestial body; and astronomers are sorely at a loss to account for this serious anomaly, which has induced M. Encke himself to embrace the highly improbable theory of a resisting medium, often discussed before."

Though the writer speaks of the "highly improbable theory of a resisting medium," yet I would ask: what else but the pressure, the resistance of the solar atmosphere, and solar, instead of universal attraction and gravitation, could have produced and could produce the effects above spoken of, if true at all? Common sense would point to a resisting medium rather than to vacant space!

I believe, however, that Encke's comet may simply exhibit to to us a "precession of the perihelion," without at all coming closer to the sun at every revolution; and this I consider far more natural than its spiral approach to the centre of our system as the result of resisting matter.

The sun, as noticed before, has an expanding, a rarefying, a dispersing, a dispelling and repulsive power; hence it is said that he repels the envelope of the comet into a tail; that, though he ATTRACTS THE COMET, he does not only not leave alone, but actually, by an opposite force, BRUSHES BACK its hair with enormous rapidity at the same time—"being" in the words of Sir John Herschell, "under the influence of a REPULSIVE force directed FROM the sun!" On the other hand, with a strange, and in astronomers not rare, inconsistency, the sun is said to attract—and consequently to CONTRACT the WATERY envelope of the earth, more than the earth itself, by raising a tide !

It is said also, that the nucleus of the comet itself possesses and exercises a repelling power upon its envelope; but if so, why should this power act backward, and not equally on all sides, or on that side which faces the sun, and which is exposed to his dissipating influence?

These anomalies cannot be overcome except by the simple adoption of a resistance of the the solar atmosphere.

This resistance also prevents the restoration of the balance of the comet's hair, its re-accumulation in front of the nucleus, and thereby causes a circulation of the nebulous matter, as of a constant emanation from the interior of the nucleus towards the sun, a kind of fountain or jet, in a direction opposite to that of the tail, but sometimes forming whirls or eddies, driven back again in wavelike rings or curves into the tail by the resisting atmosphere of the sun.

The envelope of the comet driven back into a tail, will naturally accumulate and becomes dense at the sides of the nucleus; but the brunt of resistance being borne by the forepart of the comet, the part away from the sun, the back, as it were, will be screened, and more or less remain in its normal condition; hence no accumulation of the matter of the envelope immediately behind the nucleus; and being, moreover, the shaded side of the comet, the appearance there of a dark streak, a seemingly hollow cone between the concentrated volumes and more enlightened side of the comet's tail, will be a necessary consequence.

The wavelike motion of the tails of comets, which by several has been observed, can only be ascribed to currents, or an agitated state of the solar atmosphere.

The hornlike, or curved, appearance of the projections, protuberances, or cusps, of the solar photospheres, observed during the last
eclipse of the sun, (1860) is no doubt also due to the resisting
atmosphere of the sun, within and with which he is rotating on
his own axis.

If our earth and the planets had no further envelope than an
atmosphere of air, or that the higher rings of oxygen, hydrogen,
or whatever else it may be, were not of so highly cohesive and
elastic a nature, the earth and planets would not appear as spheres
to our eyes; for, the rapid progression of the earth through the
resisting medium of the solar atmosphere would drive back into a
tail our ocean of air and water, already so easily acted upon and
lashed into fury by a slight pressure of the moon, and by a mere
accumulation of thunderclouds, as we shall see hereafter, and thus
also make it present the appearance of a comet-like body. It
therefore, is not only probable, also, but absolutely necessary, that

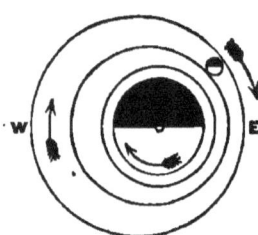

(as shown in the diagram), superior envelopes of our earth should exist, in order
to guard us against exterior pressure,
whether caused by the rapidity of the
earth's progress in the solar atmosphere, or
by passing comets or other heavenly bodies,
should they come within our vicinity.

But what, I may be questioned, about Jupiter's envelope, and
"Lexell's comet," passing twice among his satellites?

In the year 1770 a comet was discovered by Messier, but its
course and periodical return in five years and-a-half so EXACTLY
calculated by Lexell, that in consequence it bears his name.

From this calculation it appears, that not only in 1767, but also
in 1779, the same comet passed among the moons of Jupiter, without exercising the slightest effect upon them; of all the astronomers
then existing, not one could detect the least perturbation. Sir John

Herschell, in his "600 pages of OUTLINES of Astronomy," says at page 346; "It is worthy of notice that by this rencontre with the system of Jupiter's satellites, none of THEIR motions suffered any perceptible derangement,—a sufficient proof of the smallness of its (the comet's) mass."

On the first rencontre, in 1767, the comet came from a very distant errand, not to be ascertained, but, approaching Jupiter in its way, it proved such an attraction to the planet, that its former unknown course was changed into an ellipse of five-and-a-half years' duration.

The second rencontre, according to Sir John, as above, took place "about the the 23rd of August, 1779, when by a SINGULAR COINCIDENCE, the comet again approached Jupiter, whose attraction at the time was 200 times greater than that of the sun himself;" but strange enough, instead of—as in 1767—narrowing still more its calculated ellipse or course, it turned out repulsive to the planet, which so flung it off, that again no one knows, or is able to calculate, in what direction, or to what distance, it has retired.

Why the excess of Jupiter's attraction over that of the sun should not narrow the path of the comet, instead of—according to Sir John—" DEFLECTING its orbit into a curve, not one of whose elements would have the least resemblance to those of the ellipse of Lexell," I, as an amateur, cannot be expected to understand ;* I should rather have thought, that from that time the comet would have commenced coursing round

---

* Nor can I understand the VERY OPPOSITE statement of Sir John, § 530, why "At the distance at which the moon really is from us, its gravity towards the earth is ACTUALLY LESS than towards the sun." And yet the moon courses round THE EARTH !—according to Sir John, in § 529, for the FOLLOWING reason: "The moon, as we have seen, is about 60 radii of the earth distant from the centre of the latter. Its proximity, therefore, to its centre of attraction, thus estimated, is much greater than that of the planets to the sun; . . . . It is owing to the proximity,"—THOUGH THE SUN ATTRACTS HER MORE—" that the moon remains attached to the earth as a satellite."

D

Jupiter as its centre, or at least like the comet of Encke, continued narrowing its orbit instead of returning into the more distant regions of the solar system and revolving again round its primitive centre, the sun. If, however, Jupiter really sent off the comet into the wide world, there is still some hope of Encke's comet not falling into the sun, and thus escaping the threatened conflagration.

"Its observation on its first return in 1776 was rendered IMPOSSIBLE by the relative situation of the perihelion and of the earth at the time." Sir John Herchell as above.

When first I heard, and afterwards read, of these extraordinary facts concerning Lexell's comet, I can scarcely say what I felt, when in my mind I beheld this celestial body, majestically, but quietly, twice passing among the satellites of Jupiter, as if to salute and bring them tidings of other worlds, from members of the same communion and system ; and though unknown by one another, yet all warmed and held together by one common centre, to whom the messenger was going to fetch, and bring back, a fresh store of graces and vitality of which the distant flock was now in need of; and again, I felt strange, when after the second visit, like an uncertain sporadic meteor, I saw the flaming wanderer pushed off and disappear like one excommunicated and an outlaw!—But wonderment of one kind gave way to wonderment of another, when I found, that though so confidently spoken of as to the years 1767, 1776, and "about the 23rd of August, 1779," this brilliant comet WAS NEVER SEEN BEFORE 1770, when Messier discovered it, NOR EVER AFTER.

This amazing discovery destroyed the charm of previously supposed FACTS, and the TWICE UNSEEN approach to, and departure from, Jupiter and his moons, their UNDETECTED perturbation, and the IMPOSSIBLE observation of a comet NOT IN SIGHT, reminded me of the story of a mercantile house writing to one of its debtors:

" We are much astonished at not having as yet received your remittances;" and the reply : " I should have still more been astonished than yourselves if the remittance had arrived, because I never sent it."

What if the comet of 1770 never returned again at all, like so many others? Or if it be one of long period like that of Halley's? Would not, in that case, the comet also have been astonished to see itself twice observed among Jupiter's satellites, when perhaps, it was in the very opposite part of the solar system, and at all events beyond a visible distance? And would not the moons of Jupiter have been astonished too if they had seen the comet and been disturbed by him when he never came?

When of all the comets, the periodic time of THREE ONLY has been ascertained and verified from observation, that of Halley completing its passage round the sun in about 75½ years, that of Biela in about 6¾ years, and that of Encke in about 3½ years; but when these even do not turn up regularly? For, in his " *Introduction to Astronomy*," page 117, Mr. Hind says of the first one of these three comets, that Halley " foretold its appearance in 1759, which actually took place AFTER A RETARDATION OF BETWEEN ONE AND TWO YEARS through the ATTRACTION OF JUPITER AND SATURN !" What, then, can be said in favour of the correctness of Lexell's calculation, generally accepted, against the probability, if not certainty, of my suppositions? Having been acknowledged by the whole astronomical world, the calculations of Lexell ought certainly to be correct, if it were not for the drawback of want of verification from observation.

To verify the periodic time of a comet, the comet is first sought to be identified with some former one as to size, appearance, and path.

When thus found that there is a similarity of results in previous with present observations; that for some time back, say for centuries, at an interval of from about 75 to 75 years, a little more or, less either way, a similar comet had made its appearance; that therefore it might reasonably be predicted that in about 75 years hence, the same comet would turn up again : the identity of the heavenly body is tried to be verified by adapting the elements of calculation to its probable orbit, whether of long or short duration, and we have seen with anything but a certain and definite result, though nearly all those comets whose periodic time is said to have been verified, and whose number is stated by Sir John Herschell to be 36,—a great increase since the previously given number came under my notice!—move round the sun in the same direction as the planets.

The orbits, then, of these comets, ought to have been as easily and correctly (?) calculated as those of the planets have been determined, if astronomers were as sure of THEIR THEORY as they pretend to be, and of the contrary of which the loss of many asteroids, discovered but not re-found, and the late correction* in the distance of the earth from the sun, and its narrowed orbit, are striking proofs.

As to the CONSEQUENT irregularity in the CALCULATED returns of Halley's and other comets, the periodic time of which is considered to have been truly verified, Sir John has an excellent and happy mode of accounting for the same. Astronomers are not wrong, oh dear no ! " All this APPARENT irregularity is owing to the action mainly of Jupiter, which is a general disturber of

---

* Among the communications made at the last meeting of the Astronomical Society, was one from Mr. Stone, principal assistant at the Royal Observatory, Greenwich. Mr. Stone had completed the calculations of the mean horizontal parallax of the sun, as deduced from observations made at Greenwich on the planet Mars at his recent opposition, compared with other similar observations made in Australia. The result is that the heretofore-received mean distance of the earth from the sun must be diminished by about three millions of miles ! What a corollary this on the dogmatical and supercilious assurance with which some astronomers lay down the law to lay thinkers on the subject of astronomical science ! *Public Opinion*, 16 May, 1863.

comets " and of learned men, "and gives a vast deal of trouble
to calculators; and Saturn " as Mr. Hind already told us, "is not
without a finger in the pie."—"Jupiter, in fact, is a regular
stumbling block in the way of comets," and of Sir John too I
suppose.

If, then, it were not for this Jupiter, no doubt all the calculations concerning comets would be as true as the Gospel; but, as
this is not the case, as no comet as yet has ever proximately been
calculated correctly, unless seen four or five times and more; and
as in the absence of verification from observation, for which "the
poor unhappy comet" of Lexell, unfortunately, declined giving
the opportunity, I must beg to differ from the calculation as well
as from the observation on which it is based.

How futile even calculations based upon observation, will
appear from the following extract from the *Times* of 18th September,
1861, relating to a sitting of the Academy of Sciences.

"Mr. Biot communicated two letters from Mr. Valz, honorary
director of the observatory at Marseilles, in which he repeats his
assertion, that the earth passed through the comet's tail on the
30th of June last, and in support of this opinion computes the
position of the tail, the situation of the earth, &c. It is a curious
fact, that on such a mathematical question as the inclination of a
comet's tail, and the position of that body and the earth at a given
moment, there should be two, and even three, opinions; for our
readers must recollect, that Messrs. Hind, Liais, and Valz, all
admit that the comet's tail swept the earth, but neither agree as to
the hour or THE DAY; whilst Leverrier has cast great doubts upon
the occurrence."

How ridiculous, then, the authoritative tone and dicta of so
many astronomers, when among themselves they are at variance
about a question, which, one should say, could not possibly admit
of an error, either of observation or calculation; and if in a case

like this, the comet almost within the reach of the hand, such important differences exist, no agreement even about THE DAY which took us through the tail of the comet: what weight can we attach to observations and calculations of more distant and smaller bodies, and to calculations unsupported by observation? to "the SINGULAR COINCIDENCE of an UNSEEN COMET being said to have passed among Jupiter's satellites about the 23rd August, 1779?"

Notwithstanding, the correctness of Lexell's calculation is and has, in the meanwhile, been as readily taught and taken for granted, as if it were based upon infallible authority, though, certainly, things still less possible than the impossible observation of an invisible comet, have been and are seriously believed and propagated, and of which an instance or two, connected with cometary influence, will be found later on.

The CALCULATED FACTS respecting Lexell's comet, SUPPOSING them to be truly verified from observation, would, however, just as beautifully illustrate my new theory, as their absence would leave it intact ; they would establish the moon-embracing envelope of Jupiter, warding off the comet from the solid body, and by its elasticity throwing it out again after having come to take a dip into the superior oceans of the planet, as a swallow would take a dip into the water whilst flying over the lake.

This, in fact, could not be otherwise; for, what could have prevented the comet from circulating round Jupiter, or from descending upon its terra firma, when the planet attracted it 200 times stronger than the sun himself? Why should not,—as meteors are said to come from beyond our sphere upon the earth,—with this overwhelming attraction, the comet have settled upon Jupiter like the bird over its nest? The envelopes of Jupiter alone could have resisted its momentum, and prevented the warming up and melting of the planet's polar regions.

It is, however, considering Lexell's calculation to be as correct as it is false, more probable, that nevertheless, at the first rencontre with Jupiter, the comet would only have passed him by, in its regular course directed by the attraction of the sun and the currents of the solar atmosphere; and that at the second rencontre —if ever it took place—it would merely have impinged upon the exterior envelope of the planet, but been repelled and driven off by its elasticity. The very brilliancy of Jupiter may be—and no doubt actually is—owing to one of its envelopes, to an enormous ocean of hydrogen surrounding him.

And may not the moon's brightness, the visible OUTLINE of her dark body when slightly in the crescent before and after new moon, though ascribed to the light she receives from the earth, the projection on her disc of an occulted star, and of solar protuber-ances at the time of an eclipse of the sun, be owing to the ocean of hydrogen somewhere above us and surrounding our satellite?— And if so, how singularly beautiful is it not then, that these oceans of hydrogen, like a floating mirror, a moving lens, at once convex and concave, surrounding everyone of the celestial bodies, should absorb, accumulate, send down and reflect upon them in myriads of rays the concentrated light of their suns? And that by the same means their daylight should be dispersed and mutually reflected by the one into the night of the other, from star to star and from planet to planet? That thus there should be a kind of communion between the whole of the heavenly host, seen and unseen, millions of orbs saluting each other in twinkling tokens of kindred origin, and as one choir in silent harmony proclaiming the one God their common Creator?

III.

POSITIVE ELECTRICITY WITHOUT, NEGATIVE ELECTRICITY WITHIN, THE
SOLID NUCLEUS OF THE HEAVENLY BODIES.—WILLIAM GILBERT.—
LIGHTNING OF EVERY KIND, THUNDERSTORMS, WATERSPOUTS, SAND-
SPOUTS, DUSTSPOUTS, CYCLONES, METEORS AND METEORIC PHENO-
MENA OF EVERY DESCRIPTION, ARE THE PRODUCTS OF THE RE-UNION
OF POSITIVE AND NEGATIVE ELECTRICITY.—EXTENT OF THE EARTH'S
ATMOSPHERE.—PROFESSOR CHALLIS.—PROBABLE ORIGIN OF THE
MOON.—THE EXPLODED PLANET OF OLBERS.—SWARMS OF SHOOTING
STARS.—THE AURORA BOREALIS; ITS CORRESPONDENCE WITH THE
AURORA AUSTRALIS; ELECTRIC CONNECTION WITHIN THE EARTH
FROM POLE TO POLE.

HYDROGEN, as the best reflector of light, the lightest or most
expanded of all known gases, is admirably calculated to form one
of the outer, if not the last, envelopes of the heavenly bodies and
our own globe, as, owing to its expansion, it is capable of that
resistance which our earth requires for protection in its rapid
course through resisting matter. Our sphere, surrounded by an
invisible web, and pressed upon all sides by the atmosphere of
the sun, we may not inaptly compare, as I have done already,
with a balloon filled, in our own instance SURROUNDED, by hydrogen
gas. In the great Vauxhall balloon it has been calculated—the
air pressing with a weight of 15 lbs. to the square inch—that
when inflated with this gas, it sustains upon its external surface
the enormous pressure of 9,122 tons. And such an envelope of
gas is actually required for our globe to resist the pressure of the
atmosphere of the sun, and to ward off from us the effect
of a progressive motion through the heavens of 68,143 miles
an hour, or 19* miles in a second,—in the mere twinkling of
an eye.

Whether, thus surrounded and protected, like the yolk by the
white of the egg, the planets, as fiery balls, rose up from the body,

---

* These figures will have to be corrected according to foot note page 36,
though the operation prove fatal to universal gravitation—or the doctors.

and with the expanding atmosphere of the labouring sun, ascending to their designated positions, there to cool, and by degrees to
form each one a world for itself, some of them generating and
throwing up in a similar manner the satellites by whom they are
accompanied ; whether, with the grand luminary that attracts
them all, and bestows upon each of them the blessings of light
and warmth, they were dropped from the bosom of the Almighty
like soap-bubbles from the mouth of the child ; or, whether as
seed sown into chaos by God, and feeding upon it, the heavenly
bodies grew into size, as the solid part of the earth, particularly
round the equator, is still increasing in bulk by feeding upon the
surrounding atmosphere : His unerring and immutable wisdom
provided them all, like the sun himself, with electro-magnetic
hearts of intense, unquenchable fire, with veins, arteries, and
blood vessels, similar to the constitution of the human body, to
fill AND WARM THE WHOLE OF THEIR FRAME, by infusing into them
from the beginning the negative element, to grow and enlarge
together like the body with the soul ; and He weighed and adjusted its quantity, in proportion to their size and density in so
admirable a manner, that when the sun began to revolve upon his
course, they also began, like so many balloons, inflated, in a
manner, by the breath of Omnipotence, to float and revolve with
mathematical precision, certainty, and regularity, in their prescribed orbits, in his track, and upon their proper level within the
atmosphere of the solar system.

Positive electricity, which I have assumed to pervade the whole
universe, accumulates round the heavenly bodies according to their
own exterior electric constitution ; they influence and repel each
other on their approach by this very accumulation, though kept
apart already at their respective distances by the position
they occupy within the solar atmosphere, according to their inflation and specific gravity ; but, by surrounding and isolating each

of them by an almost non-conducting element, by a transparent
veil, by a sheet of paper, as it were, placed between the magnet
and the magnetized object, it was alone possible to keep each
planet confined to the place and path assigned to it by God, to
restrain within their limits the two kinds of electricity, which
always tend to unite, and also to bring together the bodies from
which they emanate, or which by them are invested.

By this isolating veil our atmosphere of air, this union, this
approximation, has so beautifully been guarded against by Divine
Providence, that, by the exquisite distribution of exterior, that is,
solar heat, of water and air, just so much of negative electricity is
liberated from each planet (if our earth be a pattern of all), and just
so much of the positive fluid allowed to penetrate and accumulate
within the upper regions of the atmosphere, as by amalgamation
in the phenomenon of lightning, and its accompanying thunder,
and the aurora borealis, &c., will be a blessing to the beings by
whom the planet is inhabited, each one of the electric elements
most likely returning again to its original destination, in some
manner or other, by a compensation veiled in mystery.

I will now proceed, from natural phenomena, to prove the
material points of my new theory of the solar system, namely : the
existence and operation of positive electricity exterior to the solid
parts of the heavenly bodies, and of negative electricity confined
within them, and keeping them together.*

That these two kinds of electricity have a tendency to unite, is
a well known matter of fact.   Positive electricity becomes permanent when we arrive in the upper regions of our atmosphere,
increasing in body and intensity the nearer we approach the

---

* William Gilbert, (about 1600), was of opinion that "the Terrestrial Globe
is held together as by an electric force ; for the electric action tends to produce
the cohesion of matter;" and Poisson supposed already, as we shall see presently, the existence of electricity in the higher regions of our atmosphere.
Cosmos ii. p. 340.

sun; negative electricity is permanent in the earth and its waters, increasing in body and intensity the nearer we come to the heart of the earth; the former is densest the colder and more rarefied the air; the latter escapes more freely from the earth, the warmer and more expanded the air becomes on its surface. Hence no thunderstorms in cold regions, or in winter, because the condition of amalgamation does not exist; that is, the cold state of the atmosphere does not allow negative electricity to escape from the earth.

In this I am confirmed by Dr. Lardner, who says : " One of the earliest results of the observation of the electrical state of the air was the discovery of the fact that, in clear weather, when the natural state of the atmosphere is undisturbed by clouds, it is always charged with positive electricity,* and the surface of the earth is, on the contrary, charg edwith negative electricity.

" The negative electricity of the ground, and the positive electricity of the stratum of air contiguous to it, have a continual tendency to recombine and neutralize each other. From this cause the lowest stratum of air in clear weather, apart from disturbing causes, is found to be in its natural state. This effect extends to the height of three or four feet from the ground; above which height the positive electricity begins to be perceivable, and increases in its intensity in ascending, according to some definite law, which observation has not yet discovered." And he further states : " From the testimony of Phipps, Scoresby, Parry, Ross, and all others, who have navigated the Polar seas, it has

---

* Dr. Lardner assumes that the positive electricity of the air is supplied to it by the earth. But this cannot be true ; for, rising from the earth, and diverging as it ascends into the ever-increasing rarefying ring of atmosphere, it is not possible that, according to the same authority, it should "augment in intensity, as the height increases, to the greatest elevation to which observation is extended." Positive electricity, therefore, must be permanent above us, decreasing in intensity in the same proportion as it has to force its way down to the earth through the increasing density of the air.

been ascertained that the frequency of thunder diminishes in approaching the pole. It appears to be certain that it never thunders in north latitudes above the 75th parallel. Between 70° and 75° thunder is sometimes, but rarely heard. Below 65° storms become more common, and their frequency increases as the latitude diminishes, the inter-tropical regions being those which, in general, present the most violent and frequent manifestations of this meteor."

Thus, the body of air floating on the earth is a constant obstacle to the union of the two electricities. When, however, the rays of the sun, in certain states of the atmosphere, fall particularly strong on the earth or the sea, their accumulation and intensity cause the air to expand ; and in proportion to this accumulated heat and consequent expansion of the air, the negative fluid is perspired by, is liberated from, the earth, or disengaged from the sea; it then rises all over the more than ordinarily heated surface, and, filling the air, produces in us that lassitude of body and depression of spirits which we always complain of in sultry weather. In this I am confirmed by Richter, who found that the electricity of the positive pole augments, whilst the negative diminishes the action of life ; the pulse of the hand, he says, held a few minutes in contact with the positive pole, is strengthened; that of the one in contact with the negative is enfeebled.

Whilst this depressing liberation of negative electricity takes place on the earth or the sea, the sky becomes overcast, clouds begin to form, and the positive fluid of the upper regions, floating there loosely like its kindred below, finds in these clouds a place of refuge, into which it creeps like water into a sponge ; or, in the words of Dr. Lardner, "the cloud takes the nature of one continued conductor, and the free electricity accumulates on its surface in the same manner as on the conductor of an electrical machine." The clouds then serve as a vehicle to carry the electricity thus contracted through the otherwise opposing air, towards the object of its attraction.

During this process of concentration of electricity below and above, both kinds begin to exercise their mutual influence,—their inherent magnetic power, their property of amalgamation, their tendency to unite, having increased with their accumulation; and thus, as it were, they are brought within sight of each other, close within contact. Positive clouds begin to draw together, to unite, and to lower as they become larger and more charged; and their movements are directed, more or less, by the accumulation of negative electricity on the earth, which draws them down, as also by atmospheric currents and the face of the country. If the place of gathering be mountainous, thunderclouds will sometimes approach from opposite directions, and hover about the tops of hills and mountains, as if they would not leave them.

The electricity of the earth tending to rise, and at a certain height finding itself everywhere impeded by a denser and colder state of the air, seeks a conductor in any elevated object that may be near; and to this it leans, as it were, for assistance, and ascending into the higher strata, forms a constantly upward-flowing stream of the electricity all round. The fluid, however, will prefer such conductors as are most congenial, to which it has the greatest affinity; and the more pointed the ends of the conductors, the more readily will the stream, pushed on from below, quit its hold and emerge into the opposing atmosphere; but the rounder or larger the extremity of the conductor, the more the negative fluid will accumulate there, and the more difficult will be its passage into the resisting air.

In this manner, then, the negative electricity escaped from, and floating on, the earth or the sea, like vapours from the surface of boiling water, is directed upwards in as many volumes or streams as there are conductors, and the better the conductor, the more

concentrated or condensed will be the stream; the best and highest conductors take the lead, and in most cases absorb the lower
and lesser streams of the neighbourhood.   On the sea, islands and
ships will act as conductors, and to these the electricity floating
on the surface of the water will incline. Islands, moreover, will
not only attract and conduct the electricity of the water that
surrounds them, but will also send up their own electric exhalations in still greater abundance.

This is confirmed by Humboldt in his " *Cosmos*," and also by
Dr. Lardner.   The former says: " When, upon the whole, where
the ocean of the air rests upon a fluid bottom, the electric balance
is more rarely disturbed than in the air on land ; it is more striking to see how, in extensive seas, small groups of islands influence
the state of the atmosphere, and cause the formation of thunderstorms."   And Dr. Lardner makes the following observation:
" By comparing the recorded cases of thunderstorms on land and
at sea, it appears to be established, on grounds at least of strong
probability, that storms are more frequent on land than at sea ;
that at sea the frequency of storms diminishes as the distance
from land increases; and analogy leads to the probable conjecture
that there is a certain distance from land at which it never
thunders."   The cause of this lies in the absence of conductors to
the negative electricity liberated from the water, and the consequent formation of waterspouts instead of thunderstorms, as we
shall see a little further on.

On land, the chief conductors will be elevated buildings, particularly steeples with metal vanes, and the like ; trees, as also the human
body, and animals in an erect position, where higher and better conductors are wanting. Hence the necessity in open fields to lie down
flat on the ground in case we are overtaken by thunderstorms.*

---

* One person is a better conductor than another, and, no doubt one animal
better than another; and this most likely depends upon the stronger or feebler

As regards mountains, the fluid will creep up at the sides, and arrive at the summit more and more concentrated, like many streams merging into one; and hence its power, at this elevation, to attract and arrest electric clouds.*

For this accumulation of electricity on the summit of mountains, Dr. Lardner accounts in the following way: "Since free electricity accumulates in great intensity at prominences and points of a conducting body, the negative electricity of the earth may be expected to be most intense at mountain-summits. Clouds being in general charged with positive electricity, an attraction will consequently be exerted upon them, which, conspiring with the attraction of gravitation, will draw them round such summits." The true cause of accumulation is, however, as I have stated it.

Besides this ascent of negative electricity by the sides of mountains, vapours rising from the earth will also act as conductors; charged with the electricity through which they ascend, they will form into light clouds, and rapidly take their direction to the clouds above.

By these means the atmosphere becomes filled with streams and volumes of negative electricity, and the stronger and higher they rise, the more will they attract and hold fast the electric clouds. Negative clouds, if but lightly charged, will swell the positive ones, and the amalgamation will take place without explosion; if heavily charged, the mutual attraction is more vehement and stronger, and the point of union of the two fluids at a greater distance between them; the positive charge passing

---

process of animation, of life, upon the respective warmth or coolness, the electric-phosphoric constitution, of the body. With animals, the form of the body will probably influence the conduction of the electric fluid. If among inanimate bodies one conductor is better than another, all else being the same, it arises from the one being chemically united with more electric matter than the other.

* It will be interesting to examine what relation there exists between pine and other forests growing on the tops of mountains, and electricity and oxygen; whether wood grown on the mountain, as a rule, burns better than wood grown in the plain or in the valley.

to the negative cloud will be ignited by friction with the air through which it darts, and the union of the clouds will be accompanied by lightning.

In the same way, when both electricities have reached that point at which their union becomes irresistible,—that distance at which the force of the electricity shall overcome the resistance of the surrounding air,—the positive will precipitate itself from the cloud towards the negative columns from the earth, ignite by friction with the air it has to pass through, and cause as many of these volumes to explode in the air, down to their very base, even into the bowels of the earth if uninterrupted, as may be dense or near enough to be capable of ignition. The almost instantaneous explosion of these electric columns rising from the earth, often cut off from the earth by the wind or atmospheric currents, and drifting about in the air, produces the various forms of fork, or zig-zag lightning, which, according to Lardner, seldom flashes between two clouds, but is, in conformity with my theory, generally manifested between a cloud and some terrestrial object, playing about in the air like flames in the escaping gasses of a coal fire.

Lightning-flashes sometimes dart upwards from the cloud ; and this takes place when volumes of electricity rising from the earth ascend higher than the position of the thundercloud hanging in the air. If such a volume be struck by a flash from the cloud, it is natural that it should explode to its extreme point, whether this be the summit of a mountain or the top of a steeple. An instance of this kind is cited by Dr. Lardner (ii. 127), where seven persons were killed on a church, erected on a lofty peak in Styria, by a flash of lightning ascending from a cloud about half the elevation below the place where they stood.

In all cases, where resistance of the air to the amalgamation of the two electricities in their accumulated state is not sufficient to cause ignition,* the union will take place without the appearance of lightning, and the affected column, or volume of electricity,— the thunderbolt as it is called,—will strike with the same destructive effect, though without the fire ; no such effect will result, however, if the accumulation of the two electricities is not sufficiently dense for the purpose.

The heavier a cloud is charged with electricity, the blacker will be its appearance ; magnetism, or its property of cohesion, will hold the fluid together in as compact a body as possible, and consequently, the vapours of the cloud are contracted, with the charge they contain, into a much denser body than otherwise would be the case ; and there is no doubt upon my mind but that the rays of the sun falling upon, and being absorbed by, the cloud, must exercise some material influence on its condition and development.

In the absence of clouds, on fine, clear, and calm summer days, the electric columns from the earth will rise to a great height. On arriving in the upper, highly rarefied regions of the air, whether cut off from, or still connected with, the stream from below, they will amalgamate with the free, yet more than ordinarily attracted and consequently dense positive electricity above : this amalgamation is accompanied by an evolution of light, and, by a rapid exhaustion and supply from above and below, produces the sheet lightnings of our summer evenings. For, light alone is evolved by the two electricities when their union takes place within a rarefied atmosphere. This is proved by a very beautiful experiment described by Dr. Lardner. Let a Leyden jar be suspended under the receiver of an air-pump in a dark room, and

---

* The distance at which the explosion will take place, and its force, will depend on many circumstances ; such as the difference between the actual charges of the clouds, and the charges due to contact, the form of clouds, and the state of the intervening atmosphere. An electrical explosion, therefore, may take place between two clouds, whether they are both similarly electrified (if unequally charged), or one be electrified and the other in its natural state.—*Lardner.*

E

let the air within the receiver be slowly and gradually rarefied by
the action of the pump. When the internal fluid is sufficiently
relieved from the restraining pressure of the air, it will be seen to
overflow the mouth of the jar, and descend the sides in a cascade
of light to meet and combine with the external fluid. Fluctuations in this cascade of light alone are wanting faithfully to
represent the rapid succession of flashes of sheet lightning.

By a similar process, assisted perhaps by a superabundant supply
of supporters of combustion existing in chaos, we may readily
imagine the inconceivable longitude and latitude of the equator
of the universe to have been lighted up, as noticed page 5, when
God said : "Let there be light;" and to have influenced the
creations which it embraced.

During common thunderstorms, sheet lightning is produced
in the same manner as above ; the two electricities combine in
the higher rarefied regions of the air, without the explosion of
electric volumes taking place. If a cloud be negatively electrified, sheet lightning will be produced behind the cloud ; and hence
it arises, no doubt, that the edges only of the clouds appear
illuminated. If the cloud be positively electrified, the sheet
lightning will naturally flash downward, the positive electricity
uniting with the ascending negative electricity in its free yet
accumulated state.

Of volcanic thunderstorms, Von Humboldt says : "The hot
steam (from the waters that may have found their way into the
interior of the mountain) rising during the eruption from the
crater, and emerging into the atmosphere, will, when cooling,
form a cloud, by which the column of ashes and fire, many
thousand feet high, is surrounded. Lightning, in winding motion,
flashes from the column, and then the rolling thunder of the
volcanic thunderstorm may be clearly distinguised from the crack-
ing within the volcano. The lightning striking down from the

volcanic steam cloud killed eleven horses and two men in Iceland, on the 17th October, 1755."

The operation here is the same as before described ; electricity from the higher regions concentrates in the volcanic cloud, whilst the volcanic column does not only carry up the electricity liberated within the volcano, but serves also as a conductor. to that rising from the earth in the vicinity round about. At the eruption of Mount Vesuvius in April, 1855, Professor Palmieri, at Naples, noticed that the electric current in general appeared to follow the course of the smoke issuing from the crater.

By a process similar to the preceding one, waterspouts are formed at sea.

In a "dead calm" the sun exercises, as a matter of course, a more powerful influence on the water than he does when it is in any way agitated. By the intensity and undisturbed accumulation of his rays on the quiescent surface, the air, resting upon it, becomes unusually rarefied ; an extraordinary evaporation,—most likely assisted, if not chiefly caused, by volcanic action below,— takes place, as if the sea were steaming, and an immense quantity of negative electricity is produced or liberated. The electric fluid tending to rise, but opposed by the resistance of the air, unites with the generated vapours as the only conductor it can lay hold of. The particles of vapour being heavier than the highly expanded air, are unable to ascend and form into a cloud ; attracted, there- fore, and held together by the electricity they contain, and striving to rise, they all begin to move towards the centre, the focus of the foggy surface, until by degrees the whole mass is set in one uniform upward spiral motion, assuming the shape of a funnel turned upside down, and becoming more and more elevated as it becomes more contracted. The calm state of the air allows the quiet formation of the electric cone, which, in the absence of a conductor, like a wedge, parts and penetrates the impeding air,

through which otherwise the body of diffused electrified vapours
would not have been able to ascend, not having even the assistance
of small things like grass and shrubs, which on land facilitate the
escape of electricity from the earth into the air,

Currents will be produced where accumulation takes place, or
the balance of fluids or liquids is disturbed.   Fire expands water
and air, and by disturbing their balance, currents are set in motion.
Heat, whether from the rays of the sun, from fire or friction, disturbs the electric equilibrium of matter, and causes currents to
take place.

Whilst thus the wedge is forming by a continual access of electricity and electric vapours, the surrounding air, likewise disturbed
in its balance, begins by degrees to be slightly set in motion by the
screw-like accumulation in the centre of the steaming surface; and
this motion takes the same direction in which the wedge or screw
is winding upwards.   The air will thus gradually press more heavily
on the base of the funnel, and less and less on the elevated part
where it meets with less surface, and where the pressure is more
on the incline; and by this means it even assists the progressive
elevation of the negative spout.

The rising in a column, however, of this mass of electric vapours
would nevertheless not take place, if before, or at the time of its
formation into a spout, no cloud had been gathering in the sky,
saturated with the positive electricity of the upper regions, exercising its attractive power on the negative acccumulation below, and
fixing through it, as it were, a vertical axis.   The two kindred
elements tend to union, but have no conductor to effect it.   Thus it
is the attraction from above that makes the electric vapours, heavier
and denser than the highly expanded air, and consequently unable
to rise, assume the natural shape of a funnel-like wedge, whilst the
attraction from below, in the absence of any conductor whatsoever,
causes the same appearance in the cloud above: holding to a com-

mon axis, it is a kind of mutual magnetic suction, the spout above causing the one below, and the one below causing the other above, until, elongated within the striking-distance, within the point of amalgamation, they break down in union in the form of rain or hail, unable any longer to be held in suspense by the air which both cones contributed to set in spiral motion.

That previous to, or at the time of actual union of the spouts, no explosion takes place, may be accounted for in several ways.

Though, in the first instance, the extreme points of the cones may be seen, yet it is likely, and it is even said to have been observed, that previous to their meeting, rays, or threadlike streams of electricity, are shooting forth from cone to cone, thus rendering visible the electric axis, the magnetic bridge, that had been thrown across at the commencement from kindred to kindred. These rays of electricity, however, concentrated by the tapering shape of the spout, as the rays of the sun are concentrated by a lens, must possess a correspondingly increased magnetic power, and may easily be supposed to act as a link, as a chain, and imponderable conductor, from cone to cone, by which both are steadily and gradually drawn together.—This electric link, the axis of the spout, as we have seen before, communicates between the cones soon after their formation, and sometimes before they come in contact. Hence, the amalgamation of the two fluids, harboured by these columns of vapour, and the union of the columns themselves, is so gradual, that no ignition, no explosion can take place, there being very little, if any, friction with the air.

The absence of explosion may, in the second instance, be accounted for by the exceedingly saturated or condensed state of the vapours; they are too much like water to let the electric fluid have an easy escape, though this does not diminish their relative attraction, their tendency to unite being only somewhat fettered.

When the union of the two cones has been accomplished, and

the two fluids have been brought to amalgamate, the whole mass, being now too condensed to be held any longer in suspense by the air, will violently burst or come down in rain or hail, and, according to circumstances, with or without lightning, accompanied with great noise, but without any thunder.

All these phenomena may easily be accounted for, particularly the bursting, the rain and hail ; they depend upon the more or less rapid amalgamation or dissipation of the two electricities; if their union or dispersion is gentle and gradual, the spout will dissolve in rain; but if violent and sudden, positive electricity most likely preponderating, hail will be the result. Lightning will more or less depend upon the density of the respective cones, and their admixture with air during the amalgamation of the two electricities; and the lightning taking place within the spouts, and merely flashing out, as it is said, readily explains the rumbling noise; but the electric fluid not passing or striking through the exterior atmosphere, as is the case during thunderstorms, sufficiently accounts for the absence of thunder.

If the "dead calm" at sea extends to a great distance, several waterspouts may naturally happen at the same time, each one forming its own centre. Their motion on the water seems to depend upon the upper currents of the atmosphere, the positive cone drawing along the negative one. They will also move in the direction of a ship, or of a prominent point on land, as centres of attraction and conductors to the electricity of the surrounding place.

Waterspouts on land happen from the same causes as those at sea, from the high position of positively electrified clouds, from isolated powerful conductors of the negative electricity of the earth, or from the absence of conductors.

Sandspouts in the desert originate in the same way as those at sea and on land. Electricity disengaged by the burning sun, by evaporation accumulated on the heated waste, and attracted either by electric clouds or óther electric accumulations in the upper air, tries to rise; but having no conductor, it concentrates in one or more places, according to local causes, and the element pressing from all sides upon the centre, it assumes that spiral motion (around its own electric axis), which is witnessed throughout all nature, in the water, the air, and in the plant that winds itself round the protecting pole or the stem of a tree; and as the centre of the spout turns and rises and puts the air in motion, it gathers strength and consistency from the pressure around, taking up the sand within its powerful grasp; and like the avalanche in its fall, accumulating in bulk, electric force, and rapidity as it whirls round over the plane, it will often, like the water and landspout, change into a hurricane, tearing up everything in its way, until its career is broken and inpeded by internal exhaustion and dissolution, or by exterior obstacles and a dissipation of its electric accumulation. The aurora borealis, in the same manner, is an originator of storms.

Mr. Redfield in his *Observations on Storms*, in the transactions of the Americ. Philos. Society, 1841, offered an opinion, that generally during a gale there is, in the lower part of the atmosphere, a spiral motion inclining downwards and towards the centre; and in the higher regions a like spiral motion inclining upwards and towards the exterior.

My theory, of storms originating in electrical accumulation,— often perhaps without either vapours or sand,—and in the impeded or unimpeded amalgamation of the two electricities, seems to confirm the opinion of Mr. Redfield; for these storms must naturally partake of the motion of the positive and negative cones from above and below.

Hurricanes, revolving storms, or "cyclones," from their revolving motion, appear to belong to that class, and are said to embrace masses of air of from 50 to 500 miles in diameter.

According to the *Manual of Scientific Enquiry*, published by Authority of the Lords Commissioners of the Admiralty, 1859, " These cyclones always revolve in the same way in each respective hemisphere, though in each their rotation is in an opposite direction. In the northern hemisphere their rotation is RETROGRADE, *i.e.*, contrary to the motion of the hands of a watch laid face upwards, or in conformity with the motion of the hand in UNSCREWING a screw. In the southern their rotation is DIRECT, conformable to the hands of a watch, or to the motion of the hand screwing IN a screw into a horizontal board."○

According to my theory, the northern cyclone is rolling towards the exhausted equator, and the southern one is doing the same; they come from opposite directions, each one following the meridian to fill up a vacuum, as it were, in the tropics ; and following also, like the electric tide, the apparent course of the sun from east to west, they naturally revolve forward to the west as described above.

At a meeting of the Meteorological Society, according to the

---

* " As the phenomena of the cyclone are reversed in the opposite hemispheres, so are those of the (atmospheric) wave. In the northern hemisphere the barometer, during the passage of an atmospheric wave, rises with N.E. winds ; IN THIS INSTANCE, in the southern hemisphere, it FALLS ; in the northern hemisphere, as the N.E. winds pass they DECREASE in force ; in the southern hemisphere they appear to increase." *Manual of Scientific Enquiry.*

Owing to the greater heat at the equator, and the equatorial depression of the envelope of the earth of which I speak further on, the air is denser and colder, consequently heavier towards the north than at the equator. When, therefore, a current, a wind, comes from the N.E., it comes from a dense quarter into a rarefied one ; hence the more rarefied air becomes heavier, and the barometer rises ; the current itself becomes diffused, and therefore decreases in force. In the southern hemisphere, the N.E. current in that direction comes in reality from the equator ; its air is warmer and lighter than that into which it flows towards the south pole ; hence the colder air is warmed and rarefied, and the barometer sinks, whilst the current itself becomes compressed by driving in between a colder atmosphere, and therefore seems to increase in force.

Daily News of 11th April, 1860, a paper was read by H. Cook, M.D., on duststorms and dustcolumns, &c. of India. The author remarked : "that there are certain days in which, however hard and violently the wind may blow, little or no dust accompanies it, whilst at other times every little puff of air or current of wind raises up and carries with it clouds of dust, and at these times the individual particles of sand appear to be in such an electrified condition that they are even ready to repel each other, and are consequently disturbed from their position and carried up into the air with the slightest current. To so great an extent does this sometimes exist, that the atmosphere is completely filled with dust, and when accompanied by a strong wind nothing is visible at a few yards, and the sun at noonday is obscured. This CONDITION OF THE ATMOSPHERE IS EVIDENTLY ACCUMULATIVE, it increases by degrees until the climax is reached, when after a certain time, usually about 24 hours, the atmosphere is cleared and equanimity restored. Dustcolumns appear under a similar condition of electrical disturbance or intensity. On calm quiet days when hardly a breath is stirring, and the sun pours down his heating rays with full force, little CIRCULAR EDDIES are seen to arise in the atmosphere near the surface of the ground ; these increase in force and diameter, until a column is formed of great height and diameter, which usually remains stationary for some time, and then sweeps away across the country at great speed, and ultimately, losing the velocity of its circular movement, dissolve and disappear. The author had seen in the valley of Mingochao, which is only a few miles across, and surrounded by high hills, on a day when not a breath of air stirred, twenty of these columns. These seldom changed their places, or but slowly moved across the level track, and they never interfered with each other."

In a paper read before the Royal Society, treating of magnetic storms, eddies, currents, &c., the astronomer-royal, Mr. Airy, AS A

NOVELTY in 1864, " assumes the presence of a magnetic ether or fluid as an envelope of some feet in thickness over the surface of the whole earth, which, being affected during magnetic storms in the same way as air and water are, occasions the phenomena which have long been regarded as the most interesting in observations of terrestrial magnetism."

Some inconsistency excepted, it will strike the reader that there is nothing in the preceding assumption, either NEW or IMPORTANT.

Spouts, like the dustcolumns here spoken of, and originated only by the heat of the sun, will scarcely ever, if at all, happen during the night, unless through some exceptional state of the atmosphere; and their formation on the sea during the day is no doubt intimately connected with sub-marine geological formations and dispositions, with the greater or lesser depth of the sea, the absence of currents, &c. The same causes most likely produce also local marine fogs and clouds which in bright weather have been observed on the high sea.

" Mists are frequently met with over shoals and sand-banks . . . . . I have seen such mists to the south of Jamaica, and in the Pacific, which have shown the outline of the shoals beneath so well defined, as to be distinctly recognised from a distance; thus forming to the eye aerial images reflecting the form of the bottom of the ocean . . . . . At sea and in very clear weather, clouds are often seen suspended above the site of sand-banks or shoals, as well as over low coral or sandy islands. Their bearings may be taken from a distant ship by the compass, precisely as that of a high mountain or solitary peak." *Cosmos* ii. 30.

These mists and clouds, like icebergs, afar off indicated by the thermometer, Humboldt ascribes to the low, if not cold, temperature of the water over shoals and sand-banks, though I think they are caused by the reverse, *i.e.*, the warm state of the water.

Shallow water is always warmer than deep water. In the former case, the proximity of land to the surface of the water makes this very apparent, negative electricity is more abundantly disengaged, draws down, or attracts, the positive fluid from above, and master cold condenses the moisture of the air through which he passes, thus causing mists and clouds as before described, the same as islands produce mists and clouds and thunderstorms, which are absent on the deep open sea away from land, and the same as mists and fogs in general are caused by the positive electricity coming down lower into our atmosphere and cooling the same, rather than by the cooling of the surface of the earth and water. For, it would be strange if the earth, heated by the direct rays of the sun, were to cool sooner than the air which it warms by reflection.

We have hitherto seen how electricity rises from the earth in streams or volumes, whether by the aid of conductors, or ascending with the cone of the spout; we have seen how by these means the electricity of the earth and that of the higher region are brought into union; but we have seen also at page 48, how the electric stream often rises beyond the region of the clouds, beyond the region where negative and positive electricity are wont to unite. And these streams, ascending higher than our ordinary conceptions, are by far more numerous than those which bring us the refreshing water and the cooling influence of the thundercloud. For, thunderstorms are comparatively rare in proportion to the electric volumes rising every day from the earth into the atmosphere.

These volumes perform their noiseless journey during the day, and in warm and cloudless nights. As evening draws on, or the atmosphere cools, as a breeze springs up, or currents of air cross each other in sundry directions,—these pillars are cut off, some early, some late, some high, some low, some in one state, some in another;[*] and as they are cut off they contract, unless dissipated by wind, and form into spheres, small or large, according to circumstances. Arago found, according to Lardner, that when electric currents were transmitted through wires forming closed curves, or complete geometrical figures, iron filings, placed within the sphere of attraction of the wire, adhered to it, so as to form concentric rings upon it. The moment the connection of the wire with the piles was broken, and the current was no longer transmitted along it, the filings fell off, and all attraction disappeared. We may, therefore, well conclude, that these spheres, balls, or globes of electricity, attract metallic and other particles floating in the air, and that by the ignition and explosion of these electric metallic accumulations, fire-balls, meteors, shooting stars, and meteoric stones are produced.

That brimstone and other analogous solutions are contained in our atmosphere is evidenced by the many cases in which the lightning-flash, and thunderstorms themselves, have been accompanied by a suffocating sulphurous smell. Hence, the first thing of a meteoric character we meet with is the appearance of ball-lightning.

Whilst the electric columns, flashing into lightning, uncondensed, and in the act of ascending, had no opportunity, nor power, to attract much of the solutions filling the air, another volume cut off before coming into, or passing by, the striking distance of an electric opposite charge, had time to contract, to condense, and to form into a sphere; or, the rising volume still clinging to the earth, untouched by surrounding explosions, but resisted in its upward progress, its upper end began to accumulate into a ball. Thus, concentrated into an electric sphere of larger or smaller dimension, it more powerfully attracted the metallic substances within its

---

* Hence, also, the rarity of lightning striking into buildings, trees, &c., at night.

reach, and coming within striking distance of another electric
accumuldtion from below or from above, it shoots, a ball of fire,
with lightning rapidity to the earth, whilst the ponderable particles that compose its bulk are less quickly dissolved or solidified ;
and by this means the globular form is preserved until its final
explosion, diffusing in sulphurous smoke, and often spreading
about in fragments, the evidence of ponderable matter diffused in
the air.*

This ponderable matter—which also is proved to fill the air, by
the fact that metallic bases have been obtained in the analysis of
plants whose nutriment has been derived exclusively from the
atmosphere—will most likely be present at different times in
different proportions ; electric clouds will also attract it, and
discharges therefrom will, therefore, often be attended by a
peculiar smell ; or, the matter in question will accumulate into a
ball, or balls, according to the constitution and electric state of
the cloud. These balls, concentrating within themselves a greater
portion of electric and material accumulation, and coming within
striking distance of volumes of electricity rising from the earth ;
or, repelled from each other and thrown off from the cloud, or
sinking down by the weight of their agglomeration until friction

* "The British Ship Albemarle, according to Lardner, was struck with lightning
off Cape Cod. A mass of burning bituminous matter fell in the boat suspended
at the stern of the vessel, which diffused an odour like that of gunpowder. It
was consumed in the place where it fell, after ineffectual attempts to extinguish
it by water, or to throw it out of the boat with rods of wood."
A similar, but more awful proof of my theory is recorded in Holy Scripture,
where (Gen. xix. 24,) we are told, that "The Lord rained brimstone and fire
upon Sodom and Gomorrah, and destroyed these cities, and all the country
about, all the inhabitants of the cities, and all things that spring from the
earth." And how naturally can we not account for such a phenomenon?
But to the Christian philosopher, on religious grounds, it is far more congenial, consoling, and encouraging still to contemplate in this event the truth
and confirmation of divine revelation ; how the laws of grace and justice
govern the laws of nature; how hand in hand they go, and in due time, and
at the proper place, not only punish the guilty creature, and purge the earth
from rotten generations, but also how therein he possesses the guarantee that the
same laws of grace and nature will go hand in hand to lead him to transcendent
bliss, as they have done so many others before him.

with the air causes explosion, they will fall to the earth in the shape of balls of fire, according to the cause and circumstance of discharge from above.

That in most cases these ponderable electric balls are attracted from their position in the air or in the clouds by volumes of electricity rising from the earth is proved by the fact, that they mostly alight upon the steeples of churches or towers, and at sea, in particular, on the masts of ships. Humboldt relates (*Cosmos* i. 125.), that " the companions of Cortes saw at Cholula the ærolite which had fallen on the neighbouring PYRAMID," by which, no doubt, it was attracted; and that (note 75) a short time before the earthquake of Quito in October, 1766, "during a whole hour, the volcano of Cayamba appeared as if veiled by the number of falling stars, and the alarmed inhabitants instituted processions."

These globes of fire, then, will be produced wherever there is a sphere-like accumulation of one kind of electricity or the other, where, at the same time, ponderable and electric or magnetic substances fill the air; the light will ensue when friction with the air, or some chemical combination within the ball, causes ignition, and the explosion will take place when the intensity of action dissolves, melts, or solidifies, and repels or disperses the component parts. The combination of ponderable matter with the electric accumulation in these balls presents to us that spontaneous explosion and disappearance which we witness in the lightning-flash.

Ball-lightning of the kind we have spoken of is rare, because, happening at the time of thunderstorms, there is opportunity enough for the discharge and amalgamation of electricity in the shape of common lightning. But in calm, serene weather, the electric emanations and volumes will rise higher than the region of the clouds. Cut off from the earth by one means or the other, they will form into larger spheres than the balls before noticed;

these spheres, assisted by the quiescent state of the atmosphere, will attract more ponderable matter, and, free from the condensed moisture which saturated the fire-ball before its ignition, and removed from the attraction of electric volumes or clouds, they will float longer in the air, until their final consummation. The electric sphere, like a shell, holds the particles it has attracted, and by interior chemical excitation, or by friction with the air when in the lower strata of the atmosphere, ignites like a will-o'-the-wisp, and the agglomerated atoms amalgamate with great intensity. Impelled by its own weight, or attracted again from below, and pressed upon by the surrounding air in proportion to its velocity; having also by its presence and heat caused an accumulation of free, positive electricity,—the explosion envelopes the electric shell from without and solidifies the agglomerated mass within, and, impressing upon its surface a deep black coating as a mark of the violence of its action and evidence of its warm embrace, sends down to the earth the product of electric combustion and amalgamation—the meteoric stone. If more than one stone fall from the exploding meteor, it is clear that in this case the ponderable accumulation was not confined to, or concentrated in, the shell, the outer envelope of the ball, but that it penetrated the whole electric sphere, which, flashing into fire from within, or across, separated the body by repulsion, and dispersed it into numbers of fragments by its explosion.

That these meteoric stones are formed within our atmosphere is beyond contradiction, and, at all events, far more probable, than " that our planet has been formed from NEBULOUS RINGS, separated from the solar atmosphere, AGGLOMERATED INTO SPHERIODS, and consolidated by progressive condensation " (*Cosmos*, 275); for, it is and has been proved and confirmed over and over again, that the sulphurous smell accompanies the lightning-flash, proceeds from the lightning-ball, which, itself issuing from the clouds, fell on the masthead and the steeple, into the church and the house, and often scattering about fragments; that volumes of dust and sand are carried high and far away* into the atmosphere, and by moisture and electricity condensed and converted into showers of ærolites, the same as vapours are converted into rain and showers of hail; that clouds are formed and meteoric stones projected from them to the earth; that balls of fire, and brilliant globes of light, and even shooting stars have let them drop.† And where there is such an absolutely undefinable transition, both in appearance and result, from the lightning-flash to the shooting star, from the will-o'-the-wisp to the most brilliant meteor: who will point out the line of demarcation, the broken link, and say, the one phenomenon is of the earth (telluric), the other comes from beyond it (cosmic)? And if of cosmic origin; why should the earth ATTRACT these cosmic bodies, whilst, as we are told, Jupiter repels the cosmic comet, if the old solar theory be correct? Is there any, and in which of these two cases does there exist, a suspension of the law of universal gravitation? What makes the comet fly off from the planet whilst, as we have seen, said to be 200 times more attracted by it than by the sun, and whilst meteoric stones FROM BEYOND OUR OWN SPHERE are said to fall upon the body of the earth? that they "are small heavenly bodies, which the attraction of our planet has caused to deviate from their previous path." (*Cosmos*, p. 51.)

Not only this: but the metallic, or earthy composition of meteoric stones contains nothing that is foreign to what the crust of the earth is composed of; and the violence of electric action

---

* Dust or sand, in which Ehrenberg discovered the remains of 18 species of siliceous-shelled polygastric infusoria, often falls on ships navigating the ocean near Cape Verd Islands, at a distance of 380 geographical miles from the African coast."—*Cosmos*, p. 345.

† Numerous instances to prove all these cases will be found in every work on meteorology or natural science.

accounts sufficiently for the solidity, as the magnetic properties of the attracted particles do for the purity of meteoric metals.

That nothing so pure and solidly blended together is found in the earth seems to prove, that a less intense, a more gentle operation of electric agency was at work when chaos was sundered and the earth formed, than when fireballs or meteoric globes issue from the still existing, though purer, chaos of the air, or from the chaotic ingredients of the clouds in which they were moulded.

When, according to the state of the atmosphere, electric volumes from the earth, or spherical accumulations of electricity rise higher still than those which ordinarily produce meteoric stones, their dimensions must naturally increase with the altitude they attain, and with the rarefied air or matter within which they ascend; the same as in the case of balloons, which must be larger and more expanded in proportion to the elevation they are to reach. Hence it follows, that the higher the meteor, the greater its size; the further away from the sun, the larger or lighter the planets; the higher a bird is to soar, the wider must be its wings and the lighter its body.

Thus constituted, and no doubt frequently, if not always, more or less charged with phosphoric emanations from the earth, meteoric accumulations rise beyond our atmosphere of air, beyond the region which holds metals in solution, or though in a decreasing ratio only; and as they ascend and increase in size and aggregation of matter, interior distribution of particles or electric agency, will often, in the higher regions perhaps always, cause a mutual repulsion before explosion; the original sphere will part, as it were, into numbers of globes, which, after a longer or shorter course becoming ignited, present to us a host of shooting stars, emanating from one common point of radiation in apparently one or the other constellation of the heavens. Numbers of these meteoric globes rising from the earth, or forming in the atmosphere, and

F

resistance to their motion diminishing as their elevation increases, their glow, ignition, and final explosion become more steady and protracted; their brilliancy augments as more and more they withdraw from the ocean of air into that of oxygen, the supporter of combustion, and thence into the still higher ocean of hydrogen, combustible itself.* Thus they begin and pursue their brief but dazzling career, and separating into fragments, like the exploding rocket scattering about the drops of light it inclosed, will give us those swarms of shooting stars, which, at certain periods of the year in particular, dart through the heavens, mainly in one direction. Less composed of ponderable substances, but more of gaseous matter, and acted upon by the currents in which they float, and the velocity or ease with which they themselves proceed, they often present an undulating course, and leave behind them streams or tails of glowing vapours.

According to the distance of meteors and shooting stars from the earth, the direction of their passage will be influenced by the currents above, and become more and more uniform the higher they ascend. The difference of temperature between the poles and the equator causes at all times currents of water and of air and of moisture between them; and these currents within our atmosphere of air are often disturbed, crossed, and altered by local and other causes. These causes of altering currents do not, however, exist beyond our ocean of air, and the oceans above it partake only of the general movement from east to west.

Within these superior oceans, however, positive electricity, kept off from the earth by the non-conducting air, is at home and accumulates on the confines of our atmosphere, watching, so to

---

* The brilliancy and colour of meteors is readily explained when we bear in mind the dazzling light produced by putting a grain of sulphur into highly heated molten saltpetre, and that all the elements of these bodies are contained in the air, brought together and acted upon by electricity.

say, every opportunity to come down and unite with the negative electricity below. This union is greatest and most intense at the equator; and though we are not as yet acquainted with the law of electric compensation, whether planetary, solar, or universal, we are certain that the greatest accumulation of terrestrial and atmospheric electricity exists at the poles. If, therefore, the evaporation of negative electricity at the equator is maintained by an uninterrupted current from the interior of the polar regions, the positive electricity of the air at the equator, spent in union with the former, is supplied by currents flowing above our atmosphere of air from the poles to the equator, uninfluenced by the configuration of the land and its interior organization, contrary to the opposite element.

When these currents, *en rapport* with the earth, in annual ebb and flow pass and repass from the poles to the equator, this *rapport* touches or strikes a corresponding electric cord in the organization of the birds of passage, in harmony with which, exhilarated, they rise until they reach the respective altitude of the conductor which awakened within them their slumbering instinct, and are safely guided by it into distant regions.

When in the same manner meteoric columns or accumulations rise to the requisite altitude, and by repulsion of agglomerated particles disperse into numberless fragments; each separated fragment will again form its own sphere, and like birds of passage, as shooting stars follow the course, more or less, of these electric currents; and hence it arises, that all seem to move in the same or similar direction, and to follow the magnetic meridian of the earth, with a strong westerly deviation on account of the westward flow of the fluid matter above, caused by the eastward rotation of the earth itself. This latter circumstance will from hour to hour remove further away from the zenith the point of radiation of meteors, or shooting stars, ascended during the day or in the course of the evening, and make it appear as if the spot from which they seem to emanate were located, nay almost fixed, in the constellation in which first they were noticed. This, however, refers chiefly to those periodical swarms of shooting stars, which, more or less, last for several, even six to eight hours of the evening, the greatest number appearing generally about midnight.

The density and velocity of the aforesaid currents of positive electricity influencing the direction of shooting stars will naturally also affect their brilliancy, motion, and duration. The existence of electricity in the superior regions of our planet, if not the consequent prevalence of electric currents within and high above our atmosphere of air, was entertained by Poisson, the friend of Humboldt. In Note 63 of I. vol. of *Cosmos*, he is quoted as follows : "At a distance from the earth where the density of the atmosphere is altogether insensible, it would be difficult to attribute, as has been done, the incandescence of aerolites to friction with the molecules of the air. Could it not be supposed that the electric fluid, in a neutral state, forms a kind of atmosphere, which extends far beyond the mass of air ; which is subject to the attraction of the earth, and which, consequently, FOLLOWS OUR GLOBE IN ITS MOVEMENTS ? On this hypothesis, the bodies in question, entering into this imponderable atmosphere, by their unequal action on both electricities, would decompose the neutral fluid, and thus electrifying themselves they would become heated and incandescent."

A recent and more important approximation in favour of my theory I find in the *Reader* of 19 March, 1864 ; a letter of Mr. Haidinger, Vienna, to M. Van Benden, of the Académie Royal de Belgique, is noticed there in which he says : "The identity of the three kinds of fiery meteors, viz: shooting stars, meteorites, and star-showers seem to me undoubted ; but I do not feel justified in expressing any decided opinion on the height of the atmosphere. Mr. Quetelet has assumed that I do not share his views as to the height of the atmosphere being greater than is generally supposed. I may, however, say that I take the same view of the case as he does, both with regard to the greater height of the atmosphere, and also to the superposition of two layers of a different nature. The lower of these, the unstable atmosphere, partakes of the rotatory motion of the earth, and is subject to the influence of currents, and to other variations, whilst the upper layer—the stable atmosphere—is of much less density and is relatively at rest. There is no doubt that the latter follows the earth in its annual revolution, but whether it partakes of its diurnal motion is a point which for the present remains undecided."

All this closely harmonises with what I have stated before, and having in 1858 had the pleasure of paying a visit to the eminent physicist and being most affably received by him, I am all the more gratified of having the concurrence of so high an authority respecting the connection between meteors and shooting-stars, and of at least ONE further ocean of matter above our atmosphere of air and different from it.

The link, then, which I have shown to exist between lightning and the highest and brightest globes of light is evident proof that these phenomena occur within the sphere of the earth, whatever distance the meteor may seem to be at, and that this sphere of ours is more than likely to embrace its faithful companion—the moon herself.

As I have once before noticed, the atmosphere of the earth has been calculated to rise to an elevation of about 45 miles, but it seems that an extension, though small, and still far from the moon, has been gained ; for, in the *Athenæum* of 11 October, 1862, evidently alluding to my new theory never before entertained by any one else, Professor Challis, at the meeting of the British

Association for the advancement of Science, is reported to have said : "It has been generally supposed that the earth's atmosphere is about 70 miles high, but on no definite grounds, and the estimates of the height have been very various. Against the opinion that it extends as far as the moon, it was argued, that, as the moon would in that case attach to itself a considerable portion by its gravitation, which would necessarily have some connexion with the rest, there would be a continual DRAG on the portion more immediately surrounding the earth, and intermediately on the earth itself, which would in some degree retard the rotation on its axis. Hence, if, as there is reason to suppose, the rotation be. strictly uniform, the earth's atmosphere cannot extend to the moon."

In this argument there are no less than three fallacies to be noticed.

If, in the first instance, the moon does float within our sphere, she no more attaches to herself a portion of the element she rides in, or upon, than a balloon attaches to itself a portion of the air by which it is surrounded, or the vessel a portion of the water on which she sails.

If, in the second instance, the moon did do so : the smallness of such attachment, and her distance from the earth, would put entirely out of the question a DRAG upon the air immediately surrounding the earth, and still more so a retardation of the rotation of the earth on its axis.

If, in the third instance, all were true respecting the attachment and the drag, then the important and indisputable fact still remains : that the present rotation of the earth on its axis is ITS NORMAL CONDITION, whether the moon, according to my theory, is actually as much attached to the earth as a balloon in the air, a ship on the water, or a fish in the sea, or whether she is as distinctly separated from the earth as Mars and Venus.

Hence, it will appear a strange conclusion to the thinking

reader, that, "if the rotation of the earth be strictly uniform, its atmosphere cannot extend to the moon;" or, which is the same: "the rotation of the earth is strictly uniform, THEREFORE its atmosphere does not extend to the moon," which, nevertheless, my theory of the tides will prove it does, and that consequently, meteors of every description have their beginning and end in our own sphere, still embracing the moon.

Before suns were kindled in the firmament of heaven, most likely during the third day of creation, when the dry land was separated from the water, when the earth in labour was heaving up towering mountains, furrowing the bed of the sea and opening its bosom in numerous clefts and volcanoes, God may have willed it that, like a nascent meteor, an electric accumulation, the moon rose up from our planet, until agglomerated to repletion it ignited in the atmosphere of moisture and gases into which it had ascended, already enveloped, perhaps, by an ocean of hydrogen, and crystallizing—without leaving any thing loose or liable to drop down upon the earth—into a body full of pores and vesicular cavities, instead of fusing into a compact meteoric composition or dispersing in numberless fragments.

As molten nickel poured into water will form little hollow globes, so the electric accumulation of the future moon, under favourable combinations, may have blazoned up without consolidation or diffusion, and, though in other proportions, similar to the meteoric stones with their pores and vesicular cavities mentioned by Humboldt, *Cosmos* p. 119, been converted into a cork-like ball, whose extensive vesicles and hollows—perchance filled with the hydrogen in which I suppose our satellite to be immersed—with its vast craters and cavities, would admirably be calculated to keep her aloft on the expansive ocean on which she found her level, and on which, preserving a semi-electric condition, she continues to float like a discharged vessel, less dense and heavy above than below, within the sphere that gave her birth, in order to reflect light on the body from which she proceeded.

The planets themselves, though as organic bodies, in a similar manner, in meteoric progression, MAY have risen from the suns round whom they revolve, when these primaries, by the Word of God, had been distributed throughout creation, each one forming strata after strata of attracted matter around their own electric accumulation by feeding upon the surrounding chaos, and thus preserving their floating, and ultimately revolving capacities. This, indeed, seems to me more likely, than that, in the words of Sir David Brewster, according to the theory of Olbers, "thirty asteroids,* relics of a once mighty planet, are revolving in dissevered orbits, and warning the vain astronomer of another world that a similar fate may await his own." That a planet should be shattered in pieces, and its thirty or more solid fragments, by some abstract magic power, assume the form of globes; that these fragments should become centres, and accommodate themselves to the reputed law of universal gravitation, keeping each other apart, and yet not beyond a certain respectful distance, and that, irrespective of the impulse they received, when torn asunder, in the direction of every quarter of the universe, they should continue to revolve in the same plane and in the same orbital space in which their formerly common body is said to have been whirled by the hand of the Almighty, is to me too much to comprehend.

The interior of that mighty planet, the same as our earth about 30 miles below the surface is said to be, was, no doubt, one mass of fiery molten matter; but what became of this interior sphere of liquid fire when its shell was rent in pieces, we are not told. Whether, exploding like a rocket, in streams of lava-lightning it furrowed the heavens in every direction, dissolved and vanished

---

* Since this was written by Sir David Brewster, 50 more asteroids have been discovered, so that there are now 80 in all.

as cosmical vapour ? Or, whether, blown up into fragments, these also, by some unconscious, mysterious agency, formed into globes; and cooling and hardening were turned into LITTLE planets, coursing in the path of their parent to whose destruction they owed their diminutive existence ? Or, whether the white-heat liquid was scattered and dispersed into millions of meteoric bodies, swarming round the sun in different orbits, and in their periodical tour closely pass us by, manifesting themselves as luminous balls and shooting stars, and by explosions and the sending down of incandescent memorials remind us of their volcanic origin and condition?—About all this the annals of astronomy are silent, and perhaps all the better; for it appears, according to *Hind's Introduction of* 1863, p. 96, that "the bursting of a planet" has recently been given up, that "a fearful catastrophe of a great planet can hardly be said to be the prevailing opinion at the PRESENT TIME as to the origin of this remarkable group of planets."

The number, time, and periodicity of meteors and shooting stars will depend upon general as well as local expansion of the air by the heat of the sun, upon the thaws and frosts of the Arctic regions, upon the tranquil, dry, or moist state of the atmosphere and of the superior oceans, upon local warmth of the land or the sea, upon the evaporation of electricity from volcanoes, and upon the electric currents from the poles to the equator.

Which of these circumstances may combine, or in what manner they may be favourable or unfavourable to meteoric phenomena, I have not had time or facilities to examine; but as regards the general temperature of the air: the increase of heat, facilitating the escape of electricity from the earth, seems to bring with it an increase in the number of shooting stars. For, according to observations made at Paris, and generally in the north of Europe, the mean number of shooting stars during the first six months of the year is only three to four per hour, whilst during the last six months it rises as high as six to eight per hour, and with the departure of the sun to the south decreases again to the minimum.

The two maxima occur in August and November, the one when the heat of summer is greatest and thunderstorms on the wane, the second when a moist state of the atmosphere steps in between the rains and dryness of summer and the snows and cold of winter, and when the electric currents of the upper regions begin again to flow more strongly to supply the consumption of the equator, which in our summer chiefly, if not exclusively, is supplied from the south. That the extraordinary swarms of shooting stars which, at the aforesaid period in particular, fill us with wonder and admiration, have no cosmic origin, and by no means are annual or periodical travellers around the sun, I hope sufficiently to have established; but there is more positive evidence which unmistakingly speaks against these suppositions.

Whether these puzzling meteors are the offspring of exploded planets, or otherwise were launched by God into space and into our solar system :—in either case it would be wonderful indeed if, in their course round the sun—not only setting aside the darling law of universal, but also the intruder of solar and planetary, gravitation—this, their centre of attraction, would let go his hold upon the little things, permit them to peep down into our garden, and even allow a good many of them to drop in uninvited. It would further be truly wonderful if, with their astounding rapidity of motion of from 10 to 70 miles or more in a second, as has been calculated, some should at once stand still in their career and quietly burn out their existence; that so many should stop short in our atmosphere, and not only stand still, but perpendicularly fall, and some others perpendicularly rise, with almost lightning velocity; that others again should incline in their fall, and numbers of others dart and disappear in lines horizontal; that often they should move up and down in every direction; that,

like birds of passage getting lame or tired, some should come
down as meteoric masses of stone or iron, others explode and
dissolve into vapours, and others suddenly light up and as suddenly extinguish, and with others in their normal, but unseen and
never-seen, condition, continue their rapid flight; that their race
round the sun should bring them down so deep into our atmosphere as to be mostly no higher above the earth than from 5 to 90,
and in the rarer and exceptional instances from 120 to 180 and even
200 miles, and that notwithstanding we should see them only for a
few seconds from the time of their entrance into to that of their exit
from it; that the lower meteors should be the SMALLER AND MOST
NUMEROUS, and the higher ones the LARGERST AND FEWEST; that
shooting stars and luminous balls should be seen even at plain
noon, in the morning and evening, but in by far the greatest
number at midnight; that they should be seen at all times of the
year, particularly from summer to autumn in every zone, in the
northern and southern hemisphere, at every height above the sea,
above plains and mountains; that consequently the earth, as it
were, should constantly be clothed, completely enveloped, by these
swiftly-passing lights, mostly taking the direction of the magnetic
meridian or a more westerly course, and in the southern hemisphere most likely the same, similar to the cyclones rolling from
north and south towards the equator, and following the sun to the
west; that throughout the year, as from a regular source or
fountain, emanating from the higher region of our pole, they mostly
should seem to proceed from one or two circumpolar constellations,
encircling the dome of the earth on every side and glorifying God
as the lights more humbly do that illuminate the dome and tomb of
the Prince of the Apostles; that, if coming from cosmos and everywhere surrounding the earth—unlike rain and hail falling upon a
globe—so few should strike upon the body of the earth, but
mostly higher above pass us by, and dissolve or vanish in our

atmosphere; that in the case of those precipitated to the earth, the sun should cease to exercise his inherent right and power of attraction, whilst, notwithstanding, these bodies are said to obey the laws of gravitation: yet ALL THIS is incompatible with, and loudly protests against, their cosmic origin, whatever the kind of meteor; protests against the closed ring in which those swarms of myriads of shooting stars appearing in August and November every year are said to course round the sun, and in their passage across the earth's orbit to touch, or come within, our sphere; and against the declaration of Mr. Faye,* "that not only does our earth," as I have seen it so beautifully DRAWN ON PAPER, "pass through a meteoric ring at that time, but that the sporadic meteors seen from time to time," more or less the WHOLE YEAR ROUND, "are BORROWED FROM THE RING AND BECOME ACTUAL TERRESTRIAL SATELLITES," though of a very AIRY nature, "UNTIL THE EARTH'S ATTRACTION PROVES TOO MUCH FOR THEM;" but ALL THE CIRCUMSTANCES connected with the phenomena of meteors and shooting stars are readily accounted for by my simple theory.

In Lisbon I happened to be present in a foundry whilst melted iron of great purity was drawn from the furnace; and as iron, when chemically pure, will ignite when coming in contact with air or oxygen, so each time the flowing metal first began to run into the crucible, little sparks, but mostly invisible particles, quickly and in great numbers, as from a playing fountain, flew up into the air, and by their contact with it, suddenly broke out into light, gently exploded, and filled the space about with a shower of falling stars, each one emanating and sparkling out from its tiny nucleus, and in most instances leaving no visible fragment behind. Hence also the unfrequency of being hurt by these beautiful little meteors,—for they were nothing else—though often falling about like flakes of snow. They never enkindled until they had risen, and

---

* See *Reader*, October 3, 1863.

then, in a graceful inclined arch, exploded and vanished in their oxidation.

Never have I seen a more beautiful illustration of luminous meteors than in this instance, and it could not but confirm me in my theory, that chemically-pure iron chiefly must also be the basis in the formation of shooting stars and all kindred phenomena.

These atoms of iron, with other mineral or metallic solutions, filling the air, aggregated into spheres, or otherwise accumulated or condensed, as I have stated before, are reduced into their chemically-pure elements by the warmth and action of the electric currents more or less constantly or regularly flowing from about the zenith of the pole towards the equator; and being in contact with our superior ocean of oxygen, or coming down into, or floating within, our atmosphere of air, these agglomerations on a grander scale will likewise explode and spread about the oxidizing particles, in their rapid flight often leave tail-like traces behind, and thus present to us all the varieties and conditions of luminous bodies and shooting stars.

Whether meteoric phenomena emanate in greater number from the sea or the land, is a question to be settled by observation, though I think that but very few have their ORIGIN on the sea ; for, the same reason which prevents thunder and lightning from want of conductors, prevents also the rising of electric streams, and their subsequent formation into spheres. One instance where such a sphere was formed ON the water, confirmatory of what I have said, is recorded in the *Philosophical Transactions:* " It happened on board the Montague, on the 4th of November, 1748, in latitude 42° 48', and 9° 3' west longitude, about noon. One of the quartermasters desired the master of the vessel to look to the windward, when he observed a large ball of blue fire rolling apparently on the surface of the water, at the distance of three miles from them. It rose almost perpendicular, when it was within forty or fifty yards from the main chains of the ship; it then went off with an explosion, as if a hundred cannons had been fired at one time, and left so strong a smell of sulphur, that the ship seemed to contain nothing else. After the noise had subsided, the maintopmast was found shattered to pieces, and the mast itself was rent quite down to the keel. Five men were knocked down, and one of them greatly burnt by the explosion." The ball was said to have been as big as a millstone.

It will also be a matter of observation, whether shooting stars diminish when the aurora borealis is prevalent; whether meteors will be scarce when and where thunderstorms are frequent; whether meteors and meteoric stones fall about the time of volcanic eruptions; and whether meteoric phenomena are most frequent and numerous in the tropics.

When in the higher latitudes these meteoric emanations from the land or the sea are arrested by the cold state of the atmosphere, by clouds of frozen vapours or snow, they will give us that most magnificent of all phenomena, the—

AURORA BOREALIS.

The accumulation of electricity, like the density of the air and of oxygen, is greatest at the poles, owing to the intense cold, the rarefaction of the upper regions of the atmosphere, and the snow and ice which cover land and sea, and which form a barrier to the escape of negative electricity from the earth. The accumulation, also, is greater at the north than at the south pole, the former having more land, and the latter more water, and the one holding more electricity than the other. The evidence of this accumulation, of these reservoirs of electricity, lies in their attraction of the magnetic needle; for, the greater the quantity of electricity, the greater the amount and force of magnetic power; and the greatest magnetic attraction is that of the north pole.

According to all accounts, the aurora borealis takes place after a thaw, though volcanic action on land and on water, as it produces clouds and thunderstorms, may also be one, if not a chief, cause of its production. By this means, and in proportion to the rapidity of the thaw, an immense quantity of negative electricity confined in the earth, and accumulated on its surface under the cover of snow and ice, is liberated; and trying to rise through the air which opposes its passage, it creeps up at the sides of hills and mountains, which abound in the Artic regions, and ascends also with the rising vapours.

Electricity is the binding and dissolving element of bodies, the same as fire and water, which decompose some and solidify others : electricity gathers the vapours of the air and concentrates them into clouds, whilst within them it accumulates, and by its repulsive or expansive property prevents their aggregation into water; but no sooner has the cloud commenced to discharge its electric accumulation, than the same operation which decomposes water into hydrogen and oxygen, and reduces these gases again into water, begins to liquefy the aqueous vapours into drops of water to fall down on the land from which electricity arose to attract and collect it. If a cloud rises very high, and in its elevated position is unable to discharge its electric accumulation by communication with electric accumulations from the earth, the positive element will more and more augment and concentrate within the cloud, until overcharged, it suddenly contracts and violently combines with the vapours, at the same time breaking them up into numberless bodies, to precipitate or send them down to the earth in the form of hail, or, as in regularly cold seasons, and with a more gradual operation of positive electricity, in flakes of snow. At night, when the atmosphere has cooled down and is less expanded, clouds cannot attain the altitude which they rise to during the day; they, therefore, are not liable to be quickly transferred to a higher and colder stratum, to be there exposed to an increased accumulation of positive electricity and a sudden contraction; and hence the scarcity of hail at night.

It has been observed by Humboldt, that in fogs, and at the beginning of the fall of snow, in the plain as well as between 10 and 14,000 feet above the level of the sea, in cold as well as in tropical regions, the atmosphere quickly and frequently changes from one electrical state to the other; and this change, in the case of the clouds, seems, beyond question, to be in connection with, if not the main cause of, the formation of hail.

Wherever contraction is concerned, positive electricity seems mainly the operating agent. In winter, when all is cold, when the normal state of the atmosphere is to the highest degree positively electric, when snow and ice cover land and water, when negative electricity has retired from the vegetable creation into the bosom of the earth, all nature seems contracted. When metals are cold, retaining even the skin of the warm hand that touches them: is it, that warmth, that negative electricity, has been driven out of them, and positive electricity takes its place? Or is it that metals are chemically combined with the positive element? that in metals, perhaps, we find positive, or, as may be, negative electricity in its most compressed state? that positive electricity not only explodes meteoric accumulations, but enters into combination with the metallic solutions of our atmosphere to produce meteoric stones? That, as fire and light, an evolution of electricity on a small scale, warm and kindle by induction, and contact, and chemical decomposition, solidify and liquefy bodies: electricity chemically combined with metals, and effecting the same on a large scale, is freed and decomposed again by electric batteries? As water is a compound of oxygen and hydrogen, and absorbs air containing oxygen: are metals, as they seem to me, a compound of some element or elements with electricity, which they absorb still by contact with electrified metals or magnets? If my theory be true, it will be an interesting subject for inquiry, what relation there is between magnetization and the specific gravity or density of metals, as also, whether iron could not be saturated with some substance or other to prevent its magnetization in shipping, &c? Whether acids are not minutely condensed particles of electricity, a condensation of electric matter?

These and other questions naturally present themselves to the mind, and seem to me to be objects of the highest importance, though for the moment they lead away from the subject.

When water begins to congeal and to freeze: is it not likely that warmth, that negative electricity gives way to positive, the same as no doubt it does in the change from rain to hail and snow? that, having been contracted and taken possession of, the newly-formed ice will expand again by a continued and passive accession of the positive fluid, by nature expansive? That by these means ice will become lighter and float upon the water from which it had been formed? That ice and snow melted by the rays of the sun will dissolve into vapours, of themselves highly electric, and thus all the more be calculated to assist and promote the formation and development of the aurora borealis?

Whether vapours thus disengaged will lose or change their electric condition, whether they will be transferred into fogs, into clouds of rain, or again into snow or icy particles, will depend on the temperature and the electric condition and currents of the superior air; and this, no doubt, will determine the way, form, colour and duration of the phenomenon in the amalgamation of the two electric fluids near the poles.

From the nature of the aurora borealis, the COLD state of the upper atmosphere, and the general rapidity of the thaw, it seems to me likely that the electricity of the earth rises, both by the conduction of mountains and vapours, generally without any of

G

the electricity above coming down any distance into our atmosphere, or settling in clouds, as it does in thunder storms and waterspouts. Hence, no condensed body, or charges of positive electricity, will in this case be formed, and the fluid remains free and passive, as it were, though accumulating by the attraction from below ; and as by electric discharges in the rarefied air of a PARTIALLY exhausted receiver, the glow, which, according to Faraday, precedes the brush and actual explosion, is readily obtained, so the negative streams, or negative clouds of vapour, coming up and in contact with the former, the silent lightning, the sheet-lightnings of our summer evenings, will take place, as also has been shown before: but by the presence or vicinity of the ocean of oxygen which I believe to exist above our atmosphere of air, by the composition of the air itself, as well as by other conditions, this silent lightning, this glow, will be changed into the Northern Light as it most commonly appears.

The shape of the earth, the ring of vapours or clouds that gird the Polar regions, the atmosphere floating like an arch on the globe, may account for the curve or arch in which the aurora borealis presents itself to our view; but it seems to me more probable, that electric columns rising up at extreme points of the plane or place of evaporation, without any, at least lofty, volumes rising up in the centre or intermediate parts, at these points draw down the positive fluid nearer to the earth than it can descend in the middle ; and hence the appearance of a vaulted firework.

One of the experiments of Sir H. Davy proves that electric accumulations and currents are influenced by magnets. It is, therefore, quite evident, that the electric volumes rising from the earth have power to draw down and to draw out of their course and altitude the currents and accumulations of the positive electricity above. "He placed" (quoting from Dr. Lardner) "two pieces of charcoal in connection with the wires of a powerful voltaic battery, and, by presenting their points towards each other, at a distance varying from one to four inches, according to the density of the air in which the experiment was made, he obtained a column of electric fluid formed by the current passing through the space between the charcoal points. This current was not transmitted, as usual, along any conductor, but merely passed through the air between the points, and its presence was rendered manifest by the light evolved. When a powerful magnet was presented to this column with its pole at a very acute angle to it, the current was attracted or repelled with a rotatory motion, or made to revolve by placing the poles in different positions, in the same manner as metallic wire conducting the current would have been. The electric column was more easily affected by the magnet, and its motion was more rapid when it passed through dense than through rarefied air; and, in this case, the conducting medium, or chain of aëriform particles, was much shorter."

The same influence which in this instance was exercised by the magnet upon the electric current between the two pieces of charcoal, is, in the case of the aurora borealis, exercised by the electric volumes rising from the earth in certain places and at certain times upon the currents of the superior regions, to which their points or poles form more or less acute angles.

Mr. Biot, according Dr. Lardner, " conceives that the luminous columns composing the aurora have not in reality the position or form which they appear to the eye to have ; but that their apparent form is merely the result of perspective. He considers, and truly so, that the phenomenon is produced by an infinite number of luminous columns, parallel to the dipping-needle and to each other, arranged side by side at nearly the same height from the surface of the earth ; these systems of columns being placed at unequal distances from the eye, and seen under different angles of obliquity, are projected into various figures, which are subject to variation arising from the varying splendour of their component rays." Add to this, that clouds of some sort or other will also form in the upper regions, or be carried thither by atmospheric currents, charged perhaps with ponderable matter, and collecting and accumulating the positive electricity of the higher sphere ; that these clouds will, dispersed or united, correspond in position to the negative columns or clouds rising from the earth, as in the case of waterspouts cone corresponds to cone ; and that for want of conductors, where trees, hills, and mountains are scarce, clouds at the formation of the aurora borealis will always assume the conical figure of the waterspout ; that there may be a discharge of electricity between two clouds, one above the other ; that, itself a dark segment, an electric cloud may send up lofty cones into the upper regions, and communicate also with the space below, as sometimes is the case with thunderclouds ; that meteoric, globular accumulations from the earth may be attracted and concentrated over a certain area by a cloud or clouds, or their ascent impeded by too cold a state of the upper strata of air, and thus ignite and break out into streams of light ; that above all there is a constant overflowing of positive electricity at the poles, and returning in regular currents high above us to the equator ; that these currents alone are sufficient to account for all the electric accumulations of the higher regions, that, in fact, the aurora borealis will form itself into all shapes, if over against numerously-ascending columns of negative electricity, positive volumes or accumulations arrange themselves from the beforementioned causes : and it may easily be imagined that the amalgamation of the two electricities under various atmospheric densities and conditions will produce all those splendid displays of the phenomena which so often have dazzled and charmed the beholder.

Brilliancy and colour, as in the case of meteors, depend upon the proportionate presence of atmospheric air and its constituents, on the state and composition of clouds and vapours, and particularly on the refraction and reflection of light in the arctic regions ; the direction and motion of the aurora borealis will be influenced by the electric fluid from the earth, currents of the atmosphere, and the rotation of the earth itself; but there can be no doubt that it is itself nothing but the amalgamation of the two electricities, in a different clime, under different circumstances, and for a different end.   Almost all philosophers of natural science are agreed, that the aurora borealis is the result of electric agency ; but according to Dr. Lardner, Eberhart, Professor of Halle, and Paul Frisi at Pisa, were the first who proposed the explanation of the phenomenon being the effect of atmospheric electricities, founded on the following facts :—

" 1. Electricity transmitted through rarefied air exhibits a luminous appearance, precisely similar to that of the aurora borealis."

" 2. The strata of atmospheric air become rarefied as their altitude above the surface of the earth is increased."   Hence they argued, in the words of the Doctor, " that the aurora is nothing more than electric discharges transmitted through parts of the upper regions of the atmosphere, so rarefied as to produce that peculiar luminous appearance which they exhibit."

Whether the aurora borealis is more frequent and intense in severe or mild winters, whether it is chiefly local, and happens about the time of electric high tide, of which latter I shall treat hereafter, I have not been able to enter into and to ascertain. But it is a most conclusive proof of my theory of an electric organization of the interior of the earth, and of an exterior corresponding operation of electricity, that—difference of geographical situation being taken into consideration,—polar lights have often been observed at the same time in the northern and in the southern regions of the earth.

The last aurora, as recorded in *Nature and Revelation*, Münster, were seen at the same corresponding time, the one at Münster, Westphalia, and the other at Melbourne, Australia, on the evening of the 9th, and morning of the 10th of April, 1858, respectively ; the magnetism of the earth was violently disturbed, and the effect upon the magnetic needle simultaneously observed at Melbourne, Bonn, Göttingen, Prague, Vienna and Christiania.

Other, and later, instances of corresponding polar lights, with great disturbance of the MAGNETISM OF THE EARTH whilst the electric tension of the ATMOSPHERE WAS NOT PARTICULARLY DISTURBED, with variation of the declination and INCLINATION of the magnetic needle, magnetic intensity and effect upon the telegraph wires, are recorded in the first number of the above named work of 1860. In one case of a magnetic storm the magnetic disturbances were very great and peculiar, and appeared to Mr. Neumayer, the observer at Melbourne, as the effect of a polar light and of an earthquake. A polar light had for some time preceded this magnetic storm.

It has also been noticed, that the auroras, or polar lights, are most frequent in those countries that are nearest to and lie on the side of the magnetic pole. In North America, for instance, being on the side of the magnetic pole, they are far more frequent, and come down further south than in Europe or Asia ; there also the magnetism of the earth is far more intense than in the latter parts of the earth.

This makes it clear, that there is a correspondence between the northern and southern auroras, consequent upon an INTERIOR CORRESPONDENCE of the earth ; that, like an induction apparatus, filled with soft iron, the EARTH, filled with metallic substances from pole to pole, is the primary cause, next to the sun, of the electric phenomena of the air.

The aurora borealis, therefore, must necessarily exercise a very sensible effect upon the magnetic needle, owing to the immense quantity of electricity liberated and brought into action from a state of rest, by the time it has reached its height, or the point of culmination ; and the disturbance in the previous magnetic state of the atmosphere must obviously be greater or less, according to the intensity of the electric evaporation and the development of the aurora.

The amalgamation of the two electricities, as in the case of waterspouts, is not necessarily accompanied with thunder, though the development of the aurora borealis, according to the competent authority of Hansteen, and some others, is at times attended with a crackling noise ; and this may be accounted for by the electric sparks coming, on these occasions, from a more or less compact segment of the aurora, more or less in contact with the denser, and to us, nearer, part of the atmosphere.

When the phenomenon takes place on the confines of our atmosphere of air, it is too distant, and the air too rarefied to produce or transmit to us any sound; for, sound is best produced and propagated where the air is densest, and where the bodies or substances causing the sound, and giving character to it, are those which are the most solid. Hence, in the case of thunderstorms even, thunder will always be loud in proportion to the density of the cloud, and weak in proportion to its attenuity. Humboldt has proved also, that sounds are audible at greater distances by night than by day ; and this arises from the greater density of the air at night to what it is by day.

IV.

SOUND, ITS CAUSES AND ORIGIN.—PROFESSOR FARADAY.—THUNDER, SUBTERRANEAN GRUMBLING.

SOUND, the most charming and soul-stirring of all the wonders of a bountiful Creator, is the effect of a more or less violent friction of the air with more solid bodies; it is an agitation of the air striking against more opaque bodies than itself, modified by the substance with which the friction takes place, or by which the agitation is produced; and the agitation itself is caused by a displacement, or by a disturbance of the equilibrium, of the air in one way or another.

When, originated by the pressure of the moon or the cloud, or by the electric phenomena of land, sand and waterspouts, the tempest roars over the sea, sweeps through the forest, assails the lofty castle, and whistles and howls about our dwellings; there is a violent friction of the air, and its sound tells us of the awful power with which God invested the element of our life. When the hammer strikes upon the bell, when lightning flashes, when the waves of the sea run up against the shore, there is a violent displacement and consequent agitation of the air; when the organ swells, the trumpet sounds, and music of every kind steals upon the ear, it comes from a more gentle, but still the same friction, displacement and agitation of the air, and fills us with delight; we have it in the song of the mother, and in the laugh of the child; in the discourse of the man and the merriment of the boy; and these human voices are the tones that are sweetest and touch us most, and above all, when coming from those whom we love; but all is friction, displacement, and agitation of the air wherever utterance was given to animated creation to proclaim the glory of God. And in all these cases, the volume, power, and weakness of sound is proportioned to the force of friction or agitation, its character established by the substance acted upon and producing agitation, and its scale determined by a concentration or extension of space of displacement, a concentration or extension of agitation or vibration.

The propagation of this agitation, of sound, depends upon the nature and state of the bodies in connection, in contact with, or contiguous to the spot or place where the displacement of the air originated, as also upon their form. If these bodies are smooth or polished, particularly when trumpet-shaped, sound may be conveyed to an extraordinary distance, for, in the latter case, the waves forced out of the instrument will remain concentrated, and thus overcome resistance a long time before diverging; if, however, they are uneven, or porous, sound will travel but a short way, and lose its intensity, and this for the following reasons:—

If a stone be thrown into the water, rings will propagate all over the surface until the original force is spent, the rings die away, are intercepted, or broken at the shore; ordinarily there is nothing to impede the extension of the circle of the wave on the surface of the water. But as the stone is sinking to the bottom, the rings around it will soon have ceased; everywhere surrounded and pressed upon by the water, the displacing force of the stone neutralized and destroyed, and the water from all sides rushing in upon the space it has passed through as it goes down, no ring can form, much less propagate, after the first moment of immersion.

It is the same with sound produced in the air. Surrounded by air, it diffuses in every direction like the rays of light; and the air also rushing from all sides upon the spot where displacement and agitation began, prevents its propagation to any particular extent. But wherever the air presents any surface, and according to the nature of this surface, sound travels to a much greater distance.

This surface of the air exists wherever it leans against a solid,

liquid, or condensed fluid body; resting upon the river, the lake or'
the sea, the lower surface of the air reposes on the surface of the
water; on land the lower surface of the air rests on rough and
broken ground ; there is a surface of the air pressing against every
body in nature capable of pressure: there is a surface of the air
on the slab of the table and underneath it ; there is one against
the wall, the floor, and the ceiling ; in fact, against every tangible
object.

If these surfaces of the air lean against surfaces of bodies polished
or uninterruptedly even, sound will travel the longest distance,
because the wave originally created is propagated without impediment until it naturally dies away ; but if the surface of bodies is
rough, loose or porous, the wave of the air passing over it is everywhere and every moment impeded, broken, or absorbed, and consequently soon exhausted or destroyed. Hence it follows that sound
is conveyed by the air only, according to the surface and form which
receives the first undulation, and that it is sustained by the length
of time required to bring the agitated air into a state of rest ; and
this will last longer on smooth and rounded surfaces than on rough
and uneven or cornered bodies, as it will also depend on the oscillation and swinging of bodies, by or on which the friction of the air,
the sound has been produced, until they likewise come to a state
of rest.

When cannons are fired at sea, it is not the water that conveys
the sound to an extraordinary distance, but the displacement and
consequent wave of the air caused by the explosion, and propagated without impediment over the liquid surface. If the water is
agitated, its undulation will, as it were, give light and shade to the
sound, and make it resemble the distant, scarcely audible, rolling
of thunder. When the tramping of horses is heard at a greater
distance by putting the ear close to the earth, than it is when heard
within, or through, the body of the air; it is not the earth that

conveys the sound, but the agitated surface of the air resting upon it. Hence also the fact, that apart from the greater density of the air, sound is propagated and heard to a greater distance in winter than at any other time, because snow covers the ground, and presents an extended and smoother surface to the air-wave passing over it, particularly when the snow is not soft, but more or less frozen over.

When in the whispering gallery of St. Paul's a person speaks across the dome, he will not be heard on the opposite side, because the sound produced, meeting everywhere with resistance, and weakened by diffusion, soon dies away in the air; but let him speak close to the wall, and the wave of sound created on the surface of the air leaning against it, will immediately pass round the whole circle, because there is nothing to hinder its propagation It is not the wall, as is supposed, that conveys or deflects the sound, but the surface of the air resting against it.

If a metallic rod be made to strike against a chair, table, or other body : it is not the metallic rod through which the sound is conveyed to our ear, but by the waving air embracing its polished surface. If a beam of timber be struck, or scratched with a pin or nail at one end, and the sound comes to your ear at the other as passing through the wood itself : it is not the wood that vibrates or conveys that sound, but the wave produced by the scratch or stroke, and propagated on the surface of the air embracing the surface of the beam, and concentrated again in one focus at the opposite end.

If, as was done by Professor Wheatstone at the Polytechnic Institution, a piano or other musical instruments are placed in the cellar or souterrain of the house, and the instruments brought in communication with an upper appartment by means of staves carried through the several intervening ceilings, the music performed below will, as if present, be heard above. For this purpose the staves must be in connection with the sounding-boards of the instruments in the cellar as well as with the sounding-board of some musical instrument above.

In this case, however, it is not, as has been imagined, the wood, the staves, that convey ex- or in-teriorly the sound from below to the room upstairs, but the wave of the air passing along the staves. The staves, like all conducting bodies of sound, are entirely passive,—they are innocent of the transmission of harmony along their surface. Had the apertures by which the staves were carried from ceiling through ceiling been made to close air-tight upon them, no sound whatever would have passed. Had the sounding-boards of harps or other instruments above been removed, the sound from below would not have spread over the apartment, but only been able to be heard by listening at the extremity of the staves, from which it would escape into the free air of the room, and there be lost by speedy diffusion. Had the experiment been reversed, and the music performed upstairs instead of below, the sound would not so easily have been heard in the cellar, because the vertical descent of an air-wave, however free to move, is against the natural upward current of the atmosphere, itself the result of a constantly higher state of warmth of its lower strata.

When the bell sounds the merry peal, it is not the shaking or motion of every particle of the metal that causes and diffuses the sound, but the violent concussion of the hammer with the bell suddenly forcing the air from its place, and driving the wave all over the bell into the open atmosphere, in proportion to the intensity of the stroke. If dust has accumulated on the bell, it will change place according to the wave of air which passed over the surface, but not on account of any motion of the atoms or particles composing the metal.*

---

* If light particles of dust be on the outside of a bell when it is struck, you

Professor Faraday, in his lecture on Friday, December 26th, 1856, according to the Morning Chronicle, says: "Observe the elasticity of a bell, which causes the sound to vibrate, and produces chimes, tolling, &c.; put the bell under a slow fire for two or three minutes, the sound is dull, the elasticity is gone; let it get cold, the sound returns to it."

But this is evidently wrong. If the sound grows dull when the bell is warmed, it is not owing to a diminished elasticity of the metal, but to the warmer state of the air coming in immediate contact with the bell, as it is a well-known fact, that sound propagates louder and farther in winter than in summer, in a cold than in a warm, in a dense than in an expanded state of the air. Hence also the difficulty of the singer, and his vain exertion to produce a good tone in warm weather, in an expanded and attenuated state of the air. The warm air is thin and more elastic, consequently gives way, and offers, as it were, no volume or body to the striking hammer, receives, therefore, but a slight shock, and propagates but a feeble wave.[*] The air, when cold, as a denser body, more resists the falling stroke, the displacement is more violent, more particles of air are acted upon, and the

will, by their motion, have no doubt but that the PARTICLES OF THE METAL MOVE TOO, though not sufficiently to be visible to the naked eye. If you take a plate of glass and sprinkle on it a little fine sand, and then hold it at one corner by a pair of pliers, and pass a violin bow along one of the sides, you will see the sand arrange itself into a uniform figure; if you apply the bow to one corner, the figure will be varied; and by these means you may produce some beautiful arrangements of the sand, and almost say that you SEE sound. T. Joyce's Scientific Dialogues.

In these instances, as I have observed already, it is not the metal that moves (the small capitals above are my own), not the glass that vibrates, no more than the cement or stone when the wind is striking against and howling along the castle wall; here there is no oscillation of particles, but it is the agitated air, the waves produced by the hammer and the bow, that cause the dust and the sand to arrange themselves into regular forms and figures on the body, over whose surface the waves are passing.

[*] That in winter or in cold weather the air-wave is more voluminous and longer, and in warm weather thinner and shorter, may perhaps, like the thick and long, thin and short, string on the harp or piano, account for the sound of organ pipes giving a deeper tone in winter than in summer.

agitation or vibration becomes stronger, and the sound louder and longer, both as to time and distance; the density of the air is a multiplied propagator of the wave and agitation originally created, and sound will continue until the agitated air has subsided into rest, for which one substance or body, surface and form, is more calculated than the other. Hence also the oscillations, swelling and sinking, &c., of sound, if they are not the intentional production of the performer.   •

It follows from all this, that sound can neither penetrate through solid bodies, which are impervious to the air itself, nor be conducted by any solid bodies, not in contact with the air. If gases, water, or other seemingly-condensed or opaque bodies, convey sound without being in immediate contact with the atmosphere, it is because they are saturated with air, the air within them is concentrated, and they are, therefore, still in contact, however remotely, with the atmosphere. If there were no communication of the atmosphere with the air in the interior of a closed watch, you would not hear its ticking, the same as you do not hear a bell if placed under an exhausted receiver, unless you put your ear close to the smooth and solid body on which the receiver may rest; for the wave, however thin, will still affect the air outside along the polished surface of the said body, as the receiver may be very highly rarefied within, but can never entirely be exhausted. It is the same with a musical box put under the receiver; if it rest on a solid substance, its sound will be heard; but if placed on a bag of wool, it will become inaudible, because the bag will absorb or break the wave.

One of the most beautiful evidences, how, from the recess of inclosure, the slightest motion given to the air is conveyed to the ear, is afforded by the stethoscope, which, on being applied to the human frame, naturally penetrated with air, reveals to us the sanitary state of the lungs and heart. Their expansion, and contraction, and pulsation, produce and convey, through mouth and nostrils exteriorly, an air-wave over the surface of the body, which wave, concentrated more or less around and within the instrument, breaks into the ear. Or, the heaving and beating of lungs and heart, by and through the skin, is communicated to the air within and without the stethoscope; and, however slight, an indent, a feeble wave or motion of the air is produced, which silently, in a whisper, in an echo, as it were, strikes the listening mind; the ear rather feels than hears the inward action.

Thus, there cannot possibly be any sound where there is no air, or where the friction or concussion of solid bodies, or the vibration of any substance is not communicated to it in some way or other. Sound, therefore, though it may have many fathers, knows no other mother and conductor but the air.

And yet I am inclined to think, that there is something more concerned in the production of sound than the mere friction or concussion with the air. The great rapidity of the conveyance of sound in comparison to the motion of the air itself, to the fleetest of hurricanes; the propagation of sound through and against the strongest wind, which, one should think, would overcome, or at least neutralize, any air-wave, any vibration of the air caused by whatever means, whilst the mere use of the human voice will penetrate through the opposing storm; the sound which will reach the ear from within the—although never entirely—exhausted receiver, notwithstanding the apparently broken connection with the air outside; the increase of sound within and from the exhausted receiver if a few drops of alcohol or ether which quickly evaporate, are put in; the divergence of musical harmonies through the air in every direction, without the waves that convey them ever commingling together, but bringing each individual note or sound distinctly and at the same time to our ear; the fact of hard and solid substances, particularly metals producing the loudest tone and propagating it farthest; that friction produces both sound and light: all this makes me suppose that the essence of sound is connected with electricity; that by whatever friction with the air, perhaps with its most important and mysterious ingredient, oxygen, a certain corresponding amount of electric excitation is produced; that this electric excitation is inseparably bound up with the air; that the air is the vehicle to bring the effects of this excitation in all their unimaginable variety to our mind; and as the sun sends forth his seven-coloured rays, shows them to us in the rainbow, and again in their most wonderful and infinite variety of mixture and union in every created object and in the works of man, so the same electric element, in indissoluble alliance with the air we breathe, sends, like rays of light, in another form, to our mind and heart the seven notes of the musical scale, with all that harmony which the genius of the composer has been and will still be able to weave out by their unlimited combination in inexhaustible measure, melody, and modulation.

The air, then, remains the mother, handmaid, and conductor of sound; it is the plastic element that yields to and adopts every impression, which it receives as a unit and sub-divides into millions of undivided parts, conveying to each hearer the unit unimpaired; for, all will hear alike, all will taste alike, their hearts and minds will be filled with the unit, and yet they cannot contain or grasp it; a beautiful type, this, of the Blessed Sacrament.

Conveying every kind of impression the air receives, undulating sounds, like the rolling of distant thunder, will fall upon our ear if a waggon rolls over the pavement, which being undulated, imparts its character and nature to the wave, the sound produced.

Thunder itself is produced in a similar manner.

Dr. Lardner states, that from " the reports of military engineers, who, being placed at elevated stations on the Pyrenees, and thus enabled to observe the superior surface of strata of clouds situated below them, it appears that there is no correspondence between the state of the lower and upper surface of a stratum of thunderclouds; that when the inferior surface is perfectly even and level, the superior surface will be broken into ridges and protuberances, rising upwards to great altitudes, like the surface of the earth in an alpine district. In times of great heat, such strata were observed to send upwards lofty vertical cones, which, stretching into higher regions of the air, established by their conducting power (no doubt themselves attracted by superior accumulations of free electricity), an electrical communication between strata of the atmosphere at very different heights. This appearance was generally observed to precede a thunderstorm."

Now when lightning flashes from the cloud, the air is violently displaced, and the wave thereby created will pass with a uniformity of sound along the inferior level surface : but passing to the superior, undulated, surface of the cloud, the rolling thunder will be heard in proportion to the distance and intensity of the flash and the density of the cloud, and continue until the flux and reflux of the air-wave or waves re-subside into a state of rest. Even when near, an electric explosion behind the cloud only will give us but a faint murmur of the agitated air; but when the flash communicates with the earth, the vibration of the air will strike against more solid bodies, and the wave or waves above, often with repeated concussions, will propagate along the space through which the lightning passed; and hence the peal of thunder, at once so loud, long, and fluctuating. The violent displacement of the air may also set the cloud in a trembling, undulating motion at the inferior surface, thereby causing, or adding to, the rolling nature of the wave. The cloud is the sounding-board of the electric string; undulated and elastic, dense or

H

light, high or low, near or far, it reverberates to us in awful majesty. the power and greatness of the Creator, the same as, caused by the vibration of the string, the sounding-board of the musical instrument sends to us the exquisite harmonies, as it were, of heaven, to make us feel and appreciate the love and goodness of the Supreme Being in having made an element, more than any other in His especial service, to gratify our senses, elevate our thoughts, and dispose our minds to everything that is gentle and noble ; for,—

> " There is in souls a sympathy with sounds ;
> And as the mind is pitch'd the ear is pleased
> With melting airs, or martial, brisk or grave :
> Some chord in unison with what we hear
> Is touch'd within us, and the heart replies."—COWPER.

That clouds are capable of reflecting sound is proved by Muschenbroeck, who, according to Lardner, " states as the result of his own observations, that a cannon which being discharged when the heavens were unclouded produced only a single report, had its sound several times reverberated when discharged in the same place under a clouded sky. In the course of experiments made in 1822 to determine the velocity of sound already referred to, the same observation was made."

That, in proportion to their density, clouds press heavily upon the air, and consequently are able to produce and reverberate sound when the air is agitated on their surfaces, is proved by the smooth level which they present on the inferior surface,—showing the weight with which they rest on the air,—and by the strong winds which generally attend thunderstorms. The air, previously expanded by the heat of the sun, is rushed in upon by denser currents from other places, and the cloud floating in the atmosphere like a bridge, presses upon it with great power, and drives it away from underneath, whilst at the same time it forms an arch, through which the air will force itself with great impetuosity, as it does through the arch of a bridge; and the more bulky the bridge or the cloud the more violent will be the rush.

Without clouds there will be no thunder: hence it is but a rumbling noise that sometimes attends the dissolution of the water-spout; yet at the time of earthquakes it seems sometimes to be heard in the air when the sky is perfectly clear. Though this is no case of common thunder, the noise, or apparent thunder, proceeds, nevertheless, from the upper regions of the atmosphere, and is no acoustic deception. Places at which the phenomenon happens to occur are subject to earthquakes, and possess volcanoes, either active or apparently burnt out. When violent concussions take place within them, the air which has sunk or settled down into the craters or fissures of the mountain is displaced and becomes agitated: the waves thus created, travel up to the mouth or opening of the abyss, whence they propagate and diffuse in the superior regions of the atmosphere; and the vibration descending partially to the earth, apparent thunder will be heard by those who receive the impression of the agitated air, according to the beautiful design by which the DIVINE INTELLIGENCE modelled the human ear, so that sound from before and behind, from above and below, the air-wave falling upon the organ of hearing, should convey and indicate to the brain and soul of man and other living creatures, the direction of its coming.

When by means of fissures or openings the air on the ground comes in contact with volcanic concussions in the interior of the earth, the noise appears to proceed from its very bowels, and the wave created will travel along the surface of the ground until it dies away or is broken, preserving all the time the subterranean character imparted to it by the original cause and material of agitation. The earth is the sounding-board of the vibration caused at certain places within, and there first communicated to the air by direct or indirect contact.

"I have minutely ascertained (says Humboldt), that the great shock of the earthquake of Riobamba (4th February, 1797), one of the most terrible phenomena of the physical history of our earth, was unaccompanied by any noise."

In this instance it is evident that the concussions, or action, of the heaving material, or the electric current below, came not in contact with the air at the place of the earthquake, though it must have done so, direct or indirectly, at a great distance away from it, where eighteen or twenty minutes AFTER the actual catastrophe, a great subterranean noise was heard. From this it also follows, that shocks of earthquakes may happen without any apparent subterranean noise, and that subterranean noise may be heard without any shock being experienced.

V.

EARTHQUAKES AND VOLCANIC ERUPTIONS, ELEVATION OF MOUNTAIN CHAINS.—ST. PATRICIUS.—SEPARATION OF THE LAND FROM THE WATER.—THE DELUGE AND THE RAINBOW; REVOLUTION WITHIN AND UPON THE EARTH.—PRECIPITATION OF NITRATES FROM THE ATMOSPHERE.—CHANGE OF TEMPERATURE UPON OUR GLOBE; ITS RELATION TO PLANTS AND ANIMALS.—ELECTRICITY IN CONNECTION WITH THE GROWTH OF PLANTS.—BARKING OF TREES.—INFLUENCE OF THE MOON ON VEGETATION.—THE COLOURS OF FLOWERS.—ANILINE AND ITS COLOURS.—ELECTRICITY IN CONNECTION WITH ANIMAL LIFE.— ELECTRICITY, THE A OF CREATION.

EARTHQUAKES and volcanoes, in a remarkable manner, seem to confirm my theory of electricity, not only on account of the thunderstorms which so frequently accompany earthquakes and volcanic eruptions, but also on account of all the other phenomena by which they are more or less attended.

That earthquakes and volcanic eruptions are the effect of electric disturbances is fully established by Professor Palmieri's observations.

On the occasion of the earthquakes of Mefi, he used the mag-

netic apparatus of Lamont, and the result led him to believe that
it would no less act on the occasion of an eruption.   He observed
great changes in the electrical condition of the atmosphere ; on the
29th April, 1855, the needles of the aforesaid apparatus had been
slightly affected ; they were greatly agitated on the 30th, so as to
amount to what the professor calls a magnetic storm, and on the
following day the eruption of Mount Vesuvius broke out.   During
the eruption the magnetic vibrations continued with increased
intensity, and the electrical state of the atmosphere, quite in
correspondence with my theory, was equally remarkable, being
greater than the maximum in ordinary times ; its diurnal period
was disturbed, greater electricity having been observed during the
night than during the day ; and what was still more singular,
during the eruption of ashes the fixed conductors gave but slight
indication of negative electricity, while the moveable ones gave the
strongest possible signs of positive tension. In general the electric
current appeared to follow the course of the smoke.   A letter in
the *Athenæum* of 2nd June, 1855, gives an interesting account
of the eruption.

The electric fluid with which our earth, and every planet and
sun has been inflated, or endowed, by Divine Wisdom, and which
by the construction and enormous thickness of the crust of our
globe, as well as by its atmosphere, is kept bound and imprisoned
within its bowels, has a natural tendency to be freed, and to burst
the vault that holds it confined, and within which it dwells and
circulates like the blood in the human body, with its heart, veins
and arteries, whilst, at the same time, it pervades the minutest
parts of the whole frame ; and this tendency to break its bonds
is further aided by the rays of the sun heating the surface of the
earth and expanding the air, thus opening, as it were, a door for
the escape of the fluid, and attracting it to the heated and expanded
area from within all parts of the earth.

In this manner, to wherever the sun shines, to wherever the earth is warmed and its atmosphere expanded by the great luminary of our system, there will be an interior, uninterrupted, current of electricity. But this current is everywhere impeded by nonconducting masses, and thus, the bursting of electric veins and arteries, or the forcing of a passage, produce "horizontal oscillations of the surface of our earth, similar to the waves of an agitated sea, violent perpendicular up-liftings; so that it would seem as if repeated explosions were exerting their force against the roof of a subterranean cavern, threatening to burst it open and to blow into the air everything placed over it."[*]   In fact, if our atmosphere, pressing with a weight of 15 lb. to the square inch upon the earth, were rarefied to a very high degree, or withdrawn from it, earthquakes would be much more frequent and violent, and the earth itself might be blown and shivered into atoms.

When electric currents, to our every-day knowledge, are capable of melting rods of iron in a moment, the electric currents circulating within the earth are capable of producing that heat of our globe which increases as deeper we descend into its bowels, as we come nearer to the heart of its electric organization; the powerful spring that drives these currents from the heart of the earth through the masses of its body, operating upon its various ingredients, melting and refining its metallic constituents, and sending up fountains after fountains of healing water, cold and warm,[†] all over the earth, explains the enormous pressure which throws up

---

[*] This description of the effects of earthquakes is the language of a writer who wanted but the key of the theory to which it corresponds.

[†] Saint Patricius, probably bishop of Pertusa, as early as the third century, conceived already the fire in the cloud to be the same as that in the earth. "Fire is nourished in the interior of the earth as well as in the clouds, as you may learn both from mount Etna and another mountain near Naples. Water rises from beneath the ground as in siphons; those at a distance from the subterranean fire are colder, but those which have their source near the fire are heated by it, and bring with them to the surface which we inhabit an insupportable degree of heat." *Cosmos* 211.

stones, molten lava, &c., from a volcano, with a force equal to from 300 to 400 atmospheres, though water decomposed, or steam generated within the mountain by the intensity of the electric current, may add to the force of electricity itself. My theory explains the connection of volcanic agency between various parts of the earth, the simultaneous shocks at places far asunder, as for instance at the earthquake of Lisbon, on the 1st November, 1755, which was felt in Greenland, Africa, and even America; and it accounts for the fact, that generally either the first, or one of the first shocks, are the most violent, because the first strokes of the electric power open the passage through opposing strata of the earth.

Volcanoes are the safety-valves, the lungs, of our globe in labour; they are the conductors of the unfettered element within; and happy those places that possess them. Providence also has placed them where they are most needed, that is, chiefly in those parts of the earth that are exposed to the greatest heat of the sun and therefore more calculated to attract the electric current within, and give to it the opportunity of liberation. Volcanoes moreover, by Divine Wisdom, seem mostly placed close to the sea, not only with the view of originally raising bulwarks to the waters of the earth, AND CAUSE THE LAND TO BE SEPARATED FROM THE WATER, but also, because, near the sea, particularly at the bottom of the ocean, the electric fluid has a thinner crust to burst through than it would have in the midst of continents. Hence electric up-heavings on land are diminished, and at sea not so dangerous to man, though at times ships are engulphed in the chasms of the ocean caused by the sudden rise and fall of its bottom.

The elevation of mountain-chains is evidently due to regular currents of elastic, i.e. electric forces; for the lines of direction which characterize each group have such relations to each other, and are so symmetrically arranged over the earth's surface, that they cannot but be the result of that electrical organization of the interior of the earth of which I have spoken.   It was, no doubt, first called into extraordinary activity when the whole solid globe was still covered with water, when the earth—though darkness had been banished from the face of the deep by the creation of light, and the waters separated from the waters by a firmament of air—was still void and empty, God said: "Let the waters that are under the heavens be gathered together into one place, and let the dry land appear.   And it was done."   And thus the electric current within the earth fulfilled the word of its Creator, and by its volcanic action undulated the earth, and systematically and symmetrically raised up mountains and mountain-chains, as so many monuments of the power and wisdom of God.   The separation of land from the water was the first violent catastrophe our earth experienced for the benefit of man, who hereafter was to inhabit it; and it is only by the operation of a pent-up force of electric matter that we can satisfactorily account for the disturbance of all the strata of the crust of the earth, and their tumbling one over the other in the most extraordinary and seemingly confused positions and directions, rendering firmer and more compact even that shell, which, with its horizontal layers, could not withstand the electric element, since that time more effectually confined, and provided with the volcanic safety-valve to limit its destructive effect on the earth.

Apart from this universal revolution in the crust of the earth, other tremendous concussions have taken place, as they still from time to time do occur, confined to individual portions of the globe. The elevation of new, and the sinking of old islands in the ocean, are evidence of this subterranean working of electricity.

Dr. Lardner says: "When storms are breaking in the heavens, and sometimes long before their commencement, and when their approach has not yet been manifested by any appearances in the firmament, phenomena are observed, apparently sympathetic, proceeding from the deep recesses of the earth, and exhibited under various forms at its surface." And these phenomena proceeding from the deep recesses of the earth, as Dr. Lardner truly says, without however, suspecting the true cause, most strikingly confirm the theory I have laid down : of an accumulation and organization of negative electricity within the earth.

This negative electricity, though separated by the nonconducting air from the positive electricity of the superior regions, is never totally severed from it ; the thread, the link of union, however finely spun, however attenuately distended, is never altogether broken; like rarefied air, to what extent soever expanded, the active, though invisible and imperceptible, union with the distant accumulation of the most dense and cold atmosphere is never for one moment interrupted ; the thread of India-rubber, drawn out to its utmost length and attenuity, its particles will still retain their hold upon one another, and resume their ordinary condition as soon as the distending power or cause is removed. The two electricities in the same way will hold each other, and approach in accumulation, or retire in expansion, according to the condition of the barrier or power that in the ordinary state keeps them in distant communion, and at other times allows them in force to amalgamate.

Thus, when storms are approaching, though unperceived in the heavens, the two electricities are gradually drawing into closer communion ; that from within the earth will force its way through certain wonted channels, causing fountains to boil, springs to flow, after having been dry for a time, and often heave up the waters of rivers and lakes and the sea, roll them in furious waves and drive them over their banks, rend hills in two, open their sides and bring with it the waters of the veins it has opened in the earth ; and, at times, even, it will open awful pits and blast abysses in the bosom of the earth, and like gashing wounds force asunder and divide the waters of the mirror of the ocean.

From the day on which God separated the waters from the waters until Noah went into the ark, dews only had nourished and refreshed the vegetable creation; no rain had as yet fallen* upon the earth. Since the sun had broken out into light, hydrogen and oxygen gases receded to a still greater distance from the earth than before, and mists rose higher into the air; but as yet all the pores of the earth were open, evaporation of water and electricity, and climate, was general and even all over its surface; there were as yet no separate accumulations or concentrations of these elements in the atmosphere, and the vapours that had risen returned again as heavy dews ere they could ascend high enough to condense and fall down in drops of rain, or in the form of hail or snow.

When at last the solid body of the earth had increased in bulk and hardened so as to accumulate a greater amount of heat on its surface and correspondingly to expand the air: God gathered the rising mist into one vast sheet, into a universal cloud enveloping the whole earth; electricity from above accumulated in the aggregated vapours (no doubt also peculiarly acted upon by the rays of the sun, which they received and absorbed), and whilst moulding them into drops, and converting into water the oxygen and hydrogen gases collected in the air and existing above our atmosphere, it drew forth from its stronghold, into closer embrace, the electricity within the bowels of the earth; and thus it came that "in the 600th year of Noah's life, in the second month, the seventeenth day of the month, the same day all the foundations of the great deep were broken up, and the windows of heaven were opened."

---

* There are still countries, such as Peru, and places, where it never rains but where heavy dews supply the earth with moisture. For the last three years it never rained at the island Teneriffe, by which famine and disease were produced and multitudes obliged to leave the island. In Egypt it has not rained for 700 or 800 years.

Thus a law of nature harmonized with the moral law of God, the law of justice, and swept away a sinful world.

Unaccompanied with lightning and thunder, the discharge in rain of the accumulated vapours of the whole earth did not come down in the heavy drops of the thunder-cloud, but fell in universal and slower measure for forty days and nights, until the electricity accumulated above had again been liberated. As rain begins to fall and to increase when lightning is nigh ceasing, or when electric discharges have taken place, so it was then that the sheet of water, spread out on high over the earth, began to lighten and to disperse, that the sun again broke forth through the dissolving veil; and from that moment the windows of heaven were shut, the power had been removed that drew electricity and water from the innermost depths of the earth, and thus also the fountains of the deep were stopped up. Nor were they to be opened again; for, when the earth was dried, mists were no longer the only accumulation of moisture; the constitution of the earth had been changed, and henceforth, clouds were to be formed to bring refreshment and carry blessings all over the face of the earth. When thus, after the Deluge, Noah had left the ark and was occupied in the cultivation of the restored land, mists again began to rise; but from the altered electric organization of our globe they rose high and formed into clouds; and fearful perhaps, lest they should gather together and again send the devastating current over the earth, God took away his fear and told the second father of mankind, that He would no more waste the earth by the waters of a flood; and as a pledge for His assurance He said to him: " This is the sign of the covenant which I give between me and you. I will set my bow in the clouds, and when I shall cover the sky with clouds, my bow shall appear in the clouds." And thus it was that God, for the FIRST TIME, decorated the FIRST-BORN CLOUD, though dark and threatening, with the arch of peace, with the glorious rainbow, to us still, as it was to Noah, the seal of the covenant, that the waters of the earth, above and below the firmament of heaven, will no more be gathered into a shroud for the human race.

From the time that Noah went into the ark nothing but water once more covered the earth, nor retired from the land until a revolution of one year and ten days' time, within the earth and on its surface, marked the progress and final subsidence of the deluge. During its prevalence the subterranean agency was not idle that originally heaved up mountains and separated the dry land from the water; continents, countries, islands, and mountains, risen perhaps and fallen already during the 2,000 years that preceded the deluge, but particularly from the third day of creation to the creation of man, rose and sunk again and again, here of a sudden, there by degrees; monsters of the deep were carried up within mountains of the soil and clay upon which they had roamed and in which now they became embedded and entombed; forests and prairies with their inhabitants sank and were swept away where forests and prairies had perhaps sunk or been swept away before, to be covered again by falling hills or buried under floating masses; the waters of the flood filled every recess, every abyss, as well as the craters and openings of the highest mountains; they extinguished the fires of numberless volcanoes and stopped up the pores of the earth; and whilst they remained, ample time was given for gigantic deposits and quiet sediments of various kinds; for, once at their height, the waters of the deluge were disturbed only by the waves of an agitated surface, by the ebb and flow of the tides; and the flood settled down, perhaps, and retired from the land, in a dead calm, until its final abatement, no doubt, however, leaving vast tracts under water that hitherto had been above the level of the ocean; for, it stands to reason, that by the accumulation of the waters from below and above, the ocean itself must have been raised and increased in extent.

Since then, in proportion to the disturbance of the former electric organization of the earth, and the alluvial deposits that had been formed by the deluge, and in which alone remains of existing animals and of man are found, its climate began 'more sensibly to vary; negative electricity, the representative of warmth, was more driven back and more closely confined at certain places within the earth; and where this had been the case, and perspiration impeded, positive electricity, the representative of cold, came down in greater accumulation upon the earth; and as the former retired from the plants it had luxuriantly raised up, the latter came to embrace them in its withering grasp.

If before the Deluge the crust and soil of the earth was not more spongy, porous, or loose, than after it; and that it did not increase in thickness by further deposits and chemical precipitation, to which latter, among others, we no doubt owe the beds of nitrate of potash, of borax and of tincal in the East Indies, and the beds of nitrate of soda, and of borate of lime in Chili and Peru,* being combinations of azote, oxygen, &c.; or that the interior electric action and evaporation of the earth was not more active when in early times the climate of the earth was hotter than it is now at the torrid zone: was it perhaps that then the earth was altogether nearer to the sun than it is at present? or

---

* According to Lardner, the transmission of the electric fluid through the atmosphere produces nitric acid. In the residue of seventeen samples of rainwater, collected during or immediately after thunderstorms, Professor Liebig found nitric acid in combination with lime, or with ammonia. In the tropics, where thunderstorms are very violent and for a length of time of daily occurence, the formation of nitric acid and its precipitation with other substances must still be very great, though diminishing, no doubt, in a progressing ratio, the same as precipitation or deposition in general must have diminished since the closing of Paradise, or more probably, since the deluge. Before this inundation of the whole earth, before the production of thunderstorms, of lightning striking through the air, I do not think any precipitation of nitrate of potash, nitrate of soda, and of borates, to have taken place.

that the oscillation of the earth's axis, its so-called inclination to the plane of the earth's orbit, was not as great as at present? or that the brine of the ocean, once more covering the whole globe, saturated the ground, and thenceforward, although diluted with the rain and sweet water of the rivers, affected the growth and nutritiousness of vegetation and consequently all life upon earth?

Henceforth, however, not only plants and animals, but man also began to diminish in size and longevity; and from the permission of God to man after the deluge, to eat the flesh of animals, there seems to be no doubt, but that by a reduced emanation of warmth from the earth, or by the abstraction by precipitation from the air of an essential element of life, perhaps of oxygen, calculated both for sustaining life and adding volume to the body by oxygenation, or by the precipitation or change of some other ingredient of the atmosphere: some important alteration must have occured in the physical condition of the surface of the earth, or in the constitution of the air, for which the said permission was to compensate in a certain degree as far as the life of man was concerned.

As to plants and animals, the change became particularly striking towards the polar regions of the earth, and is analogous to the gradual diminution of all organic life from the valley to the mountain top.

In the now colder regions of the earth, negative electricity, reduced in its exhalation from the earth both as to time and quantity, does not sufficiently long and powerfully enter into the plants so as to make them grow to the height and extent to which they grow in hotter climates, nor does it in this limited state infuse sufficient warmth into animals, to make them enlarge into the size of those of the tropics. When in spring, with respect to plants, the electric element is excited within the earth by the heat of the sun, it issues from the earth, and is conducted upwards into the air by the vegetable creation ; but it also enters into the plant itself, drives it upward, and, excited by the sun, causes it to live and to attract from the air all the substances required for its growth, its leaves and flowers, according to the order and design which God had laid down and hidden within the seed. When in autumn the electric element is driven back into the heart of the plant,—the same as in cold weather our blood is driven back to the heart, and the exterior parts of our body become chilled,—its leaves wither and the branches become bare ; and when in winter the electric element is driven out of the plant altogether, the negative overcome by the positive, warmth by cold, the latent life of the sap is gone, the plant freezes, and dies.

Thus, by this simple element and operation, we see everthing grow and ripen. Living in electricity, exteriorly and interiorly, every blade of grass, every leaf of bush or tree becomes a conductor, from which it readily emerges into the air; but the flower, the berry, the ear of corn, the apple and the pear, not being of so pointed a form, afford no such ready escape into the air; their rounded shape retains it, and for this reason a certain activity is excited, and more warmth produced; and this warmth ripens flowers and fruit, which, when sufficiently saturated or combined with electricity, are repelled by the stem that bore them.

Instances are not wanting by which my theory is proved.

" The elebrated Duhamel du Monceau states," according to Dr. Lardner " that silent lightning, unaccompanied by wind or rain, called heat lightnings, have the property of breaking the ears of corn. Farmers are well acquainted with this fact. On the 3rd of September, 1771, Duhamel himself witnessed this fact : on the morning of that day there was much lightning, and he afterwards found that all the ears of corn which were ripe were broken off at the nearest knot: the only ones which remained standing were the green ones."

We here see, in a more active degree, the usual operation of electricity on plants, both exteriorly and interiorly. As long as the plant is green, every part of it is connected by the sap of its veins, and held together by the electricity by which it is pervaded. When the ear is ripening, the sap in the fruit drying up and the grain beginning to harden, the interior conducting substance is weakened, the grain by degrees peels off from the vessel that holds it; and electrified by the electricity exteriorly rising up and accumulating round the ear, the matured corn is suddenly thrown off on occasions like the above, when the electricity rising from the earth becomes particularly excited.

" These and similar effects,' Dr. Lardner continues, "indicate an influence emanating from the ground. Such effects are not confined to corn, but probably extend to all vegetable substances. The following fact, as stated in the *Bibliothèque Britannique*, of Geneva, for the year 1796, supplies an example of this:—

"'A wood of oak situated on an eminence two leagues from Geneva was barked in May, 1795. This operation can only be effected in the season of the year when the sap, moving between the wood and the bark, diminishes sufficiently the adherence of the latter to be enabled to be separated with facility from the tree. The workmen remark also, that the state of the atmosphere produces an evident influence on the process.

"'One day the wind was blowing from the north and the sky unclouded, the bark was removed with more than usual difficulty. In the afternoon, clouds rose in the west, thunder rolled, and at the same instant the bark, to the greatest astonishment of the workmen, fell spontaneously from the trees. They soon had reason to ascribe this to the state of the atmosphere, since the effect ceased when the storm passed away.'"

As long as the wind was blowing from the north, and the air cold and without any electric accumulation, the electricity residing within the tree was driven back to the heart and to the earth, as it is in autumn; and as it withdrew and became contracted, so it drew all parts of the tree closer together, assisted by the pressure of the atmosphere; and hence the difficulty of removing the bark. Without the air even becoming warmer and more filled with electricity, but an electric accumulation taking place in the sky only, causing but a temporary electric disturbance, the negative element within the earth and trees became active and excited; it panted to be set free and to unite with the positive element above, and expanding in proportion to the superior excitation and attraction, it almost by itself, without the aid of the labourer, drove off the bark from the body of the tree.

A similar, but more striking, case is recorded in the *Times*, of May 18th, 1860, by the rector of the parish of Holywell, near St. Ives, Hunts. He says:—

"A violent thunderstorm broke out this evening at five o'clock. A few hundred yards from my residence, in a farmer's field stood an oak of about 40 feet high, and 10 feet in circumference, perfectly vigorous and sound, just putting forth its leaves. The thunderbolt struck it about 15 feet from the ground, it was broken almost two-thirds off, and the top of the tree fell to the earth, clinging to the stem by the undivided fibres.

"The tree was nearly barked from the top to the root as perfectly as human labour could have stripped it. About one-fourth part of the standing stem was cleft clean out to the very root, and flung on the side opposite to that on which the top had fallen. The remainder of the standing trunk was cleft in four or five places, so that the hand could be thrust into some parts and the fingers into others. The whole heart of the tree was literally torn into thousands of splinters. The bark and pieces that seemed torn

I

from the stem of the broken top were flung in various distances into the field."

When corn and grass have been beaten down by the rain, the force of gravity has overcome the power of electricity within, and the cold of the rain has driven it back towards the ground; but when the rays of the sun fall again upon the prostrate stalk, the cold vanishes, the rain-drops dissolve, electricity is again elicited from its retreat, or freed from its contraction; it re-enters the plant, expands in the warmth of the sun, and tending to rise towards the electricity of the superior regions, it raises up the fallen legions of the field, lifts up and elevates their drooping heads, as the grace of God lifts up the fallen sinner. If, however, the corn, or grass, or flowers, be ripe, or dry, or broken, they will no more rise when prostrated by the storm, because the conducting sap has been absorbed or scorched away, because the connecting link is withered, severed, not merely bent or weakened.

In his theory of sleep, Dr. Ashburner says: " The sensitive plant exhibits, under two different conditions, the opposite states of sleep and vigilance. Its contracted leaves are in a state of spasm—grasping, active, apparent acquiescence; and when its leaves are open and expanded, it is awake." And this he very aptly compares with the two opposite states, with the sleep and vigilance of man. But what is it that produces these states in plants? It is the electricity retiring into the deeper recesses of the plant at night, contracting all its members, as was the case with the oak-trees mentioned above, and re-entering or expanding again into its extreme part as soon as the morning sun, or the warmth of day, arouses it from its sleep by its electric touch. And how beautifully is this operation not represented by the opening and shutting of the convolvulus! by the sunflower, which then even remains turned to and following the sun when clouds are passing between

But plants in their origin, growth, maturity, and fruitfulness, exhibit to us the beneficial effect of the solar emanations.

At the time of new moon, when the moon in conjunction stands between us and the sun, she witholds from the earth that vast proportion of solar rays and heat which fall upon her back and are absorbed by her body. This deprivation produces a low state of activity on the earth, the lowest during the month. If seeds or plants are then put into the ground—everything else being the same—they will have the benefit of ever-increasing light and warmth, as by degrees the moon moves away from under the sun allowing more and more rays of light and warmth to fall again on the earth until she has reached the opposition, until she is full ; the seed or plants will therefore thrive under this gradually-increasing, genial, and invigorating influence. If on the contrary, seed and plants are put into the ground at full moon, they will germ and take root at the time of the greatest terrestrial electric activity ; and as the moon will then decrease and thus begin more and more to abstract from the life of the earth, to deprive us in an increasing ratio of a proportion of solar rays and heat until she has completed her lunar month, the plants will languish for want of good food, as it were; the stimulus that attended their birth is not only not kept up but diminishing, they are stinted immediately after they have taken root or begun to live, and thus they will remain sickly.

At a meeting of horticulturists at Berlin, in 1860, the royal gardener observed, from his own attention to the subject, that apples which fell from the tree whilst the moon was on the decline, did not begin to rot on the spot which touched the ground, but those which fell when the moon was on the increase soon rotted on those parts on which they had fallen. From this it would seem to be best to gather in fruit when the moon is on the wane.

Similar observations must have been made in very early times;

as the Roman writer Varro, contemporary of Cicero, in his book on agriculture, exhorts the farmer that in cultivating the land he is to pay the greatest attention to whether the moon is on the increase or decrease; and sheep are not to be shorn, nor one's own hair to be cut when the moon is decreasing.

Singular it is, that a shark turns black when hung up in the moonshine; and I have been told that other fish, similarly exposed, also turn, and become unfit for food.

Thus, the fact of lunar, though indirect, influence on the earth cannot be denied; it is based upon the testimony of ages and of all nations, against which the individual opinion of astronomers, because it passes their observation and comprehension, is but like a few grains of sand against the desert; and if this influence has been transferred to matters of business and other objects not within the sphere of the laws of nature, it was but a natural transition with people in a state of heathenism.

At a total eclipse of the sun, when the moon almost entirely deprives us of the benefit of his rays next to the darkness produced, is the fall of the thermometer, the lowering of the temperature even to chilliness, as in July, 1860, causing the feeling of approaching night suddenly to come over the animal creation, and birds in particular, and convolvulus and other flowers to close and hang their heads.

Vegetation and animal life thus influenced by the sun, flowers and blossoms open to him their bosoms, in order that fructification may take place, that the multiplication of their seed and kind may be insured by' electric agency and electric incorporation. For, electricity, as stated in the beginning, is incorporated with every created substance, and liberated again by dissolution, or combustion; we take it with our food, imbibe it with our drink, and inhale it with our breath. It is the fire which is latent in all ponderable matter, the most closely fettered in metals, and the most loosely allied to spirit and ether. It reposes in the mineral lodes of the earth, in its fields of coal, its lakes of oil, and beds of turf, and is embodied in every tree and plant. Of alcohol, in equal purity and composition of 52.23 carbon, 13.01 hydrogen, 34.76 oxygen, produced from every variety of grain, fruit, berry, and juice, it is the essence; it exists in the oil and tallow of the animal and vegetable creation; man skilfully concentrates it in gunpowder for the destruction and chastisement of his own species; by the bee, the pattern of order and industry, it is collected, so that, hidden within the virgin wax, the lighted candle, an emblem of faith, and striking image of the Holy Trinity, it may illuminate the altar of God, burn and shine in honour of Him who Himself came to be the light of the world, and by whom it first was created; on being kindled it starts from its repose and breaks out into fire, and, amalgamating with the electricity bound up with the oxygen of the air, will continue to flash and to flame until the body from which it is evolved or from which it emanates, is dissolved, consumed, or exhausted.

We have seen already how exterior and interior electric currents drive up the plants and cause them to abstract from the air the nourishment which they require; for, attraction, magnetism, is but a property of electricity. [*]

When the seed of the plant falls, or is put, into the ground, the positive pole strikes downward to take root in the earth, whilst the negative pole strives upward into the light. Thus the tree rises, the negative crown enlarges, and like arms, stretching forth its branches, it proclaims the praise of its Creator, and draws down upon itself the dew and showers of His bounty. The crown, however, is but the corresponding part of a root positively resting

---

[*] "William Gilbert . . . . anticipated by his conjectures much of our present knowledge. He regarded magnetism and electricity as two emanations of one fundamental force pervading all matter, and therefore treated of both at once." *Cosmos* ii., *p.* 331.

in God's own foundation, or holding fast to the rock ; and, always turned to the light, the tree luxuriates in its rays, and with us here its volume and foliage is fuller, more developed and preponderating on the south side than on that of the north. This, no doubt, is the effect of the sun, towards whom the electric currents are inclining, causing a greater vitality in that direction, and proving also, that the plant feeds upon the surrounding atmosphere, and by means of the leaves absorbs nourishment from the air. The leaves, in fact, are the outward life of the plant ; the fruit of a tree will wither when they have been eaten by caterpillars ; the tree itself will cease to grow, and perish, when they are all taken away at an unseasonable time, as by cutting off all the branches of a plant, it will, as it were, bleed to death, or, at the very least, deteriorate its produce.

Flowers and fruit, owing to their more rounded forms, retain longer the exterior and interior electric currents which more readily escape from the pointed leaf; hence they are subject to greater electric activity, particularly on those spots or parts which are shone upon by the sun. This activity attracts in greater abundance from the air the substances they require, as also those which form the various beautiful colours and bloom by which we are charmed, and which so often make us adore the goodness of God and admire the beauty of His designs. And can we SUFFICIENTLY appreciate and admire His power and wisdom in producing all this by the simplest of means ?

Meteors of every description, the metallic bases of plants living entirely in the air, have shown us already that the air is filled with metallic solutions, chemically combined with oxygen, perhaps free, perhaps in both states. The analysis of meteoric stones shows nothing foreign to what is contained in the earth ; in them we find, according to Berzelius, eighteen simple elements ; viz., oxygen, sulphur, phosphor, flint, aluminum, magnesia, calcium, kali, natron, iron, nickel, cobalt, chrome, manganese, copper, zinc, and titan. All these are highly magnetic; and each one thus attracted according to the electric organization of every plant, the metallic atoms settle down upon leaf, fruit, and flower, in the order and arrangement devised, and the pattern traced and delineated thereon by God; and moistened by interior sap, by dew, and rain, and again shone upon by the sun, oxidation takes place, the solar rays corporify, yielding to our eyes the most magnificent variety of colours, such as human chemistry has not as yet been able to produce, though the newly-discovered aniline colours surpass all other colours previously made. But their beauty and intensity is accounted for by my theory, which the discovery most clearly and remarkably confirms and illustrates.

Anil is the Indian name of the plant from which indigo or indigo-blue is obtained; hence also the name of anil given to indigo itself, by the Portugese and the Spaniards, who first met with the plant in India.

From this anil a certain liquid was produced, and on that account called aniline; and owing to the presence of this aniline in indigo it is found that there is no other blue or colour equally fast and bright.

Now aniline exists in coal, deep within the bowels of the earth, and by various processes of distillation is obtained from coal-tar, when, by means of steam, it is chemically united with mineral or metallic preparations of tin, lead, mercury, manganese, etc.; and thus the new colours are produced.

Coal being a transformation of vegetable bodies, without the loss of their component parts, we here have in an artificial manner the same operation and the same substances as in living nature, namely: the aniline of vegetation, and the metallic solutions of the atmosphere; and against steam: the solar rays combined with moisture both of plant and air, with an over-abundant

supply of oxygen, nitric and nitrous acids, salts, sulphur, &c.,
likewise required in the chemical process.

The new colours, moreover, can only be dissolved with alcohol
and wood naptha, both extracted from vegetable sources, and
both concerned in the production of floral decoration.

Hence the intensity and beauty of aniline colours, their production corresponding to the operation and combination of God's
laboratory of nature.

With further progress in the production of these colours, I
think that the constitution of plants will still be found out, why
one absorbs the cobalt, another the copper, the chrome, the iron
etc., and why others again should absorb and secrete in different
proportions the arsenic or other poisonous solutions that may be
suspended in the air; why chemical operations on the plant
produce a change of colour in the plant by changing its electric
condition or constitution, its quality and power of attraction, etc.
What wonders of Divine wisdom will not then be manifested
when science shall have arrived at this eminence! and when, as
in plants, so also in animal and human life, we find electricity
the chief, if not the only ACTING agent, may the essence and
primary created cause of mortal life itself; when, arriving at the
unit of creation, we find all things passible,—even our own bodies,
—vanishing from us as nothing real, nothing self-existing, nothing
self-sustaining, nothing self-acting! Then the very climax of
science will FORCE US TO ACKNOWLEDGE, that beyond all this,
distinct from all and ABOVE ALL, there must be a Supreme
Intelligence, an omnipotent, self-existing, self-acting, real God,
manifesting Himself to us in million-fold diversity, springing
forth from a created unit, and this diversity ever and anon returning again to the unit, in the dissolution of inorganic and organic,
and of animal and human corporal existence.

The same agencies, which in plants produce their growth and

color, in animal life produce increase in size, color and instinct; and animals being endowed with living, but mortal, souls, and because mortal, unconscious of death, instinctively self-acting, they are capable of locomotion, and so transport themselves to wherever their instinct inclines or repels them, that is : they move towards the plant or food that attracts, or anything else that agreeably affects them, whilst plants attract the nourishment necessary for their existence and development.

Why, as with plants noticed above, one animal should be attracted to feed upon one, and the other upon another plant; why one should be repelled from one and another from the other, is still a mystery, but will no doubt be found out to be traceable to electric affinities and differential electric organization, the same also that one kind of food will not incorporate with one animal when it will do so with another.

As plants root in the earth, and are warmed, expanded, and fed, by means of the electricity rising into them, their interior warmth and consequent life and growth kept up by positive electricity and oxygen from the air, and in the products of wax and oil, and seed and grain of every description, re-produce again electricity in a concentrated state : so the animal world feeding upon these products, takes up the electricity they contain ; and the focus, the hearth, which God planted somewhere in every living creature, being fed and excited by the electricity and oxygen from without, from the air, both electricities combine for the combustion and dissolution of the food taken, re-producing again in the animal body the same elements in a still more purified and concentrated state ; and as none of the oxygen consumed is given up again in the gaseous form by the animal body, so there is no pure, uncombined, oxygen at any time given up by the plant either, except in the case noticed page 134, but simply the products of combustion, both the change and the surplus; and as the food taken affects in various ways the body, so in man, the wholesome and poisonous effects, morally, of meat and drink, may and do pass over into the soul.

According to the distribution and organization of animal life, one substance produces more blood, the other more fat, in the animal body. Hybernating animals, no doubt by this means chiefly, become uncommonly fat towards the end of autumn; before they retire into their hiding places, they are, in a manner, cased in fat. During the whole of their winter's sleep they take no food; but, as we know, the accumulation of fat is absorbed by degrees, so that on their awaking in spring they are quite lean and look altogether famished.

As with the departing heat of summer the circulation of the blood becomes less rapid, and consequently the perspiration of the animal gradually diminishes and ceases, the fat produced by the food they have taken must naturally accumulate upon their bodies, instead of being thrown off in evaporation by exterior and interior heat and excited motion, the same as domestic animals are fattened by being kept in close confinement, and fed with fat-producing nourishment.

From the moment they begin their winter sleep, the accumulation of fat not only ceases by their not taking any food, but, by a wonderful disposition of Divine Providence, the fat itself is turned into their life-preserving element; and that this element may not too soon be exhausted, their respiration and circulation of the blood, from the want of access of exterior air, aided, no doubt, by an accumulation of carbonic acid about them, becomes rapidly slower, until at last they appear entirely to be deprived of life.

Under these circumstances the process of life seems to be kept up by the focus, the secret hearth within, most likely some higher concentration of electricity, attached to an organization affected from without, and tenanting the unconscious, animated, but mortal, *i.e.* totally extinguishable soul, the one perishing with the other, and returning again into chaos by the extinction of the vital spark.

This living soul, with its immediate body of latent light, like the wick of a candle, burns slower and slower with a diminution in the supply of oxygen, dissolving or liberating, in a decreasing ratio, the latent light and warmth of the fat, electricity in another state, and sending it correspondingly through the veins and arteries of the body to keep it alive; but the fat, the concentrated electricity all gone, or the electricity,—as in the case of the tree—driven out and supplanted by cold, by the positive element, and all life is gone.

Here also the two electricities are the life-preserving and motion-imparting agents, as they are the acting constructors into form of all things in existence, the obedient executors of God's will, ideas and designs, every object being but the expression of the thoughts of God made manifest to us in creation.

It is, no doubt, the enormous accumulation of fat, the concentration of an immense amount of negative electricity, not only on hybernating animals, but also on those living in cold climates, on the bear, the whale, the seal, &c., which shields them against the severest frost and makes their exterior part of the body impervious, as it were, to the icy grasp, the withering influence of positive electricity, of cold itself.

And in all this behold again the wonderful simplicity of God's creation, TWO ELEMENTS, in ORIGIN ONE, PERFORMING ALL, every thing else being an accidental or auxiliary in the carrying out, the conceptions, the thoughts of God, by His persevering will PERMANENTLY IMPRESSED upon the germ of every thing of its own kind, though nothing, in fact, is actually permanent on earth except the soul of man. The human body even is in constant change; after every seven years all its former particles have disappeared and

been supplanted by new ones, composed of the same elements ;
after death it will entirely be dissolved into its original constituents of oxygen, hydrogen, nitrogen and carbonic acid gases,
but the soul remains ; it alone is a conscious intellectual, free-willed,
by God created, self-acting REALITY, impressed with His own
image, and THEREFORE ALONE worthy the ransom of Calvary,
ALONE worthy the price of redemption ; hence also the soul of
man ALONE is conscious of the death of its body of clay, therefore also ALONE conscious of an undying existence ; for, as He is
for ever, so His image will exist for ever, but everything else is
transitory and passible, subject to everlasting change, to decay
and resurrection, to disappearance and re-appearance in some
form or other, to amalgamation and evolution ; and thus the
grand scenery of this earth, the dissolving views of a Divine
Intelligence, of an Omnipotent Prime Mover, will continue in
ever youthful renovation, until the final conflagration of the world,
and the establishment, the creation of the new, the heavenly
Jerusalem, the dwelling place of PURE and happy souls,—clothed
in their former, but transfigured, form and elements,—and the
abode of eternal peace.

As all the emanations of the brain that have been written down,
from the beginning in all nations and tongues, have been expressed
and presented to the human mind by about twenty-five letters, and
these governed by only five positive sounds, called vowels, of
which vowels that of "A," the first one also pronounced by the
lisping infant, is first in rank, the basis, the womb, the cell, as it
were, of the whole alphabet ; and as all the writings, designs,
pictures and paintings of every descripton ever executed by the
hand of man, are based upon, have for foundation but ONE SINGLE
LINE, applied in various positions, curves and combinations ; and
as all the variety of colour and of musical notes and harmonies
produced and combined by human ingenuity and in existence are

reducible to seven primaries each, and again both colour and sound having but one root, radiating and elicited but from one element; so with the corporification of the thoughts, the conception of a Divine Being, with the compositions, the pictures, the tableaux, the panoramas, the decorations, the fireworks and illuminations, the musical harmonies and the thunders of creation; all emanate from one very short but simple alphabet, unknown, as yet, to be sure, but nevertheless certain. Electricity, positive and negative, two in one, and one in two, is no doubt the alpha, as it is the omega of this alphabet, joined by oxygen and hydrogen; the two former as light when in active amalgamation, and as cold and heat* when separate; and the two latter forming the vowels of solids and liquids, ALL, with all their infinite variety of mixture and combination, radiating from the first vowel, from electricity the A of creation.

The starting-point of all life itself is said to be a single cell, by degrees doubling, quadrupling, etc., until the most complete differentiation of living creatures, from the frog to the philosopher,—as it has been expressed—is developed and established.

But behind the A of creation, behind the cell?—There is the grand composer and artist and *modelleur* from eternity!—For, though wonder upon wonder manifests itself in the simplicity of

---

* Baron von Reichenbach seems most singularly to confirm my supposition respecting the nature of the two electricities in respect to warmth and cold. Speaking of the magnetic nature of crystals, he says: "Every crystal presents two points, which lie always diametrically opposite to each other; they ARE THE POLES of a primary axis of the crystal. Both poles act in the same way, but one always much more strongly than the other, and with the distinction that from one there appears (to sensitive persons) to issue a cool, from the other a softer, gentle warm (seeming) current of air. The magnet when drawn over the hand of a sensitive person, produces the same sensation of cold at one pole, and of gentle heat at the other pole. In regard to quantity the north pole is stronger and the south pole weaker; in the crystal the northward pole is stronger, and the southward pole weaker."

Joyce, page 482, says: "The thermo-electric apparatus has been employed to prove a case in which electricity produces cold: as when a feeble current passes the junction of bismuth and antimony."

the creation, when from the most minute to the most gigantic object in nature, from inorganic to animate existence, from the growth of the insect to the growth of the elephant, from the course of the fishes in the water to the passage of the birds in the air, from the song of the nightingale to the call of the vulture, from the bleating of the lamb to the roar of the lion, from the moving centre of the smallest infusoria to the pulsation of the human heart, from the electric hearth and crystal rocks of our globe and the planets to the electric nucleus and luminous garment of the sun, from the deposition of colour on the flower by his rays to the reduction again into colour of gold and silver,* we trace nothing but the simplest of means, electricity in repose, electricity free and accumulated, electricity in silent operation and amalgamation, and electricity in dissolving, dispersing and explosive action; electricity in attraction and repulsion, in contraction and expansion, in motion, and the cause of motion; yet, for all that, it is no intelligent no self-acting force, but the obedient instrument, the spring and the lever, the pen, brush and pencil of the will of an All-powerful, Self-existing Creator.

How childish, then, does it not appear, to speak of self-acting forces in nature!

VI.

OXYGEN, AND THE SOURCE OF ITS SUPPLY.—INHALATION AND EXHAL-ATION OF PLANTS.—CARBONIC ACID ; FORMATION OF DIAMONDS.—PRECIPITATIONS FROM THE ATMOSPHERE, CREATION OF THE FIRMA-MENT.—THE EARTH SUPPLIED WITH MOISTURE.—NEWTON.

IN the life of plants, of animated creation, in the composition or dissolution of almost everything that presents itself to our

---

*As the light, the rays of the sun, with metallic solutions, deposit and corporify in surpassing splendour upon the flowers of the field, so in turn, as we have seen, metals themselves are reduced again into rays of colour; gold itself is diffused into the brightest ruby and purple upon glass and porcelain, it is all but re-converted again into the red and purple rays of the sun.

notice, electricity, with its attractive power of whatever comes with more or less affinity within its reach, the metallic solutions which fill our air, and oxygen, seem to be, of all bodies, the most intimately interwoven. Nothing is more strikingly acted upon by electricity than metals, and nothing more readily oxydises, or comes under the influence of oxygen than metals. Hence we may conclude, that metals in a state of undefinable solution are always in a state of amalgamation with oxygen, and that in this state of amalgamation, electricity, as in all cases, is the binding element of the body, the cement of the cement, whether that body be in a gaseous, fluid, opaque, or solid condition, and that it depends on the degree of electric excitation and amalgamation as to what state or property certain proportions of matter, or substances, are to assume.

In feeding upon metallic solutions derived from the air by plants, and no doubt by ourselves, and by animals on land and in the water, it will be evident, according to the above, that the solutions could not be absorbed, could not be inhaled without the necessary portion of, or combination with oxygen, and that, living by the inhalation of oxygen, we must necessarily inhale some portion of metallic matter, determining, perhaps, electrical operation always united, the various state and colour of the blood, the sap and juice of plants, etc. Thus the preceding trio seems to be inseparably connected ; and having treated already at large upon electricity, and tried to prove the metallic state of the ocean of air we live in, it still remains to demonstrate, if possible, where in the economy of Divine Providence, the important element of oxygen is derived from.

At the beginning of this work I have already spoken of the certainty of oceans of different matter existing beyond our own atmosphere of air, and these oceans, or rings, which surround our globe, are but ulterior rings and strata of a system of construction which begins within the earth itself ; for the earth, as geologists have shown, is composed of rings or strata, which, as we leave the interior, become lighter and lighter in substance the nearer we come to the surface; and this surface is girt by an ocean of water, and this again covered by a lighter ocean of air. Hence we may reasonably conclude, that the higher oceans are progressing in lightness, expansion and elasticity, in proportion to their reaching the boundary of our sphere.

That of these oceans one is of oxygen, I am the more inclined to believe, as the supply of this gas has, as yet, not been accounted for; that it is the first and next in order after our own atmosphere of air, I hope to be able to establish.

Oxygen is a body, a gas, which scientific men have, as yet, not succeeded to decompose, though several beginnings have been made. In nature we meet with it in every form, but in the air we find it in its most simple state.

The air itself is composed of two gases, of which seventy-nine parts are nitrogen, or azote, which means destructive of life, and twenty-one parts of oxygen, or vital air; and these two gases are almost evenly mixed throughout the whole ocean of our airy atmosphere, of which oxygen is the leaven. Oxygen in fact, is the all-pervading life-giving matter of this our terrestrial globe; no human being, no animal, whether of the air, water, or earth, can live, nor can plants grow without it; and, like electricity, it seems to unite, or to be in a state of incorporation, more or less, with every created substance. By this incorporation, the oxidation of inanimate matter, by the process of respiration of man, animals, and plants, by consumption through fire and other causes, so much is used and withdrawn from the air, that it is truly wonderful how the supply keeps pace with the demand, and that too with a regularity hardly conceivable. It is one of the masterpieces, if we may say so, of the God of Creation, of Him who needed no counsel in the arrangement of His works.

Azote, or nitrogen gas, the great bulk of our atmosphere, science has not been able to dissolve, or decompose, no more than oxygen ; it enters into chemical combinations with oxygen in different proportions from that in which it is mixed with it as air, forming nitric acid, etc. ; but otherwise it is not known to suffer any diminution ; and if this be so, we may conclude, that it is the permanent part of our atmosphere, through and with which we inhale oxygen in that proportion which is most suitable to our existence.

Oxygen, then, is the variable, the constantly consumed and abstracted part of our atmosphere, or more correctly speaking, of the air ; and unless my theory be correct, it will as yet have to be found out, by what law it has pleased Divine Providence to produce or renew, and from what chamber of His laboratory to furnish us with this indispensable vivifying element. From appearances, we should suppose, as we shall see, that the supply comes from above, from beyond our own atmosphere, though the results of chemistry certainly seem to be against this supposition. But then, chemistry has still to divide both nitrogen and oxygen.

There are bodies which contain very large quantities of oxygen, and in particular manganese ; from these bodies it is driven off by heat, and by being heated to a high degree, others will again absorb it. By this, and other means, scientific men have been able to collect and weigh this gas ; and they find that 100 cubic inches at a medium temperature and pressure, weigh 34·4 grains, whereas an equal column of atmospheric air weighs only 31 grains ; and azote, or nitrogen, the residue of the air after being deprived of oxygen, to weigh only 30·20 grains.

Supposing these results to be correct, it would certainly seem strange, that oxygen, the heaviest of the three bodies, should form an ocean above our atmosphere ; it would be something like water floating upon oil. It must, however, be borne in mind, that

K

oxygen, unlike water and oil, is a rarefied and elastic body, and that the oxygen measured and weighed by scientific men, has, moreover, been obtained by artificial means; and the question therefore arises : is it as pure as that which floats in our atmosphere or above it, particularly seeing, that water, however pure, differs much in weight ? How will pressure and temperature HERE and ABOVE relatively influence its density ? Professor Schönbein, some years ago discovered " ozone " in the atmosphere, a gas, which by some, is considered a compound of hydrogen and oxygen, by others, oxygen in a highly excited state. Mr. Auguste Kouzeau has also succeeded in obtaining a very odoriferous gas from oxygen, said to be dangerous to respiration.

And does not this, in some degree, bear out already my theory ?* May not oxygen form some undetected combination by coming down into our atmosphere ? Indeed, it seems to me very likely, that nitrogen, the bulk of our atmosphere, is an undetected compound, of which the heaviest part unites with the oxygen coming from above, and thus facilitates its descent to the earth. But still more probable does it seem to me, that oxygen unites with the metallic solutions with which our atmosphere is charged, and thus sinks easily down to us with uninterrupted regularity; hence also its greater weight over that of the air, or nitrogen, by themselves. If, on the other hand, oxygen is really heavier in its free state than air & and azote, there will be more difficulty to account for its rise in even proportion to the greatest height against the law of gravity, than to account for its descent on my proposition. Or, if it rise like vapours from the earth, to form into clouds and come down again : whence does it rise ?

Taking, then, for granted, that, in its pure, natural and

---

* Since this was written Professor Schönbein has followed up his important discovery ; and it is now a fact, confirmed by Dr. Faraday, that oxygen is a compound of ozone and ant-ozone gases, the more perfect knowledge of which may lead to most extraordinary results.

expanded state above, it is lighter than air, the same as air is lighter than water; and that in this assumption I am borne out by the evidence which I shall bring forward, I will now try first to show, that, as a source of supply, oxygen is not contained in the earth, nor in the water, nor in the air.

As a source of supply, oxygen does certainly not seem to be contained in the earth, or to evaporate from it, in the state in which we breathe it; for, if you dig a well or pit, it will by degrees escape from the air with which it was mixed and leave nothing but azote behind, a substance, in which neither human being, nor animal, can live, or fire burn; and in thus separating from the air, it either creeps into the wall of the pit or well to form some oxidation, or it rises again into the light of day. It is the same with mines that are not worked, or badly, or not at all ventilated; and if our supply of oxygen came from the earth, this would not be the case, and the well, pit, or mine, would never become foul, and dangerous to human life.

The lower strata of the air are always, more or less, charged with deleterious matter, which, partly as miasma and contagious gases, dangerously affect the animal organization; there are fogs which give a peculiar smell, and which, according to Humboldt, remind us at certain times of the year, of those accidental mixtures of the lower air.

Now if our supply of oxygen came from the earth, though united with various metals it forms the greater part of the crust of our globe, it stands to reason, that the strata of air nearest to the earth would always be as much, if not more, charged with it, as the upper regions, notwithstanding the exhalation of other matter, and the stagnation of the air in so many places, both above and below ground; for, the oxygen would there be fresh, and by its constant escape upwards keep the air healthy and clear like a rivulet at its spring. Nor is it likely, that, with an almost

K 2

constant deposit of carbonic and other gases, and matter floating over the surface of the earth, God should have chosen to convey to us, through this dead sea of impurities, that element, on which life itself depends, and that this element should more be shrouded in corruption at its source than at the greatest, or any, distance from it. If, again, it came from the earth, it would also have a tendency to return to it, like rain and dew to the waters from which they rise, and not, in its pure natural state, ascend and remain, in greater proportion than below, higher than it has been given to man to ascend. But the earth draws its supply from above, and amalgamating with it by the process of oxidation, holds it bound, and only allows it to be freed again by force; it is more inclined to unite with solid bodies under certain conditions, than to escape from them when once it has formed the combination. Hence the oxidation of our globe, of whose ponderable matter nearly one-half consists of oxygen; and this oxidation would not have taken place, nor be going on, if, instead of absorbing this element like all living and inanimate bodies, from the air, it were to supply them all, and the air too, from its own substance, unless it could be proved, that there was an inexhaustible supply within the earth itself, and that at any time, place, or under particular conditions, it was evolved by it into the atmosphere.

It is said that plants, and flowers in particular, at times exhale this extraordinary gas; this, however, must be very little, inasmuch as, like living beings, they always absorb the oxygen of the air, or of the water, in which they grow, to be incorporated with their sap, the same as it is with our blood. If flowers over night produce such an atmosphere in a room, that a person sleeping therein will die from its effects, like from the effect of burning charcoal: it is, that the flowers, as well as the walls of the room, absorb perhaps all the oxygen of the place, leaving nitrogen, charged with the carbonic odour of the flowers, behind, just the same as a charcoal fire in a closed department. In daytime, when the air of a room is more or less changed by the opening of doors and windows, no such éffect is perceived, and hence, no doubt, has arisen the belief, that flowers exhale oxygen by day, and carbonic acid gas by night.

Speaking of "the sources from whence the atmosphere may derive carbonic acid," *Cosmos* p. 306., Humboldt says in Note 379: "I have not included in this enumeration the exhalation of carbonic acid gas by plants during the night, WHEN THEY INHALE OXYGEN, because this source of addition is fully compensated by the respiration of plants during the day."

How these results have been arrived at I have had no means to ascertain ; but it strikes me very forcibly, that if this exhalation of carbonic acid by plants during the night did take place, it would be impossible for animals to live, or human beings to sleep, in forests or in places of rich vegetation, and the oxygen evolved by plants during the day would do them no good for the night.

Oxygen promotes our breathing, fast breathing accelerates the circulation of the blood, and a rapid flow of the blood causes perspiration. And is it not the same with plants and flowers inhaling oxygen, as it were, that they perspire after rain and dew, particularly when followed by sunshine, from increased activity of life? This activity is manifested by their erected carriage, by electricity raising up their drooping heads and limbs, sustained, refreshed, and nourished by the oxygen, which in combination with metallic solutions have been attracted and absorbed.

Oxygen gas is evolved by the action of the sun's rays on the moistened leaves of trees, which, it is said, by this agency decompose the carbonic acid DIFFUSED IN THE ATMOSPHERE from various

sources, and by combining it with their carbon, flourish and
increase in size.   Now there are at most, and at times only, but
one-and-a-half per cent. of carbonic acid contained in our atmosphere, and this, if decomposed by the aforesaid process, would
not go very far to make the leaves of trees flourish and increase in
size, particularly those on which the sun never shines.   Mere
warmth will not have the effect of evolving oxygen from moistened
leaves, or from a fresh leaf put into a glass of water; but as the
rays of the sun produce this result, they must exercise a kind of
fermenting influence on leaves in that state, from which they
disengage the oxygen, causing it to appear in the shape of little
silvery bubbles, like those that rise from the bottom of a flask of
water over a fire.   The oxygen thus evolved is no doubt the
result of an excess of electric activity produced by the rays of
the sun; this emanation, however, can never be regarded as a
supply for our atmosphere, for, the process is very partial, short-
lived, and at long intervals, and there are many more leaves
not shone, than shone, upon by the sun.

From what I have said here and at the preceding page it will
appear evident to the reader, that the inhalation and exhalation
by plants of oxygen and carbonic acid must be of very little importance as to the effects thereby produced upon the atmosphere,
and the same upon plants as far as the inhalation of carbonic acid
is concerned.   For, the minute quantity of this gas in THE ATMO-
SPHERE is utterly incapable of materially, if at all, promoting
their increase in size.   For this reason there must be another
source from whence plants derive their carbon, and particularly
from whence they derived it since the third day of creation
up to the deluge, when they grew into such gigantic dimensions,
and covering, perhaps, almost every part of our globe.

Unlike oxygen, hydrogen and nitrogen, which never, or but
in small quantities, are found on the surface of, or issuing from,

the earth,* " we see steam and carbonic acid gas issue from the ground almost always free from nitrogen ";—" of gaseous emissions those of carbonic acid are, as far as we yet know, the most numerous and the most abundant." (*Cosmos* 205 and 206.) Therefore, the source of supply must be sought in the earth; but, being a compound, how did the carbonic acid get there?

The same as from the beginning by electric agency the solid portion of the earth had been precipitated from chaos, attracted by its electric nucleus; the same as still by electric action we see pure meteoric iron, stones, pebbles, and sulphurous substances precipitated from our atmosphere, as noticed in page 63; the same as we cannot help connecting the smell of gunpowder, the bituminous matter and brimstone spoken of at page 61, with the beds of saltpetre, sulphur, and of copper and iron pyrites, etc., found in the earth; the same as we must trace the beds of nitrates, chalk, limestone, marbles, and sand, to the same source from which Liebig, page 109, obtained his nitric acid, his lime and ammonia: so we must trace the existence of carbon and carbonic acid to the same source of combination and precipitation, to our atmosphere, the firmament dividing the waters from the waters.

More than this. Similar to the formation of crystals of hail in the air, so, at some early time, crystallizations, bright as if of solar rays, may have taken place above, and in the shape of glittering diamonds been dropped upon the earth; for, it is only a superior region that any thing so pure could come from, the same as man in his original purity came from the hand of God.

But, as the diamond was washed away and embedded in the

* " The absence of hydrogen in gaseous emanations from the fissures of craters and from lava currents which have not yet cooled."—*Cosmos*, p. 226. " The small quantity of nitrogen emitted opposes similar difficulties to the hypothesis of the entrance of atmospheric air into the crater."—*Cosmos*, p. 287.

alluvial soil of the earth, and man threw himself into the soil of sin and immorality, both were tarnished, save one celestial gem, who alone was preserved in its original crystalline purity immaculate; and as the defaced diamond can only be restored to its pristine brilliancy by being rubbed and scoured with its own broken substance : so the disfigured soul of man could be redeemed, could be re-invested with its pristine purity, only by the broken heart of a God made Man like himself. And as the purified diamond exposed to the light absorbs it more and more until it becomes more and more resplendent in its fulness : so the purified soul drinks in the light and grace of God to overflowing. But what a glory of light must not SHE have drank in and absorbed, who was conceived without the taint of sin, and BORE THE LIGHT OF THE WORLD WITHIN HER!—an immensity of glory and grace, whose halo fills the heavens, embraces the whole earth, shines upon all mankind, and is partaken of by all who but will be her children.

It is melancholy to reflect, that the distribution of human diamonds—particularly among those who outside are sparkling and polished, decked with precious or imitated pearls and stones, and who far and wide diffuse around them the odour of simulated virtue and artificial perfume—should be as scarce and isolated as that of carbon diamonds hidden in the ground.

. The isolated distribution of the fields of the latter over the earth, may, perhaps, find a parallel in the remarkable fall of hail, no doubt, a rare occurrence, described by Mr. Thomas Sutton in the Proceedings of the Royal Society, as related in the *Illustrated London News* of 4th January, 1863. I give the interesting article entire though in part foreign to the subject.

" The district over which the hail fell on the 7th of May last, at Headingly, near Leeds, was very narrow. The fall was preceded by a violent storm.   The hailstones did not fall in a continuous shower, but in irregular clusters.   Sometimes a field would be thickly strewn with them whilst an adjoining one escaped with hardly any ; one part of a greenhouse would be much broken, and the remainder, similarly exposed, escape uninjured. So great was the force, that in some cases, circular holes were cut in the glass without the sheet being otherwise injured. The end of a pendulous branch of beech, 12 inches long and ⅜ inch in circumference, was cut from the tree, and several larger branches from apple and lilac trees. The hailstones were of different forms and sizes, and Mr. Sutton sketched no less than forty varieties, some of a fantastic form. The heaviest weighed by Mr. Sutton was two ounces, but some weighed by other persons were said to be upwards of five ounces. One cake-shaped hailstone had a radiated mass resembling a mushroom rising out of it. Another stone was composed of five large masses of ice, quite clear, and in size like nutmegs."

Thus, in the same way, under a different early condition of the atmosphere, diamonds, no doubt, have at times been scattered from above, and therefore we must conclude, that not only carbonic acid, but the diamond itself, have their source in the air.

In Cosmos, note 376, Humboldt refers to Boussingault " on the influence of atmospheric electricity in the production of nitrate of ammonia, which is changed into carbonic acid by contact with lime ; " and having seen that Liebig obtains from the air the very elements which constitute and produce carbonic acid, we cannot but conclude, that, when in the infancy of our planet, after the separation of the land from the water, and ATMOSPHERIC ELEC-TRICITY being probably at the climax of intensity of action, those enormous volumes of carbonic acid were precipitated, which, perhaps in the atmosphere already, combining with kindred sub-stances, formed the beds of chalk, limestone and marbles before spoken of, entered into the composition of other earthy and of metallic carbonates, thereby purifying the air for future life, fer-tilizing the land, and laying the foundation of a luxuriant vegetation.

It is even possible, that when hills and valleys and plains had been clothed with verdure and towering trees, from subterranean convulsions, they sank in all their glory, or forests perished and calcined under a burning sun, and were covered by continued and far-extending precipitations of lime and other atmospheric products, similar to heavy and lasting falls of snow, which often to an incredible depth cover the ground; and this process may have occurred again and again before the atmosphere was sufficiently pure for animal life to exist in; the very spots of sunken woods and fertile lands may have been re-decorated with the richest garments of nature, likewise to be buried and again to be re-juvenized.

All this seems to be more likely than that "at the same early period of generally-distributed VOLCANIC ACTIVITY, there also ISSUED FROM THE EARTH THE ENORMOUS QUANTITY OF CARBONIC ACID WHICH, IN COMBINATION WITH LIME, HAS FORMED THE LIME-STONE ROCKS, and of which the carbon alone, in a solid form, constitutes about the eighth part of its bulk." *Cosmos*, 207.

On the other hand, Humboldt, no doubt, is correct, when he says (page 306), that "the admixture of carbonate of ammonia IN THE ATMOSPHERE may have been older than the presence of organic life on the globe," for, without this, no organic life could have existed, and for carbonic acid to issue in such abundance from the bowels of the earth, it must have been deposited there first.

That this deposition of carbonic acid took place in its fluid form, like the precipitation of water by the chemical union of oxygen and hydrogen, I do not believe. I rather think that, besides the combination of carbonic acid with lime, &c., spoken of before,—seeing such a minute quantity of this gas in the air as to be of no account in the growth of plants, and that "according to the important experiments of Saussure and Liebig (*Cosmos*, 305) traces of ammoniacal vapours which furnish to plants the azote they contain," exist in the atmosphere : there is every reason to suppose, that, by the process of inhalation by plants of the oxygen and nitrogen of the air and their decomposition by electrical agency, combined with some other constituent, the carbon of almost the whole vegetable kingdom is produced, though it may partly be ascribed to large quantities of carbonic acid taken up by the roots of plants.

When thus successive series of vegetation were buried in the womb of the earth, they coaled by the electric heat of the interior, and thus produced the carbonic acid which in such enormous quantities is emitted "in districts of extinct volcanoes and thermal springs," and which, without rising into the air, re-combines again with water and earthy matter ; and in this way lime rocks may still be formed, though they do not seem to me to have been so formed originally, and particularly not before "the presence of organic life on the globe."

The ATMOSPHERE, then, does not derive any carbonic acid from the respiration of animals or emissions from the earth, on account of its being too heavy to rise ; it must, therefore, be and have been produced in the air, and by woods and plants or vegetable matter sunk into the earth ; for, as we have seen already, the respiration of plants, whether of carbonic acid or oxygen, counts for nothing in the bulk of the atmosphere.

That plants, like all organic life, must emit, or throw off, something, is quite certain ; they throw off their leaves, but their emission of gases in quantity and quality is still very uncertain. Some recent experiments seem to show, that plants emit ozone both by day and by night, and more by night than by day ; and if this be so, the theory of emission of oxygen, and, no doubt, of carbonic acid, by plants, if true, will, at least be considerably modified, but still more clearly prove, that the source of oxygen is not to be sought in our atmosphere of air.

Oxygen, as a reservoir for the supply of our globe, does not seem to be contained in water either, from which that part of it, required for sustaining the life of the creatures and plants it contains, is abstracted or absorbed by them, as it is done by us from the air. If this were not so, water, whether in contact with the air, or not, would never become foul, inasmuch as the combination with the life-preserving oxygen would be its unfailing and unvarying condition, and keep it fresh, whatever quantity it might have to give up to our atmosphere. But water will decompose by standing for some time, simply because the oxygen is escaping from it, the same as the air of an apartment long closed will decompose by the retirement of its vivifying element.

Water is a compound of eight parts of oxygen and one part of hydrogen; and though this appears almost similar to saying that bread is a compound of eight parts of leaven and one part of flour, yet we are unable to explain the nature of the change which takes place in producing water by means of the aforesaid combination, or chemical union of gases. Oxygen thus converted into water has evidently no longer anything more in common with the oxygen we breathe—which pervades the whole atmosphere and all the waters of the earth as leaven pervades bread, and which is as necessary for the creatures living in the water as for those living on earth—than petrified wood has anything in common with the same wood when green; the difference between the two is this: that the former is converted in a moment, and the latter by very slow degrees.

If oxygen, on being changed into water, did not change its nature altogether, none of all the creatures living, and plants growing in it, would require any pure oxygen for their sustenance, nor would this gas merely in a mechanical, and not in a chemical, manner, unite with the water. But it is a well established fact, that river water in particular possesses the property of absorbing much oxygen, and that for this reason the surface of both rivers and the ocean holds more oxygen than the atmospheric air, to the amount even of 29·1 per cent.; and this it is tha' contributes to the maintenance of the respiration of fishes and the growth of aquatic plants, the same as required for man, and the animals and plants on land.

Water again, a combination of oxygen and hydrogen, will, at a low temperature, form ice; but it is not from the ice either, no more than from the oxidized crust of our globe, that we derive the oxygen we breathe. It seems, on the contrary, that in the formation of ice, the oxygen is excluded and left behind in the unfrozen liquid, to establish, as it were, an additional store for the animals underneath, the ice being a barrier against the usual supply from the air, destitute itself of atmospheric air when first dissolved, and consequently unable to sustain respiration in fishes. Ice-water, for the same reason, the absence of atmospheric air, is mawkish and insipid, but by exposure to it, it speedily absorbs a due proportion. With snow-water it is much the same, and for this reason, no doubt, it is, when drunk, so injurious to the constitution; and frost and cold are evidently as unfavourable to oxidation as warmth and heat are favourable to it. Hence also to nature the particular benefit of a warm rain, and the invigorating influence of dew, which contains still more air and oxygen than rain.

Water also, like land, is covered with vapours of various descriptions, and the same reasons which I adduced for not considering the earth as the universal reservoir of oxygen for the benefit of all things created, will likewise hold good with regard to the former.

That the atmosphere itself is not the great holder of oxygen seems very evident from the decomposition of the air, wherever it may be, unless renewed from time to time.

I have already spoken of how this gas escapes from the air of rooms kept shut for any length of time, whether under or above ground; but, if even a chimney or tower were built a thousand or more feet high, everywhere air-tight save at the top, the oxygen would soon be found to have disappeared from the air within, by either having crept into the wall, or ascended into the regions of light, towards which it is most inclined. Being, in its purely natural state, lighter than the matter with which combined it forms air, it must naturally have a tendency to rise, unless more powerfully attracted by bodies with which it chemically as well as mechanically unites; and it would, for this reason, leave the whole earth a valley of death, had a good God, who called the world into existence, not ordained that the due and rain of heaven, and the wind and agitation of the air, should always provide us with it in that nearly uniform proportion, in which almost all scientific men have found it to exist everywhere.

And whence is it likely that the movements and currents of the air should carry to us this element of life? From where else, if not from a reservoir in the regions above, from the highest of which life itself has come?

If water, as I have shown, and as the air-vessels of fishes in particular tend to demonstrate, contains air, and of necessity oxygen, it does beyond all doubt derive it from that ocean of air which fosteringly rests upon its surface; and though the process by which the water sucks in the air from above, drawing down a lighter body than itself, is not known to us, yet we may compare it to apparently solid bodies absorbing moisture, to quicklime feeding upon vapour, until saturated to that degree assigned to it by its Creator for His own purposes; it is, in a certain manner, an oxidation of the water, perhaps an electric suction.

Our atmosphere, in the same manner, absorbs its own lighter part, from a reservoir higher above than itself, and with such regularity, that no consumption, however great, takes place, but is instantly made up again by a fresh supply. As water holds a given quantity of air, with a constant tendency to rise, so air holds a given quantity of oxygen, and this again a defined quantity of still purer matter, perhaps hydrogen, until we have arrived at the limit of our earthly sphere, where a still purer invisible matter, electricity, with its inherent magnetic power, begins to fill up the spaces between the worlds of the universe, and in which they move with almost the rapidity of thought, directed and sustained by God, the Creator of all.

To facilitate the absorption and incorporation of air by the water, He ordained, that, by the varying pressure of the atmosphere, caused by the disturbing influence of the moon gliding over it for the production of the tides, and by the turbulent action of the breeze and the hurricane, it should be brought more within the wild but benificent embrace of the air; and the atmosphere itself He provided to be rocked to and fro by the companion of the earth, and to be set in almost constant motion all over the globe by its various currents, in order to promote the absorption of the animating element, and its diffusion in every quarter, and particularly at those places were the air is most liable to decomposition. And these places are unquestionably at the equator, where nature, as it were, is constantly exhausting and recovering itself, where, more than recovering, ever since the sun burst into light, the earth has constantly been increasing in bulk, so as to make the poles appear flattened against the protuberance thus produced, if such protuberance really exist; it is there that the greatest evaporation of water, the greatest evaporation of electricity, the greatest absorption of oxygen, and the greatest decomposition of air takes place.

When in our own clime, in the heat of summer, owing, no doubt, to the escape of air and oxygen from brooks and ponds, the fishes come up to the surface, apparently gasping for breath : what must not be the evaporation under the vertical rays of the sun at the equator ! What clouds of vapour must not cover the sea and the land ! What enormous quantities of electricity must and do not rise to produce thunderstorms almost every day ! What absorption of oxygen must not take place there by the vegetable creation with a constant summer and uninterrupted vegetation ! And towards the equator it is, that from the poles, where there is no decomposition of air or water, where, in their respective regions, electricity and oxygen exist in their greatest and densest proportion, they are more particularly carried by ceaseless currents.

But also the dew and rain of heaven, as we have seen, bring down to us this indispensable oxygen, which, owing to the colder state of the higher regions of our atmosphere, is there also in greater abundance than in the warmer and more expanded strata below, the same as it is present in greater quantity in winter than in summer, as evidenced by the brighter burning of the fire at the former season. They will bring it down, not merely judging by the coolness which they impart to the air, but also by the pleasure of breathing on a fine summer morning, or after the rain of a sultry summer day ; it is the oxygen that makes us breathe again and again, as if we could not breathe enough ; it is the oxygen coming down fresh from heaven that causes the perspiration, and dissolves the fragrant gases, of plants and flowers, making them fill the air with an odour, as it were, of another world, and with the Church, offer up to God, at every moment of time, from the rising of the sun to the going down of the same, the incense of adoration.

It is after thunderstorms in particular, that the air is filled with the perfume of nature ; and this, no doubt, arises from the still greater than ordinary quantity of oxygen with which the water is charged. For, apart from the usual cause of rain through the formation of vapours into clouds, the process above, on a grand scale, seems to be that of scientific men below, who have discovered how to convert oxygen into water, and thereby changing its very nature; hydrogen, like oxygen, rises in part from the earth towards the sphere from whence they have come; both are united above, and by electric agency formed into that water, to which we owe the more particular delight of an atmosphere, almost overflowing with the sweetest fragrance.

In considering this transformation of oxygen into water, I am involuntarily carried back again to that awful instance, when by this means, and the gathering into one universal cloud of the mists of the earth, the floodgates of heaven were opened, and a sinful generation was swept away.

The deluge, though vouched for by the tradition of every nation under the sun, is, to many, difficult to believe; but suppose it to be correct, that there is an ocean of oxygen above our atmosphere, as there is an ocean of water resting on the land, and an ocean of air covering land and sea; that this body of oxygen is saturated with hydrogen from an ocean still higher above, in the same manner that land is saturated with water, water with air, and the air with oxygen; that, as we have seen already, positive electricity pervades the superior parts of our atmosphere and the whole space of the universe; and how easy was, and is, it not then for God, whose laws of nature harmonize, go hand in hand with those of morals, to excite this electricity, and by its silent operation, apart from the common rain, and the gathering of the mists of the earth into one common cloud, to convert, in an instant, and in even a comprehensible manner, an ocean of animating matter into an element of chastisement and destruction. If water is but another name for oxygen, as ice, snow, hail, dew, and steam, are but other names for water; and if within the sphere of our globe, that ocean of oxygen with its component part of hydrogen, like a line of demarcation resting upon our surface of air, forms, perhaps, the

L

blue canopy of heaven and grand reflector of light*—how beautifully true do not then become the words of Moses: "And God said, Let there be a firmament (an atmosphere of air) in the midst of the waters, and let it divide the waters from the waters. And God made the firmament, and divided the waters which WERE under the firmament from the waters which were above the firmament; and it was so."

As on each day of creation, it may be supposed, God called a new law of nature into operation, so, on the second day, He issued the "fiat" of a firmament to be made.

The laws which on the first day, in the form of water covering the whole earth, had precipitated from chaos the requisite quantity of oxygen and hydrogen, still left an enormous amount of these and other gases hanging undivided over the deep; but now another law, though still an electric agency, carried out one of the most striking designs of the wisdom and providence of God.

The firmament, the future scene of life upon earth, was not to contain anything deleterious or repugnant to man in particular. Hence, chlorine, one of the most offensive substances, and for life insupportable, though otherwise in the highest degree useful; and sodium, burning with great intensity in the air, the most remarkable devourer of oxygen, of the supporter of life, were separated from chaos and in combination precipitated into the water, that, as chloride of sodium, as sea-salt, it might saturate the entire ocean then covering our globe, in order to preserve its freshness to the end of the world, to provide an abundant store of one of the greatest essentials of life, and by this means make the firmament fit for organic life to exist in.

---

* In a pure state of the air, water-gases, when fine, reflect a beautiful blue tint; this tint, however, becomes dingy and greyish with a moist and impure atmosphere; but as neither of these conditions, or states, are normal, we only can account for the permanent blue sky by the permanent ocean of hydrogen above, which gas, among all known substances, is the best reflector of light, as has been noticed before.

The precipitation of chlorine and sodium was succeeded by the precipitation of the firmament itself, forming our grand ocean of azote, or nitrogen, permeated by a fifth portion of oxygen, so as to constitute the atmosphere in which we live.

The firmament having been separated from chaos, there was left above it still a vast ocean of pure, unappropriated oxygen, permeated by hydrogen, an ocean of the purest water, when, as we have seen, electrically acted upon ; and thus the firmament divided the waters from the waters, leaving still beyond this transparent ocean of latent water above, a still vaster and more expanded ocean of hydrogen, both divinely calculated for the penetration of light and vision, and the contemplation by man of the wonders of God on high.

As noticed before, the ocean of hydrogen above forms a vast burning-glass and reflector embracing the whole earth, in order to collect and concentrate the rays of the sun, and throw light and heat with intensity upon it. When, however, at the deluge, the windows of heaven were opened, when, in addition to descending mists, so enormous a quantity of oxygen and hydrogen was precipitated in the form of water as to cause it to rain for forty days and nights without intermission : the oceans above were so dimished in bulk or density, that the hydrogen lens was greatly reduced in its previous light-concentrating and reflecting capacity ; and hence, as I believe, the real cause of the diminution of all life upon earth after the deluge. Since then the precipitation of oxygen and hydrogen has gone on, not to inundate our globe— for the rainbow still appears—but to supply us with moisture and with the corresponding blessings of heaven.

This theory is strengthened by the following observation of Sir Isaac Newton, who says : " that COMETS seemed to be required FOR THE CONSERVATION OF THE SEAS AND FLUIDS OF PLANETS, in order that, from their condensed exhalations and vapours, the water consumed in vegetables and putrefaction, may be continually replaced and supplied. For, all vegetables grow wholly from fluids, and then are, to a great extent, turned into dry earth by putrefaction, a slime perpetually settling from putrefying fluids. Hence it is that the bulk of the solid earth is continually increasing, and that its fluids, if not supplied from any other source, must constantly decrease, and at last, entirely fail. I suspect also, that the comets supply our air with that which is the smallest, and most subtle, and useful part of it, and which is required to sustain the life of everything."

Who does not see that the functions here ascribed to COMETS are more naturally performed by the oceans of oxygen and hydrogen above, yielding their supply to the solid earth until it shall please God to consume it by fire, fed by the oxygen which it received through the air.

It is also worthy of notice, that of the superincumbent oceans of air, oxygen, and hydrogen, the air, as a general rule, but always when cold, should be positively electrified; that oxygen, in its uncombined gaseous state is attracted by the positive, and hydrogen, in its uncombined gaseous state, by the negative pole of the voltaic battery; that therefore the electric state of these bodies is calculated for the functions they have to perform, oxygen being negatively, and hydrogen positively electrified. Owing to the extraordinary absorption of oxygen and hydrogen by the earth, and by vegetation and animal life, it is not to be wondered at, that the ocean above us should by degrees diminish, in bulk or density, however little, in a given time; and to this diminution, both of the oceans of oxygen and hydrogen, we may, no doubt, ascribe the moon's acceleration, that is, her gradual, though slow approach to the earth, if she is really coming down.

VII.

PENETRATION OF LIGHT THROUGH THE ATMOSPHERE.—SPECTRAL ANALYSIS.—JUPITER'S SATELLITES.—VELOCITY OF LIGHT; RECENT DISCOVERY OF ERROR IN ITS THEORY.—HUMBOLDT.—THE LIGHTING AND WARMING OF OUR EARTH AND THE PLANETS; OBJECTIONS AGAINST THE OLD THEORY; VON GUMPACH.—THE HEAVENLY BODIES INHABITED; ASTRONOMERS ON THE OTHER SIDE OF THE MOON.—THE CHURCH A FOCUS OF LIGHT AND WARMTH.—LIGHT AND VISION OF THE SOUL.—MORE STARS SEEN NOW THAN BEFORE.—PROVED: THE EXISTENCE OF AN OCEAN OF OXYGEN ABOVE, THE EXISTENCE OF POSITIVE ELECTRICITY WITHOUT, AND OF NEGATIVE ELECTRICITY WITHIN, THE SOLID NUCLEUS OF THE HEAVENLY BODIES.

THE steady absorption of mineral and metallic solutions that fill the air, along with the absorption of oxygen and hydrogen by the solid earth from its superior oceans, in a corresponding ratio facilitates also the transmission of light through our atmosphere and the atmospheres of the heavenly bodies.

This is proved by the spectral analysis, according to which MINERAL VAPOURS ABSORB THE RAYS OF THE SUN and throw dark lines upon their spectrum, whilst mineral-vapour flames by themselves throw bright and coloured tints upon their spectrum.

Thus, the sodium flame produces yellow lines, and the flame of iron vapours produces bright rays upon the spectrum, whilst, when interposed, sodium vapour absorbs the rays of the sodium-vapour flame, and iron vapours absorb the rays of the iron-vapour flame, causing dark lines to fall upon the spectrum. Therefore, metallic or mineral solutions in the atmosphere would absorb corresponding rays coming from the sun, and produce darkness upon the earth ; and hence, the precipitation of sodium and chlorine for the formation of salt, and the purification of the firmament, as well as the deposition of other elements from the atmosphere, were, and are still, necessary for the free admission of light upon the earth ; and these continual depositions of solutions and gases, so beautifully preconceived and providentially arranged by the Creator and Ruler of chaos, must influence the oceans above us.

OUTSIDE, EXTERIOR, to these oceans, to these atmospheric envelopes of our earth, I believe light to be uninterrupted, to have been and to be simul- and instantaneously transmitted or diffused to its furthest limit.

A burning candle will not shine beyond a certain distance, if it were to burn for ever, and the same holds good with any other light or fire, according to its size and luminosity.

If the immersion and emersion of the moons of Jupiter into and from his shadow, CONTRARY TO THE CALCULATED TIME, takes place SOONER when the earth in its path round the sun is placed between the sun and Jupiter, when Jupiter is said to be in opposition, and later when the earth is passing towards the opposite side of the sun, towards conjunction of Jupiter with the sun: it does not follow—the difference between SOONER and LATER than the calculated time, amounting to about fourteen minutes—that, therefore, the rays of Jupiter's moons take a quarter of an hour longer to reach the earth when near conjunction, than when it is right underneath Jupiter, between Jupiter and the sun ; or that, when the earth is most remote from Jupiter, the rays of his moons, on their immersion into his shadow, vanish from our sight a quarter of an hour later than when the earth is nearest to Jupiter, or, which is the same, that in that case the rays of his moons remain visible to us after the moons themselves have already for a quarter of an hour been totally immersed into the shadow of their primary.

Though unable satisfactorily to demonstrate the reverse, yet, seeing that so many things have been wrongfully taken for granted and calculated, that the Newtonian theory of the solar system is altogether based upon a sandy foundation, that experiments affecting the transmission of light made in our atmosphere here below are no standard, even if correct, of its transmission through

the atmosphere of the sun and the electrical medium of the space
of the whole universe : I may reasonably be permitted to question,
particularly from what in this respect will appear further on, the
conclusion and calculation of astronomers come to and made from
the eclipses of Jupiter's moons, namely, that light travels at the
rate of 192,000 miles a second, and that the light of some of the
stars may have taken thousands, if not millions of our years until
their rays fell upon the earth ; and that other stars may have been
extinct thousands, if not millions of years ago, whose rays are still
falling upon our eyes, and make us believe in their still real
existence.

All this MAY be true, but it strikes me that it looks like seeing
a man's shadow after he is gone out of the sun, or at an eclipse
of the sun our still standing within the tip of the shadow of the
moon after she has passed away from under him.

I cannot help thinking, that light and shadow begin to contract
or recede at the extremity, as soon as the primary cause is
removed, and that there may still be some other explanation of the
phenomenon of light than the theory of emanation or undulation.

Humboldt himself does not seem to think the present theories
of light to answer every difficulty. In his *Cosmos*, p. 132, he
speaks " of observations made by the acute Olbers, ' of sudden
flashings and pulsations, which, in the course of a few seconds,
vibrate throughout the whole of a comet's tail, which is seen at
the same time to lengthen several degrees, and again to contract.
As the different portions of a comet's tail, which is millions of
miles in length, are at very unequal distances from the earth, IT
IS NOT POSSIBLE, according to the law of the velocity of light, that
actual alterations in a cosmical body filling so immense a space,
should be seen by us to take place in such short intervals of time.'
These considerations . . . . should make us careful to distinguish between effects which should be referred to the cosmical

ether and to the regions of space, and those which are referrible to the terrestrial atmospheric strata through which the bodies existing in space are beheld by us . . . . . we are not able to explain completely all that takes place at the uncertain and much-contested limit of our atmosphere . . . . . The phenomena of light . . . . have not yet been successfully investigated."

This,—though the proof be not true—is most singularly confirmed by a statement where, referring to an astronomical error about the weight of the sun and the earth found out by Le Verrier, it is said: "But in the meantime M. Foucault"—by an ingenious arrangement of reflectors—"had discovered that the velocity usually assigned to light in its passage from the sun to the earth had been over-estimated: the sun is CONSEQUENTLY one-thirtieth less distant from the earth than had been supposed . . ."

I here may be allowed to ask : is the sun less distant from the earth BECAUSE the velocity of his light has been over-estimated, or, is the velocity of light less BECAUSE the earth is one-thirtieth nearer to the sun, according to Le Verrier and the calculation referred to in foot-note, page 36 ? If, however, the discovery of Foucault, as stated, be independent of this calculation : by what standard does he estimate and compare the result of his experiments in this nether world of smoke and vapours with the transmission of light through the solar and stellar spheres ?

Supposing, however, that his velocity of light be correct : what then becomes of the velocity of light from the sun to the earth and the planets, from planet to planet, and from star to star, adopted by all astronomers and yet now not true ? What of the velocity of light calculated by Romer from the eclipses of the satellites of Jupiter, see p. 150 ? What of the aberration of light calculated from "the relative velocities of light, and of the earth in her annual orbit, upon which the effect depends," if the calcu-

lations of both the velocity of light and the orbit of the earth
have been blunders ? What of Romer's calculations, endorsed by
the highest astronomical authorities, if the earth be 3,000,000,
miles less distant from the sun, its orbit round him 3,000,000
miles narrower than he adopted in his calculations ? What of the
annual parallax of the earth hitherto adopted, and everything
based upon its foundation ? What of the velocity of the solar
system in space, say 49 miles in a second, when, by the reduction
of the orbit of the earth to the extent of 18,000,000 miles, the
velocity of our own planet has been mistaken ? What greater
certainty have we than before, that these new amendments respecting the reduced distance and orbit of the earth and the
diminished velocity of light are this time to be depended upon ?
And IF THERE EXIST a lesser velocity of light than previously
calculated: may there not be more reasons even than one to
account for it, apart from the assumption of a lesser distance
between the earth and the sun, and the necessarily defective and
doubtful experiments of scientific men ? Is it not likely that we
are as far from the knowledge of the diffusion of light in space as
we are from that of the NATURE of light ?

Which, then, of these calculations and conclusions shall we take
to be true ? The new velocity of light ? The new distance of
the earth from the sun ? The new weight of either the sun or the
earth ? or both ?

In the present stage, I think, we laymen in science had better
suspend our decision and wait further result of the Newtonian
theory. If hitherto we have believed in the accuracy of astronomical observations and calculations, astronomers themselves have
destroyed the prestige and undermined our faith ; they have shown
and still do show that they are very fallible guides, that universal
gravitation FAILS IN EVERY POINT IT HAS BEEN BROUGHT TO BEAR
UPON. What, then, would appear more natural, than that men,

and above all the eminent in learning, should abandon their old idol, and freely and cheerfully launch into the deep, into the depth and vast expanse of the Simplicity of the Creation?

My theory of light, of a hydrogen lens surrounding our earth, every planet, and most likely every sun, though emanating from an amateur also liable to err, may, nevertheless, meet some of the greatest difficulties and deficiencies of present theories, if it does not altogether set the theories themselves aside.

My theory alone explains the high degree of light and warmth which we enjoy upon earth; for, as Sir David Brewster says: "convex lenses possess peculiar advantages for concentrating the sun's rays, and for conveying to an immense distance a condensed parallel beam of light." If, therefore, there were no such lens surrounding our earth, it would be but feebly lighted and warmed.

Let the rays of the sun, as in the annexed diagram, diverge into space and a portion fall upon the body of the earth, 1, simply surrounded by our atmosphere of air, and it will then be evident, that, comparatively speaking, but few rays would strike and rest upon it, and that it would scarcely make any difference in the number and force of those falling upon the equator and those falling upon the other parts of the earth; therefore, light and heat would be weak and pretty equal on every spot shone upon by the sun. Let, however, the earth, 2, be surrounded by its hydrogen lens,—extending beyond the

moon and immensely greater in proportion to our atmosphere of air than represented in the diagram—and it will be seen, that the rays of the sun THICKLY CONCENTRATED will fall upon its body, thus causing greater intensity of light and heat, and particularly at the equator, the focus of this concentration.

This also accounts for the increasing light and heat of the sun when in the morning we enter the area of our lens until we arrive in its centre, and it explains their decreasing intensity as towards evening we pass out of its limits. The concentrated action of the solar rays falls and rests upon the earth, and as we rise in the balloon, or ascend the peak of the mountain, we gradually retire from their focus and they lose in power. If this were not so, considering the size of the sun, the insignificant ball of the earth in comparison, and the little distance between them, it would make no difference, as long as we beheld the sun, whether his rays were to fall upon us perpendicularly, or more or less oblique, whether we received them in the balloon, on the mountain top, or on the deck of the vessel, in winter or in summer, at the equator or at the poles.

The objections against the old theory are very clearly set forth by Von Gumpach. In *Baby Worlds*, page 134, he says:

"The difference of atmospheric heat, or of temperature between morning or evening and noon, is, according to the prevailing theory, explained by the OBLIQUE DIRECTION of the solar rays at the two former epochs. But it must be confessed that this is, to say the least, a very oblique explanation. For, it is evident that, if the sun's calorific action decrease as the square of the distance, and extends only beyond the orbit of Neptune, the linear dimensions of the terrestrial radius=less than 4000 miles, at a distance from the sun=95,000,000 miles, can mark no APPRECIABLE difference of temperature. In other words, according to the present theory, it ought to be as hot in the morning and the evening, as it is at noon :

or else as cold at noon, as it is in the morning and the evening.
And at the epoch of the summer solstice, when the earth is at the
greatest distance from the sun, the annual temperature of the
terrestrial atmosphere should reach its minimum of heat, and its
maximum, when the earth is nearest to the sun, at the time of the
winter solstice.   The reverse of all this is the case.   Moreover,
whether we ascend, even in the regions of the equator, some
thousand feet only above the level of the sea, straight into free
space by means of a balloon, or up the inclined slopes of a lofty
mountain, we find, AT THE VERY MOMENT OF NOON, that the
nearer we are to the sun, the lower is the temperature of the atmosphere; and that, at a certain elevation the ground ceases to
produce, and the snow no longer melts.   Or, whether, leaving the
zone of the tropics, we turn towards the north or towards the
south, we find, that the mean temperature of the terrestrial atmosphere rapidly decreases, and that eternal ice and snow begin to
clothe the barren regions in the vicinity of either pole.   That, in
the face of empirical facts as striking as these, and in spite of the
wonderful progress of the natural sciences in our days, the vulgar
opinion concerning the source of terrestrial atmospheric heat
should have maintained its undisputed ascendancy, is almost as
strange a thing, as is the mutual (electro-magnetic) action between
the sun and the planets itself."

All these anomalies find their solution in my new theory and
stamp it with the seal of truth.   And if this were not so: how
could the more distant planets receive light and heat, if it were
not for the bountiful provision of God which my theory establishes?
And is it not even possible that, according to the density or volume
of the hydrogen envelopes of Mercury and Venus, these planets
should NOT MORE be heated and lighted than our own earth, notwithstanding their nearness to the sun?

That upon this supposition, as well as on the ground of gravitation, from what will appear hereafter, all the celestial bodies might be inhabited by beings like ourselves, it is not unreasonable to conclude; the belief would be innocent, and if not in harmony with, at least not against any thing taught by, revealed relegion.

I, moreover, can understand, that noble minds would cherish the idea of beings dwelling upon the celestial orbs, in their nature equal, if not superior, to man, and needing no redemption; for, nothing more beautiful or higher in the order of creation can be imagined by man, than either angels in human form, or youth or virgin blooming in original innocence; the thought would be calculated to raise heart and soul to God in the contemplation of intellectual life of a higher grade serving and adoring Him from one end to the other of creation.

Without any warrant, by some astronomers even, the orbs of heaven have been peopled with all sorts of monstrous conceptions; they may be harmless too, and not against revealed religion either, but are by no means grand and inspiring.

Dr. Maedler keeps within a more sensible medium. Imbued with the nobility of his profession,—according to *Baby Worlds*, footnote p. 133,—he "depicts the inhabitants of the moon as endowed by nature with more than eagles' eyes, an untiring frame of body, and an extraordinary elasticity of motion, and he expresses the confident opinion, that 'that most splendid observatory in the entire solar system,—the lunar hemisphere turned away from us,—cannot well have the worst set of astronomers attached to it.'"

From this I cannot differ, for there ARE astronomers on the other side of the moon, and that we know nothing of them arises entirely from their over great humility; they do not wish to be spoken of.

The immense apertures of our satellite, generally believed to be extinct craters of volcanoes, are nothing but the mouths of tubes passing from the other side of the moon to ours, and through these lunar telescopes the men of science above watch our proceedings here below, and naturally those of their *confrères*.

When thus they look down upon Dorpat and Berlin, Greenwich and Cambridge, Paris, and even Rome, not minding much the lesser constellations, their astonishment must be great at the work they find performed, as with myriads of ants everything covered with figures, through which facts cannot be seen; and though timid and retired, yet they agree with the doctor : " that the worst set of astronomers are not attached to the moon."

But they do not wish to be discovered. Therefore, whenever with us there is an eclipse of the sun, or the occultation of a star by the moon, by general consent and with one accord they close the eye-pieces of their telescopes, though they do not always succeed all at the same time.

Thus, whenever there is one or the other weary on the watch, or too late, we can see through the moon, like a prisoner in darkness through a little hole in the roof of his cell; and though this is so very natural, yet the light from above has always been mistaken for stars reflected on the disc of our satellites, an opinion which somewhere in my work I myself have tried to confirm.

However, for all that has been said, and for all the theories in the world to the contrary, the astronomers in the moon may nevertheless be exposed to the greatest cold ; for, her climate and heat do not depend upon her greater nearness to the sun when in conjunction, no more than the temperature on the peaks of the equatorial mountains, but upon the focus of that solar focus of light, into which she may be placed, or from which she may be partially, if not totally, removed or excluded.

In a similar manner to the warming and lighting of our globe, the Church is the focus of God's love upon earth, and all who enter her precincts will be divinely warmed and enlightened in proportion to their advance into its very heart.

In her garden many sweet flowers are blooming, nourished by the centre of life, and drinking from the fountain of heaven; and warblers there are without number, rising gloriously up to the sky, and songsters in humble plumage, to cheer up the sickly and drooping in the silent night of affliction; her children are ACCUSTOMED to their strains, and harbour them with tender affection; and their sorrow is great when by small birds of prey, or royal hawks and vultures, they are driven away from their peaceful abodes, and desolation is brought over some fair portion of the grounds.

Outside, the clime is less warm and congenial, and the further removed from the focus of love, the more all is stunted and pining, no flowers can bud to perfection, and their colours want brightness and purity; some lovely Nightingale may appear, grown up, as it were, in a wilderness, lonely and without kindred companions, panting away, and leaving nurses, BUT NO SISTERS, behind; with song sublimely touching, the more welcome as it is rare, the admiration of all in the country the more she is solitary, and the more cherished and adored the farther away she is from her real home.   For, she is not in the focus of the light from heaven, in which alone the heart can fully live and expand; beyond its borders these songsters and flowers are but exotic, passing strangers; to souls yearning for a better land, all is pale moonshine, a dreary spiritual waste, in which, even for the most intellectual mind, there is no altar to attract, no ever-burning lamp of the sanctuary to guide, no shining lights to cheer, nothing to follow, except, perhaps, the unsteady cone of the zodiacal light of a departed or still hidden sun, to many, I hope, rising, but mostly nothing but the flickering will-o'-the-wisp, or the phantom of the desert, shifting and fleeting at the moment of being grasped.

When light first gave the kiss of greeting to the newly created worlds and embraced them with ever-increasing fondness, it THEN

DID NOT TRAVEL, for God said, " Let there be light," . . . . . .
and " there was light."

Before then it had been veiled, but now darkness withdrew, the
curtain dropped from creation, chaos had been resolved into order,
it was gone . . . . and there WAS light upon everything that had
been separated from the light.   As yet there was no concentration
round the intended orbs of the universe, as yet no noon and
midnight alternated upon the celestial bodies, but only evening and
morning, for there only WAS LIGHT, but NO LIGHTS had as yet
been made to divide the day from the night.   When, however,
LIGHTS had been made, light had not been withdrawn, had not
retired from the heavenly bodies ; it only became more dilated,
more rarefied, less visible in space, that, more accumulated and
bursting out into brilliant rays around the centre of every system,
its blessings might be centred upon every globe, and in particular
upon our earth, through the concentrating lens of its wondrous
blue envelope.

Thus, all over the universe, there was but a transition from
light to lights, from diffusion to concentration, from evening and
morning to midnight and noon, and every luminous orb shone at
once to the end of its destination.   Beyond this they could not
shine, for, as I have said already : a burning candle will not throw
its rays upon the moon if it were to burn for ever ; a child will
not see as many lights hung up in the vault of heaven as the
adult, and the naked eye of the adult will not equal the eye
assisted by the telescope in penetrating into the depth of creation.

Light, as it were, is the transition from imponderable matter to
the instinctive animal but mortal soul, and thence to the spiritual
soul of man, composed, the concentration, of pure light in its
essence, the eternal reflection of the Eternal Light, a LIVING
LIGHT, endowed with consciousness, free will, and everlasting,
SELF-ACTING, intellectual life.   Hence, when immaculate, and

dwelling in a pure or purified body, it perceives things material and immaterial, whether in exstacy or out of it. In that pure state, in the perfect state in which Adam and Eve were created, but more so in the state of the immaculate Virgin, over and above also FULL OF GRACE, the soul is the living image of its Creator; and as He beholds and embraces, at every instant, the whole universe, All in All, its past, present, and future : so, as an evidence of being made according to His own likeness, the clean, undefiled image of God is made to share in the attributes and prerogatives° of its Eternal Type, and at one time is privileged to ·see the past, at another the present, and again the future unfolded and revealed.

The mystic eye of such a soul, the living light itself, penetrates the veil of the body, as the light of the heavenly bodies penetrates the atmospheric veils within which they are clothed ; and as the thoughts emanating from a soul, from a living light of such transcendent purity, penetrate in one instance through the whole expanse of the universe, wing their way into heaven and there take up their abode, wander into the infernal regions and at the same moment shrink back again from their horrors : so light, in its pure state and intensity, is next in rapidity to human thoughts in traversing the whole of creation, bringing blessings wherever it is admitted.

As vice of any kind, evil thoughts, nay mere conceit, drown the light of the soul, obscure its vision, lame its wings for flights into loftier regions, keep it grovelling in the dust, bound in chains and wallowing in the mire : so mists, fogs and smoke, or an otherwise defiled atmosphere, impede the vision of our bodily eye, prevent the admission of superior light; even an atmosphere

---

* As instanced by Moses and the prophets, by a long line of saints of the new law, and by many holy souls of our own day, like sister Catharina Emmerich and numbers of others.

transparent with brightness, yet hides from our view heavenly hosts without number, stars untold, perhaps seen already by Adam and Eve, as noticed page 18., when unspotted they walked in the garden of paradise, and God with them. For, clean of heart, with unclouded, unimpaired vision of soul and body, the LIVING IMAGE OF GOD SAW GOD, and contemplated in all their splendour the glories of the heavens, though an atmosphere less clarified and transparent than at present rested upon our globe ; but the shadow of death having spread over the soul and body of man, spiritual and corporal vision became dull and confined ; sin, the substance of. the shadow of death, like a dark veil, placed itself between God and man, the image of God had been overlaid with the film of corruption, and no more saw its Eternal Type, the beauties of creation diminished, and thousands of stars in the dome of heaven disappeared to the fallen creature.

As, however, according to Sir David Brewster, "on the summit of the highest mountains, where their light has to pass through a much less extent of air, a much greater number of stars is visible to the eye than in the plains below," so by the continual growth of the earth through the constant absorption of matter from the air, this element itself became gradually clearer and brighter, until by this means, with the lapse of time, many stars re-appeared, silently, unnoticed, like hidden virtue, as it were, to the descendants of Adam, so that generation after generation beheld some celestial body or bodies, that had been hid from their predecessors, the same as thousands of stars were hid from our more immediate forefathers, which now WE see through a more purified medium, without the assistance of telescopes.   These space-penetrating instruments only make perceptible to our impaired vision the light of those millions of orbs, which from its first outburst already broke in and settled upon the borders of the atmospheres of the heavenly bodies, though unable to penetrate to our naked perception upon the earth, except by the progressive accumulation of its nucleus, particularly round the equator, occasioned, as repeatedly noticed, by the absorption from the air of its more ponderable constituents, of its oxygen, its mineral and metallic solutions, and the consequent clarification and increasing transparency of the atmosphere.

It seems natural, that, by the manifold operations in nature, by the various pressure of the moon floating on the oceans above us, by storms, and winds, and phenomena of different kinds, the presence of oxygen in the air should not be as even everywhere as has generally been believed; there are too many causes to operate against this permanently even distribution. Humboldt also states, that from observations made by Levy it has become probable, that according to seasons, the situation of places in the sea or in the interior of continents, the presence of oxygen varies a little, but sufficiently so as to be noticeable. He also states his own view to that effect, and mentions, that Martin, at a height of 8,226 feet, did not find the air he had collected poorer in oxygen than the air of Paris. Gay-Lussac, in one of his balloon ascents, brought down air from a height of 22,000 feet, or more than four miles, with the same result.

The two instances cited above are not exactly adapted to show the variableness of oxygen in different parts of the air, but they are the best proof that this element exists in the air in undiminished proportion at the highest altitude that has been ascended to, and one of the best evidences in favour of my theory.

Electricity, as it has the power to convert oxygen and hydrogen into water, so it possesses the power to evolve both oxygen and hydrogen from water. But all those instances in which these gases are liberated from solid or liquid bodies, and rise into the air, are exceptions to the general rule; and it cannot be said that by these means our atmosphere is provided with the one and the other; we then must necessarily go higher for our supply.

Meteors, and in particular the aurora borealis, of which I have treated already, and which has been calculated to rise from 50 to 600 miles, and the one simultaneously seen at Rome and at Paris, on the 16th October, 1726, to have had an elevation of 800 miles,[*] do, no doubt, owe their brilliancy to the reservoir of oxygen, within which, or in whose vicinity, they make their appearance; their various light, intensity and duration cannot be ascribed to an electric explosion or amalgamation in vacant space or rarefied air only, there must be some other body to cause their diversified appearance, and this body can be no other than oxygen, giving lustre to electric metallic combustion and conflagration.

Thus, within the solid body of the earth, upon and exterior to it, we have seen the existence of electricity and its operation, though but clumsily represented and illustrated; yet upon the evidence brought forward, until shown to the contrary, I think my propositions must be considered as established, viz: the existence, *per se*, of positive electricity without, and negative electricity within, the solid part of the heavenly bodies, if our earth be a pattern of all. But we have a further, and still greater, proof of this separation and distribution of the two electricities in the variation and the dip of the magnetic needle.

---

[*] For many most important facts and ideas I am indebted to a work published in numbers, at Munster, in Westphalia, under the title of *Nature and Revelation* (Natur und Offenbarung), and I cannot sufficiently recommend this publication, which in England is to be had through Messrs. Williams and Norgate, London.

VIII.

ELECTRIC TIDE OF THE ATMOSPHERE AND WITHIN THE EARTH; VARIA-
TION AND DIP OF THE NEEDLE.—EFFECT OF THE AURORA BOREALIS
ON THE MAGNETIC NEEDLE.—OERSTED AND VON HOFF ON TERRES-
TRIAL MAGNETISM.—MAGNETIC IRON, COAL AND SALT WITHIN THE
BOWELS OF THE EARTH.—PROVIDENCE OF GOD.—ROTATION OF THE
SUN ROUND HIS OWN AXIS.—SEAS AND DESERTS ON THE SOLAR BODY.
—VARIABLE STARS.—OSCILLATION OF THE EARTH, CAUSE OF THE
SEASONS.—NECESSARY POLARITY OF THE EARTH AND PLANETS.—
OSCILLATION OF THE SUN; EFFECT ON CLIMATE.

WHEN treating of thunderstorms, of the aurora borealis and other
phenomena, we have seen that the sun exercises the greatest,
if not exclusive, influence on the electric condition of the heavenly
bodies moving within his sphere. Exterior to the solid body of
the earth he causes an electric tide, an apparent electric current
from east to west; and the exterior electric condition always
corresponding with the interior one, the effect is manifested in
the variation of the magnetic needle, though in many instances
caused by the substances which in different places form the crust
of the earth, facilitating or obstructing, according to circumstances,
a liberation of negative electricity.

The earth being warmed by the sun, and the sun's rays accumulating on its surface, we experience the greatest heat a few
hours after he has passed the meridian. As cold increases, so
heat, by its expansive power, decreases, or diminishes the electric
density. Thus, the solar heat reduces the positive electric density
of the air, as well as the negative density of the earth, and when
at its highest, produces electric low tide, or the local evening
minimum of positive electric intensity, with its contemporary
liberation of negative electricity, and consequent local diminution
of the electric intensity of the earth. On both, the west and
eastern sides of the sun, that is, on the sides of electric low tide,

occurring about two hours after the culmination, after mid-day, there will be a corresponding electric high tide, or the morning and evening maxima, leaving between them again, somewhat after the midnight part of the earth, a natural, or normal, electric state, a morning minimum, similar to the tides produced by the moon upon the ocean, though, owing to the absence of the sun at night, the morning minimum of electric intensity of the air causes NO CORRESPONDING DIMINUTION OF TERRESTRIAL electric intensity; and as the earth turns round, we pass from the atmospheric electric maximum to minimum, from minimum to maximum, and from that again to minimum, according to our gradual approach to, or recession from, the sun. The values of the intensity of the electric tide have been very accurately ascertained.

Dr. Lardner, speaking of the variation of the electricity of the air, says: "As the diurnal change in the position of the sun, relatively to a given place, produces a periodical variation in the electric state of the air, the change of its declination, from month to month, may be expected to be followed by some corresponding periodical effect on the mean amount of the maxima and minima values of the electricity. On comparing the mean values from month to month, it is accordingly found that the values of the daily maxima and minima undergo a progressive decrease from January to July and a progressive increase from July to January. It is found, also, that during the winter the electricity of the air increases as the thermometer falls.

"On comparing the mean values of the maxima and minima throughout the year, it is found that the morning values of each are a little less than the evening values.

"The hours at which the electricity attains its maxima and minima values are, likewise, subject to variation from month to month. The hour of the morning minimum and maximum

continually advances towards noon from winter to summer, and
undergoes the contrary change in the latter part of the year.

" The observations of Schubler indicate that the hour of the
evening minimum is invariable. From June 1811 to June 1812
it took place at Stuttgart always at 2 P.M. The hour of the
second maximum also gradually approached from summer to
winter, and receded from it again from winter to summer."

All this agrees with the theory I have laid down : positive
electricity diminishing in intensity, and withdrawing, as it were,
from the earth, according to the warming and expansion of the
air by the heat of the sun, and the consequent liberation of
negative electricity from the earth. Hence also the local variations of electricity.

In the most elevated places, and in those which are best
insulated, heat cannot accumulate ; and accordingly, Saussure has
shown, that THERE the positive electricity of the air has the
greatest intensity. In the midst of squares and other open spaces
in cities, on the quays, but especially on bridges, it is even more
intense than in an open flat country, as Dr. Lardner informs us,
because here likewise the air cannot become warm on account of
its constant change and motion. In particular localities, such as
Genoa, where fogs prevail which lie low and are not converted
into rain, the positive electricity, according to the same authority,
is not intense, but also, because the rays of the sun not falling and
accumulating there upon the earth, the air can neither become
warm, nor can negative electricity be liberated from the earth,
whilst positive electricity concentrates in the vapours which compose the fog. On the other hand, in the interior of buildings,
under trees, in the streets, courts, and other inclosed and sheltered
parts of towns, no free electricity is found in the air, because
the air is warmed either by the sun, or by animal heat and the
fires of domestic life, without its being renewed by atmospheric

currents, in fact, by ventilation, to supply the positive electricity which has been dissipated, or to counteract the negative electricity rising from the earth, floating on the surface, settling in buildings and congregating under trees as its natural conductors, and which, as we have seen already, enfeebles and depresses the human frame, whilst positive electricity promotes and strengthens the action of life.   Hence, apart from the supply of oxygen, ventilation is of the utmost importance for keeping the frame of the human body in a vigorous condition.

The correspondence between the diurnal and mensual variation of the electricity of the air and of the earth, and the diurnal and mensual variation of the magnetic needle, will easily be perceived.

From observation it appears, that in a given place, the north pole of the needle begins to turn westward at seven or eight o'clock in the morning, and continues to deviate in that direction till about two o'clock, when it becomes stationary, and soon begins to turn eastward, arriving at the position it had in the morning, at the same hour in the evening.   Canton's observations showed that the amount of this deviation varied· from seven to thirteen or fourteen minutes, being greatest at midsummer, and increasing and decreasing gradually between these seasons.

From this it will be seen, that the times of greatest eastward and westward variation correspond nearly to the times of morning maximum and evening minimum ; but there are no effects exhibited by the needle corresponding to the evening maximum and morning minimum.

According to what has been said before, we know that the electric tide, the greatest electric intensity of the air, precedes and follows the course of the sun.   Thus, when at the morning maximum the north pole of the magnetic needle · begins to turn westward, it follows the maximum, or high tide, which precedes the sun; but the moment that the evening minimum (two o'clock), which likewise is the time of greatest liberation of negative electricity from the earth, has passed, the evening maximum gains upon the needle, which, having just escaped the grasp of the morning maximum, gradually returns eastward to meet again the morning maximum, unaffected by the morning minimum. This however does not, in reality, seem to be the true statement of the case.

The electric condition of the atmosphere, upon the whole, corresponds all over the earth with that of the interior of the earth; yet, the electric tide caused by the sun in the air does not, as we have seen already, effect an exactly corresponding change, or tide, within the earth; the change produced during the day amounts only to the liberation of negative electricity from the earth by the sun, and the consequent diminution of its interior electric intensity at the place of liberation, and may therefore be called a single interior tide only, against the double, though very unequal, electric tide of the air. Thus, the magnetic needle is not in reality influenced by the electric tide of the air, though it appears so; the evening maximum and morning minimum of the electric tide of the air not in any way affecting it, must be evident proof that its variation does not proceed from the electric state of the air, but from that of the earth.

This state of the earth, as we have seen, undergoes little, if any, change at night; but in the day the electric intensity is diminished; and as the electric density of the air as well as that of the earth is greatest when it is coldest, the needle, disturbed, as in the case of the aurora borealis, by the electric evaporation caused by the solar heat, turns alternately to the interior, colder and more intensely electric part on either side of the sun. After the evening minimum, the eastern side of the earth is coldest and most intensely electric and magnetic, and after the morning maximum the western side is coldest and most intensely electric and magnetic.

The diurnal changes of the magnetic needle are greater and more irregular in the northern regions, because the intensity of electricity is there greatest, and the disturbance of its balance consequently more powerful, while towards the line the variations gradually diminish, and at length disappear, because the density of terrestrial electricity becomes less and less, until it ceases to affect or to attract the needle as we approach the equator, owing to the greater, more regular, and uninterrupted liberation of negative electricity from the earth by the heat of the sun.

The mensual change is an alteration in the direction, according to the season of the year, by which the needle, during the months between the vernal equinox and the summer solstice retrogrades towards the east; and during the remaining nine months preserves its general tendency towards the west. And this change agrees with the progressive decrease and increase of the intensity of the electric tide from winter to summer and from summer to winter, as well as with the hours of maxima and minima, which continually advance towards noon from winter to summer, and recede from summer to winter, in proportion to our days getting longer or shorter, warmer or colder.

The eleven-yearly magnetic period which has been observed, is unquestionably due to the change in the solar spots which Schwabe discovered to take place within about that time, and which naturally must affect the climate of the earth.

Besides these comparatively regular, daily, monthly, and eleven-yearly variations, there are others of an irregular character, which are due to adscititious causes, present only under peculiar circumstances. Chief among these disturbing causes is the aurora borealis, during the prevalence of which the needle is considerably disturbed. Mr. Becquérel says : "Generally, the declination increases before the aurora, and often even until the phenomenon has attained a certain degree of intensity; then the great oscillations commence; afterwards the needle returns towards the east very regularly; it passes beyond its normal position, which, providing no new aurora disturbs its progress, it regains again, although not until some hours have elapsed."

The accumulation of electricity, both positive and negative, required for the formation of the aurora borealis, sufficiently accounts for the increasing declination of the needle before the aurora has commenced; the needle is attracted by the increased density of the electricity of the earth, and liberated from its hold by the amalgamation of the two electricities, by the neutralization of their power; and this neutralization, the explosion of the electric elements, produces an exhaustion of electric intensity, both of the air and the earth, which in turn causes the needle to be attracted from a greater distance by the electricity of the earth, and makes it pass beyond its normal position until the usual electric equilibrium has been restored, and the needle retired to its original direction.

In confirmation of what I have said, I may add, that Hansteen found by experience, that shortly before the commencement of the aurora borealis the magnetism of the earth is of an uncommon force or intensity, which, however, soon after the beginning of the aurora borealis decreases, and sinks down under the usual force. And on the occasion of an aurora borealis seen on the 19th December, 1829, at Alford, in Aberdeenshire, a most striking disturbance in the direction of the declination of the magnetic needle was simultaneously observed, not only on the earth at Berlin, Petersburg, Kasan, and Nicolajew, but also deep under the earth at Freiberg, in Saxony.

Whether the aurora borealis itself is in any way connected with the electric tide, I have not been able to ascertain.

That earthquakes and volcanic eruptions disturb the needle we have already seen before.

Owing to the undulated character of the surface of our earth, the construction of its crust, the various materials which here and there lie uppermost, even the position of its seemingly confused strata, the liberation of negative electricity from the earth varies in almost every place ; hence also, in the words of Dr. Lardner, there is a given declination, or deviation, from the true north, or a given direction of the needle, proper to each spot of the earth. That this DECLINATION is influenced by place, *i.e.*, by change of latitude, or longitude, or both, is acknowledged by Lardner, who says further on this point : " It is found by experience, that this relative declination bears no regular ratio to the change in latitude and longitude, but is governed by other laws ; and so regular is the rate of variation, that it is not easy to foresee the precise effect of a change of place ; so that nothing but actual observation avails for the construction of tables showing the declination in different places ; or at least, no calculations can be confided in, unless well confirmed by observation. Navigators and travellers in former days, and with them philosophers at the present moment, have accumulated their various observations ; and from these data have been constructed magnetic charts, which should present, at one view, the declination of the needle for all parts of the globe."

These charts are of the highest interest, and may easily be procured ; and it is to be hoped, that the local variations there indicated will be traced to their true source, to those other laws which as yet are unknown, but which I trust my new theory of the interior electric organization of the earth will help, if not be the means, to unfold.

That the electric state of the earth, and not that of the air, is the cause of the variation of the magnetic needle, is proved also by Cassini, who observed the diurnal variation at Paris, and who found, that neither the solar heat, nor light, influenced it; for it was the same in the deep caves constructed under the Observatory in Paris, where a sensibly constant temperature is preserved, and from which light is excluded at the surface. But the solar heat influences the electric condition of the earth, and this influences the motion of the magnetic needle.

In the absence of thunderstorms, sand, land, and waterspouts, and the aurora borealis, positive electricity in the air is not sufficiently concentrated to affect the magnetic needle, unmistakingly attracted by the more condensed or concentrated element within the interior of the earth ; and nothing points out to us this fact more clearly than the very DIP of the needle.

The discoverer of the DIP (quoting from Dr. Lardner) found that at London a magnetic needle, free to move on an axis perpendicular to the magnetic meridian, presented its north pole downwards, forming an angle of above 71°. If the instrument be carried northward, it is found that the dip generally increases ; and on reaching a certain region near the pole, the needle would become vertical, the dip being then 90°, and its north pole pointing downwards. At such a place the common compass-needle, moving on a vertical support, would lose its directive power, and rest indifferently in any position. A place where these effects would be produced is called a NORTHERN MAGNETIC POLE.

If, on the other hand, the dipping-needle were carried to the equator, the magnitude of the DIP would be gradually diminished, until, on arriving at a certain region near the equator, the needle would become horizontal, and the dip would become nothing ; and if the dipping-needle could be carried round the globe, always following such a course as would allow it to retain its horizontal position, its course traced on the globe would be the MAGNETIC EQUATOR.

The magnetic equator does not coincide with the equator of the globe, nor is it a great circle of the earth. It never departs from the equator, however, more than twelve or thirteen degrees; all which agrees with my theory.

If, after passing the magnetic equator, the dipping-needle be carried southwards; its south pole will dip or be directed downwards, and this dip will increase in magnitude as the needle approaches the south pole. A place near that pole where the needle becomes vertical, is a SOUTHERN MAGNETIC POLE.

The dip also, like the variation of the needle and the intensity of the electricity of the earth, is subject to variation according to place, time, and local causes.

To account for the phenomena of terrestrial magnetism, Oersted proposes the hypothesis of currents of electricity circulating round the terrestrial globe, from east to west, in planes at right angles to the direction of the dipping-needle; but it is evident that such currents floating on the earth and surrounding or immersing the needle, could not have the described effect either upon its variation or its dip. These currents, then, do not exist besides the electric tide I have spoken of, besides those superior electric currents from the poles to the equator, and their general deflection westward round the globe in proportion to their nearing the equator, owing to the revolution of the earth in the opposite direction. The magnetic needle too significantly points to the poles and to the interior of the earth to allow of one moment's doubt as to the interior electric organization of our globe.

If there is, as I believe, any current that produces the dip of the needle, that current exists within the interior of the earth; and as a wire, along which an electric current is passed, will have a tendency to turn a magnetic needle at right angles to it: so that interior electric current, according to its intensity, attracts and produces the dip of the magnetic needle.

One of the greatest evidences of this subterranean agency is supplied to us by Von Hoff. In the Report of the British Association for 1850, he says : "More remarkable, however, are the changes in the direction of the dip-and-variation needles, which take place (not only in the vicinity of volcanoes as we have seen by Professor Palmieri's report,—see page 100, but) AT A DISTANCE from the place where the earthquake was observed, and at a place where the shock itself is not perceptible ; as, for instance, in Paris, on the 19th February and 31st May, 1822, simultaneously with an earthquake which occured in Savoy and some of the southern parts of France. If this observation should be established by others carefully made, the existence could not be denied of a connection between terrestrial vulcanism and terrestrial magnetism."

It has been noticed, that severe earthquakes and volcanic eruptions have followed upon severe winters. M. de Tschihatcoff relates, that on the 16th February, 1755, the Golden Horn of Constantinople was covered with a sheet of ice which bore foot-passengers. Upon this followed the memorable earthquake of 1st November, 1755, as upon the strong winter of the beginning of 1855 followed the eruption of Vesuvius; in May, the earthquakes at Broussa and in the western parts of Europe. The rigorous winter of 1669 was followed by one of the most remarkable eruptions of Etna.

These severe winters shut up the evaporation of the electricity of the earth and keep it pent up, when afterwards it finds vent again by sudden shocks and by volcanic eruptions of more than ordinary violence.

Besides these proofs of an interior electric inflation, or constitution and organization, of the earth, we have still to call to witness a certain species of iron ore, or oxide of iron, abundantly found at Roslagen, in Sweden. On being taken from the mine (see Lardner) where it has lain for ages exposed to the influence of the earth's magnetism, it is itself found endowed with permanent magnetic power. It is, in fact, the loadstone. But this magnetism of the earth is its interior electric element, the current coursing through the various strata of the earth, and causing more or less all the phenomena I have just recorded.[*]

And what, moreover, has caused our enormous beds of coal and salt deep within the bowels of the earth?

When after the third day of creation until the creation of man, by the continued, though decreasing, violent action of the electricity of the earth, the configuration of our globe was altered again and again, tracts of land with their overgrown gigantic vegetation ever and anon were engulphed, forests crushed and doubled up within its depths, hills and mountains raised and their vaults and cavities filled by the in-rushing sea: it was the heat of the electric blood of the earth that scorched and coked and coaled the wood and plants swallowed up and entombed in the chambers,

---

[*] Travelling in some of the mining districts in Germany, I have often expressed my belief to gentlemen interested in mining, as well as in science, the same as I have done to various friends in England, that the time would come when certain electric phenomena would guide us in finding out the mineral wealth buried in the earth. It often struck me, that some valleys and hills in the same district, in the same neighbourhood, and in similar direction, very differently remained covered with fogs, whilst others were quite free from them ; and I could not help but ascribing this to the interior condition of the respective areas. All at once, in reading the Journal of Mayence (*Mainzer Journal*) of 18th October, 1857, I find both my new theory and my supposition respecting the discovery of metalliferous spots of the earth, confirmed by the following article :—

"These many years already it had been observed, that at the time of heavy thunderstorms, the lightning always struck into the so-called Katzenberg (cat's-mountain), in the Gemarkung (district) of Volkartshain, in the Grand Duchy of Hesse, though several still higher mountains lie around it. Attention being directed to the attraction of this mountain, search was made a few weeks since, and ironstone discovered. The quality is said to be an excellent one."

If nature points ont the way, science will not neglect to follow it up ; it will devise instruments to notice electric exhalations, and so fix upon the coveted deposits within the earth.

Since this was written, Professor Von Cotta, at the session of 4th October, 1859 of the *Mining Union* at Freiberg, Saxony, exhibited an instrument for that purpose, called, the "Iron Seeker," and which was said to be successfully applied in Finland for finding magnetic iron-ore beds.

kilns and cauldrons below ; it was the heat of the electric element that, from the colossal masses of the vegetable kingdom heaped together and enclosed within the retorts of the earth, not un-attended with explosions and eruptions, drove off the gas and oil still confined in our subterranean hollows; it was the same element again that evaporated the hidden reservoirs of sea water, and in the salt mines and springs of various countries left the evidence of a former presence of the ocean.

But in what inexpressible grandeur, above all these stupendous manifestations of electrical action, do we not see displayed here the evidence of God's love and care in regard to man ! Not only that with the utmost splendour He prepares and surrounds the introduction of our first parents into the paradise of this world, and in abundance bestows upon them everything that could make them happy; but He foresees and provides, even ere man was created—not to speak of redemption—against the wants of generation after generation unto the remotest future, storing up treasures without which now we could not exist, and for which, alas ! we scarcely ever think of being grateful!

As to the physical condition of our globe, its interior electric organization leads us to still greater results than those previously noticed, it even leads us to the cause of revolution of our globe round its own axis.

" Let us," in the words of Dr. Lardner, " consider the case of the attraction or repulsion which is apparently exerted by an electrified body, S, on a non-conducting body, S', also electrified.

" In this case, the electric fluid on S attracts or repels the electric fluid on S'. Now since the pressure of the surrounding air prevents any motion of the fluid in a direction perpendicular to the surface of S', and the non-conducting power forbids any superficial motion of the particles of fluid, no change of position *inter se* can ensue, and the shell of the electric fluid will preserve

N

its form exactly as if it were solid matter encrusting the body S'. The attraction or repulsion of the fluid on S must, therefore, cause a motion of the entire shell of fluid on S', to or from the body S; and as this preservation of the form of the electric shell necessarily requires the continuance of the body which it invests within it, that body must accompany it as it moves to or from the body S.

"This may be illustrated in the following manner. Let us suppose a sphere of cork to have its surface covered by iron dust, and imagine this dust to be pressed against the surface of the cork by a surrounding atmosphere, whose pressure is sufficient to prevent its escape from the surface. Also suppose that the roughness of the surface of the cork is sufficient to prevent the particles of iron from moving upon it, Let this sphere be placed near a powerful magnet. The iron will be strongly attracted, and if free, would leave the cork and fly to the magnet. But this is prevented by the causes just stated. The iron can neither leave the cork, nor shift its position upon it. It must, therefore, move towards the magnet in virtue of the attraction exerted on it, carrying the sphere of cork which it invests, along with it."

Now, for the sphere of cork, substitute our globe; for the iron dust, the loadstone, minerals and coal-fields of the earth, and in place of a powerful magnet, let us put the all-powerful magnet of the sun, and the cause of rotation of our earth, riding within the solar atmosphere, is explained.

But this explanation I nevertheless do not think the correct one. I think it more probable, from all I have advanced, that the electric fluid within the earth, attracted and kept in circulation by the heat of the sun, causing the variation and dip of the needle, and iron to be magnetized within the bowels of the earth, is the true cause of the revolution of our globe round its own axis, the cause of Sir Isaac Newton's supposed centrifugal, as well as centripetal, force, that holds the masses of the earth together.

As the blood of the human body flows in a prescribed direction, so God ordained, that the electric circulation of the sun and planets should proceed from west to east. The sun shining with intense heat upon the earth, thus causes the electric current to take its course round the most excited part, the belt of the equator, in the direction in which the revolving sun, as it were, precedes the earth; and the heating of one half of the earth taking place without intermission, the electric current from the cold part, to that which is warmed and expanded, flows without ceasing; but having to overcome the resisting masses of the earth within which it moves, and its power being more particularly exerted within the outer circle, within the tire, the crust of the earth, is capable of turning it round its own axis; and thus, moved by the spiral spring of negative electricity, the earth has become the divine chronometer of mankind.

When, in the foreknowledge of the faith of Joshua, God, the Father of His chosen people, applied the break of His omnipotence to the current of electricity that turns our globe, and by these means slackened the pace of its revolution, He performed that miracle which the Israelites beheld in the apparent standing still of the sun and moon, and by which the prayer and faith of the Jewish leader was rewarded. God, as He often does, did not indeed literally fulfil the request of Joshua, but granted, by an equivalent, as it were, what in His service he demanded.

The girdle of the earth, shone upon by the sun at noon, presents to us the negative, and at the opposite or midnight part, the positive side of the electric equator, whilst the earth itself, whether orange or lemon-shaped, appears like a magnetic needle in an horizontal position, with its electric poles north and south, revolving from west to east within the atmosphere of the sun, the same as all the other planets with their respective poles, revolving likewise in this direction, according to their electric inflation or constitution, the construction or density of their masses, or the circumference of their bodies.

The sun himself (and fixed stars), no doubt, revolves on his horizontal axis from the same cause which turns our globe. By most scientific men, I may say by almost all, his body is considered to consist of solid matter like our own earth, and to be surrounded by a luminous atmosphere. Dr. Elliott, as early as 1787, maintained, " that the light of the sun proceeds from a dense and universal aurora which may afford ample light to the inhabitants of the surface beneath, and yet be at such a distance aloft, as not to annoy them." And Sir David Brewster, in his *More Worlds than One*, remarks : " Sometimes by the naked eye, but frequently even by small telescopes, large black spots, many thousand miles in diameter, are seen upon its surface, and are evidently openings in the luminous atmosphere, through which we see the opaque solid nucleus, or the real body of the sun."

This confirms my new theory of accumulation of positive electricity round the heavenly bodies, of which suns are the largest.

When a magnetic needle, sufficiently light, is placed within a helix or screw of wire, so as to rest on the lower part of the wire, it will start up and place itself in the centre and remain there suspended in the air without any visible support, and against the law of gravitation, as soon as a powerful current of electricity is passed along the thread of the helix.

Like a magnetic needle within a helical current, the sun is surrounded by, and kept in suspense within, an accumulation of positive electricity, constantly in motion and action, the attracting and connecting element between the body of the sun and the negative bodies of the planets.

As the amalgamation of positive .and negative electricity with us, in the phenomena of the various descriptions of lightning, and of the aurora borealis, which has . been observed to continue sometimes for two days, take place in the upper parts of our atmosphere, so the amalgamation of the positive electricity collected and collecting round the sun, with the negative electricity with which he is inflated, takes place at a corresponding distance from the opaque body of the sun; and this constant amalgamation and consequent production of electric light and heat,—not for days, but incessantly bursting forth, perhaps from a segment,* an aggregation of matter, similar to that from which the aurora borealis is mostly evolved or developed, or forms the bearer of meteors,—acts upon his interior electric organization, the same as it does upon that of our earth.

It may be objected that this result cannot be caused, as regards the body of the sun, since he is totally surrounded by this electric photosphere, every part of which neutralizing the other, and keeping the body of the sun, supposing him to have no other momentum, in a state of repose. But it must not be forgotten, that in this fire-ring there are openings or dark spots, from a few hundred to 50,000 MILES AND MORE IN DIAMETER, which, breaking the contact, render its effect unequal on the different parts of the body of the sun.

In certain parts of our globe, neither lightning nor thunder, is ever witnessed : and it appears from this, that in these places there is no proper medium for the union of the two electricities in the air.

Humboldt says, "that in the geographical distribution of thunderstorms, the Peruvian coast-land offers the most striking contrast to the whole of the other parts of the tropics; that, whilst in these at certain times of the year, thunderstorms are formed every day

---

* Such a segment, the basis, or bearer, of the solar conflagration, would account also for detached, or seemingly-detached clouds of the solar fire-veil, of which one was distinctly noticed during the last eclipse of the sun, by Professor Airy, the Astronomer Royal.

from four to five hours after the culmination of the sun, in the former it never lightens or thunders."

This may, no doubt, be accounted for by the fact, that the coast-land of Peru is mostly arid, and covered with sands and deserts, and though it possesses many valleys, conductors in the shape of forests, or otherwise, may not be sufficiently numerous to admit of an accumulation of negative electricity in the superior atmosphere, large enough to attract the positive electricity above, whether free or harboured by clouds. Deserts, or even sandy districts, from want of moisture and of vegetation, or other causes, may not emit, or exhale perhaps, the negative element in that abundance which we witness in cultivated and highly-productive soil, and that which is so emitted has no conductors to carry it upwards. It may be also, though it seems to me less likely, that the mountain chain of the Andes, running parallel with the coast, and at no very great distance from it, down to Chili, is so powerful a conductor, as to attract the electricity emanating from the coastland as soon as it has been liberated from the earth. In either case, however, there are no two electricities to come in contact in the atmosphere in such a manner as will produce lightning or thunder.

If, therefore, the body of the sun, like the body of the earth, as is most probable, has his Egyptian plains, his deserts of Peru and Chili, of Africa and Asia, his vast oceans of water, where, according to the distance from land, it rarely or never lightens or thunders; where the isolated sand, land, and waterspout, the cradle of whirlwinds, perform the office of uniting in destructive embrace the two opposite electricities, instead of the scarcely less awful but more sublime thundercloud, the magnificent aurora borealis, and the silent lightning discharge of our summer evenings : the black spots, however vast in extent, and naturally varying in size from time to time, will then have been explained ; they break the contact of the two electricities, the veil of fire is perforated, its effect becomes unequal over the body of the sun, and the motion of the photosphere, the eddies playing around the borders of its ocean-like openings, and the rotation of the sun round his own axis, are a necessary consequence of the attraction of positive and negative electricity, their amalgamation and production of light.

The production of the solar light, if more analogous to the aurora borealis than to lightning, would lead us to suppose, that vast regions of his body must be very cold, as with us the aurora borealis is produced only in the cold regions of the earth. Whether these regions alternate with cold and warmth, as they do on our globe, will also depend on the interior electric organization and the oscillation of the sun himself. And is it not possible, nay, most probable, that these very cold regions, apart from attraction and repulsion, act upon, or are acted upon, by comets, as on their mighty sweep through the solar system, they approach the great orb of the day, lighting up his unilluminated spaces, and in his presence appear to us with increased splendour and light?

The periodical appearance and disappearance of the Mira Stella, and of about twenty other fixed stars that are said to have been observed, is, no doubt based upon a constitution of its body, similar to that of the sun*; the one side is so vast and barren, covered with water, or constituted in such a manner, that no light is produced upon it, or a light so feeble or partial, that to us it remains imperceptible.   And yet the organization, like that of the sun, must be such, that by means of its own fiery envelope, the body rotates upon its own axis, becoming invisible to us as by degrees the lightless spot is presented to us, and visible as by degrees the illuminated side returns again before our eyes.

The sun, then, by means of the black spots, the openings in his

---

* Very nearly the same theory, I find, was held by Bullialdus, in a work which appeared about 1667.

veil of fire, revolving round his own oscillating horizontal axis in one direction, the planets revolving, upon their bellies or equators, round their own horizontal axis in the same direction round and above the equator of the sun, and kept steadily there by the repelling influence of his poles on either side, whilst allowing more margin to the small asteroids on both sides of the ecliptic ; and the equatorial belt of the sun acting upon those of the planets, it follows : that the axes of all these bodies, supposing their time of the equinox happening all at the same moment, should be parallel to each other, revolve in one line and direction, and oscillate according to their constitution and their greater distance from the sun ; and thus the whole of our solar system should present to us one north and one south pole in regard to the centre of the whole universe, the same as the earth does to the sun, unless our own system itself be the centre of creation.   In this latter case, the true poles of the universe would have to be determined by the elongation of the solar axis at the time at which the axis of the earth at each equinox, runs parallel to that of the sun.

With the notion of rotation we naturally associate the idea of progression in the same direction in which the rotation itself takes place ; and if thus we consider the planets, with their axes parallel to that of the sun, we cannot but see that, as a matter of course, they ought to revolve round the sun, as a balloon would revolve round the earth, if in its construction we could but give it rotary motion round its HORIZONTAL axis.

But if this were so, it may be objected, we should not be able to account for the four seasons ; on this principle every part of a planet would always have the same season without any variation, inasmuch as in the case in question, the sun would shine equally upon every respective part the whole year round. And yet, my proposition is true after all, and the difficulty of the seasons easily solved when we bear in mind the interior electric organization of the earth, with its two magnetic poles, and its magnetic or electric centre. For as, apart from the DAILY tide of the electricity of the air and the earth, of the tide of the atmosphere and the sea, of the blood of living beings and of the sap of plants, there is also a PERIODICAL tide, an equatorial and polar flux and reflux of the air and the water, a flux and reflux of blood to and from the heart of the human body and of animals, a flux and reflux of the sap in plants : so there is also a flux and reflux of the electricity of the upper regions, and, no doubt, a corresponding spiral one of the electricity in the interior of the earth, as proved by the connection of our northern auroras with those of the southern hemisphere, (page 85) with all their attendant phenomena.

The same then, as the centre, the heart, of the earth is acted upon by the sun, so are the poles, and this in such a manner, that the balance of the electric fluid in the earth, inclining by degrees for six months to the north, the sun will attract the north pole and repel the south, negative electricity preponderating at the former and positive at the latter; and inclining for the other six months to the south, he will attract the south and repel the north pole, negative electricity preponderating at the one, and positive at the other; or which is the same: the balance of circulation, of attraction and repulsion, will cause the poles to rise and fall in turn, held, as it were, by the beam of the sun. By this gradual inclination produced by the sun, as he apparently passes through the ecliptic, the earth will undergo but two oscillations* a year, shifting, in a manner, the equator in an oblique direction north and south.

---

* If in their course round the sun, the superior planets, Jupiter, Saturn, and the rest, undergo but two oscillations during the period of one entire revolution, a change of seasons would require 12 of our years in the case of Jupiter, 29½ in the case of Saturn, 84 in that of Uranus, and 164 in the case of Neptune; everything thereon must, therefore, be very different from what it is here, and the idea of their being peopled with A LIVING CREATION SIMILAR TO OURS, can certainly not be entertained.

Whatever may be the true theory of the solar system: the polar-magnetic organization of the heavenly bodies—apart from the existing polarity of every atom in creation—was, so to speak, an absolute necessity; for, without this balance, influenced by solar atmospheric currents, interior causes, or any kind of disturbance of their equilibrium, they would, as it were, have been liable to tilt over and be the sport of the winds and waves; like a billiard or skittle ball, they would have rolled over in every direction, and regularity of motion and seasons, in fact, seasons altogether, would have been out of the question.—What a beautiful illustration this of the Divine Wisdom and Intelligence!

When, as in the following diagram, facing the south pole of the

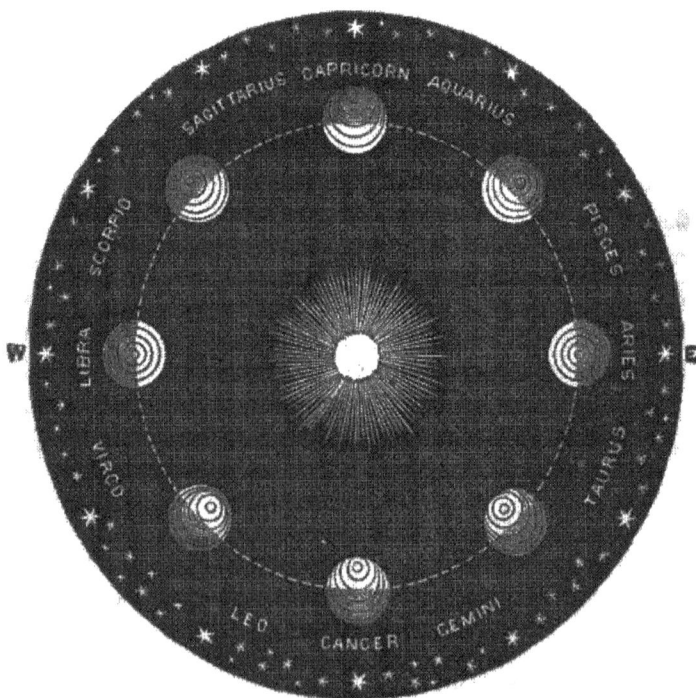

earth, the same as at page 1, facing the south poles of all our planets; and looking in the distance at the fixed stars of the northern heaven, the reader sees the earth in the sign of Libra, its axis is presented horizontally to the sun, one half of its surface is shone upon by him from pole to pole, we have the vernal equinox, our day and night are equal, we have spring. As the earth on its path advances to Scorpio and Sagittarius, the south pole is repelled by the sun, and by degrees passes into the shade, until, arrived at Capricorn, the repulsion is complete, and it then is, to a considerable distance, altogether enveloped in darkness, making it cold, winter, and short days there, whilst the north pole, attracted, and to a corresponding distance completely exposed to the light of the sun, we have it warm, have long days, and our summer here. Proceeding towards Aquarius and Pisces, the north pole is repelled by the sun, and the south pole attracted, the former pushed back or away from, and the latter drawn towards the light, until, arrived at Aries, the axis of the earth has assumed the horizontal position it had in Libra; the earth again is shone upon from pole to pole, we have the autumnal equinox, our day and night are equal, we have autumn. Passing on to Taurus and Gemini, the north pole is further repelled from, and the south pole more attracted into, the light of the sun, until, arrived at Cancer, the north pole, to a considerable distance, is completely enveloped in darkness, and the south pole, to the same corresponding distance, completely exposed to the sun; there is winter with its short days here, summer with its long days there. Progressing towards Leo and Virgo, the south pole is repelled and the north pole attracted, until spring dawns upon us again on our return to Libra.

By this oscillating motion the apparent obliquity of the ecliptic —in reality the obliquity of the earth to the ecliptic—is established, and, with the four seasons, not only accounted for in a far more simple, natural, and rational manner than hitherto they have been

explained, but the key even has been found for the DIVERSITIES of
the various seasons in different years, depending upon the greater
or lesser oscillation of our globe, caused by the oscillation of the
sun himself, and upon the periodical variation in the extent, density,
and intensity of the solar photosphere.

What are the nature and duration of the oscillations of the
sun must be left to learned astronomers to find out; but the unde-
cennial or eleven-yearly variation in the solar fire-veil arises no
doubt from this, that, like the water of the earth, it is left behind
in the rotation of the solid, or opaque, body, round its own axis,
the same also as the moon is left behind in the rotation of our
planet. And this difference of rotation between the photosphere
of the sun and his opaque body, both meeting again after about
every ten or rather eleven years in the same relative position in
which, or under the same points or constellation of the heavens,
under and from which they parted, forms a cycle like that of the
moon and the earth (and the moons and rings of the planets with
their primaries) in their united yet different speed of rotation.
And this gradual shifting of the solar fire-veil with its enormous
openings, about 35° on each side of the equator above the opaque
body of the sun,—causing perhaps to another existence an inde-
scribable spectacle of the heavens and the blessings of another order
of day and night and seasons to pass in succession over the various
portions of his surface,—both envelope and body imperceptibly
changing their mutual action and reaction, and influencing their
own condition : the intensity of the photosphere and the extent
and change of the solar spots, as the photosphere is floating by, and
its masses, eddy-like, circulating around the vacant island spaces,
as it were, above the seas and deserts of the sun, must naturally
be productive of a periodical change in the temperature and condition of the earth.

IX.

INFLUENCE OF COMETS AND PLANETS UPON THE EARTH AND ON MAN.—
COMET-STROKES.—CALUMNIES AGAINST THE CHURCH ; ARAGO, VON
·GUMPACH, GIBBON, MÆDLER AND OTHERS.—THE ANGELUS DOMINI.—
THE PHENOMENA OF NATURE TO LEAD TO GOD ; WHEN LEADING
TO HIM, NO SUPERSTITION.—THE PHARISEE IN THE "TIMES."—SUP-
POSED DANGER OF COMETS.—GALILEO, MACKINTOSH, MÆDLER AGAIN,
SIR DAVID BREWSTER, THE "E PUR SE MOUVE."—HUMBOLDT, ARIS-
TOTLE, THE CHURCH, BARBARISM AND TRUE CIVILIZATION.—THE
AGE OF LYING AND PROSTITUTION OF THE PRESS.—OUTBURSTS OF
THE SOLAR PHOTOSPHERE, THE ZODIACAL LIGHT, THEIR INFLUENCE.

COMETS and planets may indirectly, if not directly, affect the
condition of the earth, though the primary cause must again be
sought in the sun.

When a comet appears in sight, on its approach to the sun, it is
powerfully attracted by him, and this attraction increases with the
increase of proximity. By this means the body of the sun becomes
more than ordinarily excited; this excitement, in proportion to
the cometary influence, determines a corresponding intensity, or
other change, in the quantity and quality of his luminosity, in the
quantity and quality of his rays ; and this, in connexion, perhaps,
with the theory of spectral analysis and the hydrogen lens of our
sphere, operates a change in the climatic condition of the earth,
particularly when comets are large, and in their passage closely
pass us by, and most nearly approach the sun.

It does not follow from this, that this influence must manifest
itself in the same manner all over the earth; for, according to
interior and exterior organization, it may produce dryness in one
part of the world, and wet in another, thunderstorms in one, and
their absence in another. Nor does it follow, that EVERY comet
exercises such influence, as there must necessarily exist a difference

between large ones and small ones, their respective distance from, or nearness to, the sun, and the consequent excitement or non-excitement of his body or photosphere.

Comets and planets, attracted by, and moving round the centre of the sun, are attached to him by electric links which obediently they follow. If, then, it happens, that we are in conjunction with a comet or a planet, the earth with its electric link will be enveloped or embraced also by the electric link passing from the comet, or planet, to the sun; and hence the electric condition of our earth will thereby materially be affected, being exposed to a double attracting power, emanating from the sun both upon itself and upon the comet or planet; the effect will be different, again, whenever a comet or planet places itself in conjunction between us and the sun. Hence, also, an abnormal influence must be exercised by the sun upon a planet in conjunction with another, or with a comet. And that these electric influences, of which comets and planets are the mediate cause, must more or less effect the electric state of the earth, and produce a different state of fecundity in vegetable life, as they may also act upon minds whose bodies are predisposed to electrical affection,* at the right time and at the right place, for good or for bad, in the good or the wicked, for health or disease, seems to me but very reasonable to believe, and warranted by almost universal opinion and tradition, if not experience, among all nations. And we ourselves, in the train of a most splendid comet, have witnessed the blessings of an over-abundant and luxurious yield of corn and wine; we have,

---

* "In a paper on some of the phenomena of terrestrial magnetism, as observed at Greenwich, read before the Royal Society, the Astronomer-Royal shows that the revolutionary year, 1848, appears to have been one of confusion and disturbance in cosmical as well as tellurian affairs, for changes have taken place which cannot be explained on any established theory . . . . . . ".—*Extract.* To myself this is but another proof that the laws of nature are subservient to the super-natural designs of God in regard to man, and in particular to His Church, whose fetters were struck off abroad in this memorable year.

however, seen also the seed of perjury, sedition, and treason, sown by the enemy and deceiver of mankind from the beginning, fostered and quickened, and the egg of a FOUL, PRETENDED and GODLESS liberty, of premeditated revolution and aggression, hastened in its development, and hatched before the calculated time ; and we have beheld, in the horrors of war let loose against peaceful states, in the destruction of that peace about which there is something so beautiful in families and nations, the abundant harvest of Satan gathered in by his own servants and satellites. We see again and again the deceiving cuckoo, always and always cuckooing, liberty ! liberty ! liberty !—whilst laying its eggs into other birds' nests, its brood pushing out their rightful inmates and keeping them under the yoke of ITS OWN dear enlightened and precious freedom, and for the sake of greater liberty still, driving them from house and home and from their sacred places.

Nevertheless, it is not the comet which is guilty of all this iniquity, no more than wine is guilty of the vices of the drunkard, but MAN that does not resist the world, the devil, and the flesh, the excited motion of ambition, lust, avarice, and pride, and recklessly throws overboard the last vestige of christian virtue.

Apart from the physical effects which, however slightly, may, perhaps in all cases, be traced to the indirect influence of comets, so their mere appearance may strike terror into barbarous nations and uncultivated multitudes, and incite to savage warfare, as in the good it may elevate the soul to God, make it ponder over His wonderful works and prompt it to noble deeds, nay even make a prince exchange sceptre for rosary, for the glorious badge and pledge of the Lily of Nazareth, leave the palace for the cell, the rule over empire for the rule over himself.

In another form the mere sight of a comet may excite the pride of the worldly-learned at the bare idea of being able to calculate its weight and dimension, and to trace and determine its course

between large ones and small ones, their respective distance from, or nearness to, the sun, and the consequent excitement or non-excitement of his body or photosphere.

Comets and planets, attracted by, and moving round the centre of the sun, are attached to him by electric links which obediently they follow. If, then, it happens, that we are in conjunction with a comet or a planet, the earth with its electric link will be enveloped or embraced also by the electric link passing from the comet, or planet, to the sun; and hence the electric condition of our earth will thereby materially be affected, being exposed to a double attracting power, emanating from the sun both upon itself and upon the comet or planet; the effect will be different, again, whenever a comet or planet places itself in conjunction between us and the sun. Hence, also, an abnormal influence must be exercised by the sun upon a planet in conjunction with another, or with a comet. And that these electric influences, of which comets and planets are the mediate cause, must more or less effect the electric state of the earth, and produce a different state of fecundity in vegetable life, as they may also act upon minds whose bodies are predisposed to electrical affection,* at the right time and at the right place, for good or for bad, in the good or the wicked, for health or disease, seems to me but very reasonable to believe, and warranted by almost universal opinion and tradition, if not experience, among all nations. And we ourselves, in the train of a most splendid comet, have witnessed the blessings of an over-abundant and luxurious yield of corn and wine; we have,

---

* "In a paper on some of the phenomena of terrestrial magnetism, as observed at Greenwich, read before the Royal Society, the Astronomer-Royal shows that the revolutionary year, 1848, appears to have been one of confusion and disturbance in cosmical as well as tellurian affairs, for changes have taken place which cannot be explained on any established theory . . . . . ".—*Extract.* To myself this is but another proof that the laws of nature are subservient to the super-natural designs of God in regard to man, and in particular to His Church, whose fetters were struck off abroad in this memorable year.

however, seen also the seed of perjury, sedition, and treason, sown by the enemy and deceiver of mankind from the beginning, fostered and quickened, and the egg of a FOUL, PRETENDED and GODLESS liberty, of premeditated revolution and aggression, hastened in its development, and hatched before the calculated time ; and we have beheld, in the horrors of war let loose against peaceful states, in the destruction of that peace about which there is something so beautiful in families and nations, the abundant harvest of Satan gathered in by his own servants and satellites.   We see again and again the deceiving cuckoo, always and always cuckooing, liberty ! liberty ! liberty !—whilst laying its eggs into other birds' nests, its brood pushing out their rightful inmates and keeping them under the yoke of ITS OWN dear enlightened and precious freedom, and for the sake of greater liberty still, driving them from house and home and from their sacred places.

Nevertheless, it is not the comet which is guilty of all this iniquity, no more than wine is guilty of the vices of the drunkard, but MAN that does not resist the world, the devil, and the flesh, the excited motion of ambition, lust, avarice, and pride, and recklessly throws overboard the last vestige of christian virtue.

Apart from the physical effects which, however slightly, may, perhaps in all cases, be traced to the indirect influence of comets, so their mere appearance may strike terror into barbarous nations and uncultivated multitudes, and incite to savage warfare, as in the good it may elevate the soul to God, make it ponder over His wonderful works and prompt it to noble deeds, nay even make a prince exchange sceptre for rosary, for the glorious badge and pledge of the Lily of Nazareth, leave the palace for the cell, the rule over empire for the rule over himself.

In another form the mere sight of a comet may excite the pride of the worldly-learned at the bare idea of being able to calculate its weight and dimension, and to trace and determine its course

between large ones and small ones, their respective distance from, or nearness to, the sun, and the consequent excitement or non-excitement of his body or photosphere.

Comets and planets, attracted by, and moving round the centre of the sun, are attached to him by electric links which obediently they follow. If, then, it happens, that we are in conjunction with a comet or a planet, the earth with its electric link will be enveloped or embraced also by the electric link passing from the comet, or planet, to the sun; and hence the electric condition of our earth will thereby materially be affected, being exposed to a double attracting power, emanating from the sun both upon itself and upon the comet or planet; the effect will be different, again, whenever a comet or planet places itself in conjunction between us and the sun. Hence, also, an abnormal influence must be exercised by the sun upon a planet in conjunction with another, or with a comet. And that these electric influences, of which comets and planets are the mediate cause, must more or less effect the electric state of the earth, and produce a different state of fecundity in vegetable life, as they may also act upon minds whose bodies are predisposed to electrical affection,* at the right time and at the right place, for good or for bad, in the good or the wicked, for health or disease, seems to me but very reasonable to believe, and warranted by almost universal opinion and tradition, if not experience, among all nations. And we ourselves, in the train of a most splendid comet, have witnessed the blessings of an over-abundant and luxurious yield of corn and wine; we have,

---

* "In a paper on some of the phenomena of terrestrial magnetism, as observed at Greenwich, read before the Royal Society, the Astronomer-Royal shows that the revolutionary year, 1848, appears to have been one of confusion and disturbance in cosmical as well as tellurian affairs, for changes have taken place which cannot be explained on any established theory . . . . . ".—*Extract.* To myself this is but another proof that the laws of nature are subservient to the super-natural designs of God in regard to man, and in particular to His Church, whose fetters were struck off abroad in this memorable year.

however, seen also the seed of perjury, sedition, and treason, sown by the enemy and deceiver of mankind from the beginning, fostered and quickened, and the egg of a FOUL, PRETENDED and GODLESS liberty, of premeditated revolution and aggression, hastened in its development, and hatched before the calculated time ; and we have beheld, in the horrors of war let loose against peaceful states, in the destruction of that peace about which there is something so beautiful in families and nations, the abundant harvest of Satan gathered in by his own servants and satellites. We see again and again the deceiving cuckoo, always and always cuckooing, liberty ! liberty ! liberty !—whilst laying its eggs into other birds' nests, its brood pushing out their rightful inmates and keeping them under the yoke of ITS OWN dear enlightened and precious freedom, and for the sake of greater liberty still, driving them from house and home and from their sacred places.

Nevertheless, it is not the comet which is guilty of all this iniquity, no more than wine is guilty of the vices of the drunkard, but MAN that does not resist the world, the devil, and the flesh, the excited motion of ambition, lust, avarice, and pride, and recklessly throws overboard the last vestige of christian virtue.

Apart from the physical effects which, however slightly, may, perhaps in all cases, be traced to the indirect influence of comets, so their mere appearance may strike terror into barbarous nations and uncultivated multitudes, and incite to savage warfare, as in the good it may elevate the soul to God, make it ponder over His wonderful works and prompt it to noble deeds, nay even make a prince exchange sceptre for rosary, for the glorious badge and pledge of the Lily of Nazareth, leave the palace for the cell, the rule over empire for the rule over himself.

In another form the mere sight of a comet may excite the pride of the worldly-learned at the bare idea of being able to calculate its weight and dimension, and to trace and determine its course

between large ones and small ones, their respective distance from, or nearness to, the sun, and the consequent excitement or non-excitement of his body or photosphere.

Comets and planets, attracted by, and moving round the centre of the sun, are attached to him by electric links which obediently they follow. If, then, it happens, that we are in conjunction with a comet or a planet, the earth with its electric link will be enveloped or embraced also by the electric link passing from the comet, or planet, to the sun; and hence the electric condition of our earth will thereby materially be affected, being exposed to a double attracting power, emanating from the sun both upon itself and upon the comet or planet; the effect will be different, again, whenever a comet or planet places itself in conjunction between us and the sun. Hence, also, an abnormal influence must be exercised by the sun upon a planet in conjunction with another, or with a comet. And that these electric influences, of which comets and planets are the mediate cause, must more or less effect the electric state of the earth, and produce a different state of fecundity in vegetable life, as they may also act upon minds whose bodies are predisposed to electrical affection,* at the right time and at the right place, for good or for bad, in the good or the wicked, for health or disease, seems to me but very reasonable to believe, and warranted by almost universal opinion and tradition, if not experience, among all nations. And we ourselves, in the train of a most splendid comet, have witnessed the blessings of an over-abundant and luxurious yield of corn and wine; we have,

---

* "In a paper on some of the phenomena of terrestrial magnetism, as observed at Greenwich, read before the Royal Society, the Astronomer-Royal shows that the revolutionary year, 1848, appears to have been one of confusion and disturbance in cosmical as well as tellurian affairs, for changes have taken place which cannot be explained on any established theory . . . . . . ".—*Extract.* To myself this is but another proof that the laws of nature are subservient to the super-natural designs of God in regard to man, and in particular to His Church, whose fetters were struck off abroad in this memorable year.

however, seen also the seed of perjury, sedition, and treason, sown
by the enemy and deceiver of mankind from the beginning, fostered
and quickened, and the egg of a FOUL, PRETENDED and GODLESS
liberty, of premeditated revolution and aggression, hastened in its
development, and hatched before the calculated time; and we have
beheld, in the horrors of war let loose against peaceful states, in
the destruction of that peace about which there is something so
beautiful in families and nations, the abundant harvest of Satan
gathered in by his own servants and satellites. We see again and
again the deceiving cuckoo, always and always cuckooing, liberty!
liberty! liberty!—whilst laying its eggs into other birds' nests, its
brood pushing out their rightful inmates and keeping them under
the yoke of ITS OWN dear enlightened and precious freedom, and
for the sake of greater liberty still, driving them from house and
home and from their sacred places.

Nevertheless, it is not the comet which is guilty of all this
iniquity, no more than wine is guilty of the vices of the
drunkard, but MAN that does not resist the world, the devil, and
the flesh, the excited motion of ambition, lust, avarice, and pride,
and recklessly throws overboard the last vestige of christian virtue.

Apart from the physical effects which, however slightly, may,
perhaps in all cases, be traced to the indirect influence of comets,
so their mere appearance may strike terror into barbarous
nations and uncultivated multitudes, and incite to savage warfare,
as in the good it may elevate the soul to God, make it ponder
over His wonderful works and prompt it to noble deeds, nay even
make a prince exchange sceptre for rosary, for the glorious badge
and pledge of the Lily of Nazareth, leave the palace for the cell,
the rule over empire for the rule over himself.

In another form the mere sight of a comet may excite the pride
of the worldly-learned at the bare idea of being able to calculate
its weight and dimension, and to trace and determine its course

between large ones and small ones, their respective distance from, or nearness to, the sun, and the consequent excitement or non-excitement of his body or photosphere.

Comets and planets, attracted by, and moving round the centre of the sun, are attached to him by electric links which obediently they follow. If, then, it happens, that we are in conjunction with a comet or a planet, the earth with its electric link will be enveloped or embraced also by the electric link passing from the comet, or planet, to the sun; and hence the electric condition of our earth will thereby materially be affected, being exposed to a double attracting power, emanating from the sun both upon itself and upon the comet or planet; the effect will be different, again, whenever a comet or planet places itself in conjunction between us and the sun. Hence, also, an abnormal influence must be exercised by the sun upon a planet in conjunction with another, or with a comet. And that these electric influences, of which comets and planets are the mediate cause, must more or less effect the electric state of the earth, and produce a different state of fecundity in vegetable life, as they may also act upon minds whose bodies are predisposed to electrical affection,* at the right time and at the right place, for good or for bad, in the good or the wicked, for health or disease, seems to me but very reasonable to believe, and warranted by almost universal opinion and tradition, if not experience, among all nations. And we ourselves, in the train of a most splendid comet, have witnessed the blessings of an over-abundant and luxurious yield of corn and wine; we have,

---

* "In a paper on some of the phenomena of terrestrial magnetism, as observed at Greenwich, read before the Royal Society, the Astronomer-Royal shows that the revolutionary year, 1848, appears to have been one of confusion and disturbance in cosmical as well as tellurian affairs, for changes have taken place which cannot be explained on any established theory . . . . . ".—*Extract.* To myself this is but another proof that the laws of nature are subservient to the super-natural designs of God in regard to man, and in particular to His Church, whose fetters were struck off abroad in this memorable year.

however, seen also the seed of perjury, sedition, and treason, sown by the enemy and deceiver of mankind from the beginning, fostered and quickened, and the egg of a FOUL, PRETENDED and GODLESS liberty, of premeditated revolution and aggression, hastened in its development, and hatched before the calculated time ; and we have beheld, in the horrors of war let loose against peaceful states, in the destruction of that peace about which there is something so beautiful in families and nations, the abundant harvest of Satan gathered in by his own servants and satellites. We see again and again the deceiving cuckoo, always and always cuckooing, liberty ! liberty ! liberty !—whilst laying its eggs into other birds' nests, its brood pushing out their rightful inmates and keeping them under the yoke of ITS OWN dear enlightened and precious freedom, and for the sake of greater liberty still, driving them from house and home and from their sacred places.

Nevertheless, it is not the comet which is guilty of all this iniquity, no more than wine is guilty of the vices of the drunkard, but MAN that does not resist the world, the devil, and the flesh, the excited motion of ambition, lust, avarice, and pride, and recklessly throws overboard the last vestige of christian virtue.

Apart from the physical effects which, however slightly, may, perhaps in all cases, be traced to the indirect influence of comets, so their mere appearance may strike terror into barbarous nations and uncultivated multitudes, and incite to savage warfare, as in the good it may elevate the soul to God, make it ponder over His wonderful works and prompt it to noble deeds, nay even make a prince exchange sceptre for rosary, for the glorious badge and pledge of the Lily of Nazareth, leave the palace for the cell, the rule over empire for the rule over himself.

In another form the mere sight of a comet may excite the pride of the worldly-learned at the bare idea of being able to calculate its weight and dimension, and to trace and determine its course

between large ones and small ones, their respective distance from, or nearness to, the sun, and the consequent excitement or non-excitement of his body or photosphere.

Comets and planets, attracted by, and moving round the centre of the sun, are attached to him by electric links which obediently they follow. If, then, it happens, that we are in conjunction with a comet or a planet, the earth with its electric link will be enveloped or embraced also by the electric link passing from the comet, or planet, to the sun; and hence the electric condition of our earth will thereby materially be affected, being exposed to a double attracting power, emanating from the sun both upon itself and upon the comet or planet; the effect will be different, again, whenever a comet or planet places itself in conjunction between us and the sun. Hence, also, an abnormal influence must be exercised by the sun upon a planet in conjunction with another, or with a comet. And that these electric influences, of which comets and planets are the mediate cause, must more or less effect the electric state of the earth, and produce a different state of fecundity in vegetable life, as they may also act upon minds whose bodies are predisposed to electrical affection,* at the right time and at the right place, for good or for bad, in the good or the wicked, for health or disease, seems to me but very reasonable to believe, and warranted by almost universal opinion and tradition, if not experience, among all nations. And we ourselves, in the train of a most splendid comet, have witnessed the blessings of an over-abundant and luxurious yield of corn and wine; we have,

---

* "In a paper on some of the phenomena of terrestrial magnetism, as observed at Greenwich, read before the Royal Society, the Astronomer-Royal shows that the revolutionary year, 1848, appears to have been one of confusion and disturbance in cosmical as well as tellurian affairs, for changes have taken place which cannot be explained on any established theory . . . . . ".—*Extract.* To myself this is but another proof that the laws of nature are subservient to the super-natural designs of God in regard to man, and in particular to His Church, whose fetters were struck off abroad in this memorable year.

however, seen also the seed of perjury, sedition, and treason, sown by the enemy and deceiver of mankind from the beginning, fostered and quickened, and the egg of a FOUL, PRETENDED and GODLESS liberty, of premeditated revolution and aggression, hastened in its development, and hatched before the calculated time ; and we have beheld, in the horrors of war let loose against peaceful states, in the destruction of that peace about which there is something so beautiful in families and nations, the abundant harvest of Satan gathered in by his own servants and satellites. We see again and again the deceiving cuckoo, always and always cuckooing, liberty ! liberty ! liberty !—whilst laying its eggs into other birds' nests, its brood pushing out their rightful inmates and keeping them under the yoke of ITS OWN dear enlightened and precious freedom, and for the sake of greater liberty still, driving them from house and home and from their sacred places.

Nevertheless, it is not the comet which is guilty of all this iniquity, no more than wine is guilty of the vices of the drunkard, but MAN that does not resist the world, the devil, and the flesh, the excited motion of ambition, lust, avarice, and pride, and recklessly throws overboard the last vestige of christian virtue.

Apart from the physical effects which, however slightly, may, perhaps in all cases, be traced to the indirect influence of comets, so their mere appearance may strike terror into barbarous nations and uncultivated multitudes, and incite to savage warfare, as in the good it may elevate the soul to God, make it ponder over His wonderful works and prompt it to noble deeds, nay even make a prince exchange sceptre for rosary, for the glorious badge and pledge of the Lily of Nazareth, leave the palace for the cell, the rule over empire for the rule over himself.

In another form the mere sight of a comet may excite the pride of the worldly-learned at the bare idea of being able to calculate its weight and dimension, and to trace and determine its course

between large ones and small ones, their respective distance from, or nearness to, the sun, and the consequent excitement or non-excitement of his body or photosphere.

Comets and planets, attracted by, and moving round the centre of the sun, are attached to him by electric links which obediently they follow. If, then, it happens, that we are in conjunction with a comet or a planet, the earth with its electric link will be enveloped or embraced also by the electric link passing from the comet, or planet, to the sun; and hence the electric condition of our earth will thereby materially be affected, being exposed to a double attracting power, emanating from the sun both upon itself and upon the comet or planet; the effect will be different, again, whenever a comet or planet places itself in conjunction between us and the sun. Hence, also, an abnormal influence must be exercised by the sun upon a planet in conjunction with another, or with a comet. And that these electric influences, of which comets and planets are the mediate cause, must more or less effect the electric state of the earth, and produce a different state of fecundity in vegetable life, as they may also act upon minds whose bodies are predisposed to electrical affection,* at the right time and at the right place, for good or for bad, in the good or the wicked, for health or disease, seems to me but very reasonable to believe, and warranted by almost universal opinion and tradition, if not experience, among all nations. And we ourselves, in the train of a most splendid comet, have witnessed the blessings of an over-abundant and luxurious yield of corn and wine; we have,

---

* "In a paper on some of the phenomena of terrestrial magnetism, as observed at Greenwich, read before the Royal Society, the Astronomer-Royal shows that the revolutionary year, 1848, appears to have been one of confusion and disturbance in cosmical as well as tellurian affairs, for changes have taken place which cannot be explained on any established theory . . . . . . ".—Extract. To myself this is but another proof that the laws of nature are subservient to the super-natural designs of God in regard to man, and in particular to His Church, whose fetters were struck off abroad in this memorable year.

however, seen also the seed of perjury, sedition, and treason, sown
by the enemy and deceiver of mankind from the beginning, fostered
and quickened, and the egg of a FOUL, PRETENDED and GODLESS
liberty, of premeditated revolution and aggression, hastened in its
development, and hatched before the calculated time ; and we have
beheld, in the horrors of war let loose against peaceful states, in
the destruction of that peace about which there is something so
beautiful in families and nations, the abundant harvest of Satan
gathered in by his own servants and satellites. We see again and
again the deceiving cuckoo, always and always cuckooing, liberty !
liberty ! liberty !—whilst laying its eggs into other birds' nests, its
brood pushing out their rightful inmates and keeping them under
the yoke of ITS OWN dear enlightened and precious freedom, and
for the sake of greater liberty still, driving them from house and
home and from their sacred places.

Nevertheless, it is not the comet which is guilty of all this
iniquity, no more than wine is guilty of the vices of the
drunkard, but MAN that does not resist the world, the devil, and
the flesh, the excited motion of ambition, lust, avarice, and pride,
and recklessly throws overboard the last vestige of christian virtue.

Apart from the physical effects which, however slightly, may,
perhaps in all cases, be traced to the indirect influence of comets,
so their mere appearance may strike terror into barbarous
nations and uncultivated multitudes, and incite to savage warfare,
as in the good it may elevate the soul to God, make it ponder
over His wonderful works and prompt it to noble deeds, nay even
make a prince exchange sceptre for rosary, for the glorious badge
and pledge of the Lily of Nazareth, leave the palace for the cell,
the rule over empire for the rule over himself.

In another form the mere sight of a comet may excite the pride
of the worldly-learned at the bare idea of being able to calculate
its weight and dimension, and to trace and determine its course

between large ones and small ones, their respective distance from, or nearness to, the sun, and the consequent excitement or non-excitement of his body or photosphere.

Comets and planets, attracted by, and moving round the centre of the sun, are attached to him by electric links which obediently they follow. If, then, it happens, that we are in conjunction with a comet or a planet, the earth with its electric link will be enveloped or embraced also by the electric link passing from the comet, or planet, to the sun; and hence the electric condition of our earth will thereby materially be affected, being exposed to a double attracting power, emanating from the sun both upon itself and upon the comet or planet; the effect will be different, again, whenever a comet or planet places itself in conjunction between us and the sun. Hence, also, an abnormal influence must be exercised by the sun upon a planet in conjunction with another, or with a comet. And that these electric influences, of which comets and planets are the mediate cause, must more or less effect the electric state of the earth, and produce a different state of fecundity in vegetable life, as they may also act upon minds whose bodies are predisposed to electrical affection,* at the right time and at the right place, for good or for bad, in the good or the wicked, for health or disease, seems to me but very reasonable to believe, and warranted by almost universal opinion and tradition, if not experience, among all nations. And we ourselves, in the train of a most splendid comet, have witnessed the blessings of an over-abundant and luxurious yield of corn and wine; we have,

---

* "In a paper on some of the phenomena of terrestrial magnetism, as observed at Greenwich, read before the Royal Society, the Astronomer-Royal shows that the revolutionary year, 1848, appears to have been one of confusion and disturbance in cosmical as well as tellurian affairs, for changes have taken place which cannot be explained on any established theory . . . . . ".—*Extract.* To myself this is but another proof that the laws of nature are subservient to the super-natural designs of God in regard to man, and in particular to His Church, whose fetters were struck off abroad in this memorable year.

however, seen also the seed of perjury, sedition, and treason, sown by the enemy and deceiver of mankind from the beginning, fostered and quickened, and the egg of a FOUL, PRETENDED and GODLESS liberty, of premeditated revolution and aggression, hastened in its development, and hatched before the calculated time ; and we have beheld, in the horrors of war let loose against peaceful states, in the destruction of that peace about which there is something so beautiful in families and nations, the abundant harvest of Satan gathered in by his own servants and satellites.　We see again and again the deceiving cuckoo, always and always cuckooing, liberty ! liberty ! liberty !—whilst laying its eggs into other birds' nests, its brood pushing out their rightful inmates and keeping them under the yoke of ITS OWN dear enlightened and precious freedom, and for the sake of greater liberty still, driving them from house and home and from their sacred places.

Nevertheless, it is not the comet which is guilty of all this iniquity, no more than wine is guilty of the vices of the drunkard, but MAN that does not resist the world, the devil, and the flesh, the excited motion of ambition, lust, avarice, and pride, and recklessly throws overboard the last vestige of christian virtue.

Apart from the physical effects which, however slightly, may, perhaps in all cases, be traced to the indirect influence of comets, so their mere appearance may strike terror into barbarous nations and uncultivated multitudes, and incite to savage warfare, as in the good it may elevate the soul to God, make it ponder over His wonderful works and prompt it to noble deeds, nay even make a prince exchange sceptre for rosary, for the glorious badge and pledge of the Lily of Nazareth, leave the palace for the cell, the rule over empire for the rule over himself.

In another form the mere sight of a comet may excite the pride of the worldly-learned at the bare idea of being able to calculate its weight and dimension, and to trace and determine its course

between large ones and small ones, their respective distance from, or nearness to, the sun, and the consequent excitement or non-excitement of his body or photosphere.

Comets and planets, attracted by, and moving round the centre of the sun, are attached to him by electric links which obediently they follow. If, then, it happens, that we are in conjunction with a comet or a planet, the earth with its electric link will be enveloped or embraced also by the electric link passing from the comet, or planet, to the sun; and hence the electric condition of our earth will thereby materially be affected, being exposed to a double attracting power, emanating from the sun both upon itself and upon the comet or planet; the effect will be different, again, whenever a comet or planet places itself in conjunction between us and the sun. Hence, also, an abnormal influence must be exercised by the sun upon a planet in conjunction with another, or with a comet. And that these electric influences, of which comets and planets are the mediate cause, must more or less effect the electric state of the earth, and produce a different state of fecundity in vegetable life, as they may also act upon minds whose bodies are predisposed to electrical affection,* at the right time and at the right place, for good or for bad, in the good or the wicked, for health or disease, seems to me but very reasonable to believe, and warranted by almost universal opinion and tradition, if not experience, among all nations. And we ourselves, in the train of a most splendid comet, have witnessed the blessings of an over-abundant and luxurious yield of corn and wine; we have,

* "In a paper on some of the phenomena of terrestrial magnetism, as observed at Greenwich, read before the Royal Society, the Astronomer-Royal shows that the revolutionary year, 1848, appears to have been one of confusion and disturbance in cosmical as well as tellurian affairs, for changes have taken place which cannot be explained on any established theory . . . . . . ".—*Extract.* To myself this is but another proof that the laws of nature are subservient to the super-natural designs of God in regard to man, and in particular to His Church, whose fetters were struck off abroad in this memorable year.

however, seen also the seed of perjury, sedition, and treason, sown
by the enemy and deceiver of mankind from the beginning, fostered
and quickened, and the egg of a FOUL, PRETENDED and GODLESS
liberty, of premeditated revolution and aggression, hastened in its
development, and hatched before the calculated time ; and we have
beheld, in the horrors of war let loose against peaceful states, in
the destruction of that peace about which there is something so
beautiful in families and nations, the abundant harvest of Satan
gathered in by his own servants and satellites. We see again and
again the deceiving cuckoo, always and always cuckooing, liberty !
liberty ! liberty !—whilst laying its eggs into other birds' nests, its
brood pushing out their rightful inmates and keeping them under
the yoke of ITS OWN dear enlightened and precious freedom, and
for the sake of greater liberty still, driving them from house and
home and from their sacred places.

Nevertheless, it is not the comet which is guilty of all this
iniquity, no more than wine is guilty of the vices of the
drunkard, but MAN that does not resist the world, the devil, and
the flesh, the excited motion of ambition, lust, avarice, and pride,
and recklessly throws overboard the last vestige of christian virtue.

Apart from the physical effects which, however slightly, may,
perhaps in all cases, be traced to the indirect influence of comets,
so their mere appearance may strike terror into barbarous
nations and uncultivated multitudes, and incite to savage warfare,
as in the good it may elevate the soul to God, make it ponder
over His wonderful works and prompt it to noble deeds, nay even
make a prince exchange sceptre for rosary, for the glorious badge
and pledge of the Lily of Nazareth, leave the palace for the cell,
the rule over empire for the rule over himself.

In another form the mere sight of a comet may excite the pride
of the worldly-learned at the bare idea of being able to calculate
its weight and dimension, and to trace and determine its course

between large ones and small ones, their respective distance from, or nearness to, the sun, and the consequent excitement or non-excitement of his body or photosphere.

Comets and planets, attracted by, and moving round the centre of the sun, are attached to him by electric links which obediently they follow. If, then, it happens, that we are in conjunction with a comet or a planet, the earth with its electric link will be enveloped or embraced also by the electric link passing from the comet, or planet, to the sun; and hence the electric condition of our earth will thereby materially be affected, being exposed to a double attracting power, emanating from the sun both upon itself and upon the comet or planet; the effect will be different, again, when-ever a comet or planet places itself in conjunction between us and the sun. Hence, also, an abnormal influence must be exercised by the sun upon a planet in conjunction with another, or with a comet. And that these electric influences, of which comets and planets are the mediate cause, must more or less effect the electric state of the earth, and produce a different state of fecundity in vegetable life, as they may also act upon minds whose bodies are predisposed to electrical affection,* at the right time and at the right place, for good or for bad, in the good or the wicked, for health or disease, seems to me but very reasonable to believe, and warranted by almost universal opinion and tradition, if not experience, among all nations. And we ourselves, in the train of a most splendid comet, have witnessed the blessings of an over-abundant and luxurious yield of corn and wine; we have,

---

* "In a paper on some of the phenomena of terrestrial magnetism, as observed at Greenwich, read before the Royal Society, the Astronomer-Royal shows that the revolutionary year, 1848, appears to have been one of confusion and disturbance in cosmical as well as tellurian affairs, for changes have taken place which cannot be explained on any established theory . . . . . ".—*Extract.* To myself this is but another proof that the laws of nature are subservient to the super-natural designs of God in regard to man, and in particular to His Church, whose fetters were struck off abroad in this memorable year.

however, seen also the seed of perjury, sedition, and treason, sown
by the enemy and deceiver of mankind from the beginning, fostered
and quickened, and the egg of a FOUL, PRETENDED and GODLESS
liberty, of premeditated revolution and aggression, hastened in its
development, and hatched before the calculated time ; and we have
beheld, in the horrors of war let loose against peaceful states, in
the destruction of that peace about which there is something so
beautiful in families and nations, the abundant harvest of Satan
gathered in by his own servants and satellites. We see again and
again the deceiving cuckoo, always and always cuckooing, liberty !
liberty ! liberty !—whilst laying its eggs into other birds' nests, its
brood pushing out their rightful inmates and keeping them under
the yoke of ITS OWN dear enlightened and precious freedom, and
for the sake of greater liberty still, driving them from house and
home and from their sacred places.

Nevertheless, it is not the comet which is guilty of all this
iniquity, no more than wine is guilty of the vices of the
drunkard, but MAN that does not resist the world, the devil, and
the flesh, the excited motion of ambition, lust, avarice, and pride,
and recklessly throws overboard the last vestige of christian virtue.

Apart from the physical effects which, however slightly, may,
perhaps in all cases, be traced to the indirect influence of comets,
so their mere appearance may strike terror into barbarous
nations and uncultivated multitudes, and incite to savage warfare,
as in the good it may elevate the soul to God, make it ponder
over His wonderful works and prompt it to noble deeds, nay even
make a prince exchange sceptre for rosary, for the glorious badge
and pledge of the Lily of Nazareth, leave the palace for the cell,
the rule over empire for the rule over himself.

In another form the mere sight of a comet may excite the pride
of the worldly-learned at the bare idea of being able to calculate
its weight and dimension, and to trace and determine its course

between large ones and small ones, their respective distance from, or nearness to, the sun, and the consequent excitement or non-excitement of his body or photosphere.

Comets and planets, attracted by, and moving round the centre of the sun, are attached to him by electric links which obediently they follow. If, then, it happens, that we are in conjunction with a comet or a planet, the earth with its electric link will be enveloped or embraced also by the electric link passing from the comet, or planet, to the sun; and hence the electric condition of our earth will thereby materially be affected, being exposed to a double attracting power, emanating from the sun both upon itself and upon the comet or planet; the effect will be different, again, whenever a comet or planet places itself in conjunction between us and the sun. Hence, also, an abnormal influence must be exercised by the sun upon a planet in conjunction with another, or with a comet. And that these electric influences, of which comets and planets are the mediate cause, must more or less effect the electric state of the earth, and produce a different state of fecundity in vegetable life, as they may also act upon minds whose bodies are predisposed to electrical affection,[*] at the right time and at the right place, for good or for bad, in the good or the wicked, for health or disease, seems to me but very reasonable to believe, and warranted by almost universal opinion and tradition, if not experience, among all nations. And we ourselves, in the train of a most splendid comet, have witnessed the blessings of an over-abundant and luxurious yield of corn and wine; we have,

---

[*] "In a paper on some of the phenomena of terrestrial magnetism, as observed at Greenwich, read before the Royal Society, the Astronomer-Royal shows that the revolutionary year, 1848, appears to have been one of confusion and disturbance in cosmical as well as tellurian affairs, for changes have taken place which cannot be explained on any established theory . . . . . . ".—*Extract.* To myself this is but another proof that the laws of nature are subservient to the super-natural designs of God in regard to man, and in particular to His Church, whose fetters were struck off abroad in this memorable year.

however, seen also the seed of perjury, sedition, and treason, sown
by the enemy and deceiver of mankind from the beginning, fostered
and quickened, and the egg of a FOUL, PRETENDED and GODLESS
liberty, of premeditated revolution and aggression, hastened in its
development, and hatched before the calculated time ; and we have
beheld, in the horrors of war let loose against peaceful states, in
the destruction of that peace about which there is something so
beautiful in families and nations, the abundant harvest of Satan
gathered in by his own servants and satellites.   We see again and
again the deceiving cuckoo, always and always cuckooing, liberty !
liberty ! liberty !—whilst laying its eggs into other birds' nests, its
brood pushing out their rightful inmates and keeping them under
the yoke of ITS OWN dear enlightened and precious freedom, and
for the sake of greater liberty still, driving them from house and
home and from their sacred places.

Nevertheless, it is not the comet which is guilty of all this
iniquity, no more than wine is guilty of the vices of the
drunkard, but MAN that does not resist the world, the devil, and
the flesh, the excited motion of ambition, lust, avarice, and pride,
and recklessly throws overboard the last vestige of christian virtue.

Apart from the physical effects which, however slightly, may,
perhaps in all cases, be traced to the indirect influence of comets,
so their mere appearance may strike terror into barbarous
nations and uncultivated multitudes, and incite to savage warfare,
as in the good it may elevate the soul to God, make it ponder
over His wonderful works and prompt it to noble deeds, nay even
make a prince exchange sceptre for rosary, for the glorious badge
and pledge of the Lily of Nazareth, leave the palace for the cell,
the rule over empire for the rule over himself.

In another form the mere sight of a comet may excite the pride
of the worldly-learned at the bare idea of being able to calculate
its weight and dimension, and to trace and determine its course

between large ones and small ones, their respective distance from,
or nearness to, the sun, and the consequent excitement or non-
excitement of his body or photosphere.

Comets and planets, attracted by, and moving round the centre
of the sun, are attached to him by electric links which obediently
they follow. If, then, it happens, that we are in conjunction with
a comet or a planet, the earth with its electric link will be enveloped
or embraced also by the electric link passing from the comet, or
planet, to the sun; and hence the electric condition of our earth
will thereby materially be affected, being exposed to a double
attracting power, emanating from the sun both upon itself and
upon the comet or planet; the effect will be different, again, whenever a comet or planet places itself in conjunction between us and
the sun. Hence, also, an abnormal influence must be exercised
by the sun upon a planet in conjunction with another, or
with a comet. And that these electric influences, of which
comets and planets are the mediate cause, must more or less effect
the electric state of the earth, and produce a different state of
fecundity in vegetable life, as they may also act upon minds whose
bodies are predisposed to electrical affection,* at the right time
and at the right place, for good or for bad, in the good or the
wicked, for health or disease, seems to me but very reasonable to
believe, and warranted by almost universal opinion and tradition,
if not experience, among all nations. And we ourselves, in the
train of a most splendid comet, Have witnessed the blessings of an
over-abundant and luxurious yield of corn and wine; we have,

---

* "In a paper on some of the phenomena of terrestrial magnetism, as
observed at Greenwich, read before the Royal Society, the Astronomer-
Royal shows that the revolutionary year, 1848, appears to have been one of
confusion and disturbance in cosmical as well as tellurian affairs, for changes
have taken place which cannot be explained on any established theory . .
. . . ".—*Extract*. To myself this is but another proof that the laws of
nature are subservient to the super-natural designs of God in regard to man,
and in particular to His Church, whose fetters were struck off abroad in this
memorable year.

however, seen also the seed of perjury, sedition, and treason, sown by the enemy and deceiver of mankind from the beginning, fostered and quickened, and the egg of a FOUL, PRETENDED and GODLESS liberty, of premeditated revolution and aggression, hastened in its development, and hatched before the calculated time ; and we have beheld, in the horrors of war let loose against peaceful states, in the destruction of that peace about which there is something so beautiful in families and nations, the abundant harvest of Satan gathered in by his own servants and satellites.   We see again and again the deceiving cuckoo, always and always cuckooing, liberty ! liberty ! liberty !—whilst laying its eggs into other birds' nests, its brood pushing out their rightful inmates and keeping them under the yoke of ITS OWN dear enlightened and precious freedom, and for the sake of greater liberty still, driving them from house and home and from their sacred places.

Nevertheless, it is not the comet which is guilty of all this iniquity, no more than wine is guilty of the vices of the drunkard, but MAN that does not resist the world, the devil, and the flesh, the excited motion of ambition, lust, avarice, and pride, and recklessly throws overboard the last vestige of christian virtue.

Apart from the physical effects which, however slightly, may, perhaps in all cases, be traced to the indirect influence of comets, so their mere appearance may strike terror into barbarous nations and uncultivated multitudes, and incite to savage warfare, as in the good it may elevate the soul to God, make it ponder over His wonderful works and prompt it to noble deeds, nay even make a prince exchange sceptre for rosary, for the glorious badge and pledge of the Lily of Nazareth, leave the palace for the cell, the rule over empire for the rule over himself.

In another form the mere sight of a comet may excite the pride of the worldly-learned at the bare idea of being able to calculate its weight and dimension, and to trace and determine its course

between large ones and small ones, their respective distance from, or nearness to, the sun, and the consequent excitement or non-excitement of his body or photosphere.

Comets and planets, attracted by, and moving round the centre of the sun, are attached to him by electric links which obediently they follow. If, then, it happens, that we are in conjunction with a comet or a planet, the earth with its electric link will be enveloped or embraced also by the electric link passing from the comet, or planet, to the sun; and hence the electric condition of our earth will thereby materially be affected, being exposed to a double attracting power, emanating from the sun both upon itself and upon the comet or planet; the effect will be different, again, when-ever a comet or planet places itself in conjunction between us and the sun. Hence, also, an abnormal influence must be exercised by the sun upon a planet in conjunction with another, or with a comet. And that these electric influences, of which comets and planets are the mediate cause, must more or less effect the electric state of the earth, and produce a different state of fecundity in vegetable life, as they may also act upon minds whose bodies are predisposed to electrical affection,* at the right time and at the right place, for good or for bad, in the good or the wicked, for health or disease, seems to me but very reasonable to believe, and warranted by almost universal opinion and tradition, if not experience, among all nations. And we ourselves, in the train of a most splendid comet, have witnessed the blessings of an over-abundant and luxurious yield of corn and wine; we have,

---

* "In a paper on some of the phenomena of terrestrial magnetism, as observed at Greenwich, read before the Royal Society, the Astronomer-Royal shows that the revolutionary year, 1848, appears to have been one of confusion and disturbance in cosmical as well as tellurian affairs, for changes have taken place which cannot be explained on any established theory . . . . . . ".—*Extract.* To myself this is but another proof that the laws of nature are subservient to the super-natural designs of God in regard to man, and in particular to His Church, whose fetters were struck off abroad in this memorable year.

however, seen also the seed of perjury, sedition, and treason, sown
by the enemy and deceiver of mankind from the beginning, fostered
and quickened, and the egg of a FOUL, PRETENDED and GODLESS
liberty, of premeditated revolution and aggression, hastened in its
development, and hatched before the calculated time ; and we have
beheld, in the horrors of war let loose against peaceful states, in
the destruction of that peace about which there is something so
beautiful in families and nations, the abundant harvest of Satan
gathered in by his own servants and satellites.   We see again and
again the deceiving cuckoo, always and always cuckooing, liberty !
liberty ! liberty !—whilst laying its eggs into other birds' nests, its
brood pushing out their rightful inmates and keeping them under
the yoke of ITS OWN dear enlightened and precious freedom, and
for the sake of greater liberty still, driving them from house and
home and from their sacred places.

Nevertheless, it is not the comet which is guilty of all this
iniquity, no more than wine is guilty of the vices of the
drunkard, but MAN that does not resist the world, the devil, and
the flesh, the excited motion of ambition, lust, avarice, and pride,
and recklessly throws overboard the last vestige of christian virtue.

Apart from the physical effects which, however slightly, may,
perhaps in all cases, be traced to the indirect influence of comets,
so their mere appearance may strike terror into barbarous
nations and uncultivated multitudes, and incite to savage warfare,
as in the good it may elevate the soul to God, make it ponder
over His wonderful works and prompt it to noble deeds, nay even
make a prince exchange sceptre for rosary, for the glorious badge
and pledge of the Lily of Nazareth, leave the palace for the cell,
the rule over empire for the rule over himself.

In another form the mere sight of a comet may excite the pride
of the worldly-learned at the bare idea of being able to calculate
its weight and dimension, and to trace and determine its course

between large ones and small ones, their respective distance from, or nearness to, the sun, and the consequent excitement or non-excitement of his body or photosphere.

Comets and planets, attracted by, and moving round the centre of the sun, are attached to him by electric links which obediently they follow. If, then, it happens, that we are in conjunction with a comet or a planet, the earth with its electric link will be enveloped or embraced also by the electric link passing from the comet, or planet, to the sun; and hence the electric condition of our earth will thereby materially be affected, being exposed to a double attracting power, emanating from the sun both upon itself and upon the comet or planet; the effect will be different, again, whenever a comet or planet places itself in conjunction between us and the sun. Hence, also, an abnormal influence must be exercised by the sun upon a planet in conjunction with another, or with a comet. And that these electric influences, of which comets and planets are the mediate cause, must more or less effect the electric state of the earth, and produce a different state of fecundity in vegetable life, as they may also act upon minds whose bodies are predisposed to electrical affection,* at the right time and at the right place, for good or for bad, in the good or the wicked, for health or disease, seems to me but very reasonable to believe, and warranted by almost universal opinion and tradition, if not experience, among all nations. And we ourselves, in the train of a most splendid comet, Have witnessed the blessings of an over-abundant and luxurious yield of corn and wine; we have,

---

* "In a paper on some of the phenomena of terrestrial magnetism, as observed at Greenwich, read before the Royal Society, the Astronomer-Royal shows that the revolutionary year, 1848, appears to have been one of confusion and disturbance in cosmical as well as tellurian affairs, for changes have taken place which cannot be explained on any established theory . . . . . . ".—*Extract.* To myself this is but another proof that the laws of nature are subservient to the super-natural designs of God in regard to man, and in particular to His Church, whose fetters were struck off abroad in this memorable year.

however, seen also the seed of perjury, sedition, and treason, sown by the enemy and deceiver of mankind from the beginning, fostered and quickened, and the egg of a FOUL, PRETENDED and GODLESS liberty, of premeditated revolution and aggression, hastened in its development, and hatched before the calculated time ; and we have beheld, in the horrors of war let loose against peaceful states, in the destruction of that peace about which there is something so beautiful in families and nations, the abundant harvest of Satan gathered in by his own servants and satellites. We see again and again the deceiving cuckoo, always and always cuckooing, liberty ! liberty ! liberty !—whilst laying its eggs into other birds' nests, its brood pushing out their rightful inmates and keeping them under the yoke of ITS OWN dear enlightened and precious freedom, and for the sake of greater liberty still, driving them from house and home and from their sacred places.

Nevertheless, it is not the comet which is guilty of all this iniquity, no more than wine is guilty of the vices of the drunkard, but MAN that does not resist the world, the devil, and the flesh, the excited motion of ambition, lust, avarice, and pride, and recklessly throws overboard the last vestige of christian virtue.

Apart from the physical effects which, however slightly, may, perhaps in all cases, be traced to the indirect influence of comets, so their mere appearance may strike terror into barbarous nations and uncultivated multitudes, and incite to savage warfare, as in the good it may elevate the soul to God, make it ponder over His wonderful works and prompt it to noble deeds, nay even make a prince exchange sceptre for rosary, for the glorious badge and pledge of the Lily of Nazareth, leave the palace for the cell, the rule over empire for the rule over himself.

In another form the mere sight of a comet may excite the pride of the worldly-learned at the bare idea of being able to calculate its weight and dimension, and to trace and determine its course

between large ones and small ones, their respective distance from, or nearness to, the sun, and the consequent excitement or non-excitement of his body or photosphere.

Comets and planets, attracted by, and moving round the centre of the sun, are attached to him by electric links which obediently they follow. If, then, it happens, that we are in conjunction with a comet or a planet, the earth with its electric link will be enveloped or embraced also by the electric link passing from the comet, or planet, to the sun; and hence the electric condition of our earth will thereby materially be affected, being exposed to a double attracting power, emanating from the sun both upon itself and upon the comet or planet; the effect will be different, again, when-ever a comet or planet places itself in conjunction between us and the sun. Hence, also, an abnormal influence must be exercised by the sun upon a planet in conjunction with another, or with a comet. And that these electric influences, of which comets and planets are the mediate cause, must more or less effect the electric state of the earth, and produce a different state of fecundity in vegetable life, as they may also act upon minds whose bodies are predisposed to electrical affection,* at the right time and at the right place, for good or for bad, in the good or the wicked, for health or disease, seems to me but very reasonable to believe, and warranted by almost universal opinion and tradition, if not experience, among all nations. And we ourselves, in the train of a most splendid comet, Have witnessed the blessings of an over-abundant and luxurious yield of corn and wine; we have,

---

* "In a paper on some of the phenomena of terrestrial magnetism, as observed at Greenwich, read before the Royal Society, the Astronomer-Royal shows that the revolutionary year, 1848, appears to have been one of confusion and disturbance in cosmical as well as tellurian affairs, for changes have taken place which cannot be explained on any established theory . . . . . ".—*Extract.* To myself this is but another proof that the laws of nature are subservient to the super-natural designs of God in regard to man, and in particular to His Church, whose fetters were struck off abroad in this memorable year.

however, seen also the seed of perjury, sedition, and treason, sown
by the enemy and deceiver of mankind from the beginning, fostered
and quickened, and the egg of a FOUL, PRETENDED and GODLESS
liberty, of premeditated revolution and aggression, hastened in its
development, and hatched before the calculated time ; and we have
beheld, in the horrors of war let loose against peaceful states, in
the destruction of that peace about which there is something so
beautiful in families and nations, the abundant harvest of Satan
gathered in by his own servants and satellites.   We see again and
again the deceiving cuckoo, always and always cuckooing, liberty !
liberty ! liberty !—whilst laying its eggs into other birds' nests, its
brood pushing out their rightful inmates and keeping them under
the yoke of ITS OWN dear enlightened and precious freedom, and
for the sake of greater liberty still, driving them from house and
home and from their sacred places.

Nevertheless, it is not the comet which is guilty of all this
iniquity, no more than wine is guilty of the vices of the
drunkard, but MAN that does not resist the world, the devil, and
the flesh, the excited motion of ambition, lust, avarice, and pride,
and recklessly throws overboard the last vestige of christian virtue.

Apart from the physical effects which, however slightly, may,
perhaps in all cases, be traced to the indirect influence of comets,
so their mere appearance may strike terror into barbarous
nations and uncultivated multitudes, and incite to savage warfare,
as in the good it may elevate the soul to God, make it ponder
over His wonderful works and prompt it to noble deeds, nay even
make a prince exchange sceptre for rosary, for the glorious badge
and pledge of the Lily of Nazareth, leave the palace for the cell,
the rule over empire for the rule over himself.

In another form the mere sight of a comet may excite the pride
of the worldly-learned at the bare idea of being able to calculate
its weight and dimension, and to trace and determine its course

between large ones and small ones, their respective distance from, or nearness to, the sun, and the consequent excitement or non-excitement of his body or photosphere.

Comets and planets, attracted by, and moving round the centre of the sun, are attached to him by electric links which obediently they follow. If, then, it happens, that we are in conjunction with a comet or a planet, the earth with its electric link will be enveloped or embraced also by the electric link passing from the comet, or planet, to the sun; and hence the electric condition of our earth will thereby materially be affected, being exposed to a double attracting power, emanating from the sun both upon itself and upon the comet or planet; the effect will be different, again, whenever a comet or planet places itself in conjunction between us and the sun. Hence, also, an abnormal influence must be exercised by the sun upon a planet in conjunction with another, or with a comet. And that these electric influences, of which comets and planets are the mediate cause, must more or less effect the electric state of the earth, and produce a different state of fecundity in vegetable life, as they may also act upon minds whose bodies are predisposed to electrical affection,* at the right time and at the right place, for good or for bad, in the good or the wicked, for health or disease, seems to me but very reasonable to believe, and warranted by almost universal opinion and tradition, if not experience, among all nations. And we ourselves, in the train of a most splendid comet, have witnessed the blessings of an over-abundant and luxurious yield of corn and wine; we have,

---

* "In a paper on some of the phenomena of terrestrial magnetism, as observed at Greenwich, read before the Royal Society, the Astronomer-Royal shows that the revolutionary year, 1848, appears to have been one of confusion and disturbance in cosmical as well as tellurian affairs, for changes have taken place which cannot be explained on any established theory . . . . . . ".—*Extract.* To myself this is but another proof that the laws of nature are subservient to the super-natural designs of God in regard to man, and in particular to His Church, whose fetters were struck off abroad in this memorable year.

however, seen also the seed of perjury, sedition, and treason, sown by the enemy and deceiver of mankind from the beginning, fostered and quickened, and the egg of a FOUL, PRETENDED and GODLESS liberty, of premeditated revolution and aggression, hastened in its development, and hatched before the calculated time ; and we have beheld, in the horrors of war let loose against peaceful states, in the destruction of that peace about which there is something so beautiful in families and nations, the abundant harvest of Satan gathered in by his own servants and satellites. We see again and again the deceiving cuckoo, always and always cuckooing, liberty ! liberty ! liberty !—whilst laying its eggs into other birds' nests, its brood pushing out their rightful inmates and keeping them under the yoke of ITS OWN dear enlightened and precious freedom, and for the sake of greater liberty still, driving them from house and home and from their sacred places.

Nevertheless, it is not the comet which is guilty of all this iniquity, no more than wine is guilty of the vices of the drunkard, but MAN that does not resist the world, the devil, and the flesh, the excited motion of ambition, lust, avarice, and pride, and recklessly throws overboard the last vestige of christian virtue.

Apart from the physical effects which, however slightly, may, perhaps in all cases, be traced to the indirect influence of comets, so their mere appearance may strike terror into barbarous nations and uncultivated multitudes, and incite to savage warfare, as in the good it may elevate the soul to God, make it ponder over His wonderful works and prompt it to noble deeds, nay even make a prince exchange sceptre for rosary, for the glorious badge and pledge of the Lily of Nazareth, leave the palace for the cell, the rule over empire for the rule over himself.

In another form the mere sight of a comet may excite the pride of the worldly-learned at the bare idea of being able to calculate its weight and dimension, and to trace and determine its course

between large ones and small ones, their respective distance from, or nearness to, the sun, and the consequent excitement or non-excitement of his body or photosphere.

Comets and planets, attracted by, and moving round the centre of the sun, are attached to him by electric links which obediently they follow. If, then, it happens, that we are in conjunction with a comet or a planet, the earth with its electric link will be enveloped or embraced also by the electric link passing from the comet, or planet, to the sun; and hence the electric condition of our earth will thereby materially be affected, being exposed to a double attracting power, emanating from the sun both upon itself and upon the comet or planet; the effect will be different, again, whenever a comet or planet places itself in conjunction between us and the sun. Hence, also, an abnormal influence must be exercised by the sun upon a planet in conjunction with another, or with a comet. And that these electric influences, of which comets and planets are the mediate cause, must more or less effect the electric state of the earth, and produce a different state of fecundity in vegetable life, as they may also act upon minds whose bodies are predisposed to electrical affection,[*] at the right time and at the right place, for good or for bad, in the good or the wicked, for health or disease, seems to me but very reasonable to believe, and warranted by almost universal opinion and tradition, if not experience, among all nations. And we ourselves, in the train of a most splendid comet, have witnessed the blessings of an over-abundant and luxurious yield of corn and wine; we have,

---

[*] "In a paper on some of the phenomena of terrestrial magnetism, as observed at Greenwich, read before the Royal Society, the Astronomer-Royal shows that the revolutionary year, 1848, appears to have been one of confusion and disturbance in cosmical as well as tellurian affairs, for changes have taken place which cannot be explained on any established theory . . . . . . ".—*Extract.* To myself this is but another proof that the laws of nature are subservient to the super-natural designs of God in regard to man, and in particular to His Church, whose fetters were struck off abroad in this memorable year.

however, seen also the seed of perjury, sedition, and treason, sown
by the enemy and deceiver of mankind from the beginning, fostered
and quickened, and the egg of a FOUL, PRETENDED and GODLESS
liberty, of premeditated revolution and aggression, hastened in its
development, and hatched before the calculated time ; and we have
beheld, in the horrors of war let loose against peaceful states, in
the destruction of that peace about which there is something so
beautiful in families and nations, the abundant harvest of Satan
gathered in by his own servants and satellites. We see again and
again the deceiving cuckoo, always and always cuckooing, liberty !
liberty ! liberty !—whilst laying its eggs into other birds' nests, its
brood pushing out their rightful inmates and keeping them under
the yoke of ITS OWN dear enlightened and precious freedom, and
for the sake of greater liberty still, driving them from house and
home and from their sacred places.

Nevertheless, it is not the comet which is guilty of all this
iniquity, no more than wine is guilty of the vices of the
drunkard, but MAN that does not resist the world, the devil, and
the flesh, the excited motion of ambition, lust, avarice, and pride,
and recklessly throws overboard the last vestige of christian virtue.

Apart from the physical effects which, however slightly, may,
perhaps in all cases, be traced to the indirect influence of comets,
so their mere appearance may strike terror into barbarous
nations and uncultivated multitudes, and incite to savage warfare,
as in the good it may elevate the soul to God, make it ponder
over His wonderful works and prompt it to noble deeds, nay even
make a prince exchange sceptre for rosary, for the glorious badge
and pledge of the Lily of Nazareth, leave the palace for the cell,
the rule over empire for the rule over himself.

In another form the mere sight of a comet may excite the pride
of the worldly-learned at the bare idea of being able to calculate
its weight and dimension, and to trace and determine its course

between large ones and small ones, their respective distance from, or nearness to, the sun, and the consequent excitement or non-excitement of his body or photosphere.

Comets and planets, attracted by, and moving round the centre of the sun, are attached to him by electric links which obediently they follow. If, then, it happens, that we are in conjunction with a comet or a planet, the earth with its electric link will be enveloped or embraced also by the electric link passing from the comet, or planet, to the sun; and hence the electric condition of our earth will thereby materially be affected, being exposed to a double attracting power, emanating from the sun both upon itself and upon the comet or planet; the effect will be different, again, whenever a comet or planet places itself in conjunction between us and the sun. Hence, also, an abnormal influence must be exercised by the sun upon a planet in conjunction with another, or with a comet. And that these electric influences, of which comets and planets are the mediate cause, must more or less effect the electric state of the earth, and produce a different state of fecundity in vegetable life, as they may also act upon minds whose bodies are predisposed to electrical affection,* at the right time and at the right place, for good or for bad, in the good or the wicked, for health or disease, seems to me but very reasonable to believe, and warranted by almost universal opinion and tradition, if not experience, among all nations. And we ourselves, in the train of a most splendid comet, have witnessed the blessings of an over-abundant and luxurious yield of corn and wine; we have,

---

* "In a paper on some of the phenomena of terrestrial magnetism, as observed at Greenwich, read before the Royal Society, the Astronomer-Royal shows that the revolutionary year, 1848, appears to have been one of confusion and disturbance in cosmical as well as tellurian affairs, for changes have taken place which cannot be explained on any established theory . . . . . . ".—*Extract.* To myself this is but another proof that the laws of nature are subservient to the super-natural designs of God in regard to man, and in particular to His Church, whose fetters were struck off abroad in this memorable year.

however, seen also the seed of perjury, sedition, and treason, sown
by the enemy and deceiver of mankind from the beginning, fostered
and quickened, and the egg of a FOUL, PRETENDED and GODLESS
liberty, of premeditated revolution and aggression, hastened in its
development, and hatched before the calculated time ; and we have
beheld, in the horrors of war let loose against peaceful states, in
the destruction of that peace about which there is something so
beautiful in families and nations, the abundant harvest of Satan
gathered in by his own servants and satellites.   We see again and
again the deceiving cuckoo, always and always cuckooing, liberty !
liberty ! liberty !—whilst laying its eggs into other birds' nests, its
brood pushing out their rightful inmates and keeping them under
the yoke of ITS OWN dear enlightened and precious freedom, and
for the sake of greater liberty still, driving them from house and
home and from their sacred places.

Nevertheless, it is not the comet which is guilty of all this
iniquity, no more than wine is guilty of the vices of the
drunkard, but MAN that does not resist the world, the devil, and
the flesh, the excited motion of ambition, lust, avarice, and pride,
and recklessly throws overboard the last vestige of christian virtue.

Apart from the physical effects which, however slightly, may,
perhaps in all cases, be traced to the indirect influence of comets,
so their mere appearance may strike terror into barbarous
nations and uncultivated multitudes, and incite to savage warfare,
as in the good it may elevate the soul to God, make it ponder
over His wonderful works and prompt it to noble deeds, nay even
make a prince exchange sceptre for rosary, for the glorious badge
and pledge of the Lily of Nazareth, leave the palace for the cell,
the rule over empire for the rule over himself.

In another form the mere sight of a comet may excite the pride
of the worldly-learned at the bare idea of being able to calculate
its weight and dimension, and to trace and determine its course

between large ones and small ones, their respective distance from, or nearness to, the sun, and the consequent excitement or non-excitement of his body or photosphere.

Comets and planets, attracted by, and moving round the centre of the sun, are attached to him by electric links which obediently they follow. If, then, it happens, that we are in conjunction with a comet or a planet, the earth with its electric link will be enveloped or embraced also by the electric link passing from the comet, or planet, to the sun; and hence the electric condition of our earth will thereby materially be affected, being exposed to a double attracting power, emanating from the sun both upon itself and upon the comet or planet; the effect will be different, again, when-ever a comet or planet places itself in conjunction between us and the sun. Hence, also, an abnormal influence must be exercised by the sun upon a planet in conjunction with another, or with a comet. And that these electric influences, of which comets and planets are the mediate cause, must more or less effect the electric state of the earth, and produce a different state of fecundity in vegetable life, as they may also act upon minds whose bodies are predisposed to electrical affection,* at the right time and at the right place, for good or for bad, in the good or the wicked, for health or disease, seems to me but very reasonable to believe, and warranted by almost universal opinion and tradition, if not experience, among all nations. And we ourselves, in the train of a most splendid comet, Have witnessed the blessings of an over-abundant and luxurious yield of corn and wine; we have,

---

* "In a paper on some of the phenomena of terrestrial magnetism, as observed at Greenwich, read before the Royal Society, the Astronomer-Royal shows that the revolutionary year, 1848, appears to have been one of confusion and disturbance in cosmical as well as tellurian affairs, for changes have taken place which cannot be explained on any established theory . . . . . . ".—*Extract.* To myself this is but another proof that the laws of nature are subservient to the super-natural designs of God in regard to man, and in particular to His Church, whose fetters were struck off abroad in this memorable year.

however, seen also the seed of perjury, sedition, and treason, sown by the enemy and deceiver of mankind from the beginning, fostered and quickened, and the egg of a FOUL, PRETENDED and GODLESS liberty, of premeditated revolution and aggression, hastened in its development, and hatched before the calculated time ; and we have beheld, in the horrors of war let loose against peaceful states, in the destruction of that peace about which there is something so beautiful in families and nations, the abundant harvest of Satan gathered in by his own servants and satellites.   We see again and again the deceiving cuckoo, always and always cuckooing, liberty ! liberty ! liberty !—whilst laying its eggs into other birds' nests, its brood pushing out their rightful inmates and keeping them under the yoke of ITS OWN dear enlightened and precious freedom, and for the sake of greater liberty still, driving them from house and home and from their sacred places.

Nevertheless, it is not the comet which is guilty of all this iniquity, no more than wine is guilty of the vices of the drunkard, but MAN that does not resist the world, the devil, and the flesh, the excited motion of ambition, lust, avarice, and pride, and recklessly throws overboard the last vestige of christian virtue.

Apart from the physical effects which, however slightly, may, perhaps in all cases, be traced to the indirect influence of comets, so their mere appearance may strike terror into barbarous nations and uncultivated multitudes, and incite to savage warfare, as in the good it may elevate the soul to God, make it ponder over His wonderful works and prompt it to noble deeds, nay even make a prince exchange sceptre for rosary, for the glorious badge and pledge of the Lily of Nazareth, leave the palace for the cell, the rule over empire for the rule over himself.

In another form the mere sight of a comet may excite the pride of the worldly-learned at the bare idea of being able to calculate its weight and dimension, and to trace and determine its course

between large ones and small ones, their respective distance from, or nearness to, the sun, and the consequent excitement or non-excitement of his body or photosphere.

Comets and planets, attracted by, and moving round the centre of the sun, are attached to him by electric links which obediently they follow. If, then, it happens, that we are in conjunction with a comet or a planet, the earth with its electric link will be enveloped or embraced also by the electric link passing from the comet, or planet, to the sun; and hence the electric condition of our earth will thereby materially be affected, being exposed to a double attracting power, emanating from the sun both upon itself and upon the comet or planet; the effect will be different, again, when-ever a comet or planet places itself in conjunction between us and the sun. Hence, also, an abnormal influence must be exercised by the sun upon a planet in conjunction with another, or with a comet. And that these electric influences, of which comets and planets are the mediate cause, must more or less effect the electric state of the earth, and produce a different state of fecundity in vegetable life, as they may also act upon minds whose bodies are predisposed to electrical affection,* at the right time and at the right place, for good or for bad, in the good or the wicked, for health or disease, seems to me but very reasonable to believe, and warranted by almost universal opinion and tradition, if not experience, among all nations. And we ourselves, in the train of a most splendid comet, have witnessed the blessings of an over-abundant and luxurious yield of corn and wine; we have,

---

* "In a paper on some of the phenomena of terrestrial magnetism, as observed at Greenwich, read before the Royal Society, the Astronomer-Royal shows that the revolutionary year, 1848, appears to have been one of confusion and disturbance in cosmical as well as tellurian affairs, for changes have taken place which cannot be explained on any established theory . . . . . ".—*Extract.* To myself this is but another proof that the laws of nature are subservient to the super-natural designs of God in regard to man, and in particular to His Church, whose fetters were struck off abroad in this memorable year.

however, seen also the seed of perjury, sedition, and treason, sown by the enemy and deceiver of mankind from the beginning, fostered and quickened, and the egg of a FOUL, PRETENDED and GODLESS liberty, of premeditated revolution and aggression, hastened in its development, and hatched before the calculated time ; and we have beheld, in the horrors of war let loose against peaceful states, in the destruction of that peace about which there is something so beautiful in families and nations, the abundant harvest of Satan gathered in by his own servants and satellites.   We see again and again the deceiving cuckoo, always and always cuckooing, liberty ! liberty ! liberty !—whilst laying its eggs into other birds' nests, its brood pushing out their rightful inmates and keeping them under the yoke of ITS OWN dear enlightened and precious freedom, and for the sake of greater liberty still, driving them from house and home and from their sacred places.

Nevertheless, it is not the comet which is guilty of all this iniquity, no more than wine is guilty of the vices of the drunkard, but MAN that does not resist the world, the devil, and the flesh, the excited motion of ambition, lust, avarice, and pride, and recklessly throws overboard the last vestige of christian virtue.

Apart from the physical effects which, however slightly, may, perhaps in all cases, be traced to the indirect influence of comets, so their mere appearance may strike terror into barbarous nations and uncultivated multitudes, and incite to savage warfare, as in the good it may elevate the soul to God, make it ponder over His wonderful works and prompt it to noble deeds, nay even make a prince exchange sceptre for rosary, for the glorious badge and pledge of the Lily of Nazareth, leave the palace for the cell, the rule over empire for the rule over himself.

In another form the mere sight of a comet may excite the pride of the worldly-learned at the bare idea of being able to calculate its weight and dimension, and to trace and determine its course

between large ones and small ones, their respective distance from, or nearness to, the sun, and the consequent excitement or non-excitement of his body or photosphere.

Comets and planets, attracted by, and moving round the centre of the sun, are attached to him by electric links which obediently they follow. If, then, it happens, that we are in conjunction with a comet or a planet, the earth with its electric link will be enveloped or embraced also by the electric link passing from the comet, or planet, to the sun; and hence the electric condition of our earth will thereby materially be affected, being exposed to a double attracting power, emanating from the sun both upon itself and upon the comet or planet; the effect will be different, again, whenever a comet or planet places itself in conjunction between us and the sun. Hence, also, an abnormal influence must be exercised by the sun upon a planet in conjunction with another, or with a comet. And that these electric influences, of which comets and planets are the mediate cause, must more or less effect the electric state of the earth, and produce a different state of fecundity in vegetable life, as they may also act upon minds whose bodies are predisposed to electrical affection,* at the right time and at the right place, for good or for bad, in the good or the wicked, for health or disease, seems to me but very reasonable to believe, and warranted by almost universal opinion and tradition, if not experience, among all nations. And we ourselves, in the train of a most splendid comet, Have witnessed the blessings of an over-abundant and luxurious yield of corn and wine; we have,

---

* "In a paper on some of the phenomena of terrestrial magnetism, as observed at Greenwich, read before the Royal Society, the Astronomer-Royal shows that the revolutionary year, 1848, appears to have been one of confusion and disturbance in cosmical as well as tellurian affairs, for changes have taken place which cannot be explained on any established theory . . . . . . ".—*Extract*. To myself this is but another proof that the laws of nature are subservient to the super-natural designs of God in regard to man, and in particular to His Church, whose fetters were struck off abroad in this memorable year.

however, seen also the seed of perjury, sedition, and treason, sown by the enemy and deceiver of mankind from the beginning, fostered and quickened, and the egg of a FOUL, PRETENDED and GODLESS liberty, of premeditated revolution and aggression, hastened in its development, and hatched before the calculated time ; and we have beheld, in the horrors of war let loose against peaceful states, in the destruction of that peace about which there is something so beautiful in families and nations, the abundant harvest of Satan gathered in by his own servants and satellites. We see again and again the deceiving cuckoo, always and always cuckooing, liberty ! liberty ! liberty !—whilst laying its eggs into other birds' nests, its brood pushing out their rightful inmates and keeping them under the yoke of ITS OWN dear enlightened and precious freedom, and for the sake of greater liberty still, driving them from house and home and from their sacred places.

Nevertheless, it is not the comet which is guilty of all this iniquity, no more than wine is guilty of the vices of the drunkard, but MAN that does not resist the world, the devil, and the flesh, the excited motion of ambition, lust, avarice, and pride, and recklessly throws overboard the last vestige of christian virtue.

Apart from the physical effects which, however slightly, may, perhaps in all cases, be traced to the indirect influence of comets, so their mere appearance may strike terror into barbarous nations and uncultivated multitudes, and incite to savage warfare, as in the good it may elevate the soul to God, make it ponder over His wonderful works and prompt it to noble deeds, nay even make a prince exchange sceptre for rosary, for the glorious badge and pledge of the Lily of Nazareth, leave the palace for the cell, the rule over empire for the rule over himself.

In another form the mere sight of a comet may excite the pride of the worldly-learned at the bare idea of being able to calculate its weight and dimension, and to trace and determine its course

between large ones and small ones, their respective distance from, or nearness to, the sun, and the consequent excitement or non-excitement of his body or photosphere.

Comets and planets, attracted by, and moving round the centre of the sun, are attached to him by electric links which obediently they follow. If, then, it happens, that we are in conjunction with a comet or a planet, the earth with its electric link will be enveloped or embraced also by the electric link passing from the comet, or planet, to the sun; and hence the electric condition of our earth will thereby materially be affected, being exposed to a double attracting power, emanating from the sun both upon itself and upon the comet or planet; the effect will be different, again, whenever a comet or planet places itself in conjunction between us and the sun. Hence, also, an abnormal influence must be exercised by the sun upon a planet in conjunction with another, or with a comet. And that these electric influences, of which comets and planets are the mediate cause, must more or less effect the electric state of the earth, and produce a different state of fecundity in vegetable life, as they may also act upon minds whose bodies are predisposed to electrical affection,[*] at the right time and at the right place, for good or for bad, in the good or the wicked, for health or disease, seems to me but very reasonable to believe, and warranted by almost universal opinion and tradition, if not experience, among all nations. And we ourselves, in the train of a most splendid comet, Have witnessed the blessings of an over-abundant and luxurious yield of corn and wine; we have,

---

[*] "In a paper on some of the phenomena of terrestrial magnetism, as observed at Greenwich, read before the Royal Society, the Astronomer-Royal shows that the revolutionary year, 1848, appears to have been one of confusion and disturbance in cosmical as well as tellurian affairs, for changes have taken place which cannot be explained on any established theory . . . . . ".—*Extract.* To myself this is but another proof that the laws of nature are subservient to the super-natural designs of God in regard to man, and in particular to His Church, whose fetters were struck off abroad in this memorable year.

however, seen also the seed of perjury, sedition, and treason, sown by the enemy and deceiver of mankind from the beginning, fostered and quickened, and the egg of a FOUL, PRETENDED and GODLESS liberty, of premeditated revolution and aggression, hastened in its development, and hatched before the calculated time ; and we have beheld, in the horrors of war let loose against peaceful states, in the destruction of that peace about which there is something so beautiful in families and nations, the abundant harvest of Satan gathered in by his own servants and satellites.　We see again and again the deceiving cuckoo, always and always cuckooing, liberty ! liberty ! liberty !—whilst laying its eggs into other birds' nests, its brood pushing out their rightful inmates and keeping them under the yoke of ITS OWN dear enlightened and precious freedom, and for the sake of greater liberty still, driving them from house and home and from their sacred places.

Nevertheless, it is not the comet which is guilty of all this iniquity, no more than wine is guilty of the vices of the drunkard, but MAN that does not resist the world, the devil, and the flesh, the excited motion of ambition, lust, avarice, and pride, and recklessly throws overboard the last vestige of christian virtue.

Apart from the physical effects which, however slightly, may, perhaps in all cases, be traced to the indirect influence of comets, so their mere appearance may strike terror into barbarous nations and uncultivated multitudes, and incite to savage warfare, as in the good it may elevate the soul to God, make it ponder over His wonderful works and prompt it to noble deeds, nay even make a prince exchange sceptre for rosary, for the glorious badge and pledge of the Lily of Nazareth, leave the palace for the cell, the rule over empire for the rule over himself.

In another form the mere sight of a comet may excite the pride of the worldly-learned at the bare idea of being able to calculate its weight and dimension, and to trace and determine its course

between large ones and small ones, their respective distance from, or nearness to, the sun, and the consequent excitement or non-excitement of his body or photosphere.

Comets and planets, attracted by, and moving round the centre of the sun, are attached to him by electric links which obediently they follow. If, then, it happens, that we are in conjunction with a comet or a planet, the earth with its electric link will be enveloped or embraced also by the electric link passing from the comet, or planet, to the sun; and hence the electric condition of our earth will thereby materially be affected, being exposed to a double attracting power, emanating from the sun both upon itself and upon the comet or planet; the effect will be different, again, whenever a comet or planet places itself in conjunction between us and the sun. Hence, also, an abnormal influence must be exercised by the sun upon a planet in conjunction with another, or with a comet. And that these electric influences, of which comets and planets are the mediate cause, must more or less effect the electric state of the earth, and produce a different state of fecundity in vegetable life, as they may also act upon minds whose bodies are predisposed to electrical affection,* at the right time and at the right place, for good or for bad, in the good or the wicked, for health or for disease, seems to me but very reasonable to believe, and warranted by almost universal opinion and tradition, if not experience, among all nations. And we ourselves, in the train of a most splendid comet, have witnessed the blessings of an over-abundant and luxurious yield of corn and wine; we have,

---

* "In a paper on some of the phenomena of terrestrial magnetism, as observed at Greenwich, read before the Royal Society, the Astronomer-Royal shows that the revolutionary year, 1848, appears to have been one of confusion and disturbance in cosmical as well as tellurian affairs, for changes have taken place which cannot be explained on any established theory . . . . . ".—*Extract.* To myself this is but another proof that the laws of nature are subservient to the super-natural designs of God in regard to man, and in particular to His Church, whose fetters were struck off abroad in this memorable year.

however, seen also the seed of perjury, sedition, and treason, sown by the enemy and deceiver of mankind from the beginning, fostered and quickened, and the egg of a FOUL, PRETENDED and GODLESS liberty, of premeditated revolution and aggression, hastened in its development, and hatched before the calculated time ; and we have beheld, in the horrors of war let loose against peaceful states, in the destruction of that peace about which there is something so beautiful in families and nations, the abundant harvest of Satan gathered in by his own servants and satellites. We see again and again the deceiving cuckoo, always and always cuckooing, liberty ! liberty ! liberty !—whilst laying its eggs into other birds' nests, its brood pushing out their rightful inmates and keeping them under the yoke of ITS OWN dear enlightened and precious freedom, and for the sake of greater liberty still, driving them from house and home and from their sacred places.

Nevertheless, it is not the comet which is guilty of all this iniquity, no more than wine is guilty of the vices of the drunkard, but MAN that does not resist the world, the devil, and the flesh, the excited motion of ambition, lust, avarice, and pride, and recklessly throws overboard the last vestige of christian virtue.

Apart from the physical effects which, however slightly, may, perhaps in all cases, be traced to the indirect influence of comets, so their mere appearance may strike terror into barbarous nations and uncultivated multitudes, and incite to savage warfare, as in the good it may elevate the soul to God, make it ponder over His wonderful works and prompt it to noble deeds, nay even make a prince exchange sceptre for rosary, for the glorious badge and pledge of the Lily of Nazareth, leave the palace for the cell, the rule over empire for the rule over himself.

In another form the mere sight of a comet may excite the pride of the worldly-learned at the bare idea of being able to calculate its weight and dimension, and to trace and determine its course

between large ones and small ones, their respective distance from, or nearness to, the sun, and the consequent excitement or non-excitement of his body or photosphere.

Comets and planets, attracted by, and moving round the centre of the sun, are attached to him by electric links which obediently they follow. If, then, it happens, that we are in conjunction with a comet or a planet, the earth with its electric link will be enveloped or embraced also by the electric link passing from the comet, or planet, to the sun; and hence the electric condition of our earth will thereby materially be affected, being exposed to a double attracting power, emanating from the sun both upon itself and upon the comet or planet; the effect will be different, again, whenever a comet or planet places itself in conjunction between us and the sun. Hence, also, an abnormal influence must be exercised by the sun upon a planet in conjunction with another, or with a comet. And that these electric influences, of which comets and planets are the mediate cause, must more or less effect the electric state of the earth, and produce a different state of fecundity in vegetable life, as they may also act upon minds whose bodies are predisposed to electrical affection,* at the right time and at the right place, for good or for bad, in the good or the wicked, for health or disease, seems to me but very reasonable to believe, and warranted by almost universal opinion and tradition, if not experience, among all nations. And we ourselves, in the train of a most splendid comet, Have witnessed the blessings of an over-abundant and luxurious yield of corn and wine; we have,

---

* "In a paper on some of the phenomena of terrestrial magnetism, as observed at Greenwich, read before the Royal Society, the Astronomer-Royal shows that the revolutionary year, 1848, appears to have been one of confusion and disturbance in cosmical as well as tellurian affairs, for changes have taken place which cannot be explained on any established theory . . . . . ".—*Extract.* To myself this is but another proof that the laws of nature are subservient to the super-natural designs of God in regard to man, and in particular to His Church, whose fetters were struck off abroad in this memorable year.

however, seen also the seed of perjury, sedition, and treason, sown
by the enemy and deceiver of mankind from the beginning, fostered
and quickened, and the egg of a FOUL, PRETENDED and GODLESS
liberty, of premeditated revolution and aggression, hastened in its
development, and hatched before the calculated time ; and we have
beheld, in the horrors of war let loose against peaceful states, in
the destruction of that peace about which there is something so
beautiful in families and nations, the abundant harvest of Satan
gathered in by his own servants and satellites.   We see again and
again the deceiving cuckoo, always and always cuckooing, liberty !
liberty ! liberty !—whilst laying its eggs into other birds' nests, its
brood pushing out their rightful inmates and keeping them under
the yoke of ITS OWN dear enlightened and precious freedom, and
for the sake of greater liberty still, driving them from house and
home and from their sacred places.

Nevertheless, it is not the comet which is guilty of all this
iniquity, no more than wine is guilty of the vices of the
drunkard, but MAN that does not resist the world, the devil, and
the flesh, the excited motion of ambition, lust, avarice, and pride,
and recklessly throws overboard the last vestige of christian virtue.

Apart from the physical effects which, however slightly, may,
perhaps in all cases, be traced to the indirect influence of comets,
so their mere appearance may strike terror into barbarous
nations and uncultivated multitudes, and incite to savage warfare,
as in the good it may elevate the soul to God, make it ponder
over His wonderful works and prompt it to noble deeds, nay even
make a prince exchange sceptre for rosary, for the glorious badge
and pledge of the Lily of Nazareth, leave the palace for the cell,
the rule over empire for the rule over himself.

In another form the mere sight of a comet may excite the pride
of the worldly-learned at the bare idea of being able to calculate
its weight and dimension, and to trace and determine its course

between large ones and small ones, their respective distance from, or nearness to, the sun, and the consequent excitement or non-excitement of his body or photosphere.

Comets and planets, attracted by, and moving round the centre of the sun, are attached to him by electric links which obediently they follow. If, then, it happens, that we are in conjunction with a comet or a planet, the earth with its electric link will be enveloped or embraced also by the electric link passing from the comet, or planet, to the sun; and hence the electric condition of our earth will thereby materially be affected, being exposed to a double attracting power, emanating from the sun both upon itself and upon the comet or planet; the effect will be different, again, whenever a comet or planet places itself in conjunction between us and the sun. Hence, also, an abnormal influence must be exercised by the sun upon a planet in conjunction with another, or with a comet. And that these electric influences, of which comets and planets are the mediate cause, must more or less effect the electric state of the earth, and produce a different state of fecundity in vegetable life, as they may also act upon minds whose bodies are predisposed to electrical affection,* at the right time and at the right place, for good or for bad, in the good or the wicked, for health or disease, seems to me but very reasonable to believe, and warranted by almost universal opinion and tradition, if not experience, among all nations. And we ourselves, in the train of a most splendid comet, Have witnessed the blessings of an over-abundant and luxurious yield of corn and wine; we have,

---

* "In a paper on some of the phenomena of terrestrial magnetism, as observed at Greenwich, read before the Royal Society, the Astronomer-Royal shows that the revolutionary year, 1848, appears to have been one of confusion and disturbance in cosmical as well as tellurian affairs, for changes have taken place which cannot be explained on any established theory . . . . . . ".—*Extract*. To myself this is but another proof that the laws of nature are subservient to the super-natural designs of God in regard to man, and in particular to His Church, whose fetters were struck off abroad in this memorable year.

however, seen also the seed of perjury, sedition, and treason, sown
by the enemy and deceiver of mankind from the beginning, fostered
and quickened, and the egg of a FOUL, PRETENDED and GODLESS
liberty, of premeditated revolution and aggression, hastened in its
development, and hatched before the calculated time ; and we have
beheld, in the horrors of war let loose against peaceful states, in
the destruction of that peace about which there is something so
beautiful in families and nations, the abundant harvest of Satan
gathered in by his own servants and satellites.   We see again and
again the deceiving cuckoo, always and always cuckooing, liberty !
liberty ! liberty !—whilst laying its eggs into other birds' nests, its
brood pushing out their rightful inmates and keeping them under
the yoke of ITS OWN dear enlightened and precious freedom, and
for the sake of greater liberty still, driving them from house and
home and from their sacred places.

Nevertheless, it is not the comet which is guilty of all this
iniquity, no more than wine is guilty of the vices of the
drunkard, but MAN that does not resist the world, the devil, and
the flesh, the excited motion of ambition, lust, avarice, and pride,
and recklessly throws overboard the last vestige of christian virtue.

Apart from the physical effects which, however slightly, may,
perhaps in all cases, be traced to the indirect influence of comets,
so their mere appearance may strike terror into barbarous
nations and uncultivated multitudes, and incite to savage warfare,
as in the good it may elevate the soul to God, make it ponder
over His wonderful works and prompt it to noble deeds, nay even
make a prince exchange sceptre for rosary, for the glorious badge
and pledge of the Lily of Nazareth, leave the palace for the cell,
the rule over empire for the rule over himself.

In another form the mere sight of a comet may excite the pride
of the worldly-learned at the bare idea of being able to calculate
its weight and dimension, and to trace and determine its course

between large ones and small ones, their respective distance from, or nearness to, the sun, and the consequent excitement or non-excitement of his body or photosphere.

Comets and planets, attracted by, and moving round the centre of the sun, are attached to him by electric links which obediently they follow. If, then, it happens, that we are in conjunction with a comet or a planet, the earth with its electric link will be enveloped or embraced also by the electric link passing from the comet, or planet, to the sun; and hence the electric condition of our earth will thereby materially be affected, being exposed to a double attracting power, emanating from the sun both upon itself and upon the comet or planet; the effect will be different, again, when-ever a comet or planet places itself in conjunction between us and the sun. Hence, also, an abnormal influence must be exercised by the sun upon a planet in conjunction with another, or with a comet. And that these electric influences, of which comets and planets are the mediate cause, must more or less effect the electric state of the earth, and produce a different state of fecundity in vegetable life, as they may also act upon minds whose bodies are predisposed to electrical affection,* at the right time and at the right place, for good or for bad, in the good or the wicked, for health or disease, seems to me but very reasonable to believe, and warranted by almost universal opinion and tradition, if not experience, among all nations. And we ourselves, in the train of a most splendid comet, have witnessed the blessings of an over-abundant and luxurious yield of corn and wine; we have,

* "In a paper on some of the phenomena of terrestrial magnetism, as observed at Greenwich, read before the Royal Society, the Astronomer-Royal shows that the revolutionary year, 1848, appears to have been one of confusion and disturbance in cosmical as well as tellurian affairs, for changes have taken place which cannot be explained on any established theory . . . . . ".—*Extract.* To myself this is but another proof that the laws of nature are subservient to the super-natural designs of God in regard to man, and in particular to His Church, whose fetters were struck off abroad in this memorable year.

however, seen also the seed of perjury, sedition, and treason, sown by the enemy and deceiver of mankind from the beginning, fostered and quickened, and the egg of a FOUL, PRETENDED and GODLESS liberty, of premeditated revolution and aggression, hastened in its development, and hatched before the calculated time ; and we have beheld, in the horrors of war let loose against peaceful states, in the destruction of that peace about which there is something so beautiful in families and nations, the abundant harvest of Satan gathered in by his own servants and satellites. We see again and again the deceiving cuckoo, always and always cuckooing, liberty ! liberty ! liberty !—whilst laying its eggs into other birds' nests, its brood pushing out their rightful inmates and keeping them under the yoke of ITS OWN dear enlightened and precious freedom, and for the sake of greater liberty still, driving them from house and home and from their sacred places.

Nevertheless, it is not the comet which is guilty of all this iniquity, no more than wine is guilty of the vices of the drunkard, but MAN that does not resist the world, the devil, and the flesh, the excited motion of ambition, lust, avarice, and pride, and recklessly throws overboard the last vestige of christian virtue.

Apart from the physical effects which, however slightly, may, perhaps in all cases, be traced to the indirect influence of comets, so their mere appearance may strike terror into barbarous nations and uncultivated multitudes, and incite to savage warfare, as in the good it may elevate the soul to God, make it ponder over His wonderful works and prompt it to noble deeds, nay even make a prince exchange sceptre for rosary, for the glorious badge and pledge of the Lily of Nazareth, leave the palace for the cell, the rule over empire for the rule over himself.

In another form the mere sight of a comet may excite the pride of the worldly-learned at the bare idea of being able to calculate its weight and dimension, and to trace and determine its course

between large ones and small ones, their respective distance from,
or nearness to, the sun, and the consequent excitement or non-
excitement of his body or photosphere.

Comets and planets, attracted by, and moving round the centre
of the sun, are attached to him by electric links which obediently
they follow. If, then, it happens, that we are in conjunction with
a comet or a planet, the earth with its electric link will be enveloped
or embraced also by the electric link passing from the comet, or
planet, to the sun; and hence the electric condition of our earth
will thereby materially be affected, being exposed to a double
attracting power, emanating from the sun both upon itself and
upon the comet or planet; the effect will be different, again, whenever a comet or planet places itself in conjunction between us and
the sun. Hence, also, an abnormal influence must be exercised
by the sun upon a planet in conjunction with another, or
with a comet. And that these electric influences, of which
comets and planets are the mediate cause, must more or less effect
the electric state of the earth, and produce a different state of
fecundity in vegetable life, as they may also act upon minds whose
bodies are predisposed to electrical affection,* at the right time
and at the right place, for good or for bad, in the good or the
wicked, for health or disease, seems to me but very reasonable to
believe, and warranted by almost universal opinion and tradition,
if not experience, among all nations. And we ourselves, in the
train of a most splendid comet, have witnessed the blessings of an
over-abundant and luxurious yield of corn and wine; we have,

---

* "In a paper on some of the phenomena of terrestrial magnetism, as
observed at Greenwich, read before the Royal Society, the Astronomer-
Royal shows that the revolutionary year, 1848, appears to have been one of
confusion and disturbance in cosmical as well as tellurian affairs, for changes
have taken place which cannot be explained on any established theory . .
. . . ".—*Extract.* To myself this is but another proof that the laws of
nature are subservient to the super-natural designs of God in regard to man,
and in particular to His Church, whose fetters were struck off abroad in this
memorable year.

however, seen also the seed of perjury, sedition, and treason, sown by the enemy and deceiver of mankind from the beginning, fostered and quickened, and the egg of a FOUL, PRETENDED and GODLESS liberty, of premeditated revolution and aggression, hastened in its development, and hatched before the calculated time ; and we have beheld, in the horrors of war let loose against peaceful states, in the destruction of that peace about which there is something so beautiful in families and nations, the abundant harvest of Satan gathered in by his own servants and satellites.   We see again and again the deceiving cuckoo, always and always cuckooing, liberty ! liberty ! liberty !—whilst laying its eggs into other birds' nests, its brood pushing out their rightful inmates and keeping them under the yoke of ITS OWN dear enlightened and precious freedom, and for the sake of greater liberty still, driving them from house and home and from their sacred places.

Nevertheless, it is not the comet which is guilty of all this iniquity, no more than wine is guilty of the vices of the drunkard, but MAN that does not resist the world, the devil, and the flesh, the excited motion of ambition, lust, avarice, and pride, and recklessly throws overboard the last vestige of christian virtue.

Apart from the physical effects which, however slightly, may, perhaps in all cases, be traced to the indirect influence of comets, so their mere appearance may strike terror into barbarous nations and uncultivated multitudes, and incite to savage warfare, as in the good it may elevate the soul to God, make it ponder over His wonderful works and prompt it to noble deeds, nay even make a prince exchange sceptre for rosary, for the glorious badge and pledge of the Lily of Nazareth, leave the palace for the cell, the rule over empire for the rule over himself.

In another form the mere sight of a comet may excite the pride of the worldly-learned at the bare idea of being able to calculate its weight and dimension, and to trace and determine its course

between large ones and small ones, their respective distance from, or nearness to, the sun, and the consequent excitement or non-excitement of his body or photosphere.

Comets and planets, attracted by, and moving round the centre of the sun, are attached to him by electric links which obediently they follow. If, then, it happens, that we are in conjunction with a comet or a planet, the earth with its electric link will be enveloped or embraced also by the electric link passing from the comet, or planet, to the sun; and hence the electric condition of our earth will thereby materially be affected, being exposed to a double attracting power, emanating from the sun both upon itself and upon the comet or planet; the effect will be different, again, whenever a comet or planet places itself in conjunction between us and the sun. Hence, also, an abnormal influence must be exercised by the sun upon a planet in conjunction with another, or with a comet. And that these electric influences, of which comets and planets are the mediate cause, must more or less effect the electric state of the earth, and produce a different state of fecundity in vegetable life, as they may also act upon minds whose bodies are predisposed to electrical affection,* at the right time and at the right place, for good or for bad, in the good or the wicked, for health or disease, seems to me but very reasonable to believe, and warranted by almost universal opinion and tradition, if not experience, among all nations. And we ourselves, in the train of a most splendid comet, have witnessed the blessings of an over-abundant and luxurious yield of corn and wine; we have,

---

* "In a paper on some of the phenomena of terrestrial magnetism, as observed at Greenwich, read before the Royal Society, the Astronomer-Royal shows that the revolutionary year, 1848, appears to have been one of confusion and disturbance in cosmical as well as tellurian affairs, for changes have taken place which cannot be explained on any established theory . . . . . ".—*Extract.* To myself this is but another proof that the laws of nature are subservient to the super-natural designs of God in regard to man, and in particular to His Church, whose fetters were struck off abroad in this memorable year.

however, seen also the seed of perjury, sedition, and treason, sown by the enemy and deceiver of mankind from the beginning, fostered and quickened, and the egg of a FOUL, PRETENDED and GODLESS liberty, of premeditated revolution and aggression, hastened in its development, and hatched before the calculated time ; and we have beheld, in the horrors of war let loose against peaceful states, in the destruction of that peace about which there is something so beautiful in families and nations, the abundant harvest of Satan gathered in by his own servants and satellites. We see again and again the deceiving cuckoo, always and always cuckooing, liberty ! liberty ! liberty !—whilst laying its eggs into other birds' nests, its brood pushing out their rightful inmates and keeping them under the yoke of ITS OWN dear enlightened and precious freedom, and for the sake of greater liberty still, driving them from house and home and from their sacred places.

Nevertheless, it is not the comet which is guilty of all this iniquity, no more than wine is guilty of the vices of the drunkard, but MAN that does not resist the world, the devil, and the flesh, the excited motion of ambition, lust, avarice, and pride, and recklessly throws overboard the last vestige of christian virtue.

Apart from the physical effects which, however slightly, may, perhaps in all cases, be traced to the indirect influence of comets, so their mere appearance may strike terror into barbarous nations and uncultivated multitudes, and incite to savage warfare, as in the good it may elevate the soul to God, make it ponder over His wonderful works and prompt it to noble deeds, nay even make a prince exchange sceptre for rosary, for the glorious badge and pledge of the Lily of Nazareth, leave the palace for the cell, the rule over empire for the rule over himself.

In another form the mere sight of a comet may excite the pride of the worldly-learned at the bare idea of being able to calculate its weight and dimension, and to trace and determine its course

between large ones and small ones, their respective distance from,
or nearness to, the sun, and the consequent excitement or non-
excitement of his body or photosphere.

Comets and planets, attracted by, and moving round the centre
of the sun, are attached to him by electric links which obediently
they follow. If, then, it happens, that we are in conjunction with
a comet or a planet, the earth with its electric link will be enveloped
or embraced also by the electric link passing from the comet, or
planet, to the sun; and hence the electric condition of our earth
will thereby materially be affected, being exposed to a double
attracting power, emanating from the sun both upon itself and
upon the comet or planet; the effect will be different, again, whenever a comet or planet places itself in conjunction between us and
the sun. Hence, also, an abnormal influence must be exercised
by the sun upon a planet in conjunction with another, or
with a comet. And that these electric influences, of which
comets and planets are the mediate cause, must more or less effect
the electric state of the earth, and produce a different state of
fecundity in vegetable life, as they may also act upon minds whose
bodies are predisposed to electrical affection,* at the right time
and at the right place, for good or for bad, in the good or the
wicked, for health or disease, seems to me but very reasonable to
believe, and warranted by almost universal opinion and tradition,
if not experience, among all nations. And we ourselves, in the
train of a most splendid comet, have witnessed the blessings of an
over-abundant and luxurious yield of corn and wine; we have,

---

* "In a paper on some of the phenomena of terrestrial magnetism, as
observed at Greenwich, read before the Royal Society, the Astronomer-
Royal shows that the revolutionary year, 1848, appears to have been one of
confusion and disturbance in cosmical as well as tellurian affairs, for changes
have taken place which cannot be explained on any established theory . .
. . . ".—*Extract*. To myself this is but another proof that the laws of
nature are subservient to the super-natural designs of God in regard to man,
and in particular to His Church, whose fetters were struck off abroad in this
memorable year.

however, seen also the seed of perjury, sedition, and treason, sown
by the enemy and deceiver of mankind from the beginning, fostered
and quickened, and the egg of a FOUL, PRETENDED and GODLESS
liberty, of premeditated revolution and aggression, hastened in its
development, and hatched before the calculated time ; and we have
beheld, in the horrors of war let loose against peaceful states, in
the destruction of that peace about which there is something so
beautiful in families and nations, the abundant harvest of Satan
gathered in by his own servants and satellites. We see again and
again the deceiving cuckoo, always and always cuckooing, liberty !
liberty ! liberty !—whilst laying its eggs into other birds' nests, its
brood pushing out their rightful inmates and keeping them under
the yoke of ITS OWN dear enlightened and precious freedom, and
for the sake of greater liberty still, driving them from house and
home and from their sacred places.

Nevertheless, it is not the comet which is guilty of all this
iniquity, no more than wine is guilty of the vices of the
drunkard, but MAN that does not resist the world, the devil, and
the flesh, the excited motion of ambition, lust, avarice, and pride,
and recklessly throws overboard the last vestige of christian virtue.

Apart from the physical effects which, however slightly, may,
perhaps in all cases, be traced to the indirect influence of comets,
so their mere appearance may strike terror into barbarous
nations and uncultivated multitudes, and incite to savage warfare,
as in the good it may elevate the soul to God, make it ponder
over His wonderful works and prompt it to noble deeds, nay even
make a prince exchange sceptre for rosary, for the glorious badge
and pledge of the Lily of Nazareth, leave the palace for the cell,
the rule over empire for the rule over himself.

In another form the mere sight of a comet may excite the pride
of the worldly-learned at the bare idea of being able to calculate
its weight and dimension, and to trace and determine its course

between large ones and small ones, their respective distance from, or nearness to, the sun, and the consequent excitement or non-excitement of his body or photosphere.

Comets and planets, attracted by, and moving round the centre of the sun, are attached to him by electric links which obediently they follow. If, then, it happens, that we are in conjunction with a comet or a planet, the earth with its electric link will be enveloped or embraced also by the electric link passing from the comet, or planet, to the sun; and hence the electric condition of our earth will thereby materially be affected, being exposed to a double attracting power, emanating from the sun both upon itself and upon the comet or planet; the effect will be different, again, whenever a comet or planet places itself in conjunction between us and the sun. Hence, also, an abnormal influence must be exercised by the sun upon a planet in conjunction with another, or with a comet. And that these electric influences, of which comets and planets are the mediate cause, must more or less effect the electric state of the earth, and produce a different state of fecundity in vegetable life, as they may also act upon minds whose bodies are predisposed to electrical affection,* at the right time and at the right place, for good or for bad, in the good or the wicked, for health or disease, seems to me but very reasonable to believe, and warranted by almost universal opinion and tradition, if not experience, among all nations. And we ourselves, in the train of a most splendid comet, have witnessed the blessings of an over-abundant and luxurious yield of corn and wine; we have,

---

* "In a paper on some of the phenomena of terrestrial magnetism, as observed at Greenwich, read before the Royal Society, the Astronomer-Royal shows that the revolutionary year, 1848, appears to have been one of confusion and disturbance in cosmical as well as tellurian affairs, for changes have taken place which cannot be explained on any established theory . . . . . . ".—*Extract.* To myself this is but another proof that the laws of nature are subservient to the super-natural designs of God in regard to man, and in particular to His Church, whose fetters were struck off abroad in this memorable year.

however, seen also the seed of perjury, sedition, and treason, sown by the enemy and deceiver of mankind from the beginning, fostered and quickened, and the egg of a FOUL, PRETENDED and GODLESS liberty, of premeditated revolution and aggression, hastened in its development, and hatched before the calculated time ; and we have beheld, in the horrors of war let loose against peaceful states, in the destruction of that peace about which there is something so beautiful in families and nations, the abundant harvest of Satan gathered in by his own servants and satellites. We see again and again the deceiving cuckoo, always and always cuckooing, liberty ! liberty ! liberty !—whilst laying its eggs into other birds' nests, its brood pushing out their rightful inmates and keeping them under the yoke of ITS OWN dear enlightened and precious freedom, and for the sake of greater liberty still, driving them from house and home and from their sacred places.

Nevertheless, it is not the comet which is guilty of all this iniquity, no more than wine is guilty of the vices of the drunkard, but MAN that does not resist the world, the devil, and the flesh, the excited motion of ambition, lust, avarice, and pride, and recklessly throws overboard the last vestige of christian virtue.

Apart from the physical effects which, however slightly, may, perhaps in all cases, be traced to the indirect influence of comets, so their mere appearance may strike terror into barbarous nations and uncultivated multitudes, and incite to savage warfare, as in the good it may elevate the soul to God, make it ponder over His wonderful works and prompt it to noble deeds, nay even make a prince exchange sceptre for rosary, for the glorious badge and pledge of the Lily of Nazareth, leave the palace for the cell, the rule over empire for the rule over himself.

In another form the mere sight of a comet may excite the pride of the worldly-learned at the bare idea of being able to calculate its weight and dimension, and to trace and determine its course

between large ones and small ones, their respective distance from, or nearness to, the sun, and the consequent excitement or non-excitement of his body or photosphere.

Comets and planets, attracted by, and moving round the centre of the sun, are attached to him by electric links which obediently they follow. If, then, it happens, that we are in conjunction with a comet or a planet, the earth with its electric link will be enveloped or embraced also by the electric link passing from the comet, or planet, to the sun; and hence the electric condition of our earth will thereby materially be affected, being exposed to a double attracting power, emanating from the sun both upon itself and upon the comet or planet; the effect will be different, again, when-ever a comet or planet places itself in conjunction between us and the sun. Hence, also, an abnormal influence must be exercised by the sun upon a planet in conjunction with another, or with a comet. And that these electric influences, of which comets and planets are the mediate cause, must more or less effect the electric state of the earth, and produce a different state of fecundity in vegetable life, as they may also act upon minds whose bodies are predisposed to electrical affection,* at the right time and at the right place, for good or for bad, in the good or the wicked, for health or disease, seems to me but very reasonable to believe, and warranted by almost universal opinion and tradition, if not experience, among all nations. And we ourselves, in the train of a most splendid comet, have witnessed the blessings of an over-abundant and luxurious yield of corn and wine; we have,

---

* "In a paper on some of the phenomena of terrestrial magnetism, as observed at Greenwich, read before the Royal Society, the Astronomer-Royal shows that the revolutionary year, 1848, appears to have been one of confusion and disturbance in cosmical as well as tellurian affairs, for changes have taken place which cannot be explained on any established theory . . . . . . ".—*Extract.* To myself this is but another proof that the laws of nature are subservient to the super-natural designs of God in regard to man, and in particular to His Church, whose fetters were struck off abroad in this memorable year.

however, seen also the seed of perjury, sedition, and treason, sown
by the enemy and deceiver of mankind from the beginning, fostered
and quickened, and the egg of a FOUL, PRETENDED and GODLESS
liberty, of premeditated revolution and aggression, hastened in its
development, and hatched before the calculated time ; and we have
beheld, in the horrors of war let loose against peaceful states, in
the destruction of that peace about which there is something so
beautiful in families and nations, the abundant harvest of Satan
gathered in by his own servants and satellites.   We see again and
again the deceiving cuckoo, always and always cuckooing, liberty !
liberty ! liberty !—whilst laying its eggs into other birds' nests, its
brood pushing out their rightful inmates and keeping them under
the yoke of ITS OWN dear enlightened and precious freedom, and
for the sake of greater liberty still, driving them from house and
home and from their sacred places.

Nevertheless, it is not the comet which is guilty of all this
iniquity, no more than wine is guilty of the vices of the
drunkard, but MAN that does not resist the world, the devil, and
the flesh, the excited motion of ambition, lust, avarice, and pride,
and recklessly throws overboard the last vestige of christian virtue.

Apart from the physical effects which, however slightly, may,
perhaps in all cases, be traced to the indirect influence of comets,
so their mere appearance may strike terror into barbarous
nations and uncultivated multitudes, and incite to savage warfare,
as in the good it may elevate the soul to God, make it ponder
over His wonderful works and prompt it to noble deeds, nay even
make a prince exchange sceptre for rosary, for the glorious badge
and pledge of the Lily of Nazareth, leave the palace for the cell,
the rule over empire for the rule over himself.

In another form the mere sight of a comet may excite the pride
of the worldly-learned at the bare idea of being able to calculate
its weight and dimension, and to trace and determine its course

between large ones and small ones, their respective distance from, or nearness to, the sun, and the consequent excitement or non-excitement of his body or photosphere.

Comets and planets, attracted by, and moving round the centre of the sun, are attached to him by electric links which obediently they follow. If, then, it happens, that we are in conjunction with a comet or a planet, the earth with its electric link will be enveloped or embraced also by the electric link passing from the comet, or planet, to the sun; and hence the electric condition of our earth will thereby materially be affected, being exposed to a double attracting power, emanating from the sun both upon itself and upon the comet or planet; the effect will be different, again, whenever a comet or planet places itself in conjunction between us and the sun. Hence, also, an abnormal influence must be exercised by the sun upon a planet in conjunction with another, or with a comet. And that these electric influences, of which comets and planets are the mediate cause, must more or less effect the electric state of the earth, and produce a different state of fecundity in vegetable life, as they may also act upon minds whose bodies are predisposed to electrical affection,* at the right time and at the right place, for good or for bad, in the good or the wicked, for health or disease, seems to me but very reasonable to believe, and warranted by almost universal opinion and tradition, if not experience, among all nations. And we ourselves, in the train of a most splendid comet, have witnessed the blessings of an over-abundant and luxurious yield of corn and wine; we have,

---

* "In a paper on some of the phenomena of terrestrial magnetism, as observed at Greenwich, read before the Royal Society, the Astronomer-Royal shows that the revolutionary year, 1848, appears to have been one of confusion and disturbance in cosmical as well as tellurian affairs, for changes have taken place which cannot be explained on any established theory . . . . . ."—*Extract.* To myself this is but another proof that the laws of nature are subservient to the super-natural designs of God in regard to man, and in particular to His Church, whose fetters were struck off abroad in this memorable year.

however, seen also the seed of perjury, sedition, and treason, sown by the enemy and deceiver of mankind from the beginning, fostered and quickened, and the egg of a FOUL, PRETENDED and GODLESS liberty, of premeditated revolution and aggression, hastened in its development, and hatched before the calculated time ; and we have beheld, in the horrors of war let loose against peaceful states, in the destruction of that peace about which there is something so beautiful in families and nations, the abundant harvest of Satan gathered in by his own servants and satellites. We see again and again the deceiving cuckoo, always and always cuckooing, liberty ! liberty ! liberty !—whilst laying its eggs into other birds' nests, its brood pushing out their rightful inmates and keeping them under the yoke of ITS OWN dear enlightened and precious freedom, and for the sake of greater liberty still, driving them from house and home and from their sacred places.

Nevertheless, it is not the comet which is guilty of all this iniquity, no more than wine is guilty of the vices of the drunkard, but MAN that does not resist the world, the devil, and the flesh, the excited motion of ambition, lust, avarice, and pride, and recklessly throws overboard the last vestige of christian virtue.

Apart from the physical effects which, however slightly, may, perhaps in all cases, be traced to the indirect influence of comets, so their mere appearance may strike terror into barbarous nations and uncultivated multitudes, and incite to savage warfare, as in the good it may elevate the soul to God, make it ponder over His wonderful works and prompt it to noble deeds, nay even make a prince exchange sceptre for rosary, for the glorious badge and pledge of the Lily of Nazareth, leave the palace for the cell, the rule over empire for the rule over himself.

In another form the mere sight of a comet may excite the pride of the worldly-learned at the bare idea of being able to calculate its weight and dimension, and to trace and determine its course

between large ones and small ones, their respective distance from, or nearness to, the sun, and the consequent excitement or non-excitement of his body or photosphere.

Comets and planets, attracted by, and moving round the centre of the sun, are attached to him by electric links which obediently they follow. If, then, it happens, that we are in conjunction with a comet or a planet, the earth with its electric link will be enveloped or embraced also by the electric link passing from the comet, or planet, to the sun; and hence the electric condition of our earth will thereby materially be affected, being exposed to a double attracting power, emanating from the sun both upon itself and upon the comet or planet; the effect will be different, again, when-ever a comet or planet places itself in conjunction between us and the sun. Hence, also, an abnormal influence must be exercised by the sun upon a planet in conjunction with another, or with a comet. And that these electric influences, of which comets and planets are the mediate cause, must more or less effect the electric state of the earth, and produce a different state of fecundity in vegetable life, as they may also act upon minds whose bodies are predisposed to electrical affection,* at the right time and at the right place, for good or for bad, in the good or the wicked, for health or disease, seems to me but very reasonable to believe, and warranted by almost universal opinion and tradition, if not experience, among all nations. And we ourselves, in the train of a most splendid comet, have witnessed the blessings of an over-abundant and luxurious yield of corn and wine; we have,

---

* "In a paper on some of the phenomena of terrestrial magnetism, as observed at Greenwich, read before the Royal Society, the Astronomer-Royal shows that the revolutionary year, 1848, appears to have been one of confusion and disturbance in cosmical as well as tellurian affairs, for changes have taken place which cannot be explained on any established theory . . . . . . ".—*Extract.* To myself this is but another proof that the laws of nature are subservient to the super-natural designs of God in regard to man, and in particular to His Church, whose fetters were struck off abroad in this memorable year.

however, seen also the seed of perjury, sedition, and treason, sown by the enemy and deceiver of mankind from the beginning, fostered and quickened, and the egg of a FOUL, PRETENDED and GODLESS liberty, of premeditated revolution and aggression, hastened in its development, and hatched before the calculated time ; and we have beheld, in the horrors of war let loose against peaceful states, in the destruction of that peace about which there is something so beautiful in families and nations, the abundant harvest of Satan gathered in by his own servants and satellites. We see again and again the deceiving cuckoo, always and always cuckooing, liberty ! liberty ! liberty !—whilst laying its eggs into other birds' nests, its brood pushing out their rightful inmates and keeping them under the yoke of ITS OWN dear enlightened and precious freedom, and for the sake of greater liberty still, driving them from house and home and from their sacred places.

Nevertheless, it is not the comet which is guilty of all this iniquity, no more than wine is guilty of the vices of the drunkard, but MAN that does not resist the world, the devil, and the flesh, the excited motion of ambition, lust, avarice, and pride, and recklessly throws overboard the last vestige of christian virtue.

Apart from the physical effects which, however slightly, may, perhaps in all cases, be traced to the indirect influence of comets, so their mere appearance may strike terror into barbarous nations and uncultivated multitudes, and incite to savage warfare, as in the good it may elevate the soul to God, make it ponder over His wonderful works and prompt it to noble deeds, nay even make a prince exchange sceptre for rosary, for the glorious badge and pledge of the Lily of Nazareth, leave the palace for the cell, the rule over empire for the rule over himself.

In another form the mere sight of a comet may excite the pride of the worldly-learned at the bare idea of being able to calculate its weight and dimension, and to trace and determine its course

between large ones and small ones, their respective distance from, or nearness to, the sun, and the consequent excitement or non-excitement of his body or photosphere.

Comets and planets, attracted by, and moving round the centre of the sun, are attached to him by electric links which obediently they follow. If, then, it happens, that we are in conjunction with a comet or a planet, the earth with its electric link will be enveloped or embraced also by the electric link passing from the comet, or planet, to the sun; and hence the electric condition of our earth will thereby materially be affected, being exposed to a double attracting power, emanating from the sun both upon itself and upon the comet or planet; the effect will be different, again, whenever a comet or planet places itself in conjunction between us and the sun. Hence, also, an abnormal influence must be exercised by the sun upon a planet in conjunction with another, or with a comet. And that these electric influences, of which comets and planets are the mediate cause, must more or less effect the electric state of the earth, and produce a different state of fecundity in vegetable life, as they may also act upon minds whose bodies are predisposed to electrical affection,* at the right time and at the right place, for good or for bad, in the good or the wicked, for health or disease, seems to me but very reasonable to believe, and warranted by almost universal opinion and tradition, if not experience, among all nations. And we ourselves, in the train of a most splendid comet, have witnessed the blessings of an over-abundant and luxurious yield of corn and wine; we have,

---

* "In a paper on some of the phenomena of terrestrial magnetism, as observed at Greenwich, read before the Royal Society, the Astronomer-Royal shows that the revolutionary year, 1848, appears to have been one of confusion and disturbance in cosmical as well as tellurian affairs, for changes have taken place which cannot be explained on any established theory . . . . . ".—*Extract*. To myself this is but another proof that the laws of nature are subservient to the super-natural designs of God in regard to man, and in particular to His Church, whose fetters were struck off abroad in this memorable year.

however, seen also the seed of perjury, sedition, and treason, sown by the enemy and deceiver of mankind from the beginning, fostered and quickened, and the egg of a FOUL, PRETENDED and GODLESS liberty, of premeditated revolution and aggression, hastened in its development, and hatched before the calculated time ; and we have beheld, in the horrors of war let loose against peaceful states, in the destruction of that peace about which there is something so beautiful in families and nations, the abundant harvest of Satan gathered in by his own servants and satellites. We see again and again the deceiving cuckoo, always and always cuckooing, liberty ! liberty ! liberty !—whilst laying its eggs into other birds' nests, its brood pushing out their rightful inmates and keeping them under the yoke of ITS OWN dear enlightened and precious freedom, and for the sake of greater liberty still, driving them from house and home and from their sacred places.

Nevertheless, it is not the comet which is guilty of all this iniquity, no more than wine is guilty of the vices of the drunkard, but MAN that does not resist the world, the devil, and the flesh, the excited motion of ambition, lust, avarice, and pride, and recklessly throws overboard the last vestige of christian virtue.

Apart from the physical effects which, however slightly, may, perhaps in all cases, be traced to the indirect influence of comets, so their mere appearance may strike terror into barbarous nations and uncultivated multitudes, and incite to savage warfare, as in the good it may elevate the soul to God, make it ponder over His wonderful works and prompt it to noble deeds, nay even make a prince exchange sceptre for rosary, for the glorious badge and pledge of the Lily of Nazareth, leave the palace for the cell, the rule over empire for the rule over himself.

In another form the mere sight of a comet may excite the pride of the worldly-learned at the bare idea of being able to calculate its weight and dimension, and to trace and determine its course

even through unknown and unfathomable regions.   Instances are not wanting to prove all this.

Pope Calixtus III., in the year 1455, immediately after his elevation to the chair of St. Peter, renewed the encyclical letter of his predecessor Nicholas V., proclaiming a general crusade against the Turks, whilst at the same time he sent out legates into all christian countries, to settle existing differences among princes, and to induce them to enter into the war against the common foe of christendom, in which, however, he did not succeed.   Numerous preachers of the crusade, among whom also John Capistran, went about in Germany, France, Spain and Hungary, and exhorted the people, not only to make contributions and take a personal share in the war, but also to offer up public and private prayers for the success of the christian arms.   From this invitation dates the ringing of bells for prayer THREE times a day in the whole of western christendom, being introduced into France by Louis XI., in the year 1472 ; but a Synod of Paris, in 1346, already, expressly calls the evening bell a regulation of Pope John XXII., whilst, according to the general opinion, Pope Urban II. introduced the morning as well as the evening bell of the " Angelus Domini."   Raynaldus, in his *Annals of the Church*, 1753, says :—" That according to the historian Nicolas, of Para, not only the trumpet of the above-named Capistran was sounded for the crusade against the Turks, but that heaven itself, as it were, summoned to it by the appearance of two comets, which, before the arrival of the Turks in June, 1456, were seen in the mornings and evenings in the form of trumpets."

This last passage is the only existing historical record relating to the grand comets of 1456, which, as indicated before, must have very differently affected the christians and the fanatical Turks.

However, ignorantly or wilfully, there are, and always will be, writers siding with

> Heathen, Turk, or Atheist,
> Anyone, but a Papist ;

and thus the simple fact above narrated has equally been perverted against Pope and Papist.

"Lorcequ' en 1456, on vit paraître l'éclatante comète, qui doit revenir en 1835, le pape Calixte en fut si effrayé, qu'il ordonna pour un certain tems des prières publiques, dans lesquelles, au milieu de chaque jour, on excommuniait à la fois la comète et les Turcs ; et afin que personne ne manquât à ce devoir, il établit l'usage, qui dépuis s'est conservé, de sonner à midi les cloches des églises."

Thus Arago, the illustrious astronomer, the intelligent (!) propagator of this fable, in his popular "Notices sur les comètes" published in 1832.

As far as I am aware, he does not state whence he got his valuable information, but in the year 1835 he again made the best of the "ingenious device" against the Church ; and as in each of these years a comet appeared, they cannot but have affected his mind in the way I have indicated, and this so much so, that even, after having had his attention drawn to the falsity of his statement, he nevertheless republished it in his later writings.

I can scarcely bring myself to believe that Arago, on the one hand so learned, should, on the other, have been so ignorant as not to know better, or so "borné" as to believe the story ; and unless at the sight of the comet he was so blinded by pride as not to see the silliness of his anecdote, or by its publication thought to add greater lustre to the halo of his name, there remains but one reason for him to have adhered to it, and this is : he may have felt having committed himself, but finding the tale so easily believed, and not wishing to appear foolish, or having so grossly played the fool with his friends and those whom he addressed, as also those who with so much simplicity had copied it into their works, he thought it better not to retract.

But that things like these, so eminently stupid and absurd, and so palpably untrue, should, in sober earnest, be propagated and

o

believed and repeated, even by men high in literature, science and station, with but very few exceptions, is a sad sign of modern enlightenment and civilization; it is indeed the splendid tail of the comet, DRIFTED AWAY FROM THE NUCLEUS, making a great stir in the world, all bright outside, but dark and hollow within, the boasted progress of learned ignorance.

Less precise in the narrative of the story than Arago, the writer of the article on "astrology" in *Knight's Cyclopædia* says: "The astrology of comets, which is hardly yet out of date, has even been recognised by a Pope: in the fifteenth century Calixtus III. directed prayers and anathemas against a comet which had either assisted in or predicted the success of the Turks against the Christians."

With Arago, the writer ventures to set down this in utter ignorance about excommunication, anathema, and historical fact, quite forgetful, moreover, of what he had written A FEW MOMENTS BEFORE, namely: the truly historical event, that " the Church, in its public capacity, condemned the art (of astrology) in the first councils of Braga and Toledo, and in the Decretals (cited by Vossius)."

What confidence, then, is to be placed in these writers?

Johannes Von Gumpach—equally at sea in things which these 1800 years have been familiar to the faithful—in his recent, otherwise excellent and profound work, called *Baby Worlds*, introduces the subject of the comet of 1456 as follows:

" The days are gone by . . . . . . when comets caused Roman bishops to exorcise their threatening flames by means of public prayers, which to this day ring in the Catholic's ear." HE, then, at least, in a foot-note, gives the source of his information, which is nothing more than the notorious infidel Gibbon, and in the following words relates the exorcism of " FLAMES," by " PUBLIC PRAYERS," and the " RINGING OF PRAYERS " in our ears.

" The appearance of Halley's comet, on its return to perihelion in 1456, was, by the Turks, then preparing for a new invasion, hailed as a favourable omen; while the Christian nations, scourged by a pestilence, looked upon it as the messenger of their approaching destruction. 'In this extremity,' GIBBON TELLS US, 'and placed between two unavoidable dangers, either of being destroyed by a barbarous enemy, who knew no pity, or by the comet, which in their opinion proceeded directly from hell, they returned imploringly to higher powers, distrusting their own resources. Calixtus III., at whose feet almost all Europe lay prostrated in the deepest humility, had the condescension to hear their prayers. At his intervention, a prayer was read every day at noon in all Churches, in which the comet, as well as the Turks and the destructive plague, were solemnly exorcised and excommunicated. But that no one should neglect this important and solemn act, the custom was introduced of giving the signal for the general assembly, at the moment of noon, by the Church bell, a custom that has been continued to the present time.' "

This article of Gibbon is so miserably senseless in every line, that I am surprised how any one, with the commonest under-standing, and with the slightest consideration of its contents, can for a moment refer to it except as the aberration and actual imbecility of an infidel and totally perverted mind; in fact, it seems, generally, as if the intellect of the opponents of the Church were failing and giving way whenever they try their heads against her. Do but analyze and compare the three different versions of the comet, the great Pope Calixtus III., and the hallowed " Angelus Domini ! "

Yes, the " ANGELUS BELL," with thousands and thousands of silvery tongues, is still ringing in the Catholic's ear and calling, and every morning, noon, and evening, the CATHOLIC'S PRAYER is still ascending ; daily still its endless triple garland is twining all around this beautiful earth, and without intermission and to the end of the world the Salutation will continue to be said and to be heard :

"The angel of the Lord declared unto Mary and she conceived of the Holy Ghost. Hail Mary, full of grace, the Lord is with thee, blessed art thou among women, and blessed is the fruit of thy womb, Jesus. Holy Mary, Mother of God, pray for us sinners, NOW, and AT THE HOUR OF OUR DEATH, Amen.

"Behold I am the handmaid of the Lord, be it done to me according to thy word. Hail Mary, &c.

"And the Word was made flesh and dwelt amongst us. Hail Mary, &c.

Yes, to this scriptural prayer, to that public act of faith in the incarnation, the bell is still inviting, and may you who read this,—if a stranger still to the Church,—may you one day join into it with head uncovered; you will then find how sweet it is, —not three times a day, but three times three and more,—to say with the angel : Ave Maria ! Hail Mary!

Alas ! that I am obliged to return again to the sorrowful task of exposing the errors and perversions of anti-Catholic writers, in matters, too, so old, so palpable, so well-known and clear to every child of faith, and themselves therein so lamentably deficient and ignorant !

Mädler, the director of the observatory at Dorpat, seems to have laboured under the effects of the comet of 1858; for, in an article of his on comets, to be found in a work of Dr. Peters,* he speaks, in scripture phraseology, of a Catholic priest as a "Darkling," an "Obscurantist," simply because he was (falsely) said to have predicted the destruction of the world by this comet, and thereby set millions of people into fear and terror.

A SO-CALLED Rev. Presbyterian Dr., whose writings abound in diatribes and falsehoods against the Church, counting a large body of—even aristocratic—dupes and admirers, not long ago predicted the end of the world in three years ; and whilst his work, written in flaming language, chiefly directed against the very old antichrist

---

* See *Nature and Revelation*, vol. v., 335.

at Rome, was taken and read with avidity, and I believe, selling in several editions—he quietly took a new lease of a house for twenty-one years.

But what, if in noble opposition to this speculation, the Catholic priest had actually made use of the appearance of the comet to remind men of their duty and make them better and humble?

Saint Mary Magdalene of Pazzi one day culled a flower in the garden of her cloister and broke out in rapture: "Oh, my good God, from eternity Thou hast loved me, from eternity Thou knewest that this flower would delight me!"

If, then, to the Christian mind, nature is always calculated to turn his thoughts to God: how much more those phenomena that are rare and awe-inspiring?

Does not the moon affect the epileptic? Have not comets influenced the minds of men, learned and unlearned, the godly and godless? Has not the earthquake shaken down the palaces of the rich and the huts of the poor, and hurried into eternity the guilty and the innocent? Did not a solar eclipse beneficially influence HEATHENS when it led to the conclusion of war between the Medes and Lydians just as they were going to battle? Did not the lightning of grace convert Saul the persecutor of the christians? Has not the lightning of justice more than once struck the blasphemer in the very act of defying the Almighty? Did not the sign of the cross in the heavens tell Constantine and his army: "In hoc signo vinces:" "In this sign thou shalt conquer?" Did not the star of the Magi lead these truly wise men, and astronomers too, to the far brighter Star of the Sea, and to the very Light of the World in whose glory it shone? Do we forget the awful obscuration of the sun with the attendant convulsion of the earth at the crucifixion of our Lord? "Either the God of nature is dying or the machinery of the world is being broken in pieces!"

exclaimed Dionysius the Areopagite, at the time it occurred, though distant from the spot ; he believed the former and was converted to the religion of Him who then DID die.

And though nature will have it, yet fault is found and superstition it is called, when poor mortals, and great sinners moreover, are told to beware, and mind the warning of Him in whose hands they are, who carries the lightning, guides the hurricane, rules the waves, moves the heavens, and makes the earth tremble ! No, I say ; nay, blessed that superstition, if superstition it be, that leads the soul to God, makes the proud one humble, the libertine chaste, the angry meek, the miser give alms, the infidel to pray, the scoffer to adore, the calcumniator to retract, the over-weaning to pause, scientific conceit to see the hollowness of its mere secular knowledge ; which at the contemplation of a comet—and beautiful, if true—makes an imperial sinner repent and abdicate his throne, and, another Magi of wisdom, lay down his crown at the feet of the Blessed Virgin-Mother and Child, to exchange it for a crown in heaven ! Blessed superstition if these be thy effects !

But there is a real superstition which NEVER FOUND A RESTING PLACE in the Church, nor in ANY of her moderately-well-instructed PRACTICAL children, which ONLY and EXCLUSIVELY is found outside her precincts, more or less gross according to the departure from her doctrine and teaching. Solomon even became an idolator when he had left the Church of God. Outside to her, and CONDEMNED by her, there is a superstition that LEADS TO EVIL, everywhere and always at the heel of and in opposition to good, as falsehood is opposed to truth, as the shadow clings to light, and night, with its incoherent forms, follows day.

There is the superstition of spiritualism and all its kindred, and of heresy in general, from manichæism down to our latest revivals ; there is that satanic superstition, witnessed in Protestant Sweden a few years ago, which forcibly breaking the ranks of the soldiers,

greedily seized upon the blood of executed MURDERERS to find a cure for diseases of men and cattle; there is an abundant crop of real superstition in this kingdom, and next to it that depraved tendency of the mind which collects the relics of criminals high and low, and blasphemes and sins under the gallows. And who would not rather join that superstition, if so you will have it, that collects and venerates the blood and relics of the MARTYRS, of men and women died in the service of their heavenly Master and for the corporal and spiritual welfare of the human race? And who would not be moved by that superstition which teaches and MAKES an assembled crowd fall down on their knees and in awful silence pray for the expiring culprit?

But there is that departure from moral perception, FOSTERED by apostates and the robbers of the Church and their descendants, which cannot distinguish the true from the false, deprecating what leads to God, and exalting what keeps and leads away from Him; that blindness which uses "superstition" without knowing what it is, unless by reference to Catholic authority.

As a striking proof I need only refer to an article of the *Saturday Review* of 14th May, 1864, on *Superstition and Education*. Though the paper is one of the cleverest of the day, yet, in the article in question, there is a development of ignorance and confusion on the subject which I have seldom or ever witnessed.

But there is even Sir John Herschel, so learned and acute in astronomical science, and, no doubt, a model to the society he moves in, whom I must class among the blind, and whose ignorance on superstition I must deplore.

"I have not heard"—he says, in *Good Words!*—"that it was necessary to liquefy the blood of St. Januarius, or to carry his bones about the streets, on account of any of these later great comets."

The introduction of this paragraph by Sir John is quite gratui-

exclaimed Dionysius the Areopagite, at the time it occurred, though distant from the spot; he believed the former and was converted to the religion of Him who then DID die.

And though nature will have it, yet fault is found and superstition it is called, when poor mortals, and great sinners moreover, are told to beware, and mind the warning of Him in whose hands they are, who carries the lightning, guides the hurricane, rules the waves, moves the heavens, and makes the earth tremble! No, I say; nay, blessed that superstition, if superstition it be, that leads the soul to God, makes the proud one humble, the libertine chaste, the angry meek, the miser give alms, the infidel to pray, the scoffer to adore, the calcumniator to retract, the over-weaning to pause, scientific conceit to see the hollowness of its mere secular knowledge; which at the contemplation of a comet—and beautiful, if true—makes an imperial sinner repent and abdicate his throne, and, another Magi of wisdom, lay down his crown at the feet of the Blessed Virgin-Mother and Child, to exchange it for a crown in heaven! Blessed superstition if these be thy effects!

But there is a real superstition which NEVER FOUND A RESTING PLACE in the Church, nor in ANY of her moderately-well-instructed PRACTICAL children, which ONLY and EXCLUSIVELY is found outside her precincts, more or less gross according to the departure from her doctrine and teaching. Solomon even became an idolator when he had left the Church of God. Outside to her, and CONDEMNED by her, there is a superstition that LEADS TO EVIL, everywhere and always at the heel of and in opposition to good, as falsehood is opposed to truth, as the shadow clings to light, and night, with its incoherent forms, follows day.

There is the superstition of spiritualism and all its kindred, and of heresy in general, from manichæism down to our latest revivals; there is that satanic superstition, witnessed in Protestant Sweden a few years ago, which forcibly breaking the ranks of the soldiers,

greedily seized upon the blood of executed MURDERERS to find a
cure for diseases of men and cattle; there is an abundant crop of
real superstition in this kingdom, and next to it that depraved
tendency of the mind which collects the relics of criminals high
and low, and blasphemes and sins under the gallows. And who
would not rather join that superstition, if so you will have it,
that collects and venerates the blood and relics of the MARTYRS,
of men and women died in the service of their heavenly Master
and for the corporal and spiritual welfare of the human race?
And who would not be moved by that superstition which teaches
and MAKES an assembled crowd fall down on their knees and in
awful silence pray for the expiring culprit?

But there is that departure from moral perception, FOSTERED
by apostates and the robbers of the Church and their descendants,
which cannot distinguish the true from the false, deprecating
what leads to God, and exalting what keeps and leads away
from Him; that blindness which uses "superstition" without
knowing what it is, unless by reference to Catholic authority.

As a striking proof I need only refer to an article of the
Saturday Review of 14th May, 1864, on *Superstition and Education.*
Though the paper is one of the cleverest of the day, yet, in
the article in question, there is a development of ignorance and
confusion on the subject which I have seldom or ever witnessed.

But there is even Sir John Herschel, so learned and acute
in astronomical science, and, no doubt, a model to the society
he moves in, whom I must class among the blind, and whose
ignorance on superstition I must deplore.

"I have not heard"—he says, in *Good Words!*—"that it was
necessary to liquefy the blood of St. Januarius, or to carry his
bones about the streets, on account of any of these later great
comets."

The introduction of this paragraph by Sir John is quite gratui-

exclaimed Dionysius the Areopagite, at the time it occurred, though distant from the spot; he believed the former and was converted to the religion of Him who then DID die.

And though nature will have it, yet fault is found and superstition it is called, when poor mortals, and great sinners moreover, are told to beware, and mind the warning of Him in whose hands they are, who carries the lightning, guides the hurricane, rules the waves, moves the heavens, and makes the earth tremble! No, I say; nay, blessed that superstition, if superstition it be, that leads the soul to God, makes the proud one humble, the libertine chaste, the angry meek, the miser give alms, the infidel to pray, the scoffer to adore, the calcumniator to retract, the over-weaning to pause, scientific conceit to see the hollowness of its mere secular knowledge; which at the contemplation of a comet—and beautiful, if true—makes an imperial sinner repent and abdicate his throne, and, another Magi of wisdom, lay down his crown at the feet of the Blessed Virgin-Mother and Child, to exchange it for a crown in heaven! Blessed superstition if these be thy effects!

But there is a real superstition which NEVER FOUND A RESTING PLACE in the Church, nor in ANY of her moderately-well-instructed PRACTICAL children, which ONLY and EXCLUSIVELY is found outside her precincts, more or less gross according to the departure from her doctrine and teaching. Solomon even became an idolator when he had left the Church of God. Outside to her, and CONDEMNED by her, there is a superstition that LEADS TO EVIL, everywhere and always at the heel of and in opposition to good, as falsehood is opposed to truth, as the shadow clings to light, and night, with its incoherent forms, follows day.

There is the superstition of spiritualism and all its kindred, and of heresy in general, from manichæism down to our latest revivals; there is that satanic superstition, witnessed in Protestant Sweden a few years ago, which forcibly breaking the ranks of the soldiers,

greedily seized upon the blood of executed MURDERERS to find a cure for diseases of men and cattle; there is an abundant crop of real superstition in this kingdom, and next to it that depraved tendency of the mind which collects the relics of criminals high and low, and blasphemes and sins under the gallows. And who would not rather join that superstition, if so you will have it, that collects and venerates the blood and relics of the MARTYRS, of men and women died in the service of their heavenly Master and for the corporal and spiritual welfare of the human race? And who would not be moved by that superstition which teaches and MAKES an assembled crowd fall down on their knees and in awful silence pray for the expiring culprit?

But there is that departure from moral perception, FOSTERED by apostates and the robbers of the Church and their descendants, which cannot distinguish the true from the false, deprecating what leads to God, and exalting what keeps and leads away from Him; that blindness which uses "superstition" without knowing what it is, unless by reference to Catholic authority.

As a striking proof I need only refer to an article of the *Saturday Review* of 14th May, 1864, on *Superstition and Education*. Though the paper is one of the cleverest of the day, yet, in the article in question, there is a development of ignorance and confusion on the subject which I have seldom or ever witnessed.

But there is even Sir John Herschel, so learned and acute in astronomical science, and, no doubt, a model to the society he moves in, whom I must class among the blind, and whose ignorance on superstition I must deplore.

"I have not heard"—he says, in *Good Words!*—"that it was necessary to liquefy the blood of St. Januarius, or to carry his bones about the streets, on account of any of these later great comets."

The introduction of this paragraph by Sir John is quite gratui-

exclaimed Dionysius the Areopagite, at the time it occurred, though distant from the spot; he believed the former and was converted to the religion of Him who then DID die.

And though nature will have it, yet fault is found and superstition it is called, when poor mortals, and great sinners moreover, are told to beware, and mind the warning of Him in whose hands they are, who carries the lightning, guides the hurricane, rules the waves, moves the heavens, and makes the earth tremble! No, I say; nay, blessed that superstition, if superstition it be, that leads the soul to God, makes the proud one humble, the libertine chaste, the angry meek, the miser give alms, the infidel to pray, the scoffer to adore, the calcumniator to retract, the over-weaning to pause, scientific conceit to see the hollowness of its mere secular knowledge; which at the contemplation of a comet—and beautiful, if true—makes an imperial sinner repent and abdicate his throne, and, another Magi of wisdom, lay down his crown at the feet of the Blessed Virgin-Mother and Child, to exchange it for a crown in heaven! Blessed superstition if these be thy effects!

But there is a real superstition which NEVER FOUND A RESTING-PLACE in the Church, nor in ANY of her moderately-well-instructed PRACTICAL children, which ONLY and EXCLUSIVELY is found outside her precincts, more or less gross according to the departure from her doctrine and teaching. Solomon even became an idolator when he had left the Church of God. Outside to her, and CONDEMNED by her, there is a superstition that LEADS TO EVIL, everywhere and always at the heel of and in opposition to good, as falsehood is opposed to truth, as the shadow clings to light, and night, with its incoherent forms, follows day.

There is the superstition of spiritualism and all its kindred, and of heresy in general, from manichæism down to our latest revivals; there is that satanic superstition, witnessed in Protestant Sweden a few years ago, which forcibly breaking the ranks of the soldiers,

greedily seized upon the blood of executed MURDERERS to find a
cure for diseases of men and cattle; there is an abundant crop of
real superstition in this kingdom, and next to it that depraved
tendency of the mind which collects the relics of criminals high
and low, and blasphemes and sins under the gallows. And who
would not rather join that superstition, if so you will have it,
that collects and venerates the blood and relics of the MARTYRS,
of men and women died in the service of their heavenly Master
and for the corporal and spiritual welfare of the human race?
And who would not be moved by that superstition which teaches
and MAKES an assembled crowd fall down on their knees and in
awful silence pray for the expiring culprit?

But there is that departure from moral perception, FOSTERED
by apostates and the robbers of the Church and their descendants,
which cannot distinguish the true from the false, deprecating
what leads to God, and exalting what keeps and leads away
from Him; that blindness which uses "superstition" without
knowing what it is, unless by reference to Catholic authority.

As a striking proof I need only refer to an article of the
Saturday Review of 14th May, 1864, on *Superstition and Education.*
Though the paper is one of the cleverest of the day, yet, in
the article in question, there is a development of ignorance and
confusion on the subject which I have seldom or ever witnessed.

But there is even Sir John Herschel, so learned and acute
in astronomical science, and, no doubt, a model to the society
he moves in, whom I must class among the blind, and whose
ignorance on superstition I must deplore.

"I have not heard"—he says, in *Good Words!*—"that it was
necessary to liquefy the blood of St. Januarius, or to carry his
bones about the streets, on account of any of these later great
comets."

The introduction of this paragraph by Sir John is quite gratui-

exclaimed Dionysius the Areopagite, at the time it occurred, though distant from the spot; he believed the former and was converted to the religion of Him who then DID die.

And though nature will have it, yet fault is found and superstition it is called, when poor mortals, and great sinners moreover, are told to beware, and mind the warning of Him in whose hands they are, who carries the lightning, guides the hurricane, rules the waves, moves the heavens, and makes the earth tremble! No, I say; nay, blessed that superstition, if superstition it be, that leads the soul to God, makes the proud one humble, the libertine chaste, the angry meek, the miser give alms, the infidel to pray, the scoffer to adore, the calcumniator to retract, the over-weaning to pause, scientific conceit to see the hollowness of its mere secular knowledge; which at the contemplation of a comet—and beautiful, if true—makes an imperial sinner repent and abdicate his throne, and, another Magi of wisdom, lay down his crown at the feet of the Blessed Virgin-Mother and Child, to exchange it for a crown in heaven! Blessed superstition if these be thy effects!

But there is a real superstition which NEVER FOUND A RESTING PLACE in the Church, nor in ANY of her moderately-well-instructed PRACTICAL children, which ONLY and EXCLUSIVELY is found outside her precincts, more or less gross according to the departure from her doctrine and teaching. Solomon even became an idolator when he had left the Church of God. Outside to her, and CONDEMNED by her, there is a superstition that LEADS TO EVIL, everywhere and always at the heel of and in opposition to good, as falsehood is opposed to truth, as the shadow clings to light, and night, with its incoherent forms, follows day.

There is the superstition of spiritualism and all its kindred, and of heresy in general, from manichæism down to our latest revivals; there is that satanic superstition, witnessed in Protestant Sweden a few years ago, which forcibly breaking the ranks of the soldiers,

greedily seized upon the blood of executed MURDERERS to find a cure for diseases of men and cattle; there is an abundant crop of real superstition in this kingdom, and next to it that depraved tendency of the mind which collects the relics of criminals high and low, and blasphemes and sins under the gallows. And who would not rather join that superstition, if so you will have it, that collects and venerates the blood and relics of the MARTYRS, of men and women died in the service of their heavenly Master and for the corporal and spiritual welfare of the human race? And who would not be moved by that superstition which teaches and MAKES an assembled crowd fall down on their knees and in awful silence pray for the expiring culprit?

But there is that departure from moral perception, FOSTERED by apostates and the robbers of the Church and their descendants, which cannot distinguish the true from the false, deprecating what leads to God, and exalting what keeps and leads away from Him; that blindness which uses "superstition" without knowing what it is, unless by reference to Catholic authority.

As a striking proof I need only refer to an article of the *Saturday Review* of 14th May, 1864, on *Superstition and Education*. Though the paper is one of the cleverest of the day, yet, in the article in question, there is a development of ignorance and confusion on the subject which I have seldom or ever witnessed.

But there is even Sir John Herschel, so learned and acute in astronomical science, and, no doubt, a model to the society he moves in, whom I must class among the blind, and whose ignorance on superstition I must deplore.

"I have not heard"—he says, in *Good Words!*—"that it was necessary to liquefy the blood of St. Januarius, or to carry his bones about the streets, on account of any of these later great comets."

The introduction of this paragraph by Sir John is quite gratui-

exclaimed Dionysius the Areopagite, at the time it occurred, though distant from the spot; he believed the former and was converted to the religion of Him who then DID die.

And though nature will have it, yet fault is found and superstition it is called, when poor mortals, and great sinners moreover, are told to beware, and mind the warning of Him in whose hands they are, who carries the lightning, guides the hurricane, rules the waves, moves the heavens, and makes the earth tremble! No, I say; nay, blessed that superstition, if superstition it be, that leads the soul to God, makes the proud one humble, the libertine chaste, the angry meek, the miser give alms, the infidel to pray, the scoffer to adore, the calcumniator to retract, the over-weaning to pause, scientific conceit to see the hollowness of its mere secular knowledge; which at the contemplation of a comet—and beautiful, if true—makes an imperial sinner repent and abdicate his throne, and, another Magi of wisdom, lay down his crown at the feet of the Blessed Virgin-Mother and Child, to exchange it for a crown in heaven! Blessed superstition if these be thy effects!

But there is a real superstition which NEVER FOUND A RESTING PLACE in the Church, nor in ANY of her moderately-well-instructed PRACTICAL children, which ONLY and EXCLUSIVELY is found outside her precincts, more or less gross according to the departure from her doctrine and teaching. Solomon even became an idolator when he had left the Church of God. Outside to her, and CONDEMNED by her, there is a superstition that LEADS TO EVIL, everywhere and always at the heel of and in opposition to good, as falsehood is opposed to truth, as the shadow clings to light, and night, with its incoherent forms, follows day.

There is the superstition of spiritualism and all its kindred, and of heresy in general, from manichæism down to our latest revivals; there is that satanic superstition, witnessed in Protestant Sweden a few years ago, which forcibly breaking the ranks of the soldiers,

greedily seized upon the blood of executed MURDERERS to find a cure for diseases of men and cattle; there is an abundant crop of real superstition in this kingdom, and next to it that depraved tendency of the mind which collects the relics of criminals high and low, and blasphemes and sins under the gallows. And who would not rather join that superstition, if so you will have it, that collects and venerates the blood and relics of the MARTYRS, of men and women died in the service of their heavenly Master and for the corporal and spiritual welfare of the human race? And who would not be moved by that superstition which teaches and MAKES an assembled crowd fall down on their knees and in awful silence pray for the expiring culprit?

But there is that departure from moral perception, FOSTERED by apostates and the robbers of the Church and their descendants, which cannot distinguish the true from the false, deprecating what leads to God, and exalting what keeps and leads away from Him; that blindness which uses "superstition" without knowing what it is, unless by reference to Catholic authority.

As a striking proof I need only refer to an article of the *Saturday Review* of 14th May, 1864, on *Superstition and Education.* Though the paper is one of the cleverest of the day, yet, in the article in question, there is a development of ignorance and confusion on the subject which I have seldom or ever witnessed.

But there is even Sir John Herschel, so learned and acute in astronomical science, and, no doubt, a model to the society he moves in, whom I must class among the blind, and whose ignorance on superstition I must deplore.

"I have not heard"—he says, in *Good Words!*—"that it was necessary to liquefy the blood of St. Januarius, or to carry his bones about the streets, on account of any of these later great comets."

The introduction of this paragraph by Sir John is quite gratui-

exclaimed Dionysius the Areopagite, at the time it occurred, though distant from the spot; he believed the former and was converted to the religion of Him who then DID die.

And though nature will have it, yet fault is found and superstition it is called, when poor mortals, and great sinners moreover, are told to beware, and mind the warning of Him in whose hands they are, who carries the lightning, guides the hurricane, rules the waves, moves the heavens, and makes the earth tremble! No, I say; nay, blessed that superstition, if superstition it be, that leads the soul to God, makes the proud one humble, the libertine chaste, the angry meek, the miser give alms, the infidel to pray, the scoffer to adore, the calcumniator to retract, the over-weaning to pause, scientific conceit to see the hollowness of its mere secular knowledge; which at the contemplation of a comet—and beautiful, if true—makes an imperial sinner repent and abdicate his throne, and, another Magi of wisdom, lay down his crown at the feet of the Blessed Virgin-Mother and Child, to exchange it for a crown in heaven! Blessed superstition if these be thy effects!

But there is a real superstition which NEVER FOUND A RESTING-PLACE in the Church, nor in ANY of her moderately-well-instructed PRACTICAL children, which ONLY and EXCLUSIVELY is found outside her precincts, more or less gross according to the departure from her doctrine and teaching. Solomon even became an idolator when he had left the Church of God. Outside to her, and CONDEMNED by her, there is a superstition that LEADS TO EVIL, everywhere and always at the heel of and in opposition to good, as falsehood is opposed to truth, as the shadow clings to light, and night, with its incoherent forms, follows day.

There is the superstition of spiritualism and all its kindred, and of heresy in general, from manichæism down to our latest revivals; there is that satanic superstition, witnessed in Protestant Sweden a few years ago, which forcibly breaking the ranks of the soldiers,

greedily seized upon the blood of executed MURDERERS to find a cure for diseases of men and cattle; there is an abundant crop of real superstition in this kingdom, and next to it that depraved tendency of the mind which collects the relics of criminals high and low, and blasphemes and sins under the gallows. And who would not rather join that superstition, if so you will have it, that collects and venerates the blood and relics of the MARTYRS, of men and women died in the service of their heavenly Master and for the corporal and spiritual welfare of the human race? And who would not be moved by that superstition which teaches and MAKES an assembled crowd fall down on their knees and in awful silence pray for the expiring culprit?

But there is that departure from moral perception, FOSTERED by apostates and the robbers of the Church and their descendants, which cannot distinguish the true from the false, deprecating what leads to God, and exalting what keeps and leads away from Him; that blindness which uses "superstition" without knowing what it is, unless by reference to Catholic authority.

As a striking proof I need only refer to an article of the *Saturday Review* of 14th May, 1864, on *Superstition and Education.* Though the paper is one of the cleverest of the day, yet, in the article in question, there is a development of ignorance and confusion on the subject which I have seldom or ever witnessed.

But there is even Sir John Herschel, so learned and acute in astronomical science, and, no doubt, a model to the society he moves in, whom I must class among the blind, and whose ignorance on superstition I must deplore.

"I have not heard"—he says, in *Good Words!*—"that it was necessary to liquefy the blood of St. Januarius, or to carry his bones about the streets, on account of any of these later great comets."

The introduction of this paragraph by Sir John is quite gratui-

exclaimed Dionysius the Areopagite, at the time it occurred, though distant from the spot; he believed the former and was converted to the religion of Him who then DID die.

And though nature will have it, yet fault is found and superstition it is called, when poor mortals, and great sinners moreover, are told to beware, and mind the warning of Him in whose hands they are, who carries the lightning, guides the hurricane, rules the waves, moves the heavens, and makes the earth tremble! No, I say; nay, blessed that superstition, if superstition it be, that leads the soul to God, makes the proud one humble, the libertine chaste, the angry meek, the miser give alms, the infidel to pray, the scoffer to adore, the calcumniator to retract, the over-weaning to pause, scientific conceit to see the hollowness of its mere secular knowledge; which at the contemplation of a comet—and beautiful, if true—makes an imperial sinner repent and abdicate his throne, and, another Magi of wisdom, lay down his crown at the feet of the Blessed Virgin-Mother and Child, to exchange it for a crown in heaven! Blessed superstition if these be thy effects!

But there is a real superstition which NEVER FOUND A RESTING PLACE in the Church, nor in ANY of her moderately-well-instructed PRACTICAL children, which ONLY and EXCLUSIVELY is found outside her precincts, more or less gross according to the departure from her doctrine and teaching. Solomon even became an idolator when he had left the Church of God. Outside to her, and CONDEMNED by her, there is a superstition that LEADS TO EVIL, everywhere and always at the heel of and in opposition to good, as falsehood is opposed to truth, as the shadow clings to light, and night, with its incoherent forms, follows day.

There is the superstition of spiritualism and all its kindred, and of heresy in general, from manichæism down to our latest revivals; there is that satanic superstition, witnessed in Protestant Sweden a few years ago, which forcibly breaking the ranks of the soldiers,

greedily seized upon the blood of executed MURDERERS to find a cure for diseases of men and cattle; there is an abundant crop of real superstition in this kingdom, and next to it that depraved tendency of the mind which collects the relics of criminals high and low, and blasphemes and sins under the gallows. And who would not rather join that superstition, if so you will have it, that collects and venerates the blood and relics of the MARTYRS, of men and women died in the service of their heavenly Master and for the corporal and spiritual welfare of the human race? And who would not be moved by that superstition which teaches and MAKES an assembled crowd fall down on their knees and in awful silence pray for the expiring culprit?

But there is that departure from moral perception, FOSTERED by apostates and the robbers of the Church and their descendants, which cannot distinguish the true from the false, deprecating what leads to God, and exalting what keeps and leads away from Him; that blindness which uses "superstition" without knowing what it is, unless by reference to Catholic authority.

As a striking proof I need only refer to an article of the *Saturday Review* of 14th May, 1864, on *Superstition and Education.* Though the paper is one of the cleverest of the day, yet, in the article in question, there is a development of ignorance and confusion on the subject which I have seldom or ever witnessed.

But there is even Sir John Herschel, so learned and acute in astronomical science, and, no doubt, a model to the society he moves in, whom I must class among the blind, and whose ignorance on superstition I must deplore.

"I have not heard"—he says, in *Good Words!*—"that it was necessary to liquefy the blood of St. Januarius, or to carry his bones about the streets, on account of any of these later great comets."

The introduction of this paragraph by Sir John is quite gratui-

exclaimed Dionysius the Areopagite, at the time it occurred, though distant from the spot; he believed the former and was converted to the religion of Him who then DID die.

And though nature will have it, yet fault is found and super-stition it is called, when poor mortals, and great sinners moreover, are told to beware, and mind the warning of Him in whose hands they are, who carries the lightning, guides the hurricane, rules the waves, moves the heavens, and makes the earth tremble! No, I say; nay, blessed that superstition, if superstition it be, that leads the soul to God, makes the proud one humble, the libertine chaste, the angry meek, the miser give alms, the infidel to pray, the scoffer to adore, the calcumniator to retract, the over-weaning to pause, scientific conceit to see the hollowness of its mere secular knowledge; which at the contemplation of a comet—and beauti-ful, if true—makes an imperial sinner repent and abdicate his throne, and, another Magi of wisdom, lay down his crown at the feet of the Blessed Virgin-Mother and Child, to exchange it for a crown in heaven! Blessed superstition if these be thy effects!

But there is a real superstition which NEVER FOUND A RESTING PLACE in the Church, nor in ANY of her moderately-well-instructed PRACTICAL children, which ONLY and EXCLUSIVELY is found outside her precincts, more or less gross according to the depar-ture from her doctrine and teaching. Solomon even became an idolator when he had left the Church of God. Outside to her, and CONDEMNED by her, there is a superstition that LEADS TO EVIL, everywhere and always at the heel of and in opposition to good, as falsehood is opposed to truth, as the shadow clings to light, and night, with its incoherent forms, follows day.

There is the superstition of spiritualism and all its kindred, and of heresy in general, from manichæism down to our latest revivals; there is that satanic superstition, witnessed in Protestant Sweden a few years ago, which forcibly breaking the ranks of the soldiers,

greedily seized upon the blood of executed MURDERERS to find a cure for diseases of men and cattle; there is an abundant crop of real superstition in this kingdom, and next to it that depraved tendency of the mind which collects the relics of criminals high and low, and blasphemes and sins under the gallows. And who would not rather join that superstition, if so you will have it, that collects and venerates the blood and relics of the MARTYRS, of men and women died in the service of their heavenly Master and for the corporal and spiritual welfare of the human race? And who would not be moved by that superstition which teaches and MAKES an assembled crowd fall down on their knees and in awful silence pray for the expiring culprit?

But there is that departure from moral perception, FOSTERED by apostates and the robbers of the Church and their descendants, which cannot distinguish the true from the false, deprecating what leads to God, and exalting what keeps and leads away from Him; that blindness which uses "superstition" without knowing what it is, unless by reference to Catholic authority.

As a striking proof I need only refer to an article of the *Saturday Review* of 14th May, 1864, on *Superstition and Education.* Though the paper is one of the cleverest of the day, yet, in the article in question, there is a development of ignorance and confusion on the subject which I have seldom or ever witnessed.

But there is even Sir John Herschel, so learned and acute in astronomical science, and, no doubt, a model to the society he moves in, whom I must class among the blind, and whose ignorance on superstition I must deplore.

"I have not heard"—he says, in *Good Words!*—" that it was necessary to liquefy the blood of St. Januarius, or to carry his bones about the streets, on account of any of these later great comets."

The introduction of this paragraph by Sir John is quite gratui-

exclaimed Dionysius the Areopagite, at the time it occurred, though distant from the spot; he believed the former and was converted to the religion of Him who then DID die.

And though nature will have it, yet fault is found and superstition it is called, when poor mortals, and great sinners moreover, are told to beware, and mind the warning of Him in whose hands they are, who carries the lightning, guides the hurricane, rules the waves, moves the heavens, and makes the earth tremble! No, I say; nay, blessed that superstition, if superstition it be, that leads the soul to God, makes the proud one humble, the libertine chaste, the angry meek, the miser give alms, the infidel to pray, the scoffer to adore, the calcumniator to retract, the over-weaning to pause, scientific conceit to see the hollowness of its mere secular knowledge; which at the contemplation of a comet—and beautiful, if true—makes an imperial sinner repent and abdicate his throne, and, another Magi of wisdom, lay down his crown at the feet of the Blessed Virgin-Mother and Child, to exchange it for a crown in heaven! Blessed superstition if these be thy effects!

But there is a real superstition which NEVER FOUND A RESTING PLACE in the Church, nor in ANY of her moderately-well-instructed PRACTICAL children, which ONLY and EXCLUSIVELY is found outside her precincts, more or less gross according to the departure from her doctrine and teaching. Solomon even became an idolator when he had left the Church of God. Outside to her, and CONDEMNED by her, there is a superstition that LEADS TO EVIL, everywhere and always at the heel of and in opposition to good, as falsehood is opposed to truth, as the shadow clings to light, and night, with its incoherent forms, follows day.

There is the superstition of spiritualism and all its kindred, and of heresy in general, from manichæism down to our latest revivals; there is that satanic superstition, witnessed in Protestant Sweden a few years ago, which forcibly breaking the ranks of the soldiers,

greedily seized upon the blood of executed MURDERERS to find a cure for diseases of men and cattle; there is an abundant crop of real superstition in this kingdom, and next to it that depraved tendency of the mind which collects the relics of criminals high and low, and blasphemes and sins under the gallows. And who would not rather join that superstition, if so you will have it, that collects and venerates the blood and relics of the MARTYRS, of men and women died in the service of their heavenly Master and for the corporal and spiritual welfare of the human race? And who would not be moved by that superstition which teaches and MAKES an assembled crowd fall down on their knees and in awful silence pray for the expiring culprit?

But there is that departure from moral perception, FOSTERED by apostates and the robbers of the Church and their descendants, which cannot distinguish the true from the false, deprecating what leads to God, and exalting what keeps and leads away from Him; that blindness which uses "superstition" without knowing what it is, unless by reference to Catholic authority.

As a striking proof I need only refer to an article of the *Saturday Review* of 14th May, 1864, on *Superstition and Education.* Though the paper is one of the cleverest of the day, yet, in the article in question, there is a development of ignorance and confusion on the subject which I have seldom or ever witnessed.

But there is even Sir John Herschel, so learned and acute in astronomical science, and, no doubt, a model to the society he moves in, whom I must class among the blind, and whose ignorance on superstition I must deplore.

"I have not heard"—he says, in *Good Words!*—"that it was necessary to liquefy the blood of St. Januarius, or to carry his bones about the streets, on account of any of these later great comets."

The introduction of this paragraph by Sir John is quite gratui-

exclaimed Dionysius the Areopagite, at the time it occurred, though distant from the spot; he believed the former and was converted to the religion of Him who then DID die.

And though nature will have it, yet fault is found and superstition it is called, when poor mortals, and great sinners moreover, are told to beware, and mind the warning of Him in whose hands they are, who carries the lightning, guides the hurricane, rules the waves, moves the heavens, and makes the earth tremble! No, I say; nay, blessed that superstition, if superstition it be, that leads the soul to God, makes the proud one humble, the libertine chaste, the angry meek, the miser give alms, the infidel to pray, the scoffer to adore, the calumniator to retract, the over-weaning to pause, scientific conceit to see the hollowness of its mere secular knowledge; which at the contemplation of a comet—and beautiful, if true—makes an imperial sinner repent and abdicate his throne, and, another Magi of wisdom, lay down his crown at the feet of the Blessed Virgin-Mother and Child, to exchange it for a crown in heaven! Blessed superstition if these be thy effects!

But there is a real superstition which NEVER FOUND A RESTING PLACE in the Church, nor in ANY of her moderately-well-instructed PRACTICAL children, which ONLY and EXCLUSIVELY is found outside her precincts, more or less gross according to the departure from her doctrine and teaching. Solomon even became an idolator when he had left the Church of God. Outside to her, and CONDEMNED by her, there is a superstition that LEADS TO EVIL, everywhere and always at the heel of and in opposition to good, as falsehood is opposed to truth, as the shadow clings to light, and night, with its incoherent forms, follows day.

There is the superstition of spiritualism and all its kindred, and of heresy in general, from manichæism down to our latest revivals; there is that satanic superstition, witnessed in Protestant Sweden a few years ago, which forcibly breaking the ranks of the soldiers,

greedily seized upon the blood of executed MURDERERS to find a cure for diseases of men and cattle; there is an abundant crop of real superstition in this kingdom, and next to it that depraved tendency of the mind which collects the relics of criminals high and low, and blasphemes and sins under the gallows. And who would not rather join that superstition, if so you will have it, that collects and venerates the blood and relics of the MARTYRS, of men and women died in the service of their heavenly Master and for the corporal and spiritual welfare of the human race? And who would not be moved by that superstition which teaches and MAKES an assembled crowd fall down on their knees and in awful silence pray for the expiring culprit?

But there is that departure from moral perception, FOSTERED by apostates and the robbers of the Church and their descendants, which cannot distinguish the true from the false, deprecating what leads to God, and exalting what keeps and leads away from Him; that blindness which uses "superstition" without knowing what it is, unless by reference to Catholic authority.

As a striking proof I need only refer to an article of the *Saturday Review* of 14th May, 1864, on *Superstition and Education.* Though the paper is one of the cleverest of the day, yet, in the article in question, there is a development of ignorance and confusion on the subject which I have seldom or ever witnessed.

But there is even Sir John Herschel, so learned and acute in astronomical science, and, no doubt, a model to the society he moves in, whom I must class among the blind, and whose ignorance on superstition I must deplore.

"I have not heard"—he says, in *Good Words*!—"that it was necessary to liquefy the blood of St. Januarius, or to carry his bones about the streets, on account of any of these later great comets."

The introduction of this paragraph by Sir John is quite gratui-

exclaimed Dionysius the Areopagite, at the time it occurred, though distant from the spot ; he believed the former and was converted to the religion of Him who then DID die.

And though nature will have it, yet fault is found and superstition it is called, when poor mortals, and great sinners moreover, are told to beware, and mind the warning of Him in whose hands they are, who carries the lightning, guides the hurricane, rules the waves, moves the heavens, and makes the earth tremble! No, I say ; nay, blessed that superstition, if superstition it be, that leads the soul to God, makes the proud one humble, the libertine chaste, the angry meek, the miser give alms, the infidel to pray, the scoffer to adore, the calcumniator to retract, the over-weaning to pause, scientific conceit to see the hollowness of its mere secular knowledge ; which at the contemplation of a comet—and beautiful, if true—makes an imperial sinner repent and abdicate his throne, and, another Magi of wisdom, lay down his crown at the feet of the Blessed Virgin-Mother and Child, to exchange it for a crown in heaven! Blessed superstition if these be thy effects!

But there is a real superstition which NEVER FOUND A RESTING PLACE in the Church, nor in ANY of her moderately-well-instructed PRACTICAL children, which ONLY and EXCLUSIVELY is found outside her precincts, more or less gross according to the departure from her doctrine and teaching. Solomon even became an idolator when he had left the Church of God. Outside to her, and CONDEMNED by her, there is a superstition that LEADS TO EVIL, everywhere and always at the heel of and in opposition to good, as falsehood is opposed to truth, as the shadow clings to light, and night, with its incoherent forms, follows day.

There is the superstition of spiritualism and all its kindred, and of heresy in general, from manichæism down to our latest revivals ; there is that satanic superstition, witnessed in Protestant Sweden a few years ago, which forcibly breaking the ranks of the soldiers,

greedily seized upon the blood of executed MURDERERS to find a cure for diseases of men and cattle; there is an abundant crop of real superstition in this kingdom, and next to it that depraved tendency of the mind which collects the relics of criminals high and low, and blasphemes and sins under the gallows. And who would not rather join that superstition, if so you will have it, that collects and venerates the blood and relics of the MARTYRS, of men and women died in the service of their heavenly Master and for the corporal and spiritual welfare of the human race? And who would not be moved by that superstition which teaches and MAKES an assembled crowd fall down on their knees and in awful silence pray for the expiring culprit?

But there is that departure from moral perception, FOSTERED by apostates and the robbers of the Church and their descendants, which cannot distinguish the true from the false, deprecating what leads to God, and exalting what keeps and leads away from Him; that blindness which uses "superstition" without knowing what it is, unless by reference to Catholic authority.

As a striking proof I need only refer to an article of the *Saturday Review* of 14th May, 1864, on *Superstition and Education.* Though the paper is one of the cleverest of the day, yet, in the article in question, there is a development of ignorance and confusion on the subject which I have seldom or ever witnessed.

But there is even Sir John Herschel, so learned and acute in astronomical science, and, no doubt, a model to the society he moves in, whom I must class among the blind, and whose ignorance on superstition I must deplore.

"I have not heard"—he says, in *Good Words!*—"that it was necessary to liquefy the blood of St. Januarius, or to carry his bones about the streets, on account of any of these later great comets."

The introduction of this paragraph by Sir John is quite gratui-

exclaimed Dionysius the Areopagite, at the time it occurred, though distant from the spot; he believed the former and was converted to the religion of Him who then DID die.

And though nature will have it, yet fault is found and superstition it is called, when poor mortals, and great sinners moreover, are told to beware, and mind the warning of Him in whose hands they are, who carries the lightning, guides the hurricane, rules the waves, moves the heavens, and makes the earth tremble! No, I say; nay, blessed that superstition, if superstition it be, that leads the soul to God, makes the proud one humble, the libertine chaste, the angry meek, the miser give alms, the infidel to pray, the scoffer to adore, the calcumniator to retract, the over-weaning to pause, scientific conceit to see the hollowness of its mere secular knowledge; which at the contemplation of a comet—and beautiful, if true—makes an imperial sinner repent and abdicate his throne, and, another Magi of wisdom, lay down his crown at the feet of the Blessed Virgin-Mother and Child, to exchange it for a crown in heaven! Blessed superstition if these be thy effects!

But there is a real superstition which NEVER FOUND A RESTING-PLACE in the Church, nor in ANY of her moderately-well-instructed PRACTICAL children, which ONLY and EXCLUSIVELY is found outside her precincts, more or less gross according to the departure from her doctrine and teaching. Solomon even became an idolator when he had left the Church of God. Outside to her, and CONDEMNED by her, there is a superstition that LEADS TO EVIL, everywhere and always at the heel of and in opposition to good, as falsehood is opposed to truth, as the shadow clings to light, and night, with its incoherent forms, follows day.

There is the superstition of spiritualism and all its kindred, and of heresy in general, from manichæism down to our latest revivals; there is that satanic superstition, witnessed in Protestant Sweden a few years ago, which forcibly breaking the ranks of the soldiers,

greedily seized upon the blood of executed MURDERERS to find a cure for diseases of men and cattle; there is an abundant crop of real superstition in this kingdom, and next to it that depraved tendency of the mind which collects the relics of criminals high and low, and blasphemes and sins under the gallows. And who would not rather join that superstition, if so you will have it, that collects and venerates the blood and relics of the MARTYRS, of men and women died in the service of their heavenly Master and for the corporal and spiritual welfare of the human race? And who would not be moved by that superstition which teaches and MAKES an assembled crowd fall down on their knees and in awful silence pray for the expiring culprit?

But there is that departure from moral perception, FOSTERED by apostates and the robbers of the Church and their descendants, which cannot distinguish the true from the false, deprecating what leads to God, and exalting what keeps and leads away from Him; that blindness which uses "superstition" without knowing what it is, unless by reference to Catholic authority.

As a striking proof I need only refer to an article of the *Saturday Review* of 14th May, 1864, on *Superstition and Education*. Though the paper is one of the cleverest of the day, yet, in the article in question, there is a development of ignorance and confusion on the subject which I have seldom or ever witnessed.

But there is even Sir John Herschel, so learned and acute in astronomical science, and, no doubt, a model to the society he moves in, whom I must class among the blind, and whose ignorance on superstition I must deplore.

"I have not heard"—he says, in *Good Words*!—"that it was necessary to liquefy the blood of St. Januarius, or to carry his bones about the streets, on account of any of these later great comets."

The introduction of this paragraph by Sir John is quite gratui-

exclaimed Dionysius the Areopagite, at the time it occurred, though distant from the spot ; he believed the former and was converted to the religion of Him who then DID die.

And though nature will have it, yet fault is found and super-stition it is called, when poor mortals, and great sinners moreover, are told to beware, and mind the warning of Him in whose hands they are, who carries the lightning, guides the hurricane, rules the waves, moves the heavens, and makes the earth tremble ! No, I say ; nay, blessed that superstition, if superstition it be, that leads the soul to God, makes the proud one humble, the libertine chaste, the angry meek, the miser give alms, the infidel to pray, the scoffer to adore, the calcumniator to retract, the over-weaning to pause, scientific conceit to see the hollowness of its mere secular knowledge ; which at the contemplation of a comet—and beauti-ful, if true—makes an imperial sinner repent and abdicate his throne, and, another Magi of wisdom, lay down his crown at the feet of the Blessed Virgin-Mother and Child, to exchange it for a crown in heaven ! Blessed superstition if these be thy effects !

But there is a real superstition which NEVER FOUND A RESTING PLACE in the Church, nor in ANY of her moderately-well-instructed PRACTICAL children, which ONLY and EXCLUSIVELY is found outside her precincts, more or less gross according to the depar-ture from her doctrine and teaching. Solomon even became an idolator when he had left the Church of God. Outside to her, and CONDEMNED by her, there is a superstition that LEADS TO EVIL, everywhere and always at the heel of and in opposition to good, as falsehood is opposed to truth, as the shadow clings to light, and night, with its incoherent forms, follows day.

There is the superstition of spiritualism and all its kindred, and of heresy in general, from manichæism down to our latest revivals ; there is that satanic superstition, witnessed in Protestant Sweden a few years ago, which forcibly breaking the ranks of the soldiers,

greedily seized upon the blood of executed MURDERERS to find a cure for diseases of men and cattle; there is an abundant crop of real superstition in this kingdom, and next to it that depraved tendency of the mind which collects the relics of criminals high and low, and blasphemes and sins under the gallows. And who would not rather join that superstition, if so you will have it, that collects and venerates the blood and relics of the MARTYRS, of men and women died in the service of their heavenly Master and for the corporal and spiritual welfare of the human race? And who would not be moved by that superstition which teaches and MAKES an assembled crowd fall down on their knees and in awful silence pray for the expiring culprit?

But there is that departure from moral perception, FOSTERED by apostates and the robbers of the Church and their descendants, which cannot distinguish the true from the false, deprecating what leads to God, and exalting what keeps and leads away from Him; that blindness which uses "superstition" without knowing what it is, unless by reference to Catholic authority.

As a striking proof I need only refer to an article of the *Saturday Review* of 14th May, 1864, on *Superstition and Education.* Though the paper is one of the cleverest of the day, yet, in the article in question, there is a development of ignorance and confusion on the subject which I have seldom or ever witnessed.

But there is even Sir John Herschel, so learned and acute in astronomical science, and, no doubt, a model to the society he moves in, whom I must class among the blind, and whose ignorance on superstition I must deplore.

"I have not heard"—he says, in *Good Words!*—"that it was necessary to liquefy the blood of St. Januarius, or to carry his bones about the streets, on account of any of these later great comets."

The introduction of this paragraph by Sir John is quite gratui-

exclaimed Dionysius the Areopagite, at the time it occurred, though distant from the spot; he believed the former and was converted to the religion of Him who then DID die.

And though nature will have it, yet fault is found and superstition it is called, when poor mortals, and great sinners moreover, are told to beware, and mind the warning of Him in whose hands they are, who carries the lightning, guides the hurricane, rules the waves, moves the heavens, and makes the earth tremble! No, I say; nay, blessed that superstition, if superstition it be, that leads the soul to God, makes the proud one humble, the libertine chaste, the angry meek, the miser give alms, the infidel to pray, the scoffer to adore, the calcumniator to retract, the over-weaning to pause, scientific conceit to see the hollowness of its mere secular knowledge; which at the contemplation of a comet—and beautiful, if true—makes an imperial sinner repent and abdicate his throne, and, another Magi of wisdom, lay down his crown at the feet of the Blessed Virgin-Mother and Child, to exchange it for a crown in heaven! Blessed superstition if these be thy effects!

But there is a real superstition which NEVER FOUND A RESTING PLACE in the Church, nor in ANY of her moderately-well-instructed PRACTICAL children, which ONLY and EXCLUSIVELY is found outside her precincts, more or less gross according to the departure from her doctrine and teaching. Solomon even became an idolator when he had left the Church of God. Outside to her, and CONDEMNED by her, there is a superstition that LEADS TO EVIL, everywhere and always at the heel of and in opposition to good, as falsehood is opposed to truth, as the shadow clings to light, and night, with its incoherent forms, follows day.

There is the superstition of spiritualism and all its kindred, and of heresy in general, from manichæism down to our latest revivals; there is that satanic superstition, witnessed in Protestant Sweden a few years ago, which forcibly breaking the ranks of the soldiers,

greedily seized upon the blood of executed MURDERERS to find a cure for diseases of men and cattle; there is an abundant crop of real superstition in this kingdom, and next to it that depraved tendency of the mind which collects the relics of criminals high and low, and blasphemes and sins under the gallows. And who would not rather join that superstition, if so you will have it, that collects and venerates the blood and relics of the MARTYRS, of men and women died in the service of their heavenly Master and for the corporal and spiritual welfare of the human race? And who would not be moved by that superstition which teaches and MAKES an assembled crowd fall down on their knees and in awful silence pray for the expiring culprit?

But there is that departure from moral perception, FOSTERED by apostates and the robbers of the Church and their descendants, which cannot distinguish the true from the false, deprecating what leads to God, and exalting what keeps and leads away from Him; that blindness which uses "superstition" without knowing what it is, unless by reference to Catholic authority.

As a striking proof I need only refer to an article of the *Saturday Review* of 14th May, 1864, on *Superstition and Education.* Though the paper is one of the cleverest of the day, yet, in the article in question, there is a development of ignorance and confusion on the subject which I have seldom or ever witnessed.

But there is even Sir John Herschel, so learned and acute in astronomical science, and, no doubt, a model to the society he moves in, whom I must class among the blind, and whose ignorance on superstition I must deplore.

"I have not heard"—he says, in *Good Words!*—"that it was necessary to liquefy the blood of St. Januarius, or to carry his bones about the streets, on account of any of these later great comets."

The introduction of this paragraph by Sir John is quite gratui-

exclaimed Dionysius the Areopagite, at the time it occurred, though
distant from the spot; he believed the former and was converted
to the religion of Him who then DID die.

And though nature will have it, yet fault is found and superstition it is called, when poor mortals, and great sinners moreover,
are told to beware, and mind the warning of Him in whose hands
they are, who carries the lightning, guides the hurricane, rules
the waves, moves the heavens, and makes the earth tremble! No,
I say; nay, blessed that superstition, if superstition it be, that
leads the soul to God, makes the proud one humble, the libertine
chaste, the angry meek, the miser give alms, the infidel to pray,
the scoffer to adore, the calcumniator to retract, the over-weaning
to pause, scientific conceit to see the hollowness of its mere secular
knowledge; which at the contemplation of a comet—and beautiful, if true—makes an imperial sinner repent and abdicate his
throne, and, another Magi of wisdom, lay down his crown at the
feet of the Blessed Virgin-Mother and Child, to exchange it for a
crown in heaven! Blessed superstition if these be thy effects!

But there is a real superstition which NEVER FOUND A RESTING
PLACE in the Church, nor in ANY of her moderately-well-instructed
PRACTICAL children, which ONLY and EXCLUSIVELY is found
outside her precincts, more or less gross according to the departure from her doctrine and teaching. Solomon even became an
idolator when he had left the Church of God. Outside to her,
and CONDEMNED by her, there is a superstition that LEADS TO EVIL,
everywhere and always at the heel of and in opposition to good,
as falsehood is opposed to truth, as the shadow clings to light,
and night, with its incoherent forms, follows day.

There is the superstition of spiritualism and all its kindred, and
of heresy in general, from manichæism down to our latest revivals;
there is that satanic superstition, witnessed in Protestant Sweden
a few years ago, which forcibly breaking the ranks of the soldiers,

greedily seized upon the blood of executed MURDERERS to find a cure for diseases of men and cattle; there is an abundant crop of real superstition in this kingdom, and next to it that depraved tendency of the mind which collects the relics of criminals high and low, and blasphemes and sins under the gallows. And who would not rather join that superstition, if so you will have it, that collects and venerates the blood and relics of the MARTYRS, of men and women died in the service of their heavenly Master and for the corporal and spiritual welfare of the human race? And who would not be moved by that superstition which teaches and MAKES an assembled crowd fall down on their knees and in awful silence pray for the expiring culprit?

But there is that departure from moral perception, FOSTERED by apostates and the robbers of the Church and their descendants, which cannot distinguish the true from the false, deprecating what leads to God, and exalting what keeps and leads away from Him; that blindness which uses "superstition" without knowing what it is, unless by reference to Catholic authority.

As a striking proof I need only refer to an article of the *Saturday Review* of 14th May, 1864, on *Superstition and Education.* Though the paper is one of the cleverest of the day, yet, in the article in question, there is a development of ignorance and confusion on the subject which I have seldom or ever witnessed.

But there is even Sir John Herschel, so learned and acute in astronomical science, and, no doubt, a model to the society he moves in, whom I must class among the blind, and whose ignorance on superstition I must deplore.

"I have not heard"—he says, in *Good Words!*—"that it was necessary to liquefy the blood of St. Januarius, or to carry his bones about the streets, on account of any of these later great comets."

The introduction of this paragraph by Sir John is quite gratui-

exclaimed Dionysius the Areopagite, at the time it occurred, though distant from the spot; he believed the former and was converted to the religion of Him who then DID die.

And though nature will have it, yet fault is found and superstition it is called, when poor mortals, and great sinners moreover, are told to beware, and mind the warning of Him in whose hands they are, who carries the lightning, guides the hurricane, rules the waves, moves the heavens, and makes the earth tremble! No, I say; nay, blessed that superstition, if superstition it be, that leads the soul to God, makes the proud one humble, the libertine chaste, the angry meek, the miser give alms, the infidel to pray, the scoffer to adore, the calumniator to retract, the over-weaning to pause, scientific conceit to see the hollowness of its mere secular knowledge; which at the contemplation of a comet—and beautiful, if true—makes an imperial sinner repent and abdicate his throne, and, another Magi of wisdom, lay down his crown at the feet of the Blessed Virgin-Mother and Child, to exchange it for a crown in heaven! Blessed superstition if these be thy effects!

But there is a real superstition which NEVER FOUND A RESTING PLACE in the Church, nor in ANY of her moderately-well-instructed PRACTICAL children, which ONLY and EXCLUSIVELY is found outside her precincts, more or less gross according to the departure from her doctrine and teaching. Solomon even became an idolator when he had left the Church of God. Outside to her, and CONDEMNED by her, there is a superstition that LEADS TO EVIL, everywhere and always at the heel of and in opposition to good, as falsehood is opposed to truth, as the shadow clings to light, and night, with its incoherent forms, follows day.

There is the superstition of spiritualism and all its kindred, and of heresy in general, from manichæism down to our latest revivals; there is that satanic superstition, witnessed in Protestant Sweden a few years ago, which forcibly breaking the ranks of the soldiers,

greedily seized upon the blood of executed MURDERERS to find a cure for diseases of men and cattle; there is an abundant crop of real superstition in this kingdom, and next to it that depraved tendency of the mind which collects the relics of criminals high and low, and blasphemes and sins under the gallows. And who would not rather join that superstition, if so you will have it, that collects and venerates the blood and relics of the MARTYRS, of men and women died in the service of their heavenly Master and for the corporal and spiritual welfare of the human race? And who would not be moved by that superstition which teaches and MAKES an assembled crowd fall down on their knees and in awful silence pray for the expiring culprit?

But there is that departure from moral perception, FOSTERED by apostates and the robbers of the Church and their descendants, which cannot distinguish the true from the false, deprecating what leads to God, and exalting what keeps and leads away from Him; that blindness which uses "superstition" without knowing what it is, unless by reference to Catholic authority.

As a striking proof I need only refer to an article of the *Saturday Review* of 14th May, 1864, on *Superstition and Education.* Though the paper is one of the cleverest of the day, yet, in the article in question, there is a development of ignorance and confusion on the subject which I have seldom or ever witnessed.

But there is even Sir John Herschel, so learned and acute in astronomical science, and, no doubt, a model to the society he moves in, whom I must class among the blind, and whose ignorance on superstition I must deplore.

"I have not heard"—he says, in *Good Words!*—"that it was necessary to liquefy the blood of St. Januarius, or to carry his bones about the streets, on account of any of these later great comets."

The introduction of this paragraph by Sir John is quite gratui-

exclaimed Dionysius the Areopagite, at the time it occurred, though distant from the spot ; he believed the former and was converted to the religion of Him who then DID die.

And though nature will have it, yet fault is found and superstition it is called, when poor mortals, and great sinners moreover, are told to beware, and mind the warning of Him in whose hands they are, who carries the lightning, guides the hurricane, rules the waves, moves the heavens, and makes the earth tremble ! No, I say ; nay, blessed that superstition, if superstition it be, that leads the soul to God, makes the proud one humble, the libertine chaste, the angry meek, the miser give alms, the infidel to pray, the scoffer to adore, the calcumniator to retract, the over-weaning to pause, scientific conceit to see the hollowness of its mere secular knowledge ; which at the contemplation of a comet—and beautiful, if true—makes an imperial sinner repent and abdicate his throne, and, another Magi of wisdom, lay down his crown at the feet of the Blessed Virgin-Mother and Child, to exchange it for a crown in heaven ! Blessed superstition if these be thy effects !

But there is a real superstition which NEVER FOUND A RESTING PLACE in the Church, nor in ANY of her moderately-well-instructed PRACTICAL children, which ONLY and EXCLUSIVELY is found outside her precincts, more or less gross according to the departure from her doctrine and teaching. Solomon even became an idolator when he had left the Church of God. Outside to her, and CONDEMNED by her, there is a superstition that LEADS TO EVIL, everywhere and always at the heel of and in opposition to good, as falsehood is opposed to truth, as the shadow clings to light, and night, with its incoherent forms, follows day.

There is the superstition of spiritualism and all its kindred, and of heresy in general, from manichæism down to our latest revivals ; there is that satanic superstition, witnessed in Protestant Sweden a few years ago, which forcibly breaking the ranks of the soldiers,

greedily seized upon the blood of executed MURDERERS to find a cure for diseases of men and cattle; there is an abundant crop of real superstition in this kingdom, and next to it that depraved tendency of the mind which collects the relics of criminals high and low, and blasphemes and sins under the gallows. And who would not rather join that superstition, if so you will have it, that collects and venerates the blood and relics of the MARTYRS, of men and women died in the service of their heavenly Master and for the corporal and spiritual welfare of the human race? And who would not be moved by that superstition which teaches and MAKES an assembled crowd fall down on their knees and in awful silence pray for the expiring culprit?

But there is that departure from moral perception, FOSTERED by apostates and the robbers of the Church and their descendants, which cannot distinguish the true from the false, deprecating what leads to God, and exalting what keeps and leads away from Him; that blindness which uses "superstition" without knowing what it is, unless by reference to Catholic authority.

As a striking proof I need only refer to an article of the *Saturday Review* of 14th May, 1864, on *Superstition and Education.* Though the paper is one of the cleverest of the day, yet, in the article in question, there is a development of ignorance and confusion on the subject which I have seldom or ever witnessed.

But there is even Sir John Herschel, so learned and acute in astronomical science, and, no doubt, a model to the society he moves in, whom I must class among the blind, and whose ignorance on superstition I must deplore.

"I have not heard"—he says, in *Good Words!*—"that it was necessary to liquefy the blood of St. Januarius, or to carry his bones about the streets, on account of any of these later great comets."

The introduction of this paragraph by Sir John is quite gratui-

exclaimed Dionysius the Areopagite, at the time it occurred, though
distant from the spot ; he believed the former and was converted
to the religion of Him who then DID die.

And though nature will have it, yet fault is found and superstition it is called, when poor mortals, and great sinners moreover,
are told to beware, and mind the warning of Him in whose hands
they are, who carries the lightning, guides the hurricane, rules
the waves, moves the heavens, and makes the earth tremble ! No,
I say ; nay, blessed that superstition, if superstition it be, that
leads the soul to God, makes the proud one humble, the libertine
chaste, the angry meek, the miser give alms, the infidel to pray,
the scoffer to adore, the calcumniator to retract, the over-weaning
to pause, scientific conceit to see the hollowness of its mere secular
knowledge ; which at the contemplation of a comet—and beautiful, if true—makes an imperial sinner repent and abdicate his
throne, and, another Magi of wisdom, lay down his crown at the
feet of the Blessed Virgin-Mother and Child, to exchange it for a
crown in heaven ! Blessed superstition if these be thy effects !

But there is a real superstition which NEVER FOUND A RESTING
PLACE in the Church, nor in ANY of her moderately-well-instructed
PRACTICAL children, which ONLY and EXCLUSIVELY is found
outside her precincts, more or less gross according to the departure from her doctrine and teaching. Solomon even became an
idolator when he had left the Church of God. Outside to her,
and CONDEMNED by her, there is a superstition that LEADS TO EVIL,
everywhere and always at the heel of and in opposition to good,
as falsehood is opposed to truth, as the shadow clings to light,
and night, with its incoherent forms, follows day.

There is the superstition of spiritualism and all its kindred, and
of heresy in general, from manichæism down to our latest revivals ;
there is that satanic superstition, witnessed in Protestant Sweden
a few years ago, which forcibly breaking the ranks of the soldiers,

greedily seized upon the blood of executed MURDERERS to find a cure for diseases of men and cattle; there is an abundant crop of real superstition in this kingdom, and next to it that depraved tendency of the mind which collects the relics of criminals high and low, and blasphemes and sins under the gallows. And who would not rather join that superstition, if so you will have it, that collects and venerates the blood and relics of the MARTYRS, of men and women died in the service of their heavenly Master and for the corporal and spiritual welfare of the human race? And who would not be moved by that superstition which teaches and MAKES an assembled crowd fall down on their knees and in awful silence pray for the expiring culprit?

But there is that departure from moral perception, FOSTERED by apostates and the robbers of the Church and their descendants, which cannot distinguish the true from the false, deprecating what leads to God, and exalting what keeps and leads away from Him; that blindness which uses "superstition" without knowing what it is, unless by reference to Catholic authority.

As a striking proof I need only refer to an article of the *Saturday Review* of 14th May, 1864, on *Superstition and Education*. Though the paper is one of the cleverest of the day, yet, in the article in question, there is a development of ignorance and confusion on the subject which I have seldom or ever witnessed.

But there is even Sir John Herschel, so learned and acute in astronomical science, and, no doubt, a model to the society he moves in, whom I must class among the blind, and whose ignorance on superstition I must deplore.

"I have not heard"—he says, in *Good Words!*—"that it was necessary to liquefy the blood of St. Januarius, or to carry his bones about the streets, on account of any of these later great comets."

The introduction of this paragraph by Sir John is quite gratui-

exclaimed Dionysius the Areopagite, at the time it occurred, though
distant from the spot; he believed the former and was converted
to the religion of Him who then DID die.

And though nature will have it, yet fault is found and superstition it is called, when poor mortals, and great sinners moreover,
are told to beware, and mind the warning of Him in whose hands
they are, who carries the lightning, guides the hurricane, rules
the waves, moves the heavens, and makes the earth tremble! No,
I say; nay, blessed that superstition, if superstition it be, that
leads the soul to God, makes the proud one humble, the libertine
chaste, the angry meek, the miser give alms, the infidel to pray,
the scoffer to adore, the calcumniator to retract, the over-weaning
to pause, scientific conceit to see the hollowness of its mere secular
knowledge; which at the contemplation of a comet—and beautiful, if true—makes an imperial sinner repent and abdicate his
throne, and, another Magi of wisdom, lay down his crown at the
feet of the Blessed Virgin-Mother and Child, to exchange it for a
crown in heaven! Blessed superstition if these be thy effects!

But there is a real superstition which NEVER FOUND A RESTING
PLACE in the Church, nor in ANY of her moderately-well-instructed
PRACTICAL children, which ONLY and EXCLUSIVELY is found
outside her precincts, more or less gross according to the departure from her doctrine and teaching. Solomon even became an
idolator when he had left the Church of God. Outside to her,
and CONDEMNED by her, there is a superstition that LEADS TO EVIL,
everywhere and always at the heel of and in opposition to good,
as falsehood is opposed to truth, as the shadow clings to light,
and night, with its incoherent forms, follows day.

There is the superstition of spiritualism and all its kindred, and
of heresy in general, from manichæism down to our latest revivals;
there is that satanic superstition, witnessed in Protestant Sweden
a few years ago, which forcibly breaking the ranks of the soldiers,

greedily seized upon the blood of executed MURDERERS to find a cure for diseases of men and cattle; there is an abundant crop of real superstition in this kingdom, and next to it that depraved tendency of the mind which collects the relics of criminals high and low, and blasphemes and sins under the gallows. And who would not rather join that superstition, if so you will have it, that collects and venerates the blood and relics of the MARTYRS, of men and women died in the service of their heavenly Master and for the corporal and spiritual welfare of the human race? And who would not be moved by that superstition which teaches and MAKES an assembled crowd fall down on their knees and in awful silence pray for the expiring culprit?

But there is that departure from moral perception, FOSTERED by apostates and the robbers of the Church and their descendants, which cannot distinguish the true from the false, deprecating what leads to God, and exalting what keeps and leads away from Him; that blindness which uses "superstition" without knowing what it is, unless by reference to Catholic authority.

As a striking proof I need only refer to an article of the *Saturday Review* of 14th May, 1864, on *Superstition and Education.* Though the paper is one of the cleverest of the day, yet, in the article in question, there is a development of ignorance and confusion on the subject which I have seldom or ever witnessed.

But there is even Sir John Herschel, so learned and acute in astronomical science, and, no doubt, a model to the society he moves in, whom I must class among the blind, and whose ignorance on superstition I must deplore.

"I have not heard"—he says, in *Good Words*!—"that it was necessary to liquefy the blood of St. Januarius, or to carry his bones about the streets, on account of any of these later great comets."

The introduction of this paragraph by Sir John is quite gratui-

exclaimed Dionysius the Areopagite, at the time it occurred, though distant from the spot; he believed the former and was converted to the religion of Him who then DID die.

And though nature will have it, yet fault is found and superstition it is called, when poor mortals, and great sinners moreover, are told to beware, and mind the warning of Him in whose hands they are, who carries the lightning, guides the hurricane, rules the waves, moves the heavens, and makes the earth tremble! No, I say; nay, blessed that superstition, if superstition it be, that leads the soul to God, makes the proud one humble, the libertine chaste, the angry meek, the miser give alms, the infidel to pray, the scoffer to adore, the calcumniator to retract, the over-weaning to pause, scientific conceit to see the hollowness of its mere secular knowledge; which at the contemplation of a comet—and beautiful, if true—makes an imperial sinner repent and abdicate his throne, and, another Magi of wisdom, lay down his crown at the feet of the Blessed Virgin-Mother and Child, to exchange it for a crown in heaven! Blessed superstition if these be thy effects!

But there is a real superstition which NEVER FOUND A RESTING PLACE in the Church, nor in ANY of her moderately-well-instructed PRACTICAL children, which ONLY and EXCLUSIVELY is found outside her precincts, more or less gross according to the departure from her doctrine and teaching. Solomon even became an idolator when he had left the Church of God. Outside to her, and CONDEMNED by her, there is a superstition that LEADS TO EVIL, everywhere and always at the heel of and in opposition to good, as falsehood is opposed to truth, as the shadow clings to light, and night, with its incoherent forms, follows day.

There is the superstition of spiritualism and all its kindred, and of heresy in general, from manichæism down to our latest revivals; there is that satanic superstition, witnessed in Protestant Sweden a few years ago, which forcibly breaking the ranks of the soldiers,

greedily seized upon the blood of executed MURDERERS to find a
cure for diseases of men and cattle; there is an abundant crop of
real superstition in this kingdom, and next to it that depraved
tendency of the mind which collects the relics of criminals high
and low, and blasphemes and sins under the gallows. And who
would not rather join that superstition, if so you will have it,
that collects and venerates the blood and relics of the MARTYRS,
of men and women died in the service of their heavenly Master
and for the corporal and spiritual welfare of the human race?
And who would not be moved by that superstition which teaches
and MAKES an assembled crowd fall down on their knees and in
awful silence pray for the expiring culprit?

But there is that departure from moral perception, FOSTERED
by apostates and the robbers of the Church and their descendants,
which cannot distinguish the true from the false, deprecating
what leads to God, and exalting what keeps and leads away
from Him; that blindness which uses "superstition" without
knowing what it is, unless by reference to Catholic authority.

As a striking proof I need only refer to an article of the
Saturday Review of 14th May, 1864, on *Superstition and Education.*
Though the paper is one of the cleverest of the day, yet, in
the article in question, there is a development of ignorance and
confusion on the subject which I have seldom or ever witnessed.

But there is even Sir John Herschel, so learned and acute
in astronomical science, and, no doubt, a model to the society
he moves in, whom I must class among the blind, and whose
ignorance on superstition I must deplore.

"I have not heard"—he says, in *Good Words*!—"that it was
necessary to liquefy the blood of St. Januarius, or to carry his
bones about the streets, on account of any of these later great
comets."

The introduction of this paragraph by Sir John is quite gratui-

exclaimed Dionysius the Areopagite, at the time it occurred, though distant from the spot; he believed the former and was converted to the religion of Him who then DID die.

And though nature will have it, yet fault is found and superstition it is called, when poor mortals, and great sinners moreover, are told to beware, and mind the warning of Him in whose hands they are, who carries the lightning, guides the hurricane, rules the waves, moves the heavens, and makes the earth tremble! No, I say; nay, blessed that superstition, if superstition it be, that leads the soul to God, makes the proud one humble, the libertine chaste, the angry meek, the miser give alms, the infidel to pray, the scoffer to adore, the calcumniator to retract, the over-weaning to pause, scientific conceit to see the hollowness of its mere secular knowledge; which at the contemplation of a comet—and beautiful, if true—makes an imperial sinner repent and abdicate his throne, and, another Magi of wisdom, lay down his crown at the feet of the Blessed Virgin-Mother and Child, to exchange it for a crown in heaven! Blessed superstition if these be thy effects!

But there is a real superstition which NEVER FOUND A RESTING PLACE in the Church, nor in ANY of her moderately-well-instructed PRACTICAL children, which ONLY and EXCLUSIVELY is found outside her precincts, more or less gross according to the departure from her doctrine and teaching. Solomon even became an idolator when he had left the Church of God. Outside to her, and CONDEMNED by her, there is a superstition that LEADS TO EVIL, everywhere and always at the heel of and in opposition to good, as falsehood is opposed to truth, as the shadow clings to light, and night, with its incoherent forms, follows day.

There is the superstition of spiritualism and all its kindred, and of heresy in general, from manichæism down to our latest revivals; there is that satanic superstition, witnessed in Protestant Sweden a few years ago, which forcibly breaking the ranks of the soldiers,

greedily seized upon the blood of executed MURDERERS to find a cure for diseases of men and cattle; there is an abundant crop of real superstition in this kingdom, and next to it that depraved tendency of the mind which collects the relics of criminals high and low, and blasphemes and sins under the gallows. And who would not rather join that superstition, if so you will have it, that collects and venerates the blood and relics of the MARTYRS, of men and women died in the service of their heavenly Master and for the corporal and spiritual welfare of the human race? And who would not be moved by that superstition which teaches and MAKES an assembled crowd fall down on their knees and in awful silence pray for the expiring culprit?

But there is that departure from moral perception, FOSTERED by apostates and the robbers of the Church and their descendants, which cannot distinguish the true from the false, deprecating what leads to God, and exalting what keeps and leads away from Him; that blindness which uses "superstition" without knowing what it is, unless by reference to Catholic authority.

As a striking proof I need only refer to an article of the *Saturday Review* of 14th May, 1864, on *Superstition and Education.* Though the paper is one of the cleverest of the day, yet, in the article in question, there is a development of ignorance and confusion on the subject which I have seldom or ever witnessed.

But there is even Sir John Herschel, so learned and acute in astronomical science, and, no doubt, a model to the society he moves in, whom I must class among the blind, and whose ignorance on superstition I must deplore.

"I have not heard"—he says, in *Good Words!*—"that it was necessary to liquefy the blood of St. Januarius, or to carry his bones about the streets, on account of any of these later great comets."

The introduction of this paragraph by Sir John is quite gratui-

exclaimed Dionysius the Areopagite, at the time it occurred, though distant from the spot; he believed the former and was converted to the religion of Him who then DID die.

And though nature will have it, yet fault is found and super-stition it is called, when poor mortals, and great sinners moreover, are told to beware, and mind the warning of Him in whose hands they are, who carries the lightning, guides the hurricane, rules the waves, moves the heavens, and makes the earth tremble! No, I say; nay, blessed that superstition, if superstition it be, that leads the soul to God, makes the proud one humble, the libertine chaste, the angry meek, the miser give alms, the infidel to pray, the scoffer to adore, the calcumniator to retract, the over-weaning to pause, scientific conceit to see the hollowness of its mere secular knowledge; which at the contemplation of a comet—and beauti-ful, if true—makes an imperial sinner repent and abdicate his throne, and, another Magi of wisdom, lay down his crown at the feet of the Blessed Virgin-Mother and Child, to exchange it for a crown in heaven! Blessed superstition if these be thy effects!

But there is a real superstition which NEVER FOUND A RESTING PLACE in the Church, nor in ANY of her moderately-well-instructed PRACTICAL children, which ONLY and EXCLUSIVELY is found outside her precincts, more or less gross according to the depar-ture from her doctrine and teaching. Solomon even became an idolator when he had left the Church of God. Outside to her, and CONDEMNED by her, there is a superstition that LEADS TO EVIL, everywhere and always at the heel of and in opposition to good, as falsehood is opposed to truth, as the shadow clings to light, and night, with its incoherent forms, follows day.

There is the superstition of spiritualism and all its kindred, and of heresy in general, from manichæism down to our latest revivals; there is that satanic superstition, witnessed in Protestant Sweden a few years ago, which forcibly breaking the ranks of the soldiers,

greedily seized upon the blood of executed MURDERERS to find a
cure for diseases of men and cattle; there is an abundant crop of
real superstition in this kingdom, and next to it that depraved
tendency of the mind which collects the relics of criminals high
and low, and blasphemes and sins under the gallows. And who
would not rather join that superstition, if so you will have it,
that collects and venerates the blood and relics of the MARTYRS,
of men and women died in the service of their heavenly Master
and for the corporal and spiritual welfare of the human race?
And who would not be moved by that superstition which teaches
and MAKES an assembled crowd fall down on their knees and in
awful silence pray for the expiring culprit?

But there is that departure from moral perception, FOSTERED
by apostates and the robbers of the Church and their descendants,
which cannot distinguish the true from the false, deprecating
what leads to God, and exalting what keeps and leads away
from Him; that blindness which uses "superstition" without
knowing what it is, unless by reference to Catholic authority.

As a striking proof I need only refer to an article of the
Saturday Review of 14th May, 1864, on *Superstition and Education*.
Though the paper is one of the cleverest of the day, yet, in
the article in question, there is a development of ignorance and
confusion on the subject which I have seldom or ever witnessed.

But there is even Sir John Herschel, so learned and acute
in astronomical science, and, no doubt, a model to the society
he moves in, whom I must class among the blind, and whose
ignorance on superstition I must deplore.

" I have not heard"—he says, in *Good Words!*—" that it was
necessary to liquefy the blood of St. Januarius, or to carry his
bones about the streets, on account of any of these later great
comets."

The introduction of this paragraph by Sir John is quite gratui-

exclaimed Dionysius the Areopagite, at the time it occurred, though
distant from the spot ; he believed the former and was converted
to the religion of Him who then DID die.

And though nature will have it, yet fault is found and superstition it is called, when poor mortals, and great sinners moreover,
are told to beware, and mind the warning of Him in whose hands
they are, who carries the lightning, guides the hurricane, rules
the waves, moves the heavens, and makes the earth tremble ! No,
I say ; nay, blessed that superstition, if superstition it be, that
leads the soul to God, makes the proud one humble, the libertine
chaste, the angry meek, the miser give alms, the infidel to pray,
the scoffer to adore, the calcumniator to retract, the over-weaning
to pause, scientific conceit to see the hollowness of its mere secular
knowledge ; which at the contemplation of a comet—and beautiful, if true—makes an imperial sinner repent and abdicate his
throne, and, another Magi of wisdom, lay down his crown at the
feet of the Blessed Virgin-Mother and Child, to exchange it for a
crown in heaven ! Blessed superstition if these be thy effects !

But there is a real superstition which NEVER FOUND A RESTING-
PLACE in the Church, nor in ANY of her moderately-well-instructed
PRACTICAL children, which ONLY and EXCLUSIVELY is found
outside her precincts, more or less gross according to the departure from her doctrine and teaching. Solomon even became an
idolator when he had left the Church of God. Outside to her,
and CONDEMNED by her, there is a superstition that LEADS TO EVIL,
everywhere and always at the heel of and in opposition to good,
as falsehood is opposed to truth, as the shadow clings to light,
and night, with its incoherent forms, follows day.

There is the superstition of spiritualism and all its kindred, and
of heresy in general, from manichæism down to our latest revivals ;
there is that satanic superstition, witnessed in Protestant Sweden
a few years ago, which forcibly breaking the ranks of the soldiers,

greedily seized upon the blood of executed MURDERERS to find a cure for diseases of men and cattle; there is an abundant crop of real superstition in this kingdom, and next to it that depraved tendency of the mind which collects the relics of criminals high and low, and blasphemes and sins under the gallows. And who would not rather join that superstition, if so you will have it, that collects and venerates the blood and relics of the MARTYRS, of men and women died in the service of their heavenly Master and for the corporal and spiritual welfare of the human race? And who would not be moved by that superstition which teaches and MAKES an assembled crowd fall down on their knees and in awful silence pray for the expiring culprit?

But there is that departure from moral perception, FOSTERED by apostates and the robbers of the Church and their descendants, which cannot distinguish the true from the false, deprecating what leads to God, and exalting what keeps and leads away from Him; that blindness which uses " superstition " without knowing what it is, unless by reference to Catholic authority.

As a striking proof I need only refer to an article of the *Saturday Review* of 14th May, 1864, on *Superstition and Education.* Though the paper is one of the cleverest of the day, yet, in the article in question, there is a development of ignorance and confusion on the subject which I have seldom or ever witnessed.

But there is even Sir John Herschel, so learned and acute in astronomical science, and, no doubt, a model to the society he moves in, whom I must class among the blind, and whose ignorance on superstition I must deplore.

" I have not heard"—he says, in *Good Words!*—" that it was necessary to liquefy the blood of St. Januarius, or to carry his bones about the streets, on account of any of these later great comets."

The introduction of this paragraph by Sir John is quite gratui-

exclaimed Dionysius the Areopagite, at the time it occurred, though distant from the spot; he believed the former and was converted to the religion of Him who then DID die.

And though nature will have it, yet fault is found and superstition it is called, when poor mortals, and great sinners moreover, are told to beware, and mind the warning of Him in whose hands they are, who carries the lightning, guides the hurricane, rules the waves, moves the heavens, and makes the earth tremble! No, I say; nay, blessed that superstition, if superstition it be, that leads the soul to God, makes the proud one humble, the libertine chaste, the angry meek, the miser give alms, the infidel to pray, the scoffer to adore, the calcumniator to retract, the over-weaning to pause, scientific conceit to see the hollowness of its mere secular knowledge; which at the contemplation of a comet—and beautiful, if true—makes an imperial sinner repent and abdicate his throne, and, another Magi of wisdom, lay down his crown at the feet of the Blessed Virgin-Mother and Child, to exchange it for a crown in heaven! Blessed superstition if these be thy effects!

But there is a real superstition which NEVER FOUND A RESTING PLACE in the Church, nor in ANY of her moderately-well-instructed PRACTICAL children, which ONLY and EXCLUSIVELY is found outside her precincts, more or less gross according to the departure from her doctrine and teaching. Solomon even became an idolator when he had left the Church of God. Outside to her, and CONDEMNED by her, there is a superstition that LEADS TO EVIL, everywhere and always at the heel of and in opposition to good, as falsehood is opposed to truth, as the shadow clings to light, and night, with its incoherent forms, follows day.

There is the superstition of spiritualism and all its kindred, and of heresy in general, from manichæism down to our latest revivals; there is that satanic superstition, witnessed in Protestant Sweden a few years ago, which forcibly breaking the ranks of the soldiers,

greedily seized upon the blood of executed MURDERERS to find a cure for diseases of men and cattle; there is an abundant crop of real superstition in this kingdom, and next to it that depraved tendency of the mind which collects the relics of criminals high and low, and blasphemes and sins under the gallows. And who would not rather join that superstition, if so you will have it, that collects and venerates the blood and relics of the MARTYRS, of men and women died in the service of their heavenly Master and for the corporal and spiritual welfare of the human race? And who would not be moved by that superstition which teaches and MAKES an assembled crowd fall down on their knees and in awful silence pray for the expiring culprit?

But there is that departure from moral perception, FOSTERED by apostates and the robbers of the Church and their descendants, which cannot distinguish the true from the false, deprecating what leads to God, and exalting what keeps and leads away from Him; that blindness which uses "superstition" without knowing what it is, unless by reference to Catholic authority.

As a striking proof I need only refer to an article of the *Saturday Review* of 14th May, 1864, on *Superstition and Education.* Though the paper is one of the cleverest of the day, yet, in the article in question, there is a development of ignorance and confusion on the subject which I have seldom or ever witnessed.

But there is even Sir John Herschel, so learned and acute in astronomical science, and, no doubt, a model to the society he moves in, whom I must class among the blind, and whose ignorance on superstition I must deplore.

" I have not heard"—he says, in *Good Words*!—" that it was necessary to liquefy the blood of St. Januarius, or to carry his bones about the streets, on account of any of these later great comets."

The introduction of this paragraph by Sir John is quite gratui-

exclaimed Dionysius the Areopagite, at the time it occurred, though distant from the spot; he believed the former and was converted to the religion of Him who then DID die.

And though nature will have it, yet fault is found and superstition it is called, when poor mortals, and great sinners moreover, are told to beware, and mind the warning of Him in whose hands they are, who carries the lightning, guides the hurricane, rules the waves, moves the heavens, and makes the earth tremble! No, I say; nay, blessed that superstition, if superstition it be, that leads the soul to God, makes the proud one humble, the libertine chaste, the angry meek, the miser give alms, the infidel to pray, the scoffer to adore, the calcumniator to retract, the over-weaning to pause, scientific conceit to see the hollowness of its mere secular knowledge; which at the contemplation of a comet—and beautiful, if true—makes an imperial sinner repent and abdicate his throne, and, another Magi of wisdom, lay down his crown at the feet of the Blessed Virgin-Mother and Child, to exchange it for a crown in heaven! Blessed superstition if these be thy effects!

But there is a real superstition which NEVER FOUND A RESTING PLACE in the Church, nor in ANY of her moderately-well-instructed PRACTICAL children, which ONLY and EXCLUSIVELY is found outside her precincts, more or less gross according to the departure from her doctrine and teaching. Solomon even became an idolator when he had left the Church of God. Outside to her, and CONDEMNED by her, there is a superstition that LEADS TO EVIL, everywhere and always at the heel of and in opposition to good, as falsehood is opposed to truth, as the shadow clings to light, and night, with its incoherent forms, follows day.

There is the superstition of spiritualism and all its kindred, and of heresy in general, from manichæism down to our latest revivals; there is that satanic superstition, witnessed in Protestant Sweden a few years ago, which forcibly breaking the ranks of the soldiers,

greedily seized upon the blood of executed MURDERERS to find a cure for diseases of men and cattle; there is an abundant crop of real superstition in this kingdom, and next to it that depraved tendency of the mind which collects the relics of criminals high and low, and blasphemes and sins under the gallows. And who would not rather join that superstition, if so you will have it, that collects and venerates the blood and relics of the MARTYRS, of men and women died in the service of their heavenly Master and for the corporal and spiritual welfare of the human race? And who would not be moved by that superstition which teaches and MAKES an assembled crowd fall down on their knees and in awful silence pray for the expiring culprit?

But there is that departure from moral perception, FOSTERED by apostates and the robbers of the Church and their descendants, which cannot distinguish the true from the false, deprecating what leads to God, and exalting what keeps and leads away from Him; that blindness which uses "superstition" without knowing what it is, unless by reference to Catholic authority.

As a striking proof I need only refer to an article of the *Saturday Review* of 14th May, 1864, on *Superstition and Education.* Though the paper is one of the cleverest of the day, yet, in the article in question, there is a development of ignorance and confusion on the subject which I have seldom or ever witnessed.

But there is even Sir John Herschel, so learned and acute in astronomical science, and, no doubt, a model to the society he moves in, whom I must class among the blind, and whose ignorance on superstition I must deplore.

"I have not heard"—he says, in *Good Words!*—" that it was necessary to liquefy the blood of St. Januarius, or to carry his bones about the streets, on account of any of these later great comets."

The introduction of this paragraph by Sir John is quite gratui-

exclaimed Dionysius the Areopagite, at the time it occurred, though
distant from the spot; he believed the former and was converted
to the religion of Him who then DID die.

And though nature will have it, yet fault is found and superstition it is called, when poor mortals, and great sinners moreover,
are told to beware, and mind the warning of Him in whose hands
they are, who carries the lightning, guides the hurricane, rules
the waves, moves the heavens, and makes the earth tremble! No,
I say; nay, blessed that superstition, if superstition it be, that
leads the soul to God, makes the proud one humble, the libertine
chaste, the angry meek, the miser give alms, the infidel to pray,
the scoffer to adore, the calcumniator to retract, the over-weaning
to pause, scientific conceit to see the hollowness of its mere secular
knowledge; which at the contemplation of a comet—and beautiful, if true—makes an imperial sinner repent and abdicate his
throne, and, another Magi of wisdom, lay down his crown at the
feet of the Blessed Virgin-Mother and Child, to exchange it for a
crown in heaven! Blessed superstition if these be thy effects!

But there is a real superstition which NEVER FOUND A RESTING
PLACE in the Church, nor in ANY of her moderately-well-instructed
PRACTICAL children, which ONLY and EXCLUSIVELY is found
outside her precincts, more or less gross according to the departure from her doctrine and teaching. Solomon even became an
idolator when he had left the Church of God. Outside to her,
and CONDEMNED by her, there is a superstition that LEADS TO EVIL,
everywhere and always at the heel of and in opposition to good,
as falsehood is opposed to truth, as the shadow clings to light,
and night, with its incoherent forms, follows day.

There is the superstition of spiritualism and all its kindred, and
of heresy in general, from manichæism down to our latest revivals;
there is that satanic superstition, witnessed in Protestant Sweden
a few years ago, which forcibly breaking the ranks of the soldiers,

greedily seized upon the blood of executed MURDERERS to find a cure for diseases of men and cattle; there is an abundant crop of real superstition in this kingdom, and next to it that depraved tendency of the mind which collects the relics of criminals high and low, and blasphemes and sins under the gallows. And who would not rather join that superstition, if so you will have it, that collects and venerates the blood and relics of the MARTYRS, of men and women died in the service of their heavenly Master and for the corporal and spiritual welfare of the human race? And who would not be moved by that superstition which teaches and MAKES an assembled crowd fall down on their knees and in awful silence pray for the expiring culprit?

But there is that departure from moral perception, FOSTERED by apostates and the robbers of the Church and their descendants, which cannot distinguish the true from the false, deprecating what leads to God, and exalting what keeps and leads away from Him; that blindness which uses "superstition" without knowing what it is, unless by reference to Catholic authority.

As a striking proof I need only refer to an article of the *Saturday Review* of 14th May, 1864, on *Superstition and Education*. Though the paper is one of the cleverest of the day, yet, in the article in question, there is a development of ignorance and confusion on the subject which I have seldom or ever witnessed.

But there is even Sir John Herschel, so learned and acute in astronomical science, and, no doubt, a model to the society he moves in, whom I must class among the blind, and whose ignorance on superstition I must deplore.

"I have not heard"—he says, in *Good Words!*—"that it was necessary to liquefy the blood of St. Januarius, or to carry his bones about the streets, on account of any of these later great comets."

The introduction of this paragraph by Sir John is quite gratui-

exclaimed Dionysius the Areopagite, at the time it occurred, though distant from the spot; he believed the former and was converted to the religion of Him who then DID die.

And though nature will have it, yet fault is found and superstition it is called, when poor mortals, and great sinners moreover, are told to beware, and mind the warning of Him in whose hands they are, who carries the lightning, guides the hurricane, rules the waves, moves the heavens, and makes the earth tremble! No, I say; nay, blessed that superstition, if superstition it be, that leads the soul to God, makes the proud one humble, the libertine chaste, the angry meek, the miser give alms, the infidel to pray, the scoffer to adore, the calcumniator to retract, the over-weaning to pause, scientific conceit to see the hollowness of its mere secular knowledge; which at the contemplation of a comet—and beautiful, if true—makes an imperial sinner repent and abdicate his throne, and, another Magi of wisdom, lay down his crown at the feet of the Blessed Virgin-Mother and Child, to exchange it for a crown in heaven! Blessed superstition if these be thy effects!

But there is a real superstition which NEVER FOUND A RESTING PLACE in the Church, nor in ANY of her moderately-well-instructed PRACTICAL children, which ONLY and EXCLUSIVELY is found outside her precincts, more or less gross according to the departure from her doctrine and teaching. Solomon even became an idolator when he had left the Church of God. Outside to her, and CONDEMNED by her, there is a superstition that LEADS TO EVIL, everywhere and always at the heel of and in opposition to good, as falsehood is opposed to truth, as the shadow clings to light, and night, with its incoherent forms, follows day.

There is the superstition of spiritualism and all its kindred, and of heresy in general, from manichæism down to our latest revivals; there is that satanic superstition, witnessed in Protestant Sweden a few years ago, which forcibly breaking the ranks of the soldiers,

greedily seized upon the blood of executed MURDERERS to find a cure for diseases of men and cattle; there is an abundant crop of real superstition in this kingdom, and next to it that depraved tendency of the mind which collects the relics of criminals high and low, and blasphemes and sins under the gallows. And who would not rather join that superstition, if so you will have it, that collects and venerates the blood and relics of the MARTYRS, of men and women died in the service of their heavenly Master and for the corporal and spiritual welfare of the human race? And who would not be moved by that superstition which teaches and MAKES an assembled crowd fall down on their knees and in awful silence pray for the expiring culprit?

But there is that departure from moral perception, FOSTERED by apostates and the robbers of the Church and their descendants, which cannot distinguish the true from the false, deprecating what leads to God, and exalting what keeps and leads away from Him; that blindness which uses "superstition" without knowing what it is, unless by reference to Catholic authority.

As a striking proof I need only refer to an article of the *Saturday Review* of 14th May, 1864, on *Superstition and Education*. Though the paper is one of the cleverest of the day, yet, in the article in question, there is a development of ignorance and confusion on the subject which I have seldom or ever witnessed.

But there is even Sir John Herschel, so learned and acute in astronomical science, and, no doubt, a model to the society he moves in, whom I must class among the blind, and whose ignorance on superstition I must deplore.

"I have not heard"—he says, in *Good Words*!—"that it was necessary to liquefy the blood of St. Januarius, or to carry his bones about the streets, on account of any of these later great comets."

The introduction of this paragraph by Sir John is quite gratui-

exclaimed Dionysius the Areopagite, at the time it occurred, though distant from the spot; he believed the former and was converted to the religion of Him who then DID die.

And though nature will have it, yet fault is found and super-stition it is called, when poor mortals, and great sinners moreover, are told to beware, and mind the warning of Him in whose hands they are, who carries the lightning, guides the hurricane, rules the waves, moves the heavens, and makes the earth tremble! No, I say; nay, blessed that superstition, if superstition it be, that leads the soul to God, makes the proud one humble, the libertine chaste, the angry meek, the miser give alms, the infidel to pray, the scoffer to adore, the calcumniator to retract, the over-weaning to pause, scientific conceit to see the hollowness of its mere secular knowledge; which at the contemplation of a comet—and beauti-ful, if true—makes an imperial sinner repent and abdicate his throne, and, another Magi of wisdom, lay down his crown at the feet of the Blessed Virgin-Mother and Child, to exchange it for a crown in heaven! Blessed superstition if these be thy effects!

But there is a real superstition which NEVER FOUND A RESTING PLACE in the Church, nor in ANY of her moderately-well-instructed PRACTICAL children, which ONLY and EXCLUSIVELY is found outside her precincts, more or less gross according to the depar-ture from her doctrine and teaching. Solomon even became an idolator when he had left the Church of God. Outside to her, and CONDEMNED by her, there is a superstition that LEADS TO EVIL, everywhere and always at the heel of and in opposition to good, as falsehood is opposed to truth, as the shadow clings to light, and night, with its incoherent forms, follows day.

There is the superstition of spiritualism and all its kindred, and of heresy in general, from manichæism down to our latest revivals; there is that satanic superstition, witnessed in Protestant Sweden a few years ago, which forcibly breaking the ranks of the soldiers,

greedily seized upon the blood of executed MURDERERS to find a
cure for diseases of men and cattle; there is an abundant crop of
real superstition in this kingdom, and next to it that depraved
tendency of the mind which collects the relics of criminals high
and low, and blasphemes and sins under the gallows. And who
would not rather join that superstition, if so you will have it,
that collects and venerates the blood and relics of the MARTYRS,
of men and women died in the service of their heavenly Master
and for the corporal and spiritual welfare of the human race?
And who would not be moved by that superstition which teaches
and MAKES an assembled crowd fall down on their knees and in
awful silence pray for the expiring culprit?

But there is that departure from moral perception, FOSTERED
by apostates and the robbers of the Church and their descendants,
which cannot distinguish the true from the false, deprecating
what leads to God, and exalting what keeps and leads away
from Him; that blindness which uses "superstition" without
knowing what it is, unless by reference to Catholic authority.

As a striking proof I need only refer to an article of the
Saturday Review of 14th May, 1864, on Superstition and Education.
Though the paper is one of the cleverest of the day, yet, in
the article in question, there is a development of ignorance and
confusion on the subject which I have seldom or ever witnessed.

But there is even Sir John Herschel, so learned and acute
in astronomical science, and, no doubt, a model to the society
he moves in, whom I must class among the blind, and whose
ignorance on superstition I must deplore.

"I have not heard"—he says, in Good Words!—"that it was
necessary to liquefy the blood of St. Januarius, or to carry his
bones about the streets, on account of any of these later great
comets."

The introduction of this paragraph by Sir John is quite gratui-

exclaimed Dionysius the Areopagite, at the time it occurred, though
distant from the spot; he believed the former and was converted
to the religion of Him who then DID die.

And though nature will have it, yet fault is found and superstition it is called, when poor mortals, and great sinners moreover,
are told to beware, and mind the warning of Him in whose hands
they are, who carries the lightning, guides the hurricane, rules
the waves, moves the heavens, and makes the earth tremble! No,
I say; nay, blessed that superstition, if superstition it be, that
leads the soul to God, makes the proud one humble, the libertine
chaste, the angry meek, the miser give alms, the infidel to pray,
the scoffer to adore, the calcumniator to retract, the over-weaning
to pause, scientific conceit to see the hollowness of its mere secular
knowledge; which at the contemplation of a comet—and beautiful, if true—makes an imperial sinner repent and abdicate his
throne, and, another Magi of wisdom, lay down his crown at the
feet of the Blessed Virgin-Mother and Child, to exchange it for a
crown in heaven! Blessed superstition if these be thy effects!

But there is a real superstition which NEVER FOUND A RESTING
PLACE in the Church, nor in ANY of her moderately-well-instructed
PRACTICAL children, which ONLY and EXCLUSIVELY is found
outside her precincts, more or less gross according to the departure from her doctrine and teaching. Solomon even became an
idolator when he had left the Church of God. Outside to her,
and CONDEMNED by her, there is a superstition that LEADS TO EVIL,
everywhere and always at the heel of and in opposition to good,
as falsehood is opposed to truth, as the shadow clings to light,
and night, with its incoherent forms, follows day.

There is the superstition of spiritualism and all its kindred, and
of heresy in general, from manichæism down to our latest revivals;
there is that satanic superstition, witnessed in Protestant Sweden
a few years ago, which forcibly breaking the ranks of the soldiers,

greedily seized upon the blood of executed MURDERERS to find a cure for diseases of men and cattle; there is an abundant crop of real superstition in this kingdom, and next to it that depraved tendency of the mind which collects the relics of criminals high and low, and blasphemes and sins under the gallows. And who would not rather join that superstition, if so you will have it, that collects and venerates the blood and relics of the MARTYRS, of men and women died in the service of their heavenly Master and for the corporal and spiritual welfare of the human race? And who would not be moved by that superstition which teaches and MAKES an assembled crowd fall down on their knees and in awful silence pray for the expiring culprit?

But there is that departure from moral perception, FOSTERED by apostates and the robbers of the Church and their descendants, which cannot distinguish the true from the false, deprecating what leads to God, and exalting what keeps and leads away from Him; that blindness which uses "superstition" without knowing what it is, unless by reference to Catholic authority.

As a striking proof I need only refer to an article of the *Saturday Review* of 14th May, 1864, on *Superstition and Education*. Though the paper is one of the cleverest of the day, yet, in the article in question, there is a development of ignorance and confusion on the subject which I have seldom or ever witnessed.

But there is even Sir John Herschel, so learned and acute in astronomical science, and, no doubt, a model to the society he moves in, whom I must class among the blind, and whose ignorance on superstition I must deplore.

"I have not heard"—he says, in *Good Words!*—"that it was necessary to liquefy the blood of St. Januarius, or to carry his bones about the streets, on account of any of these later great comets."

The introduction of this paragraph by Sir John is quite gratui-

exclaimed Dionysius the Areopagite, at the time it occurred, though
distant from the spot; he believed the former and was converted
to the religion of Him who then DID die.

And though nature will have it, yet fault is found and superstition it is called, when poor mortals, and great sinners moreover,
are told to beware, and mind the warning of Him in whose hands
they are, who carries the lightning, guides the hurricane, rules
the waves, moves the heavens, and makes the earth tremble! No,
I say; nay, blessed that superstition, if superstition it be, that
leads the soul to God, makes the proud one humble, the libertine
chaste, the angry meek, the miser give alms, the infidel to pray,
the scoffer to adore, the calcumniator to retract, the over-weaning
to pause, scientific conceit to see the hollowness of its mere secular
knowledge; which at the contemplation of a comet—and beautiful, if true—makes an imperial sinner repent and abdicate his
throne, and, another Magi of wisdom, lay down his crown at the
feet of the Blessed Virgin-Mother and Child, to exchange it for a
crown in heaven! Blessed superstition if these be thy effects!

But there is a real superstition which NEVER FOUND A RESTING
PLACE in the Church, nor in ANY of her moderately-well-instructed
PRACTICAL children, which ONLY and EXCLUSIVELY is found
outside her precincts, more or less gross according to the departure from her doctrine and teaching. Solomon even became an
idolator when he had left the Church of God. Outside to her,
and CONDEMNED by her, there is a superstition that LEADS TO EVIL,
everywhere and always at the heel of and in opposition to good,
as falsehood is opposed to truth, as the shadow clings to light,
and night, with its incoherent forms, follows day.

There is the superstition of spiritualism and all its kindred, and
of heresy in general, from manichæism down to our latest revivals;
there is that satanic superstition, witnessed in Protestant Sweden
a few years ago, which forcibly breaking the ranks of the soldiers,

greedily seized upon the blood of executed MURDERERS to find a cure for diseases of men and cattle; there is an abundant crop of real superstition in this kingdom, and next to it that depraved tendency of the mind which collects the relics of criminals high and low, and blasphemes and sins under the gallows. And who would not rather join that superstition, if so you will have it, that collects and venerates the blood and relics of the MARTYRS, of men and women died in the service of their heavenly Master and for the corporal and spiritual welfare of the human race? And who would not be moved by that superstition which teaches and MAKES an assembled crowd fall down on their knees and in awful silence pray for the expiring culprit?

But there is that departure from moral perception, FOSTERED by apostates and the robbers of the Church and their descendants, which cannot distinguish the true from the false, deprecating what leads to God, and exalting what keeps and leads away from Him; that blindness which uses "superstition" without knowing what it is, unless by reference to Catholic authority.

As a striking proof I need only refer to an article of the *Saturday Review* of 14th May, 1864, on *Superstition and Education.* Though the paper is one of the cleverest of the day, yet, in the article in question, there is a development of ignorance and confusion on the subject which I have seldom or ever witnessed.

But there is even Sir John Herschel, so learned and acute in astronomical science, and, no doubt, a model to the society he moves in, whom I must class among the blind, and whose ignorance on superstition I must deplore.

"I have not heard"—he says, in *Good Words!*—"that it was necessary to liquefy the blood of St. Januarius, or to carry his bones about the streets, on account of any of these later great comets."

The introduction of this paragraph by Sir John is quite gratui-

tous; but if, what he says, he had heard to have taken place on account of a comet, it would have been as wonderful as if he had seen Lexell's comet paying a visit to the satellites of that stumbling block Jupiter, which it never did. But there are a GREAT MANY REAL THINGS which he never heard either, no more than the true history of what he sneers at, what with the accumulated science, learning, intellect and cleverness, of the WHOLE PROTESTANT, HEATHEN AND INFIDEL WORLD, he can neither explain nor reproduce, no more than any of the authenticated miracles which the Catholic Church believes and teaches; and if, of the inexhaustible treasure of works of the Catholic Church, ignored by an anti-Catholic and infidel press, and therefore unknown to the non-Catholic masses, he were to read but the most elementary treatise, so to speak, he would find immensely more than he ever "dreamt of in HIS philosophy," and our younger seminarists could instruct him with more precision in things divine than he ever attained in his most successful astronomical calculations, dissertations and theories.

However excusable the above may be, it cannot be pardoned in a learned man, and a christian, I suppose, when he says that it was "the MOST ABJECT SUPERSTITION THAT DETACHED HIS (Charles V.) THOUGHTS FROM EARTHLY THINGS AND FIXED THEM ON HIS ETERNAL INTERESTS!" Had the emperor not done penance had he had the misfortune to be an unrepenting, excommunicated, licentious, sacrilegious and perfidious robber, the head of military assassins slaying without declaration of war, without notice or warning: I am sincerely afraid that, with many others, from fear of superstition that leads to God, from hatred of the Church, grounded on ignorance and prejudice, he too would have called him an enlightened ruler, an "Imperatore galantuómo," though, without even knowing that the CHURCH THROWS AWAY NO PEARL, that no good children ever leave a true mother, I cannot harbour

the uncharitable supposition, that, like many courtiers and others, he would so far have forgotten his dignity, as gratuitously to press his polluted hand or kiss the seam of his dirty garment had he had the opportunity of doing so.

Besides ignorance and prejudice in matters of the Church presiding over many a noble and excellent heart with which we have to contend, there is also the canting pharisee, the artful scribe, and the whitened sepulchre; for "When cities abandon themselves to revelry, and are driven by terror-stricken consciences to the altar, then, while the lights are burning, a cloud of incense is rising, and EVERY KNEE IS BENDED, the walls totter and gape, and down falls the ponderous vault upon a thousand sinners." But "Here in these cooler climes, with more reasonable temperaments, and a purer faith," where its meaning is unknown, no one OUT OF THE CHURCH able to MAKE an act of faith, where mere opinion is substituted for belief, and where the Apostle's and the Nicene creeds are empty words, and, I hope in ALL CASES, an UNCONSCIOUS falsehood in the mouth of the reciter, "it is hoped that we do not need this awful language. The ALMIGHTY foot-fall is soft here, even in the earthquake and the storm."

Thus, according to the national Leader and presumptuous Instructor in the *Times*, of 8th October, 1863, there is no handwriting against the country, no "social evil," alias sink of iniquity, no infanticides, no crimes, no misery, no starvation cases unparalleled in civilized Europe, no criminal treatment of the poor, no emaciated and depopulated step-sister-island, no still idolatrous India; the slight earthquake° just experienced need excite no apprehension, the ring of Polycrates will for ever return, our SUPER-STRATAL happiness and prosperity will not leave us; why then pray, why the "mea culpa," why the "miserere," though danger seems nigh!—yet the penitent thief on the cross was saved

---

* It occurred on the 6th October, 1863, at 2.30 p.m.

\

tous; but if, what he says, he had heard to have taken place
on account of a comet, it would have been as wonderful as if
he had seen Lexell's comet paying a visit to the satellites of
that stumbling block Jupiter, which it never did. But there
are a GREAT MANY REAL THINGS which he never heard either,
no more than the true history of what he sneers at, what with
the accumulated science, learning, intellect and cleverness, of
the WHOLE PROTESTANT, HEATHEN AND INFIDEL WORLD, he can
neither explain nor reproduce, no more than any of the authenticated miracles which the Catholic Church believes and teaches;
and if, of the inexhaustible treasure of works of the Catholic
Church, ignored by an anti-Catholic and infidel press, and therefore unknown to the non-Catholic masses, he were to read but
the most elementary treatise, so to speak, he would find immensely more than he ever "dreamt of in HIS philosophy," and
our younger seminarists could instruct him with more precision
in things divine than he ever attained in his most successful
astronomical calculations, dissertations and theories.

However excusable the above may be, it cannot be pardoned
in a learned man, and a christian, I suppose, when he says that
it was "the MOST ABJECT SUPERSTITION THAT DETACHED HIS
(Charles V.) THOUGHTS FROM EARTHLY THINGS AND FIXED THEM
ON HIS ETERNAL INTERESTS!" Had the emperor not done penance
had he had the misfortune to be an unrepenting, excommunicated,
licentious, sacrilegious and perfidious robber, the head of military
assassins slaying without declaration of war, without notice or
warning: I am sincerely afraid that, with many others, from fear
of superstition that leads to God, from hatred of the Church,
grounded on ignorance and prejudice, he too would have called
him an enlightened ruler, an "Imperatore galantuómo," though,
without even knowing that the CHURCH THROWS AWAY NO PEARL,
that no good children ever leave a true mother, I cannot harbour

the uncharitable supposition, that, like many courtiers and others, he would so far have forgotten his dignity, as gratuitously to press his polluted hand or kiss the seam of his dirty garment had he had the opportunity of doing so.

Besides ignorance and prejudice in matters of the Church presiding over many a noble and excellent heart with which we have to contend, there is also the canting pharisee, the artful scribe, and the whitened sepulchre; for " When cities abandon themselves to revelry, and are driven by terror-stricken consciences to the altar, then, while the lights are burning, a cloud of incense is rising, and EVERY KNEE IS BENDED, the walls totter and gape, and down falls the ponderous vault upon a thousand sinners." But " Here in these cooler climes, with more reasonable temperaments, and a purer faith," where its meaning is unknown, no one OUT OF THE CHURCH able to MAKE an act of faith, where mere opinion is substituted for belief, and where the Apostle's and the Nicene creeds are empty words, and, I hope in ALL CASES, an UNCONSCIOUS falsehood in the mouth of the reciter, " it is hoped that we do not need this awful language. The ALMIGHTY foot-fall is soft here, even in the earthquake and the storm."

Thus, according to the national Leader and presumptuous Instructor in the *Times*, of 8th October, 1863, there is no handwriting against the country, no "social evil," alias sink of iniquity, no infanticides, no crimes, no misery, no starvation cases unparalleled in civilized Europe, no criminal treatment of the poor, no emaciated and depopulated step-sister-island, no still idolatrous India; the slight earthquake° just experienced need excite no apprehension, the ring of Polycrates will for ever return, our SUPER-STRATAL happiness and prosperity will not leave us ; why then pray, why the " mea culpa," why the " miserere," though danger seems nigh !—yet the penitent thief on the cross was saved

---

* It occurred on the 6th October, 1863, at 2.30 p.m.

tous; but if, what he says, he had heard to have taken place
on account of a comet, it would have been as wonderful as if
he had seen Lexell's comet paying a visit to the satellites of
that stumbling block Jupiter, which it never did. But there
are a GREAT MANY REAL THINGS which he never heard either,
no more than the true history of what he sneers at, what with
the accumulated science, learning, intellect and cleverness, of
the WHOLE PROTESTANT, HEATHEN AND INFIDEL WORLD, he can
neither explain nor reproduce, no more than any of the authenticated miracles which the Catholic Church believes and teaches;
and if, of the inexhaustible treasure of works of the Catholic
Church, ignored by an anti-Catholic and infidel press, and therefore unknown to the non-Catholic masses, he were to read but
the most elementary treatise, so to speak, he would find immensely more than he ever "dreamt of in HIS philosophy," and
our younger seminarists could instruct him with more precision
in things divine than he ever attained in his most successful
astronomical calculations, dissertations and theories.

However excusable the above may be, it cannot be pardoned
in a learned man, and a christian, I suppose, when he says that
it was "the MOST ABJECT SUPERSTITION THAT DETACHED HIS
(Charles V.) THOUGHTS FROM EARTHLY THINGS AND FIXED THEM
ON HIS ETERNAL INTERESTS!" Had the emperor not done penance
had he had the misfortune to be an unrepenting, excommunicated,
licentious, sacrilegious and perfidious robber, the head of military
assassins slaying without declaration of war, without notice or
warning: I am sincerely afraid that, with many others, from fear
of superstition that leads to God, from hatred of the Church,
grounded on ignorance and prejudice, he too would have called
him an enlightened ruler, an "Imperatore galantuómo," though,
without even knowing that the CHURCH THROWS AWAY NO PEARL,
that no good children ever leave a true mother, I cannot harbour

the uncharitable supposition, that, like many courtiers and others, he would so far have forgotten his dignity, as gratuitously to press his polluted hand or kiss the seam of his dirty garment had he had the opportunity of doing so.

Besides ignorance and prejudice in matters of the Church presiding over many a noble and excellent heart with which we have to contend, there is also the canting pharisee, the artful scribe, and the whitened sepulchre; for "When cities abandon themselves to revelry, and are driven by terror-stricken consciences to the altar, then, while the lights are burning, a cloud of incense is rising, and EVERY KNEE IS BENDED, the walls totter and gape, and down falls the ponderous vault upon a thousand sinners." But "Here in these cooler climes, with more reasonable temperaments, and a purer faith," where its meaning is unknown, no one OUT OF THE CHURCH able to MAKE an act of faith, where mere opinion is substituted for belief, and where the Apostle's and the Nicene creeds are empty words, and, I hope in ALL CASES, an UNCONSCIOUS falsehood in the mouth of the reciter, "it is hoped that we do not need this awful language. The ALMIGHTY foot-fall is soft here, even in the earthquake and the storm."

Thus, according to the national Leader and presumptuous Instructor in the *Times*, of 8th October, 1863, there is no hand-writing against the country, no "social evil," alias sink of iniquity, no infanticides, no crimes, no misery, no starvation cases unparalleled in civilized Europe, no criminal treatment of the poor, no emaciated and depopulated step-sister-island, no still idolatrous India; the slight earthquake° just experienced need excite no apprehension, the ring of Polycrates will for ever return, our SUPER-STRATAL happiness and prosperity will not leave us; why then pray, why the "mea culpa," why the "miserere," though danger seems nigh!—yet the penitent thief on the cross was saved

---

° It occurred on the 6th October, 1863, at 2.30 p.m.

tous; but if, what he says, he had heard to have taken place on account of a comet, it would have been as wonderful as if he had seen Lexell's comet paying a visit to the satellites of that stumbling block Jupiter, which it never did. But there are a GREAT MANY REAL THINGS which he never heard either, no more than the true history of what he sneers at, what with the accumulated science, learning, intellect and cleverness, of the WHOLE PROTESTANT, HEATHEN AND INFIDEL WORLD, he can neither explain nor reproduce, no more than any of the authenticated miracles which the Catholic Church believes and teaches; and if, of the inexhaustible treasure of works of the Catholic Church, ignored by an anti-Catholic and infidel press, and therefore unknown to the non-Catholic masses, he were to read but the most elementary treatise, so to speak, he would find immensely more than he ever "dreamt of in HIS philosophy," and our younger seminarists could instruct him with more precision in things divine than he ever attained in his most successful astronomical calculations, dissertations and theories.

However excusable the above may be, it cannot be pardoned in a learned man, and a christian, I suppose, when he says that it was "the MOST ABJECT SUPERSTITION THAT DETACHED HIS (Charles V.) THOUGHTS FROM EARTHLY THINGS AND FIXED THEM ON HIS ETERNAL INTERESTS!" Had the emperor not done penance had he had the misfortune to be an unrepenting, excommunicated, licentious, sacrilegious and perfidious robber, the head of military assassins slaying without declaration of war, without notice or warning: I am sincerely afraid that, with many others, from fear of superstition that leads to God, from hatred of the Church, grounded on ignorance and prejudice, he too would have called him an enlightened ruler, an "Imperatore galantuómo," though, without even knowing that the CHURCH THROWS AWAY NO PEARL, that no good children ever leave a true mother, I cannot harbour

the uncharitable supposition, that, like many courtiers and others, he would so far have forgotten his dignity, as gratuitously to press his polluted hand or kiss the seam of his dirty garment had he had the opportunity of doing so.

Besides ignorance and prejudice in matters of the Church presiding over many a noble and excellent heart with which we have to contend, there is also the canting pharisee, the artful scribe, and the whitened sepulchre; for "When cities abandon themselves to revelry, and are driven by terror-stricken consciences to the altar, then, while the lights are burning, a cloud of incense is rising, and EVERY KNEE IS BENDED, the walls totter and gape, and down falls the ponderous vault upon a thousand sinners." But "Here in these cooler climes, with more reasonable temperaments, and a purer faith," where its meaning is unknown, no one OUT OF THE CHURCH able to MAKE an act of faith, where mere opinion is substituted for belief, and where the Apostle's and the Nicene creeds are empty words, and, I hope in ALL CASES, an UNCONSCIOUS falsehood in the mouth of the reciter, "it is hoped that we do not need this awful language. The ALMIGHTY foot-fall is soft here, even in the earthquake and the storm."

Thus, according to the national Leader and presumptuous Instructor in the *Times*, of 8th October, 1863, there is no handwriting against the country, no "social evil," alias sink of iniquity, no infanticides, no crimes, no misery, no starvation cases unparalleled in civilized Europe, no criminal treatment of the poor, no emaciated and depopulated step-sister-island, no still idolatrous India; the slight earthquake* just experienced need excite no apprehension, the ring of Polycrates will for ever return, our SUPER-STRATAL happiness and prosperity will not leave us; why then pray, why the "mea culpa," why the "miserere," though danger seems nigh!—yet the penitent thief on the cross was saved

---

* It occurred on the 6th October, 1863, at 2.30 p.m.

tous; but if, what he says, he had heard to have taken place on account of a comet, it would have been as wonderful as if he had seen Lexell's comet paying a visit to the satellites of that stumbling block Jupiter, which it never did. But there are a GREAT MANY REAL THINGS which he never heard either, no more than the true history of what he sneers at, what with the accumulated science, learning, intellect and cleverness, of the WHOLE PROTESTANT, HEATHEN AND INFIDEL WORLD, he can neither explain nor reproduce, no more than any of the authenticated miracles which the Catholic Church believes and teaches; and if, of the inexhaustible treasure of works of the Catholic Church, ignored by an anti-Catholic and infidel press, and therefore unknown to the non-Catholic masses, he were to read but the most elementary treatise, so to speak, he would find immensely more than he ever "dreamt of in HIS philosophy," and our younger seminarists could instruct him with more precision in things divine than he ever attained in his most successful astronomical calculations, dissertations and theories.

However excusable the above may be, it cannot be pardoned in a learned man, and a christian, I suppose, when he says that it was "the MOST ABJECT SUPERSTITION THAT DETACHED HIS (Charles V.) THOUGHTS FROM EARTHLY THINGS AND FIXED THEM ON HIS ETERNAL INTERESTS!" Had the emperor not done penance had he had the misfortune to be an unrepenting, excommunicated, licentious, sacrilegious and perfidious robber, the head of military assassins slaying without declaration of war, without notice or warning: I am sincerely afraid that, with many others, from fear of superstition that leads to God, from hatred of the Church, grounded on ignorance and prejudice, he too would have called him an enlightened ruler, an "Imperatore galantuómo," though, without even knowing that the CHURCH THROWS AWAY NO PEARL, that no good children ever leave a true mother, I cannot harbour

the uncharitable supposition, that, like many courtiers and others, he would so far have forgotten his dignity, as gratuitously to press his polluted hand or kiss the seam of his dirty garment had he had the opportunity of doing so.

Besides ignorance and prejudice in matters of the Church presiding over many a noble and excellent heart with which we have to contend, there is also the canting pharisee, the artful scribe, and the whitened sepulchre; for "When cities abandon themselves to revelry, and are driven by terror-stricken consciences to the altar, then, while the lights are burning, a cloud of incense is rising, and EVERY KNEE IS BENDED, the walls totter and gape, and down falls the ponderous vault upon a thousand sinners." But "Here in these cooler climes, with more reasonable temperaments, and a purer faith," where its meaning is unknown, no one OUT OF THE CHURCH able to MAKE an act of faith, where mere opinion is substituted for belief, and where the Apostle's and the Nicene creeds are empty words, and, I hope in ALL CASES, an UNCONSCIOUS falsehood in the mouth of the reciter, "it is hoped that we do not need this awful language. The ALMIGHTY foot-fall is soft here, even in the earthquake and the storm."

Thus, according to the national Leader and presumptuous Instructor in the *Times,* of 8th October, 1863, there is no handwriting against the country, no "social evil," alias sink of iniquity, no infanticides, no crimes, no misery, no starvation cases unparalleled in civilized Europe, no criminal treatment of the poor, no emaciated and depopulated step-sister-island, no still idolatrous India; the slight earthquake° just experienced need excite no apprehension, the ring of Polycrates will for ever return, our SUPER-STRATAL happiness and prosperity will not leave us ; why then pray, why the "mea culpa," why the "miserere," though danger seems nigh !—yet the penitent thief on the cross was saved

---

* It occurred on the 6th October, 1863, at 2.30 p.m.

tous; but if, what he says, he had heard to have taken place on account of a comet, it would have been as wonderful as if he had seen Lexell's comet paying a visit to the satellites of that stumbling block Jupiter, which it never did. But there are a GREAT MANY REAL THINGS which he never heard either, no more than the true history of what he sneers at, what with the accumulated science, learning, intellect and cleverness, of the WHOLE PROTESTANT, HEATHEN AND INFIDEL WORLD, he can neither explain nor reproduce, no more than any of the authenticated miracles which the Catholic Church believes and teaches; and if, of the inexhaustible treasure of works of the Catholic Church, ignored by an anti-Catholic and infidel press, and therefore unknown to the non-Catholic masses, he were to read but the most elementary treatise, so to speak, he would find immensely more than he ever "dreamt of in HIS philosophy," and our younger seminarists could instruct him with more precision in things divine than he ever attained in his most successful astronomical calculations, dissertations and theories.

However excusable the above may be, it cannot be pardoned in a learned man, and a christian, I suppose, when he says that it was "the MOST ABJECT SUPERSTITION THAT DETACHED HIS (Charles V.) THOUGHTS FROM EARTHLY THINGS AND FIXED THEM ON HIS ETERNAL INTERESTS!" Had the emperor not done penance had he had the misfortune to be an unrepenting, excommunicated, licentious, sacrilegious and perfidious robber, the head of military assassins slaying without declaration of war, without notice or warning: I am sincerely afraid that, with many others, from fear of superstition that leads to God, from hatred of the Church, grounded on ignorance and prejudice, he too would have called him an enlightened ruler, an "Imperatore galantuómo," though, without even knowing that the CHURCH THROWS AWAY NO PEARL, that no good children ever leave a true mother, I cannot harbour

the uncharitable supposition, that, like many courtiers and others, he would so far have forgotten his dignity, as gratuitously to press his polluted hand or kiss the seam of his dirty garment had he had the opportunity of doing so.

Besides ignorance and prejudice in matters of the Church presiding over many a noble and excellent heart with which we have to contend, there is also the canting pharisee, the artful scribe, and the whitened sepulchre; for "When cities abandon themselves to revelry, and are driven by terror-stricken consciences to the altar, then, while the lights are burning, a cloud of incense is rising, and EVERY KNEE IS BENDED, the walls totter and gape, and down falls the ponderous vault upon a thousand sinners." But "Here in these cooler climes, with more reasonable temperaments, and a purer faith," where its meaning is unknown, no one OUT OF THE CHURCH able to MAKE an act of faith, where mere opinion is substituted for belief, and where the Apostle's and the Nicene creeds are empty words, and, I hope in ALL CASES, an UNCONSCIOUS falsehood in the mouth of the reciter, "it is hoped that we do not need this awful language. The ALMIGHTY foot-fall is soft here, even in the earthquake and the storm."

Thus, according to the national Leader and presumptuous Instructor in the *Times*, of 8th October, 1863, there is no hand-writing against the country, no "social evil," alias sink of iniquity, no infanticides, no crimes, no misery, no starvation cases unparalleled in civilized Europe, no criminal treatment of the poor, no emaciated and depopulated step-sister-island, no still idolatrous India; the slight earthquake⁰ just experienced need excite no apprehension, the ring of Polycrates will for ever return, our SUPER-STRATAL happiness and prosperity will not leave us; why then pray, why the "mea culpa," why the "miserere," though danger seems nigh!—yet the penitent thief on the cross was saved

---

* It occurred on the 6th October, 1863, at 2.30 p.m.

tous; but if, what he says, he had heard to have taken place on account of a comet, it would have been as wonderful as if he had seen Lexell's comet paying a visit to the satellites of that stumbling block Jupiter, which it never did. But there are a GREAT MANY REAL THINGS which he never heard either, no more than the true history of what he sneers at, what with the accumulated science, learning, intellect and cleverness, of the WHOLE PROTESTANT, HEATHEN AND INFIDEL WORLD, he can neither explain nor reproduce, no more than any of the authenticated miracles which the Catholic Church believes and teaches; and if, of the inexhaustible treasure of works of the Catholic Church, ignored by an anti-Catholic and infidel press, and therefore unknown to the non-Catholic masses, he were to read but the most elementary treatise, so to speak, he would find immensely more than he ever "dreamt of in HIS philosophy," and our younger seminarists could instruct him with more precision in things divine than he ever attained in his most successful astronomical calculations, dissertations and theories.

However excusable the above may be, it cannot be pardoned in a learned man, and a christian, I suppose, when he says that it was "the MOST ABJECT SUPERSTITION THAT DETACHED HIS (Charles V.) THOUGHTS FROM EARTHLY THINGS AND FIXED THEM ON HIS ETERNAL INTERESTS!" Had the emperor not done penance had he had the misfortune to be an unrepenting, excommunicated, licentious, sacrilegious and perfidious robber, the head of military assassins slaying without declaration of war, without notice or warning: I am sincerely afraid that, with many others, from fear of superstition that leads to God, from hatred of the Church, grounded on ignorance and prejudice, he too would have called him an enlightened ruler, an "Imperatore galantuómo," though, without even knowing that the CHURCH THROWS AWAY NO PEARL, that no good children ever leave a true mother, I cannot harbour

the uncharitable supposition, that, like many courtiers and others, he would so far have forgotten his dignity, as gratuitously to press his polluted hand or kiss the seam of his dirty garment had he had the opportunity of doing so.

Besides ignorance and prejudice in matters of the Church presiding over many a noble and excellent heart with which we have to contend, there is also the canting pharisee, the artful scribe, and the whitened sepulchre; for "When cities abandon themselves to revelry, and are driven by terror-stricken consciences to the altar, then, while the lights are burning, a cloud of incense is rising, and EVERY KNEE IS BENDED, the walls totter and gape, and down falls the ponderous vault upon a thousand sinners." But "Here in these cooler climes, with more reasonable temperaments, and a purer faith," where its meaning is unknown, no one OUT OF THE CHURCH able to MAKE an act of faith, where mere opinion is substituted for belief, and where the Apostle's and the Nicene creeds are empty words, and, I hope in ALL CASES, an UNCONSCIOUS falsehood in the mouth of the reciter, " it is hoped that we do not need this awful language. The ALMIGHTY foot-fall is soft here, even in the earthquake and the storm."

Thus, according to the national Leader and presumptuous Instructor in the *Times*, of 8th October, 1863, there is no handwriting against the country, no "social evil," alias sink of iniquity, no infanticides, no crimes, no misery, no starvation cases unparalleled in civilized Europe, no criminal treatment of the poor, no emaciated and depopulated step-sister-island, no still idolatrous India; the slight earthquake° just experienced need excite no apprehension, the ring of Polycrates will for ever return, our SUPER-STRATAL happiness and prosperity will not leave us ; why then pray, why the "mea culpa," why the "miserere," though danger seems nigh !—yet the penitent thief on the cross was saved

---

* It occurred on the 6th October, 1863, at 2.80 p.m.

tous; but if, what he says, he had heard to have taken place on account of a comet, it would have been as wonderful as if he had seen Lexell's comet paying a visit to the satellites of that stumbling block Jupiter, which it never did. But there are a GREAT MANY REAL THINGS which he never heard either, no more than the true history of what he sneers at, what with the accumulated science, learning, intellect and cleverness, of the WHOLE PROTESTANT, HEATHEN AND INFIDEL WORLD, he can neither explain nor reproduce, no more than any of the authenticated miracles which the Catholic Church believes and teaches; and if, of the inexhaustible treasure of works of the Catholic Church, ignored by an anti-Catholic and infidel press, and therefore unknown to the non-Catholic masses, he were to read but the most elementary treatise, so to speak, he would find immensely more than he ever "dreamt of in HIS philosophy," and our younger seminarists could instruct him with more precision in things divine than he ever attained in his most successful astronomical calculations, dissertations and theories.

However excusable the above may be, it cannot be pardoned in a learned man, and a christian, I suppose, when he says that it was "the MOST ABJECT SUPERSTITION THAT DETACHED HIS (Charles V.) THOUGHTS FROM EARTHLY THINGS AND FIXED THEM ON HIS ETERNAL INTERESTS!" Had the emperor not done penance had he had the misfortune to be an unrepenting, excommunicated, licentious, sacrilegious and perfidious robber, the head of military assassins slaying without declaration of war, without notice or warning: I am sincerely afraid that, with many others, from fear of superstition that leads to God, from hatred of the Church, grounded on ignorance and prejudice, he too would have called him an enlightened ruler, an "Imperatore galantuómo," though, without even knowing that the CHURCH THROWS AWAY NO PEARL, that no good children ever leave a true mother, I cannot harbour

the uncharitable supposition, that, like many courtiers and others, he would so far have forgotten his dignity, as gratuitously to press his polluted hand or kiss the seam of his dirty garment had he had the opportunity of doing so.

Besides ignorance and prejudice in matters of the Church presiding over many a noble and excellent heart with which we have to contend, there is also the canting pharisee, the artful scribe, and the whitened sepulchre; for "When cities abandon themselves to revelry, and are driven by terror-stricken consciences to the altar, then, while the lights are burning, a cloud of incense is rising, and EVERY KNEE IS BENDED, the walls totter and gape, and down falls the ponderous vault upon a thousand sinners." But "Here in these cooler climes, with more reasonable temperaments, and a purer faith," where its meaning is unknown, no one OUT OF THE CHURCH able to MAKE an act of faith, where mere opinion is substituted for belief, and where the Apostle's and the Nicene creeds are empty words, and, I hope in ALL CASES, an UNCONSCIOUS falsehood in the mouth of the reciter, "it is hoped that we do not need this awful language. The ALMIGHTY foot-fall is soft here, even in the earthquake and the storm."

Thus, according to the national Leader and presumptuous Instructor in the *Times*, of 8th October, 1863, there is no handwriting against the country, no "social evil," alias sink of iniquity, no infanticides, no crimes, no misery, no starvation cases unparalleled in civilized Europe, no criminal treatment of the poor, no emaciated and depopulated step-sister-island, no still idolatrous India; the slight earthquake* just experienced need excite no apprehension, the ring of Polycrates will for ever return, our SUPER-STRATAL happiness and prosperity will not leave us ; why then pray, why the "mea culpa," why the "miserere," though danger seems nigh !—yet the penitent thief on the cross was saved

---

* It occurred on the 6th October, 1863, at 2.30 p.m.

tous; but if, what he says, he had heard to have taken place on account of a comet, it would have been as wonderful as if he had seen Lexell's comet paying a visit to the satellites of that stumbling block Jupiter, which it never did. But there are a GREAT MANY REAL THINGS which he never heard either, no more than the true history of what he sneers at, what with the accumulated science, learning, intellect and cleverness, of the WHOLE PROTESTANT, HEATHEN AND INFIDEL WORLD, he can neither explain nor reproduce, no more than any of the authenticated miracles which the Catholic Church believes and teaches; and if, of the inexhaustible treasure of works of the Catholic Church, ignored by an anti-Catholic and infidel press, and therefore unknown to the non-Catholic masses, he were to read but the most elementary treatise, so to speak, he would find immensely more than he ever "dreamt of in HIS philosophy," and our younger seminarists could instruct him with more precision in things divine than he ever attained in his most successful astronomical calculations, dissertations and theories.

However excusable the above may be, it cannot be pardoned in a learned man, and a christian, I suppose, when he says that it was "the MOST ABJECT SUPERSTITION THAT DETACHED HIS (Charles V.) THOUGHTS FROM EARTHLY THINGS AND FIXED THEM ON HIS ETERNAL INTERESTS!" Had the emperor not done penance had he had the misfortune to be an unrepenting, excommunicated, licentious, sacrilegious and perfidious robber, the head of military assassins slaying without declaration of war, without notice or warning: I am sincerely afraid that, with many others, from fear of superstition that leads to God, from hatred of the Church, grounded on ignorance and prejudice, he too would have called him an enlightened ruler, an "Imperatore galantuómo," though, without even knowing that the CHURCH THROWS AWAY NO PEARL, that no good children ever leave a true mother, I cannot harbour

the uncharitable supposition, that, like many courtiers and others, he would so far have forgotten his dignity, as gratuitously to press his polluted hand or kiss the seam of his dirty garment had he had the opportunity of doing so.

Besides ignorance and prejudice in matters of the Church presiding over many a noble and excellent heart with which we have to contend, there is also the canting pharisee, the artful scribe, and the whitened sepulchre; for "When cities abandon themselves to revelry, and are driven by terror-stricken consciences to the altar, then, while the lights are burning, a cloud of incense is rising, and EVERY KNEE IS BENDED, the walls totter and gape, and down falls the ponderous vault upon a thousand sinners." But "Here in these cooler climes, with more reasonable temperaments, and a purer faith," where its meaning is unknown, no one OUT OF THE CHURCH able to MAKE an act of faith, where mere opinion is substituted for belief, and where the Apostle's and the Nicene creeds are empty words, and, I hope in ALL CASES, an UNCONSCIOUS falsehood in the mouth of the reciter, "it is hoped that we do not need this awful language. The ALMIGHTY foot-fall is soft here, even in the earthquake and the storm."

Thus, according to the national Leader and presumptuous Instructor in the *Times*, of 8th October, 1863, there is no handwriting against the country, no "social evil," alias sink of iniquity, no infanticides, no crimes, no misery, no starvation cases unparalleled in civilized Europe, no criminal treatment of the poor, no emaciated and depopulated step-sister-island, no still idolatrous India; the slight earthquake° just experienced need excite no apprehension, the ring of Polycrates will for ever return, our SUPER-STRATAL happiness and prosperity will not leave us; why then pray, why the "mea culpa," why the "miserere," though danger seems nigh!—yet the penitent thief on the cross was saved

---

* It occurred on the 6th October, 1863, at 2.30 p.m.

tous; but if, what he says, he had heard to have taken place
on account of a comet, it would have been as wonderful as if
he had seen Lexell's comet paying a visit to the satellites of
that stumbling block Jupiter, which it never did. But there
are a GREAT MANY REAL THINGS which he never heard either,
no more than the true history of what he sneers at, what with
the accumulated science, learning, intellect and cleverness, of
the WHOLE PROTESTANT, HEATHEN AND INFIDEL WORLD, he can
neither explain nor reproduce, no more than any of the authenticated miracles which the Catholic Church believes and teaches;
and if, of the inexhaustible treasure of works of the Catholic
Church, ignored by an anti-Catholic and infidel press, and therefore unknown to the non-Catholic masses, he were to read but
the most elementary treatise, so to speak, he would find immensely more than he ever "dreamt of in HIS philosophy," and
our younger seminarists could instruct him with more precision
in things divine than he ever attained in his most successful
astronomical calculations, dissertations and theories.

However excusable the above may be, it cannot be pardoned
in a learned man, and a christian, I suppose, when he says that
it was "the MOST ABJECT SUPERSTITION THAT DETACHED HIS
(Charles V.) THOUGHTS FROM EARTHLY THINGS AND FIXED THEM
ON HIS ETERNAL INTERESTS!" Had the emperor not done penance
had he had the misfortune to be an unrepenting, excommunicated,
licentious, sacrilegious and perfidious robber, the head of military
assassins slaying without declaration of war, without notice or
warning: I am sincerely afraid that, with many others, from fear
of superstition that leads to God, from hatred of the Church,
grounded on ignorance and prejudice, he too would have called
him an enlightened ruler, an "Imperatore galantuómo," though,
without even knowing that the CHURCH THROWS AWAY NO PEARL,
that no good children ever leave a true mother, I cannot harbour

the uncharitable supposition, that, like many courtiers and others, he would so far have forgotten his dignity, as gratuitously to press his polluted hand or kiss the seam of his dirty garment had he had the opportunity of doing so.

Besides ignorance and prejudice in matters of the Church presiding over many a noble and excellent heart with which we have to contend, there is also the canting pharisee, the artful scribe, and the whitened sepulchre; for "When cities abandon themselves to revelry, and are driven by terror-stricken consciences to the altar, then, while the lights are burning, a cloud of incense is rising, and EVERY KNEE IS BENDED, the walls totter and gape, and down falls the ponderous vault upon a thousand sinners." But "Here in these cooler climes, with more reasonable temperaments, and a purer faith," where its meaning is unknown, no one OUT OF THE CHURCH able to MAKE an act of faith, where mere opinion is substituted for belief, and where the Apostle's and the Nicene creeds are empty words, and, I hope in ALL CASES, an UNCONSCIOUS falsehood in the mouth of the reciter, "it is hoped that we do not need this awful language. The ALMIGHTY foot-fall is soft here, even in the earthquake and the storm."

Thus, according to the national Leader and presumptuous Instructor in the *Times*, of 8th October, 1863, there is no handwriting against the country, no "social evil," alias sink of iniquity, no infanticides, no crimes, no misery, no starvation cases unparalleled in civilized Europe, no criminal treatment of the poor, no emaciated and depopulated step-sister-island, no still idolatrous India; the slight earthquake* just experienced need excite no apprehension, the ring of Polycrates will for ever return, our SUPER-STRATAL happiness and prosperity will not leave us; why then pray, why the "mea culpa," why the "miserere," though danger seems nigh!—yet the penitent thief on the cross was saved

---

* It occurred on the 6th October, 1863, at 2.30 p.m.

tous; but if, what he says, he had heard to have taken place on account of a comet, it would have been as wonderful as if he had seen Lexell's comet paying a visit to the satellites of that stumbling block Jupiter, which it never did. But there are a GREAT MANY REAL THINGS which he never heard either, no more than the true history of what he sneers at, what with the accumulated science, learning, intellect and cleverness, of the WHOLE PROTESTANT, HEATHEN AND INFIDEL WORLD, he can neither explain nor reproduce, no more than any of the authenticated miracles which the Catholic Church believes and teaches; and if, of the inexhaustible treasure of works of the Catholic Church, ignored by an anti-Catholic and infidel press, and therefore unknown to the non-Catholic masses, he were to read but the most elementary treatise, so to speak, he would find immensely more than he ever "dreamt of in HIS philosophy," and our younger seminarists could instruct him with more precision in things divine than he ever attained in his most successful astronomical calculations, dissertations and theories.

However excusable the above may be, it cannot be pardoned in a learned man, and a christian, I suppose, when he says that it was "the MOST ABJECT SUPERSTITION THAT DETACHED HIS (Charles V.) THOUGHTS FROM EARTHLY THINGS AND FIXED THEM ON HIS ETERNAL INTERESTS!" Had the emperor not done penance had he had the misfortune to be an unrepenting, excommunicated, licentious, sacrilegious and perfidious robber, the head of military assassins slaying without declaration of war, without notice or warning: I am sincerely afraid that, with many others, from fear of superstition that leads to God, from hatred of the Church, grounded on ignorance and prejudice, he too would have called him an enlightened ruler, an "Imperatore galantuómo," though, without even knowing that the CHURCH THROWS AWAY NO PEARL, that no good children ever leave a true mother, I cannot harbour

the uncharitable supposition, that, like many courtiers and others, he would so far have forgotten his dignity, as gratuitously to press his polluted hand or kiss the seam of his dirty garment had he had the opportunity of doing so.

Besides ignorance and prejudice in matters of the Church presiding over many a noble and excellent heart with which we have to contend, there is also the canting pharisee, the artful scribe, and the whitened sepulchre; for " When cities abandon themselves to revelry, and are driven by terror-stricken consciences to the altar, then, while the lights are burning, a cloud of incense is rising, and EVERY KNEE IS BENDED, the walls totter and gape, and down falls the ponderous vault upon a thousand sinners." But "Here in these cooler climes, with more reasonable temperaments, and a purer faith," where its meaning is unknown, no one OUT OF THE CHURCH able to MAKE an act of faith, where mere opinion is substituted for belief, and where the Apostle's and the Nicene creeds are empty words, and, I hope in ALL CASES, an UNCONSCIOUS falsehood in the mouth of the reciter, "it is hoped that we do not need this awful language. The ALMIGHTY foot-fall is soft here, even in the earthquake and the storm."

Thus, according to the national Leader and presumptuous Instructor in the *Times*, of 8th October, 1863, there is no hand-writing against the country, no "social evil," alias sink of iniquity, no infanticides, no crimes, no misery, no starvation cases unparalleled in civilized Europe, no criminal treatment of the poor, no emaciated and depopulated step-sister-island, no still idolatrous India; the slight earthquake* just experienced need excite no apprehension, the ring of Polycrates will for ever return, our SUPER-STRATAL happiness and prosperity will not leave us; why then pray, why the "mea culpa," why the "miserere," though danger seems nigh!—yet the penitent thief on the cross was saved

---

* It occurred on the 6th October, 1863, at 2.30 p.m.

tous; but if, what he says, he had heard to have taken place
on account of a comet, it would have been as wonderful as if
he had seen Lexell's comet paying a visit to the satellites of
that stumbling block Jupiter, which it never did. But there
are a GREAT MANY REAL THINGS which he never heard either,
no more than the true history of what he sneers at, what with
the accumulated science, learning, intellect and cleverness, of
the WHOLE PROTESTANT, HEATHEN AND INFIDEL WORLD, he can
neither explain nor reproduce, no more than any of the authenticated miracles which the Catholic Church believes and teaches;
and if, of the inexhaustible treasure of works of the Catholic
Church, ignored by an anti-Catholic and infidel press, and therefore unknown to the non-Catholic masses, he were to read but
the most elementary treatise, so to speak, he would find immensely more than he ever "dreamt of in HIS philosophy," and
our younger seminarists could instruct him with more precision
in things divine than he ever attained in his most successful
astronomical calculations, dissertations and theories.

However excusable the above may be, it cannot be pardoned
in a learned man, and a christian, I suppose, when he says that
it was "the MOST ABJECT SUPERSTITION THAT DETACHED HIS
(Charles V.) THOUGHTS FROM EARTHLY THINGS AND FIXED THEM
ON HIS ETERNAL INTERESTS!" Had the emperor not done penance
had he had the misfortune to be an unrepenting, excommunicated,
licentious, sacrilegious and perfidious robber, the head of military
assassins slaying without declaration of war, without notice or
warning: I am sincerely afraid that, with many others, from fear
of superstition that leads to God, from hatred of the Church,
grounded on ignorance and prejudice, he too would have called
him an enlightened ruler, an "Imperatore galantuómo," though,
without even knowing that the CHURCH THROWS AWAY NO PEARL,
that no good children ever leave a true mother, I cannot harbour

the uncharitable supposition, that, like many courtiers and others, he would so far have forgotten his dignity, as gratuitously to press his polluted hand or kiss the seam of his dirty garment had he had the opportunity of doing so.

Besides ignorance and prejudice in matters of the Church presiding over many a noble and excellent heart with which we have to contend, there is also the canting pharisee, the artful scribe, and the whitened sepulchre; for "When cities abandon themselves to revelry, and are driven by terror-stricken consciences to the altar, then, while the lights are burning, a cloud of incense is rising, and EVERY KNEE IS BENDED, the walls totter and gape, and down falls the ponderous vault upon a thousand sinners." But "Here in these cooler climes, with more reasonable temperaments, and a purer faith," where its meaning is unknown, no one OUT OF THE CHURCH able to MAKE an act of faith, where mere opinion is substituted for belief, and where the Apostle's and the Nicene creeds are empty words, and, I hope in ALL CASES, an UNCONSCIOUS falsehood in the mouth of the reciter, "it is hoped that we do not need this awful language. The ALMIGHTY foot-fall is soft here, even in the earthquake and the storm."

Thus, according to the national Leader and presumptuous Instructor in the *Times*, of 8th October, 1863, there is no handwriting against the country, no "social evil," alias sink of iniquity, no infanticides, no crimes, no misery, no starvation cases unparalleled in civilized Europe, no criminal treatment of the poor, no emaciated and depopulated step-sister-island, no still idolatrous India; the slight earthquake* just experienced need excite no apprehension, the ring of Polycrates will for ever return, our SUPER-STRATAL happiness and prosperity will not leave us; why then pray, why the "mea culpa," why the "miserere," though danger seems nigh !—yet the penitent thief on the cross was saved

---

* It occurred on the 6th October, 1863, at 2.30 p.m.

tous; but if, what he says, he had heard to have taken place on account of a comet, it would have been as wonderful as if he had seen Lexell's comet paying a visit to the satellites of that stumbling block Jupiter, which it never did. But there are a GREAT MANY REAL THINGS which he never heard either, no more than the true history of what he sneers at, what with the accumulated science, learning, intellect and cleverness, of the WHOLE PROTESTANT, HEATHEN AND INFIDEL WORLD, he can neither explain nor reproduce, no more than any of the authenticated miracles which the Catholic Church believes and teaches; and if, of the inexhaustible treasure of works of the Catholic Church, ignored by an anti-Catholic and infidel press, and therefore unknown to the non-Catholic masses, he were to read but the most elementary treatise, so to speak, he would find immensely more than he ever "dreamt of in HIS philosophy," and our younger seminarists could instruct him with more precision in things divine than he ever attained in his most successful astronomical calculations, dissertations and theories.

However excusable the above may be, it cannot be pardoned in a learned man, and a christian, I suppose, when he says that it was "the MOST ABJECT SUPERSTITION THAT DETACHED HIS (Charles V.) THOUGHTS FROM EARTHLY THINGS AND FIXED THEM ON HIS ETERNAL INTERESTS!" Had the emperor not done penance had he had the misfortune to be an unrepenting, excommunicated, licentious, sacrilegious and perfidious robber, the head of military assassins slaying without declaration of war, without notice or warning: I am sincerely afraid that, with many others, from fear of superstition that leads to God, from hatred of the Church, grounded on ignorance and prejudice, he too would have called him an enlightened ruler, an "Imperatore galantuómo," though, without even knowing that the CHURCH THROWS AWAY NO PEARL, that no good children ever leave a true mother, I cannot harbour

the uncharitable supposition, that, like many courtiers and others, he would so far have forgotten his dignity, as gratuitously to press his polluted hand or kiss the seam of his dirty garment had he had the opportunity of doing so.

Besides ignorance and prejudice in matters of the Church presiding over many a noble and excellent heart with which we have to contend, there is also the canting pharisee, the artful scribe, and the whitened sepulchre; for " When cities abandon themselves to revelry, and are driven by terror-stricken consciences to the altar, then, while the lights are burning, a cloud of incense is rising, and EVERY KNEE IS BENDED, the walls totter and gape, and down falls the ponderous vault upon a thousand sinners." But "Here in these cooler climes, with more reasonable temperaments, and a purer faith," where its meaning is unknown, no one OUT OF THE CHURCH able to MAKE an act of faith, where mere opinion is substituted for belief, and where the Apostle's and the Nicene creeds are empty words, and, I hope in ALL CASES, an UNCONSCIOUS falsehood in the mouth of the reciter, "it is hoped that we do not need this awful language. The ALMIGHTY foot-fall is soft here, even in the earthquake and the storm."

Thus, according to the national Leader and presumptuous Instructor in the *Times*, of 8th October, 1863, there is no handwriting against the country, no "social evil," alias sink of iniquity, no infanticides, no crimes, no misery, no starvation cases unparalleled in civilized Europe, no criminal treatment of the poor, no emaciated and depopulated step-sister-island, no still idolatrous India; the slight earthquake* just experienced need excite no apprehension, the ring of Polycrates will for ever return, our SUPER-STRATAL happiness and prosperity will not leave us ; why then pray, why the "mea culpa," why the "miserere," though danger seems nigh !—yet the penitent thief on the cross was saved

---

* It occurred on the 6th October, 1863, at 2.30 p.m.

tous; but if, what he says, he had heard to have taken place on account of a comet, it would have been as wonderful as if he had seen Lexell's comet paying a visit to the satellites of that stumbling block Jupiter, which it never did. But there are a GREAT MANY REAL THINGS which he never heard either, no more than the true history of what he sneers at, what with the accumulated science, learning, intellect and cleverness, of the WHOLE PROTESTANT, HEATHEN AND INFIDEL WORLD, he can neither explain nor reproduce, no more than any of the authenticated miracles which the Catholic Church believes and teaches; and if, of the inexhaustible treasure of works of the Catholic Church, ignored by an anti-Catholic and infidel press, and therefore unknown to the non-Catholic masses, he were to read but the most elementary treatise, so to speak, he would find immensely more than he ever "dreamt of in HIS philosophy," and our younger seminarists could instruct him with more precision in things divine than he ever attained in his most successful astronomical calculations, dissertations and theories.

However excusable the above may be, it cannot be pardoned in a learned man, and a christian, I suppose, when he says that it was "the MOST ABJECT SUPERSTITION THAT DETACHED HIS (Charles V.) THOUGHTS FROM EARTHLY THINGS AND FIXED THEM ON HIS ETERNAL INTERESTS!" Had the emperor not done penance had he had the misfortune to be an unrepenting, excommunicated, licentious, sacrilegious and perfidious robber, the head of military assassins slaying without declaration of war, without notice or warning: I am sincerely afraid that, with many others, from fear of superstition that leads to God, from hatred of the Church, grounded on ignorance and prejudice, he too would have called him an enlightened ruler, an "Imperatore galantuómo," though, without even knowing that the CHURCH THROWS AWAY NO PEARL, that no good children ever leave a true mother, I cannot harbour

the uncharitable supposition, that, like many courtiers and others, he would so far have forgotten his dignity, as gratuitously to press his polluted hand or kiss the seam of his dirty garment had he had the opportunity of doing so.

Besides ignorance and prejudice in matters of the Church presiding over many a noble and excellent heart with which we have to contend, there is also the canting pharisee, the artful scribe, and the whitened sepulchre; for "When cities abandon themselves to revelry, and are driven by terror-stricken consciences to the altar, then, while the lights are burning, a cloud of incense is rising, and EVERY KNEE IS BENDED, the walls totter and gape, and down falls the ponderous vault upon a thousand sinners." But "Here in these cooler climes, with more reasonable temperaments, and a purer faith," where its meaning is unknown, no one OUT OF THE CHURCH able to MAKE an act of faith, where mere opinion is substituted for belief, and where the Apostle's and the Nicene creeds are empty words, and, I hope in ALL CASES, an UNCONSCIOUS falsehood in the mouth of the reciter, "it is hoped that we do not need this awful language. The ALMIGHTY foot-fall is soft here, even in the earthquake and the storm."

Thus, according to the national Leader and presumptuous Instructor in the *Times*, of 8th October, 1863, there is no handwriting against the country, no "social evil," alias sink of iniquity, no infanticides, no crimes, no misery, no starvation cases unparalleled in civilized Europe, no criminal treatment of the poor, no emaciated and depopulated step-sister-island, no still idolatrous India; the slight earthquake* just experienced need excite no apprehension, the ring of Polycrates will for ever return, our SUPER-STRATAL happiness and prosperity will not leave us; why then pray, why the "mea culpa," why the "miserere," though danger seems nigh!—yet the penitent thief on the cross was saved

* It occurred on the 6th October, 1863, at 2.30 p.m.

tous; but if, what be says, he had heard to have taken place on account of a comet, it would have been as wonderful as if he had seen Lexell's comet paying a visit to the satellites of that stumbling block Jupiter, which it never did. But there are a GREAT MANY REAL THINGS which he never heard either, no more than the true history of what he sneers at, what with the accumulated science, learning, intellect and cleverness, of the WHOLE PROTESTANT, HEATHEN AND INFIDEL WORLD, he can neither explain nor reproduce, no more than any of the authenticated miracles which the Catholic Church believes and teaches; and if, of the inexhaustible treasure of works of the Catholic Church, ignored by an anti-Catholic and infidel press, and therefore unknown to the non-Catholic masses, he were to read but the most elementary treatise, so to speak, he would find immensely more than he ever "dreamt of in HIS philosophy," and our younger seminarists could instruct him with more precision in things divine than he ever attained in his most successful astronomical calculations, dissertations and theories.

However excusable the above may be, it cannot be pardoned in a learned man, and a christian, I suppose, when he says that it was "the MOST ABJECT SUPERSTITION THAT DETACHED HIS (Charles V.) THOUGHTS FROM EARTHLY THINGS AND FIXED THEM ON HIS ETERNAL INTERESTS!'" Had the emperor not done penance had he had the misfortune to be an unrepenting, excommunicated, licentious, sacrilegious and perfidious robber, the head of military assassins slaying without declaration of war, without notice or warning: I am sincerely afraid that, with many others, from fear of superstition that leads to God, from hatred of the Church, grounded on ignorance and prejudice, he too would have called him an enlightened ruler, an "Imperatore galantuómo," though, without even knowing that the CHURCH THROWS AWAY NO PEARL, that no good children ever leave a true mother, I cannot harbour

the uncharitable supposition, that, like many courtiers and others, he would so far have forgotten his dignity, as gratuitously to press his polluted hand or kiss the seam of his dirty garment had he had the opportunity of doing so.

Besides ignorance and prejudice in matters of the Church presiding over many a noble and excellent heart with which we have to contend, there is also the canting pharisee, the artful scribe, and the whitened sepulchre; for "When cities abandon themselves to revelry, and are driven by terror-stricken consciences to the altar, then, while the lights are burning, a cloud of incense is rising, and EVERY KNEE IS BENDED, the walls totter and gape, and down falls the ponderous vault upon a thousand sinners." But "Here in these cooler climes, with more reasonable temperaments, and a purer faith," where its meaning is unknown, no one OUT OF THE CHURCH able to MAKE an act of faith, where mere opinion is substituted for belief, and where the Apostle's and the Nicene creeds are empty words, and, I hope in ALL CASES, an UNCONSCIOUS falsehood in the mouth of the reciter, "it is hoped that we do not need this awful language. The ALMIGHTY foot-fall is soft here, even in the earthquake and the storm."

Thus, according to the national Leader and presumptuous Instructor in the *Times*, of 8th October, 1863, there is no handwriting against the country, no "social evil," alias sink of iniquity, no infanticides, no crimes, no misery, no starvation cases unparalleled in civilized Europe, no criminal treatment of the poor, no emaciated and depopulated step-sister-island, no still idolatrous India; the slight earthquake* just experienced need excite no apprehension, the ring of Polycrates will for ever return, our SUPER-STRATAL happiness and prosperity will not leave us; why then pray, why the "mea culpa," why the "miserere," though danger seems nigh!—yet the penitent thief on the cross was saved

---

* It occurred on the 6th October, 1863, at 2.30 p.m.

tous; but if, what he says, he had heard to have taken place on account of a comet, it would have been as wonderful as if he had seen Lexell's comet paying a visit to the satellites of that stumbling block Jupiter, which it never did. But there are a GREAT MANY REAL THINGS which he never heard either, no more than the true history of what he sneers at, what with the accumulated science, learning, intellect and cleverness, of the WHOLE PROTESTANT, HEATHEN AND INFIDEL WORLD, he can neither explain nor reproduce, no more than any of the authenticated miracles which the Catholic Church believes and teaches; and if, of the inexhaustible treasure of works of the Catholic Church, ignored by an anti-Catholic and infidel press, and therefore unknown to the non-Catholic masses, he were to read but the most elementary treatise, so to speak, he would find immensely more than he ever "dreamt of in HIS philosophy," and our younger seminarists could instruct him with more precision in things divine than he ever attained in his most successful astronomical calculations, dissertations and theories.

However excusable the above may be, it cannot be pardoned in a learned man, and a christian, I suppose, when he says that it was "the MOST ABJECT SUPERSTITION THAT DETACHED HIS (Charles V.) THOUGHTS FROM EARTHLY THINGS AND FIXED THEM ON HIS ETERNAL INTERESTS!" Had the emperor not done penance had he had the misfortune to be an unrepenting, excommunicated, licentious, sacrilegious and perfidious robber, the head of military assassins slaying without declaration of war, without notice or warning: I am sincerely afraid that, with many others, from fear of superstition that leads to God, from hatred of the Church, grounded on ignorance and prejudice, he too would have called him an enlightened ruler, an "Imperatore galantuómo," though, without even knowing that the CHURCH THROWS AWAY NO PEARL, that no good children ever leave a true mother, I cannot harbour

the uncharitable supposition, that, like many courtiers and others, he would so far have forgotten his dignity, as gratuitously to press his polluted hand or kiss the seam of his dirty garment had he had the opportunity of doing so.

Besides ignorance and prejudice in matters of the Church presiding over many a noble and excellent heart with which we have to contend, there is also the canting pharisee, the artful scribe, and the whitened sepulchre; for " When cities abandon themselves to revelry, and are driven by terror-stricken consciences to the altar, then, while the lights are burning, a cloud of incense is rising, and EVERY KNEE IS BENDED, the walls totter and gape, and down falls the ponderous vault upon a thousand sinners." But " Here in these cooler climes, with more reasonable temperaments, and a purer faith," where its meaning is unknown, no one OUT OF THE CHURCH able to MAKE an act of faith, where mere opinion is substituted for belief, and where the Apostle's and the Nicene creeds are empty words, and, I hope in ALL CASES, an UNCONSCIOUS falsehood in the mouth of the reciter, " it is hoped that we do not need this awful language. The ALMIGHTY foot-fall is soft here, even in the earthquake and the storm."

Thus, according to the national Leader and presumptuous Instructor in the *Times*, of 8th October, 1863, there is no handwriting against the country, no "social evil," alias sink of iniquity, no infanticides, no crimes, no misery, no starvation cases unparalleled in civilized Europe, no criminal treatment of the poor, no emaciated and depopulated step-sister-island, no still idolatrous India; the slight earthquake* just experienced need excite no apprehension, the ring of Polycrates will for ever return, our SUPER-STRATAL happiness and prosperity will not leave us; why then pray, why the "mea culpa," why the "miserere," though danger seems nigh !—yet the penitent thief on the cross was saved

* It occurred on the 6th October, 1863, at 2.30 p.m.

tous; but if, what he says, he had heard to have taken place on account of a comet, it would have been as wonderful as if he had seen Lexell's comet paying a visit to the satellites of that stumbling block Jupiter, which it never did. But there are a GREAT MANY REAL THINGS which he never heard either, no more than the true history of what he sneers at, what with the accumulated science, learning, intellect and cleverness, of the WHOLE PROTESTANT, HEATHEN AND INFIDEL WORLD, he can neither explain nor reproduce, no more than any of the authenticated miracles which the Catholic Church believes and teaches; and if, of the inexhaustible treasure of works of the Catholic Church, ignored by an anti-Catholic and infidel press, and therefore unknown to the non-Catholic masses, he were to read but the most elementary treatise, so to speak, he would find immensely more than he ever "dreamt of in HIS philosophy," and our younger seminarists could instruct him with more precision in things divine than he ever attained in his most successful astronomical calculations, dissertations and theories.

However excusable the above may be, it cannot be pardoned in a learned man, and a christian, I suppose, when he says that it was "the MOST ABJECT SUPERSTITION THAT DETACHED HIS (Charles V.) THOUGHTS FROM EARTHLY THINGS AND FIXED THEM ON HIS ETERNAL INTERESTS!" Had the emperor not done penance had he had the misfortune to be an unrepenting, excommunicated, licentious, sacrilegious and perfidious robber, the head of military assassins slaying without declaration of war, without notice or warning: I am sincerely afraid that, with many others, from fear of superstition that leads to God, from hatred of the Church, grounded on ignorance and prejudice, he too would have called him an enlightened ruler, an "Imperatore galantuómo," though, without even knowing that the CHURCH THROWS AWAY NO PEARL, that no good children ever leave a true mother, I cannot harbour

the uncharitable supposition, that, like many courtiers and others, he would so far have forgotten his dignity, as gratuitously to press his polluted hand or kiss the seam of his dirty garment had he had the opportunity of doing so.

Besides ignorance and prejudice in matters of the Church presiding over many a noble and excellent heart with which we have to contend, there is also the canting pharisee, the artful scribe, and the whitened sepulchre; for " When cities abandon themselves to revelry, and are driven by terror-stricken consciences to the altar, then, while the lights are burning, a cloud of incense is rising, and EVERY KNEE IS BENDED, the walls totter and gape, and down falls the ponderous vault upon a thousand sinners." But " Here in these cooler climes, with more reasonable temperaments, and a purer faith," where its meaning is unknown, no one OUT OF THE CHURCH able to MAKE an act of faith, where mere opinion is substituted for belief, and where the Apostle's and the Nicene creeds are empty words, and, I hope in ALL CASES, an UNCONSCIOUS falsehood in the mouth of the reciter, " it is hoped that we do not need this awful language. The ALMIGHTY foot-fall is soft here, even in the earthquake and the storm."

Thus, according to the national Leader and presumptuous Instructor in the *Times*, of 8th October, 1863, there is no hand-writing against the country, no "social evil," alias sink of iniquity, no infanticides, no crimes, no misery, no starvation cases unparalleled in civilized Europe, no criminal treatment of the poor, no emaciated and depopulated step-sister-island, no still idolatrous India; the slight earthquake° just experienced need excite no apprehension, the ring of Polycrates will for ever return, our SUPER-STRATAL happiness and prosperity will not leave us ; why then pray, why the " mea culpa," why the " miserere," though danger seems nigh !—yet the penitent thief on the cross was saved

---

* It occurred on the 6th October, 1863, at 2.30 p.m.

tous; but if, what he says, he had heard to have taken place
on account of a comet, it would have been as wonderful as if
he had seen Lexell's comet paying a visit to the satellites of
that stumbling block Jupiter, which it never did. But there
are a GREAT MANY REAL THINGS which he never heard either,
no more than the true history of what he sneers at, what with
the accumulated science, learning, intellect and cleverness, of
the WHOLE PROTESTANT, HEATHEN AND INFIDEL WORLD, he can
neither explain nor reproduce, no more than any of the authenticated miracles which the Catholic Church believes and teaches;
and if, of the inexhaustible treasure of works of the Catholic
Church, ignored by an anti-Catholic and infidel press, and therefore unknown to the non-Catholic masses, he were to read but
the most elementary treatise, so to speak, he would find immensely more than he ever "dreamt of in HIS philosophy," and
our younger seminarists could instruct him with more precision
in things divine than he ever attained in his most successful
astronomical calculations, dissertations and theories.

However excusable the above may be, it cannot be pardoned
in a learned man, and a christian, I suppose, when he says that
it was "the MOST ABJECT SUPERSTITION THAT DETACHED HIS
(Charles V.) THOUGHTS FROM EARTHLY THINGS AND FIXED THEM
ON HIS ETERNAL INTERESTS!" Had the emperor not done penance
had he had the misfortune to be an unrepenting, excommunicated,
licentious, sacrilegious and perfidious robber, the head of military
assassins slaying without declaration of war, without notice or
warning: I am sincerely afraid that, with many others, from fear
of superstition that leads to God, from hatred of the Church,
grounded on ignorance and prejudice, he too would have called
him an enlightened ruler, an "Imperatore galantuómo," though,
without even knowing that the CHURCH THROWS AWAY NO PEARL,
that no good children ever leave a true mother, I cannot harbour

the uncharitable supposition, that, like many courtiers and others, he would so far have forgotten his dignity, as gratuitously to press his polluted hand or kiss the seam of his dirty garment had he had the opportunity of doing so.

Besides ignorance and prejudice in matters of the Church presiding over many a noble and excellent heart with which we have to contend, there is also the canting pharisee, the artful scribe, and the whitened sepulchre; for " When cities abandon themselves to revelry, and are driven by terror-stricken consciences to the altar, then, while the lights are burning, a cloud of incense is rising, and EVERY KNEE IS BENDED, the walls totter and gape, and down falls the ponderous vault upon a thousand sinners." But "Here in these cooler climes, with more reasonable temperaments, and a purer faith," where its meaning is unknown, no one OUT OF THE CHURCH able to MAKE an act of faith, where mere opinion is substituted for belief, and where the Apostle's and the Nicene creeds are empty words, and, I hope in ALL CASES, an UNCONSCIOUS falsehood in the mouth of the reciter, " it is hoped that we do not need this awful language. The ALMIGHTY foot-fall is soft here, even in the earthquake and the storm."

Thus, according to the national Leader and presumptuous Instructor in the *Times*, of 8th October, 1863, there is no handwriting against the country, no "social evil," alias sink of iniquity, no infanticides, no crimes, no misery, no starvation cases unparalleled in civilized Europe, no criminal treatment of the poor, no emaciated and depopulated step-sister-island, no still idolatrous India; the slight earthquake* just experienced need excite no apprehension, the ring of Polycrates will for ever return, our SUPER-STRATAL happiness and prosperity will not leave us; why then pray, why the "mea culpa," why the "miserere," though danger seems nigh !—yet the penitent thief on the cross was saved

---

* It occurred on the 6th October, 1863, at 2.30 p.m.

tous; but if, what he says, he had heard to have taken place on account of a comet, it would have been as wonderful as if he had seen Lexell's comet paying a visit to the satellites of that stumbling block Jupiter, which it never did.   But there are a GREAT MANY REAL THINGS which he never heard either, no more than the true history of what he sneers at, what with the accumulated science, learning, intellect and cleverness, of the WHOLE PROTESTANT, HEATHEN AND INFIDEL WORLD, he can neither explain nor reproduce, no more than any of the authenticated miracles which the Catholic Church believes and teaches; and if, of the inexhaustible treasure of works of the Catholic Church, ignored by an anti-Catholic and infidel press, and therefore unknown to the non-Catholic masses, he were to read but the most elementary treatise, so to speak, he would find immensely more than he ever "dreamt of in HIS philosophy," and our younger seminarists could instruct him with more precision in things divine than he ever attained in his most successful astronomical calculations, dissertations and theories.

However excusable the above may be, it cannot be pardoned in a learned man, and a christian, I suppose, when he says that it was "the MOST ABJECT SUPERSTITION THAT DETACHED HIS (Charles V.) THOUGHTS FROM EARTHLY THINGS AND FIXED THEM ON HIS ETERNAL INTERESTS!" Had the emperor not done penance had he had the misfortune to be an unrepenting, excommunicated, licentious, sacrilegious and perfidious robber, the head of military assassins slaying without declaration of war, without notice or warning: I am sincerely afraid that, with many others, from fear of superstition that leads to God, from hatred of the Church, grounded on ignorance and prejudice, he too would have called him an enlightened ruler, an "Imperatore galantuómo," though, without even knowing that the CHURCH THROWS AWAY NO PEARL, that no good children ever leave a true mother, I cannot harbour

the uncharitable supposition, that, like many courtiers and others, he would so far have forgotten his dignity, as gratuitously to press his polluted hand or kiss the seam of his dirty garment had he had the opportunity of doing so.

Besides ignorance and prejudice in matters of the Church presiding over many a noble and excellent heart with which we have to contend, there is also the canting pharisee, the artful scribe, and the whitened sepulchre; for " When cities abandon themselves to revelry, and are driven by terror-stricken consciences to the altar, then, while the lights are burning, a cloud of incense is rising, and EVERY KNEE IS BENDED, the walls totter and gape, and down falls the ponderous vault upon a thousand sinners." But " Here in these cooler climes, with more reasonable temperaments, and a purer faith," where its meaning is unknown, no one OUT OF THE CHURCH able to MAKE an act of faith, where mere opinion is substituted for belief, and where the Apostle's and the Nicene creeds are empty words, and, I hope in ALL CASES, an UNCONSCIOUS falsehood in the mouth of the reciter, " it is hoped that we do not need this awful language. The ALMIGHTY foot-fall is soft here, even in the earthquake and the storm."

Thus, according to the national Leader and presumptuous Instructor in the *Times*, of 8th October, 1863, there is no handwriting against the country, no "social evil," alias sink of iniquity, no infanticides, no crimes, no misery, no starvation cases unparalleled in civilized Europe, no criminal treatment of the poor, no emaciated and depopulated step-sister-island, no still idolatrous India; the slight earthquake* just experienced need excite no apprehension, the ring of Polycrates will for ever return, our SUPER-STRATAL happiness and prosperity will not leave us; why then pray, why the "mea culpa," why the "miserere," though danger seems nigh !—yet the penitent thief on the cross was saved

---

* It occurred on the 6th October, 1863, at 2.30 p.m.

tous; but if, what he says, he had heard to have taken place on account of a comet, it would have been as wonderful as if he had seen Lexell's comet paying a visit to the satellites of that stumbling block Jupiter, which it never did. But there are a GREAT MANY REAL THINGS which he never heard either, no more than the true history of what he sneers at, what with the accumulated science, learning, intellect and cleverness, of the WHOLE PROTESTANT, HEATHEN AND INFIDEL WORLD, he can neither explain nor reproduce, no more than any of the authenticated miracles which the Catholic Church believes and teaches; and if, of the inexhaustible treasure of works of the Catholic Church, ignored by an anti-Catholic and infidel press, and therefore unknown to the non-Catholic masses, he were to read but the most elementary treatise, so to speak, he would find immensely more than he ever "dreamt of in HIS philosophy," and our younger seminarists could instruct him with more precision in things divine than he ever attained in his most successful astronomical calculations, dissertations and theories.

However excusable the above may be, it cannot be pardoned in a learned man, and a christian, I suppose, when he says that it was "the MOST ABJECT SUPERSTITION THAT DETACHED HIS (Charles V.) THOUGHTS FROM EARTHLY THINGS AND FIXED THEM ON HIS ETERNAL INTERESTS!" Had the emperor not done penance had he had the misfortune to be an unrepenting, excommunicated, licentious, sacrilegious and perfidious robber, the head of military assassins slaying without declaration of war, without notice or warning: I am sincerely afraid that, with many others, from fear of superstition that leads to God, from hatred of the Church, grounded on ignorance and prejudice, he too would have called him an enlightened ruler, an "Imperatore galantuómo," though, without even knowing that the CHURCH THROWS AWAY NO PEARL, that no good children ever leave a true mother, I cannot harbour

the uncharitable supposition, that, like many courtiers and others, he would so far have forgotten his dignity, as gratuitously to press his polluted hand or kiss the seam of his dirty garment had he had the opportunity of doing so.

Besides ignorance and prejudice in matters of the Church presiding over many a noble and excellent heart with which we have to contend, there is also the canting pharisee, the artful scribe, and the whitened sepulchre; for "When cities abandon themselves to revelry, and are driven by terror-stricken consciences to the altar, then, while the lights are burning, a cloud of incense is rising, and EVERY KNEE IS BENDED, the walls totter and gape, and down falls the ponderous vault upon a thousand sinners." But "Here in these cooler climes, with more reasonable temperaments, and a purer faith," where its meaning is unknown, no one OUT OF THE CHURCH able to MAKE an act of faith, where mere opinion is substituted for belief, and where the Apostle's and the Nicene creeds are empty words, and, I hope in ALL CASES, an UNCONSCIOUS falsehood in the mouth of the reciter, "it is hoped that we do not need this awful language. The ALMIGHTY foot-fall is soft here, even in the earthquake and the storm."

Thus, according to the national Leader and presumptuous Instructor in the *Times*, of 8th October, 1863, there is no hand-writing against the country, no "social evil," alias sink of iniquity, no infanticides, no crimes, no misery, no starvation cases unparalleled in civilized Europe, no criminal treatment of the poor, no emaciated and depopulated step-sister-island, no still idolatrous India; the slight earthquake* just experienced need excite no apprehension, the ring of Polycrates will for ever return, our SUPER-STRATAL happiness and prosperity will not leave us; why then pray, why the "mea culpa," why the "miserere," though danger seems nigh!—yet the penitent thief on the cross was saved

---

* It occurred on the 6th October, 1863, at 2.30 p.m.

tous; but if, what he says, he had heard to have taken place on account of a comet, it would have been as wonderful as if he had seen Lexell's comet paying a visit to the satellites of that stumbling block Jupiter, which it never did. But there are a GREAT MANY REAL THINGS which he never heard either, no more than the true history of what he sneers at, what with the accumulated science, learning, intellect and cleverness, of the WHOLE PROTESTANT, HEATHEN AND INFIDEL WORLD, he can neither explain nor reproduce, no more than any of the authenticated miracles which the Catholic Church believes and teaches; and if, of the inexhaustible treasure of works of the Catholic Church, ignored by an anti-Catholic and infidel press, and therefore unknown to the non-Catholic masses, he were to read but the most elementary treatise, so to speak, he would find immensely more than he ever "dreamt of in HIS philosophy," and our younger seminarists could instruct him with more precision in things divine than he ever attained in his most successful astronomical calculations, dissertations and theories.

However excusable the above may be, it cannot be pardoned in a learned man, and a christian, I suppose, when he says that it was "the MOST ABJECT SUPERSTITION THAT DETACHED HIS (Charles V.) THOUGHTS FROM EARTHLY THINGS AND FIXED THEM ON HIS ETERNAL INTERESTS!" Had the emperor not done penance had he had the misfortune to be an unrepenting, excommunicated, licentious, sacrilegious and perfidious robber, the head of military assassins slaying without declaration of war, without notice or warning: I am sincerely afraid that, with many others, from fear of superstition that leads to God, from hatred of the Church, grounded on ignorance and prejudice, he too would have called him an enlightened ruler, an "Imperatore galantuómo," though, without even knowing that the CHURCH THROWS AWAY NO PEARL, that no good children ever leave a true mother, I cannot harbour

the uncharitable supposition, that, like many courtiers and others, he would so far have forgotten his dignity, as gratuitously to press his polluted hand or kiss the seam of his dirty garment had he had the opportunity of doing so.

Besides ignorance and prejudice in matters of the Church presiding over many a noble and excellent heart with which we have to contend, there is also the canting pharisee, the artful scribe, and the whitened sepulchre; for " When cities abandon themselves to revelry, and are driven by terror-stricken consciences to the altar, then, while the lights are burning, a cloud of incense is rising, and EVERY KNEE IS BENDED, the walls totter and gape, and down falls the ponderous vault upon a thousand sinners." But " Here in these cooler climes, with more reasonable temperaments, and a purer faith," where its meaning is unknown, no one OUT OF THE CHURCH able to MAKE an act of faith, where mere opinion is substituted for belief, and where the Apostle's and the Nicene creeds are empty words, and, I hope in ALL CASES, an UNCONSCIOUS falsehood in the mouth of the reciter, " it is hoped that we do not need this awful language. The ALMIGHTY foot-fall is soft here, even in the earthquake and the storm."

Thus, according to the national Leader and presumptuous Instructor in the *Times*, of 8th October, 1863, there is no handwriting against the country, no "social evil," alias sink of iniquity, no infanticides, no crimes, no misery, no starvation cases unparalleled in civilized Europe, no criminal treatment of the poor, no emaciated and depopulated step-sister-island, no still idolatrous India; the slight earthquake⊙ just experienced need excite no apprehension, the ring of Polycrates will for ever return, our SUPER-STRATAL happiness and prosperity will not leave us; why then pray, why the " mea culpa," why the " miserere," though danger seems nigh !—yet the penitent thief on the cross was saved

---

* It occurred on the 6th October, 1863, at 2.30 p.m.

tous; but if, what he says, he had heard to have taken place
on account of a comet, it would have been as wonderful as if
he had seen Lexell's comet paying a visit to the satellites of
that stumbling block Jupiter, which it never did. But there
are a GREAT MANY REAL THINGS which he never heard either,
no more than the true history of what he sneers at, what with
the accumulated science, learning, intellect and cleverness, of
the WHOLE PROTESTANT, HEATHEN AND INFIDEL WORLD, he can
neither explain nor reproduce, no more than any of the authenticated miracles which the Catholic Church believes and teaches;
and if, of the inexhaustible treasure of works of the Catholic
Church, ignored by an anti-Catholic and infidel press, and therefore unknown to the non-Catholic masses, he were to read but
the most elementary treatise, so to speak, he would find immensely more than he ever "dreamt of in HIS philosophy," and
our younger seminarists could instruct him with more precision
in things divine than he ever attained in his most successful
astronomical calculations, dissertations and theories.

However excusable the above may be, it cannot be pardoned
in a learned man, and a christian, I suppose, when he says that
it was "the MOST ABJECT SUPERSTITION THAT DETACHED HIS
(Charles V.)THOUGHTS FROM EARTHLY THINGS AND FIXED THEM
ON HIS ETERNAL INTERESTS!" Had the emperor not done penance
had he had the misfortune to be an unrepenting, excommunicated,
licentious, sacrilegious and perfidious robber, the head of military
assassins slaying without declaration of war, without notice or
warning: I am sincerely afraid that, with many others, from fear
of superstition that leads to God, from hatred of the Church,
grounded on ignorance and prejudice, he too would have called
him an enlightened ruler, an "Imperatore galantuómo," though,
without even knowing that the CHURCH THROWS AWAY NO PEARL,
that no good children ever leave a true mother, I cannot harbour

the uncharitable supposition, that, like many courtiers and others, he would so far have forgotten his dignity, as gratuitously to press his polluted hand or kiss the seam of his dirty garment had he had the opportunity of doing so.

Besides ignorance and prejudice in matters of the Church presiding over many a noble and excellent heart with which we have to contend, there is also the canting pharisee, the artful scribe, and the whitened sepulchre; for " When cities abandon themselves to revelry, and are driven by terror-stricken consciences to the altar, then, while the lights are burning, a cloud of incense is rising, and EVERY KNEE IS BENDED, the walls totter and gape, and down falls the ponderous vault upon a thousand sinners." But " Here in these cooler climes, with more reasonable temperaments, and a purer faith," where its meaning is unknown, no one OUT OF THE CHURCH able to MAKE an act of faith, where mere opinion is substituted for belief, and where the Apostle's and the Nicene creeds are empty words, and, I hope in ALL CASES, an UNCONSCIOUS falsehood in the mouth of the reciter, " it is hoped that we do not need this awful language. The ALMIGHTY foot-fall is soft here, even in the earthquake and the storm."

Thus, according to the national Leader and presumptuous Instructor in the *Times*, of 8th October, 1863, there is no handwriting against the country, no "social evil," alias sink of iniquity, no infanticides, no crimes, no misery, no starvation cases unparalleled in civilized Europe, no criminal treatment of the poor, no emaciated and depopulated step-sister-island, no still idolatrous India; the slight earthquake* just experienced need excite no apprehension, the ring of Polycrates will for ever return, our SUPER-STRATAL happiness and prosperity will not leave us ; why then pray, why the " mea culpa," why the " miserere," though danger seems nigh !—yet the penitent thief on the cross was saved

---

* It occurred on the 6th October, 1863, at 2.30 p.m.

the very moment the wall fell that separated him from eternity—and from his God! And what good christian would not rather be surprised and perish with the sinner in Church, than quietly rest with the sinner and pharisee in bed? And who would not say with me: "Thou hypocrite, first cast out the beam out of thine own eye; and THEN shalt thou see clearly to cast out the mote out of thy brother's eye!"

"Initium sapientiæ timor Domini." "The fear of the Lord is the beginning of wisdom." If then the phenomena of nature have swayed and do sway the moral and physical life of mankind, and in bold relief set up in the history of man the indelible evidence of the law of grace and justice going hand-in-hand with that of nature: that BLESSED superstition will continue to the end of the world, "per omnia sæcula sæculorum," to teach man that salutary fear of God which will bring him to ETERNAL LIFE, whenever the pregnant cloud hovers above his head, whenever the flaming comet reminds him of the flaming sword that drove, and still keeps him out of a forfeited, but to be re-conquered, paradise; whenever, like a thief in the night—taught so by super-wise men of science themselves—the comet brings with it the chance of breaking up the earth, or destroying all life upon it, whilst, in reality, there is more to be dreaded from the wicked and ignorant who, at that very time, under the high-sounding names of progress and civilization of which they do not know the meaning, deride the idea and practice of praying against danger of whatever kind, real or imaginary, not knowing, or ignoring, that prayer will always be heard and in some way or other rewarded.

In his *Outlines*, p. 341, Sir John Herschel, speaking of Biela's comet, says: "Its orbit, by a remarkable coincidence, very nearly intersects that of the earth; and had the latter, at the time of its passage in 1832, been a month in advance of its actual place, it would have passed through the comet,—a singular rencontre,

perhaps not unattended with danger." And Mädler, falling foul
of the priest: does he not himself, in his *Popular Astronomy*
p. 405, say, according to *Baby Worlds*, p. 100: "In truth, the
chance of a collision between a comet and the earth should not
be calculated by hundreds of thousands or millions of years—IT
MAY HAPPEN OVER NIGHT; and no Laplace will save us from
destruction, if it is thus to come upon us."

With THIS DANGER OVER NIGHT, with all the "IFS" of astronomy, the kindling of houses by fragments of exploded meteors,
the precipitation of enormous masses of iron, said to come from
beyond the earth; the extensive shower of stones which attended
a falling star at Angers on the 9th June, 1822; the remains of
Olbers' shattered planet before their mind, and its molten inside,
perhaps, still dropping down somewhere in space upon some unfortunate body, or in the shape of dispersed comets flying about
to set other worlds on fire or causing them to burst: are there no
grounds for astronomers even to fear? and what wonder if
thousands of less enlightened persons of all classes in England
were really and truly frightened, and multitudes leaving London
in 1832, lest there should be a collision of the earth with Biela's
comet at the time of its appearance? And would it have been
blamable, would it have been superstitious, nay, would it not
have been beautiful, even as a sign of the pretended spread of the
Gospel, if they also had said their prayers, if, following the divine
precept, they had in their hearts and with their mouths called
again and again: "deliver us from all evil!" or been reminded
to do so?—Away, then, with your ignorant or hypocritical
aspersions if still you profess to be christians!

If Mädler and Arago, in their effusion against the Pope and the
Priest, did not know better, it only shows how conceit, inflamed
in the learned mind at the sight of a comet, can blind the understanding, and how eagerly every opportunity is caught at by other-

wise, no doubt, well informed, and in MANY CASES, highly estimable, persons, to cast a stone at a Church of which they have heard and seen nothing but a caricature, and against men, infinitely their superiors in everything, except, perhaps, the coveted opportunity of looking at the stars, the brilliant ornaments of the dome of heaven, and—unless dedicated to science and teaching—getting a knowledge of geometrical measurements and mathematical calculations to the NEGLECT OF THE FLOCKS COMMITTED TO THEIR CARE.

Apart from the effect of comet-strokes, such as noticed, astronomy has abundantly been laid hold of to blacken the Church, the greatest patron and promoter of science, whether it came from Galileo or Kepler, from Canon Copernicus or Tycho Brahe, from Friar Bacon, or Lord Bacon, from the Monk Albert the Great, or Newton, from Cardinal Cusa or Humboldt, from friend or foe; and whilst she abstains from dwelling upon facts like the persecution of Kepler by the Protestant theologians of Würtemberg, the condemnation of the Copernican system by Melanchton,—the new theory going straight against the text of the bible, their sole rule of faith;—its rejection by Tycho Brahe and Lord Bacon; the tardy acceptation of the Gregorian calendar by all Protestant nations, schismatic Russia still lagging behind—simply because it came from the head of the Church; the destruction of all Catholic art and literature wherever Protestantism predominated, and the like; whilst still less she degrades herself by inventions and misrepresentations, she herself is constantly made the object of slander.

Mackintosh, so clear-headed in the exposition of his *Electrical Theory*, informs us at page 38 and 53 :—" That Galileo was cast into a dungeon," (which is not true) " because his discoveries UPROOTED the ancient foundations of purgatory ; because they REMOVED purgatory from its ancient foundation."

This, as it was in every respect new to me, a long practised hand in the Church, will certainly be something new to all Catholic readers not acquainted with the work of Mackintosh, and expose the propounder of this extraordinary fact, if still living, to the merriment and ridicule even of our schoolboys. Or was he aware, according to *Baby Worlds*, page 72, that the ANGLICAN PARSON, contemporary and protegé of Sir Isaac Newton, the celebrated "Whiston, ... made out the comet's purpose to be that of a travelling hell, a residence of the damned? 'Wheeled from the remotest limits of the system'—thus Fergusson describes his theory—'the chilling regions of darkness and cold, the comet wafted them (the damned) into the very vicinity of the sun; and thus alternately hurried its wretched tenants to the terrifying extremes of perishing cold and devouring fire.'"

Mädler makes Galileo discover still the libration of the moon, with "THE REST OF HIS EYESIGHT RUINED IN THE PRISONS OF THE INQUISITION," in which, it is fearful to relate, HE NEVER WAS. For, if Mädler had but read the letters of Galileo himself, or *The Martyrs of Science*, by Sir David Brewster, he would have found that, in the words of the learned author, Galileo suffered only a "nominal confinement."

Unhappily, Sir David Brewster's honorable vindication of the Church and even of the Inquisition, in the generally-distorted picture of the so-called persecution of the great astronomer, is not untainted by inaccuracies and seemingly-gratuitous aspersions, evidently the result of unacquaintance with the Catholic Church.

Thus, I have NOWHERE found, that "his (Galileo's) noble discoveries . . . . . were even denounced as crimes which merited the vengeance of heaven," page 2—"Galileo must have felt himself . . . . in the possession of the fullest license to prosecute his researches and publish his discoveries, provided he avoided THAT DOGMA of the Church, which, even in the present day, it has not VENTURED to renounce," page 78. "The DOGMAS

OF THE CATHOLIC FAITH had been brought into direct collision with the deduction of science," page 84.

What are "THAT dogma and THE dogmas" here alluded to, I have not been able to make out. Whether Sir David refers to Mackintosh's removal of purgatory, to the earth's motion, or to infallibility : it is clear, to a Catholic reader, that however learned in science, he has never enquired what constitutes a dogma of the Church, nor what is the dogma of infallibility : otherwise he would have used neither of the foregoing expressions, nor said at page 94 "the presumptuous priest pronouncing infallible the decrees of his own erring judgment."

Page 235, Sir David speaks of the ROMAN CATHOLICS (of Grätz) HAVING EXCOMMUNICATED Kepler from their body, to which, as a a Protestant, he NEVER BELONGED, no more than the Turks and the comet excommunicated under the auspices of Arago, or the "threatening flames" and the "destructive plague" exorcised and excommunicated by Von Gumpach and Gibbon.   Alas! for the occasion to bring these names together! And what a pity that the goose-quill of some anti-Catholic firebrand has not yet been exorcised and excommunicated !

Of the tolerance and liberality of the Catholics, having invited the truly persecuted Kepler, as they did Tycho Brahe, and appointed him to office and professorships, not a word is said. But only think of Oxford or Cambridge choosing Catholic professors of astronomy, or the direction of the Observatory of Greenwich being entrusted to one !

Truly, persons outside the Church, and candid and sincere minds in particular, should honestly study her in HER OWN WORKS, if it were only to learn her PENNY CATECHISM first before touching upon Catholic ground ; to be more sure still, they should show themselves to the priest, or at least go to practical believers for information, instead of listening to or learning from aliens, or from bankrupts and insolvents both in religion and morality.

With the above drawbacks to a candid exposition of the " persecution of Galileo," the work of the noble-minded author cannot enough be recommended to a sincere enquirer after an historical truth.

As to the famous "E pur si muove," Sir David, page 93, quotes the following from the *Library of Useful Knowledge*, : "It has been said, BUT UPON WHAT AUTHORITY WE CANNOT STATE, that when Galileo rose from his knees, he stamped on the ground and said in a whisper to one of his friends, " E pur si muove." " It does move, though."

No doubt some clever Arago was father to this story,—in which " STAMPING ON THE GROUND AND WHISPERING " so ill agree,—but which, if true, would stamp the crime of perjury on the "high-minded " Galileo. And yet this perjury is ostentatiously proclaimed, not of the martyr' of science, but of the martyr of pride and bad theology : or, if this be not so, then, *The Whole of the Natural Sciences*, a work published in Germany, and *The Earth and its Mechanism*, by Henry Worms, published in London, bear the impression of falsehood on their forehead, by having imprinted thereon the "E pur se muove."

As to the last-named author, whose work was published only in 1862, I must, in charity, suppose, that he never heard of, much less read, Sir David Brewster's valuable work, the *Martyrs of Science*, or he would not have been so awfully shocked at the idea that Galileo, for " asserting the justness of the Copernican system was cited before the inquisition, CONDEMNED to prison, but released after some months, ..." without having been in it ! His furious sally against " the cruel intolerance of Jesuitical tribunals," the " calumniating fanaticism of a bigoted priesthood ; " is quite in keeping with the "E pur se muove," and I should not wonder if the gratuitous fit of excitement in which he must have thrown himself whilst writing the above, necessitated medical assistance, or, at least, a composing draught.

Otherwise, his laborious work triumphantly proves that a wheel goes round when set in motion.

Humboldt is grossly unjust to the Church. The following extracts from *Cosmos*, though written against us, will, to a THINK-ING and conscientious reader, almost of themselves speak in our favour.

In Vol. ii., p. 29, he gives a description of nature by St. Chrysostom which he thus concludes: " . . . . it might have seemed that eloquence had found again her element, freedom, on returning to the bosom of nature in the then forest-covered mountain districts of Syria and Asia Minor." He then continues:

"But when during the subsequent period, SO HOSTILE TO ALL INTELLECTUAL CULTIVATION, Christianity spread among the Germanic and Celtic races, who had PREVIOUSLY BEEN DEVOTED TO THE WORSHIP OF NATURE, and who honored under rude symbols its preserving and destroying powers, the close and affectionate intercourse with the external world of phenomena which we have remarked among the early Christians of Greece and Italy, as well as all endeavours to trace the action of natural forces, FELL GRADUALLY UNDER SUSPICION, as tending TOWARDS SORCERY. They were THEREFORE regarded as not less dangerous than the art of the sculptor had appeared to Tertullian, Clemens of Alexandria, and almost all the most ancient fathers of the Church. In the 12th and 13th centuries, the Councils of Tours (1163) and of Paris (1209) forbade to MONKS the SINFUL reading of WRITINGS ON PHYSICAL SCIENCE."

In Note 51 to *Cosmos* ii., the Council of Tours (1163) under Pope Alexander III., the Council of Paris of 1209, and the bull of Gregory IX., of the year 1231 are cited, and we are told that "Heavy penances were attached,"—not to the reading of writings on PHYSICAL SCIENCE as stated just before, but—" to the reading of the physical books of ARISTOTLE."

Note 338, speaking of latin translations of Aristotle from the Arabic and Greek, Humboldt says : " We may recognize an allusion to the same twofold source in the memorable letter which the Emperor Frederick II. of Hohenstaufen sent with trans- lations of Aristotle to his universities, and especially to that of Bologna in 1232," ONE YEAR AFTER THE BULL OF GREGORY IX.

Note 382. " We can only interpret the wish expressed by him (Roger Bacon) in a letter to Pope Clement VI. to ' burn the works of Aristotle, in order to stop the propagation of error among the schools,' as referring to the bad latin translations from the Arabic."

Cosmos ii., p. 30. " These intellectual fetters were first broken by the courage of Albert the Great and Roger Bacon ; when nature was pronounced pure, and reinstated in her ancient rights."

To any careful reader it will appear evident, that, with the material before him, Humboldt, with a biassed mind, did not know what he was writing about ; and if he did not care to enquire into the reasons for the measures, whether active or passive, of the Church whom he blames, that in this case, though otherwise called the " Nestor of Science," he neither displayed discrimination, nor moderately sound judgment and capacity of reasoning.

When the German and Celtic races were first visited by the Catholic Missionaries, these races had to be converted and civilized, to be made christians before being made natural philosophers ; and having previously been DEVOTED TO THE WORSHIP OF NATURE, it was necessary by all means to withdraw them from this worship, to lead them to the worship of the true God and His Eternal Son; to prevent and avoid everything that could tend TOWARDS SORCERY which they had practised,—and which, alas ! even in our own age is practised still!—to absorb their devotion to nature into the devotion of christianity—non angli sed angeli,—and not let their infant-christianity merge again into their former heathenism.

The arduous task of converting nations from barbarism and idolatry to christianity left neither time nor means to devote to science; but if this were not so: the necessity of restraining and turning aside devotion to nature among peoples who made it their idol is proved by Humboldt himself, whose christianity, and with it, nobility of mind, was wrecked in "the intellectual and affectionate, but heathenish, intercourse with the external world of phenomena," though living so much later than the conversion of his rude and idolatrous German ancestors, living in our own so highly enlightened time!

From what I have said already the reader will easily perceive that, to the first christians, converted from the worship of idols, the art of the sculptor was equally dangerous towards a relapse into idolatry from which, at the cost of the blood of so many martyrs, they had been rescued; and as we still see that the cultivation of science is so injurious to the faith and morality of so many ill-directed minds, without going back to the first ages of christianity, the Fathers of the Church were right and prudent in keeping out of sight, if not removing, in due time and season, all that could re-throw mankind into the demoralizing state of heathenism and idolatry.

With the slightest circumspection Humboldt ought to have seen this, particularly if he had but reflected upon Catholics being still so ignorantly accused of idolatry and picture worship, the very thing THEY ALONE abolished, and that their finest works of art were destroyed by the Iconoclasts and the Reformation, and of which the remains are still to be seen in museums and the galleries and collections of the rich, and in the many ruins of religious establishments, of desecrated abbeys and cathedrals,—

Where stood the altar in the ancient days,
  No priest now, but a stranger, sad and lone,
Stands on the steps of mossy stone,
  Ascended once with highest note of praise.

All's silent on the broken altar-stairs—
  While through the vacant window the blue heaven
Looks wistful in, defrauded of the prayers
  Once here in high response and answer given.

Where once the glad "Te Deum" flung abroad
  To heaven the music of its matchless song;
Where once the Miserere wailed to God,
  Joy echoing sweet, and sorrow sobbing long.

Where once the prayers were more than words could tell,
  Impatient wishes that besieged the sky;
Nor was there doubt of any miracle,
  Save that life's longings and its hopes could die.

But now, subdued by tedious toils and cares,
  Desire falls faint—hope falters on the strain;
And Time and Nature, with a deep Amen,
  Fill up the breaks and echoes of old prayers.*

For the same reason as before stated, LECTURES upon Aristotle were forbidden to abate a local and temporary evil; the SINFUL reading, in the way in which Humboldt states it, was forbidden TO MONKS UNDER OBEDIENCE; with PERMISSION any work may be read; but with many there was danger of christianity being absorbed by the Aristotelian philosophy, and that danger over, the school was open again to be canvassed. The same caution, though alas! not with the same result, was observed with the Bible, when men were tempted to make to themselves a religion, like the murmuring Israelites, a calf of their own. The evidence of the former is furnished by Humboldt himself, who not only lost his christianity, but stained his memory with the blackest ingratitude towards his patrons and benefactors, sinking below the animal grateful to the hand that feeds and strokes it; and by the evidence of the latter we are everywhere surrounded.

The prohibitions against Aristotle—to take the widest range assigned by Humboldt, without its being worth while to examine

---

* From a poem, "Amen!—in the Cathedral, St. Andrews," in *Blackwood's Edinburg Magazine*, October, 1863, by M. O. W. O.

their correctness—extend from 1163 to 1231, which is 68 years
and no more; for, "these intellectual fetters were broken by Albert
the Great and Roger Bacon," TWO MONKS,* who, with Cardinal
Cusa, Thomas-à-Kempis, Matthew Paris and others, flourished
about the very same time of 1231, which could not have been had
these fetters at all existed.

This is proved further by Humboldt's reference to Frederick
II. of Hohenstaufen, who in 1232, a year after the bull of Gregory
IX., sent—not ONE translation, but—TRANSLATIONS of Aristotle,
—not to ONE university, but—to ALL HIS UNIVERSITIES; and these
translations must have been in hand long before the time specified,
thereby showing, that there was no general interdict upon the
works of this great philosopher, though there seems to have
been one upon bad translations, as shown by Humboldt's
quotation from Bacon.

The *Penny Cyclopædiæ* is more explicit on this point, for, in the
article on Roger Bacon we read that " In 1209, a council at Paris
condemned and burnt, if not the works of Aristotle, at least the
MUTILATED AND INTERPOLATED TRANSLATIONS from the Arabic
which then existed."—And who would not on that account admire
the Church in her endeavour to preserve even the philosophy of
a heathen in its purity? And who would not admire her still
more in the preservation of the purity of the Holy Scriptures, and
her condemnation, though not with equal success, of all SPURIOUS
AND INTERPOLATED translations?†

That there was no general prohibition of READING and studying

* As the anti-Catholic press ignores our choicest literary productions, so
Humboldt studiously ignores that two monks, two Catholic priests, broke his
imaginary fetters; and Hind, in his *Introduction to Astronomy*, instead of giving
Copernicus his proper title of priest and canon of the Catholic Church, quite
coolly calls him : a "PRUSSIAN ASTRONOMER!" PRUSSIA and the time of
Copernicus too !

† See the little work on *The Origin of the Holy Scriptures, their Progress,
Transmission, Corruption and True Character*.

Aristotle or other scientific works is again proved by the quotation of Humboldt which expressly refers to MONKS; and if LECTURES were temporarily forbidden, it was the same as later with Galileo, who ALONE was prohibited to LECTURE upon the Copernican theory, though EVERY ONE ELSE might do so, and, like himself, "IN THE POSSESSION OF THE FULLEST LICENSE TO PROSECUTE HIS RESEARCHES AND PUBLISH HIS DISCOVERIES," (see page 205) study and perfect it to their heart's content.

If monks were forbidden to read Aristotle, it was for as good a reason as, according to Note 51, of *Cosmos* ii., "in the Council of Lateran, of 1139, monks were forbidden to exercise the art of medicine." But Humboldt did not see the wisdom of these measures, could not see that these monks were religious and not doctors, that they could not serve two masters! But such is human blindness.

The following are some extracts from *Cosmos* respecting the learning and enterprise in the Catholic church before the so-called Reformation.

Cosmos ii., note 385, "Il existe aussi de Pierre d'Ailly, que Don Fernando Colon nomme toujours Petro de Helico, cinq mémoires de Concordantia Astronomiæ cum Theologiæ." And page 249, Humboldt says: "I have shown elsewhere that this work," the Imago Mundi of Cardinal Alliacus (Pierre d'Ailly, Bishop of Cambray in 1410)—"was more influential on the discovery of America than the correspondence with the learned Florentine Toscanelli."

Note 433. Sebastian Cabot, "the great navigator, who gave to Great Britain almost an entire continent, and without whom the English language would perhaps not have been spoken by many millions who now inhabit America."

But these examples are isolated; *Cosmos* is full, full of eminent men before that disastrous period, and the preceding instances are only two of the most striking ones, particularly that of Cardinal Alliacus, as it will remind the Catholic reader of the *Connexion between the Sciences and Revealed Religion*, by Cardinal Wiseman, by THE CARDINAL *par excellence*, and, in real eminence, above all the learned men of our age, apart from the nobility of his ecclesiastical character and private virtues.

Besides this host of talent and learning and spirit, " . . . . Somewhat earlier," than the appearance of Luther, "Michael Angelo, Leonardo da Vinci, Titian, and Raphael flourished in Italy, and Holbein and Albert Dürer in our German country. In the year in which Columbus died, fourteen years after the discovery of the new continent, the order of the universe was discovered though not publicly announced by Copernicus."* *Cosmos*, p. 298. Yet in the face of all this splendid array of illustrious Catholic names brought before us from the beginning to the end of *Cosmos*, Humboldt says, at the same page, 298 : "In the same month in which Hernan Cortes, after the battle of Otumba, advanced to besiege Mexico, Martin Luther burnt the Papal Bull at Wittenberg, and laid the foundation of the Reformation, which promised,"—but did not give,—" to the mind of man FREEDOM AND PROGRESS IN ALMOST UNTRIED PATHS."

If Humboldt had said : freedom from confession, fasting, good works, and moral and religious restraint, thereby laying the foundation of discord and strife, of the devastation and dismemberment of his own country, and the ungodly despotism of the great majority of its princes, opening the flood-gates of revolution and civil war all over Europe, and the doors of divorce to whole nations, it would have been intelligible ; but I really cannot see what other exploits he performed, what other new fields Luther opened in ANY BRANCH OF LEARNING, SCIENCE, ART, OR PROFESSION, which, according to Humboldt's own account, were not opened long before him.

* The PRUSSIAN astronomer.

The still existing monuments of true civilization of the middle ages, and especially of the time of Luther,—a vicious literature and machinery excepted—have, as yet, not been surpassed; from the stimulus then given, the progress in experimental philosophy and empirical knowledge, was inevitable, but would have been by far greater, had it not been for the deluge of heresy, the vandalism of' the so-called Reformation, the continual and still continued oppression of the Church in almost every part of Europe, and the destruction, alienation, and perversion of her seats of learning. Yet, notwithstanding, if NOW I look round for any of the higher tests of progress without finding them outside her pale, and on the other hand to the poverty, degradation, neglect, and ignorance of the great majority of the industrious masses, to the UNCLIMATICAL immorality of society at large with all its frightful consequences ; to dishonesty and fraud in almost every profession, business, or occupation, and a perverted sense of right and wrong, to unbelief, seduction, treason, and revolution in high quarters and low ; to a living industrial mechanism in bloom, and almost everything lofty and exalted in art, science and literature—out of the Church— stunted : I see a rotten state of society, unprecedented in the annals of christianity, and far less freedom and expansion of the mind, than at all times was, and is still, possessed by the true children of the Church ever since her foundation by Christ ; she alone has been, is, and ever will remain, the fond nurse and fostering mother of all that is grand ; under the shelter of HER wings there is real progress, a " close and affectionate, BUT CHRISTIAN, intercourse with the external world of phenomena," precisely still the same, "which we have remarked among the early christians of Greece and Italy," not leading to a re-formation, a re-introduction of refined heathenism, a civilization of vice, a demoralizing industry, and the goddess of reason, but to all that is noble for heaven and earth.   The eternal city, CHRISTIAN ROME, with its long line of Pontiffs, the ill-requited civilizer of Europe, is STILL the unarmed mistress of the world, the centre of all christian nations, the seat of wisdom, learning, arts and science, education and CHARITY, to HER all resorts ; her Pontiff, what his predecessors EVER WERE, is still the protector and the champion of the slave and the oppressed, the stern, inflexible because the infallible, teacher of religion and morality, the fearless rebuker of the godless ; the Ruler of Rome and the Church, Pio Nono *felicemente regnante,* he ALONE in the world, with but the pastoral staff of his Lord, stays Atilla, calmly warns off the eagle circling over his head, still beards the tiger ready to spring, the bear with his paws upon his prey, and the lion lurking in his den ; he ALONE in the whole world holds out to every nation and tongue the olive-branch of peace, the palm of a christian's victory, and the standard of triumphant christianity.

When Spain and Portugal—whose old monuments of CATHOLIC, but with her Orders departed, GRANDEUR I have lately seen— were on the point of war respecting the line of demarcation between their respective possessions ; when, *Cosmos,* p. 282, " Neither the state of science, nor that of the imperfect instruments employed at sea in 1493, whether for measuring angles or time, were competent to the practical solution of so difficult a problem, Pope Alexander VI. PRESUMPTUOUSLY divided half the globe between two powerful states."

That such a decision—the line of demarcation to be a hundred miles westward of the Azores—should be called PRESUMPTUOUS when solicited by the contending parties—or even without, in order to avoid war ; and when Humboldt HIMSELF SHOWS that, owing to the state of science and defective instruments, such a decision was an ABSOLUTE NECESSITY, but too clearly exposes the false, because blind, reasoning of Humboldt, and the animus with which he looked upon this important act of the Pope, and the sentiments, or ignorance, by which writers against the Church are generally guided.

Goodwin, one of the authors of the notorious *Essays and Reviews*, referring to the Ptolemaic system, says at page 208 : "The Romish Church, it is presumed, adheres to the old views to the present day."

Whether this passage is the result of culpable ignorance, or of malice, I must leave to the reader to decide. However, both causes collectively and separately, combine to originate and propagate the most stupid though often most serious, stories and slanders against the Church ; and whilst in this way, to the great detriment of all religion and morality, science is made subservient to falsehood ; whilst electricity, the A of created substances, is degraded into the bearer of lying telegrams, conveying them with lightning speed from one end of Europe to the other ; and whilst the greatest talent and most ingenious mechanism is set in motion to multiply them in the press, and deluge society with all kinds of fabrications and misinformation, liberty of writing having become synonymous with liberty of lying: "lying itself," in the words of Cardinal Rauscher, "has become a science," and this so much so, from the highest ministerial to the lowest political mountebank, from the richest to the poorest stock-jobber and gambler, through trades and professions, that, knock at almost every house, and you will find it at home, were it only in the paper on the table, the book on the shelf, or the picture on the wall.   And is it, then, a wonder, that we live in an age of learned ignorance, that within this labyrinth of lying and lie-believing, real ignorance, spurious knowledge, and perverted judgment, particularly in matters of religion and morality, should have set up their empire, the empire of internal and inter- national strife and misfortune ?   And if in secular science these things happen with the LEARNED and hundreds of well-disposed men, otherwise patterns of society : what can be expected of those who confidingly take them for reliable authorities ? what of thousands who blindly follow more unscrupulous, if not mercenary, guides ?

what of the masses who, schooled, misled, and imposed upon by malevolent LEADERS, treasure up the lumber of crazy anti-Catholic minds as the most precious production of an enlightened age ?

It may be said that this digression, longer than I anticipated, has nothing to do with the " Simplicity of the Creation." But when we see that there is scarcely a single protestant work or publication, of science in particular, even *Good Words,* in which Catholics are not hurt or offended by all sorts of fables, remarks, false and unjust accusations, it may well be pardoned, if, without having recourse to the same weapons, without wilfully wronging or hurting any one that differs from us, we maintain our own strong positive Catholic stand-point, and in a work on physical science refute and confound those calumnies and misrepresentations with which works of the same class on the other side are abounding, and of which the extracts calling forth my digression are proofs as sufficient and striking as they are discreditable to their originators or propagators.

However, it is still worse, nay criminal, when extreme ignorance is displayed by Anglican Divines, by men in " priestly garments dressed ;" THEY, AT LEAST, should be expected to be intimately ac quainted with all the Catholic Church believes, and teaches, and practices, the why and the wherefore, the same, at all events, as we PRACTICAL LAYMEN, even among the humbler classes, in most cases know the doctrines and practices, and in ALL CASES the cause, the why and the wherefore, of all heresies and schisms and separations from the Church, that there are such things as pride and ambition, the sixth commandment of God, and the third in the list of mortal sins.

But the reverse is the case.

Just before the preceding portion of my work was going to press, my eye caught, in a shop window, the " Celestial Measurings and Weighings" of Sir John Herschel, on the wrapper of *Good*

Words for June, 1864. A kind Protestant friend lent me the number, as he had done some others before, and having read the article in question, to which I shall have occasion again to refer, I also noticed a communication "From Rome," by the Dean of Canterbury, in which he gives a portrait of his interior self; and such a portrait! such a Dean of Canterbury! a mournful picture for St. Augustine and St. Thomas to look upon.

This portrait of the Dean, delineated by his own hand, unhappily more than confirms whatever I have said before; and I most sincerely and earnestly invite every one of my non-Catholic readers—for we ourselves have plenty of such drawings every day set before us, and are connoisseurs at the first glance—calmly, considerately, and conscientiously, to peruse and analyze this literary production "FROM ROME" BY THE DEAN OF CANTERBURY; and I am confident that, if animated with a christian spirit, and without even having recourse to CATHOLIC works, and living CATHOLIC authorities, he will find in it neither the gentleman, nor the scholar, natural or moral philosopher, neither the Christian nor the Divine; he will simply find the Dean of Canterbury, coarse invectives and babbling, and very silly insinuations and exaggerations, in fact, the Dean of Canterbury; that is what HE will find.

To US, the article is full of misconceptions, senseless and incoherent reasoning, misrepresentation and slander; and if the Dean of Canterbury, his hand on his heart, cannot plead an excusable ignorance, an affected mind, a disordered, wandering, or excitable brain; and if, moreover, without any such excuse, with sound mind and intellect, he has read our *Penny Catechism*, but in particular the instructions on indulgences contained in the *Ursuline Manual* to which book he refers respecting the ten commandments: then, in charity, I must leave the verdict to the candid reader and enquirer, whether he is the CONSCIOUS or MIS-

TAKEN detractor of the Church of Christ and of her children, of whom he is no member. And he, at any rate, the Dean of Canterbury !

And the Rev. Norman Macleod, D.D., ONE OF HER MAJESTY'S CHAPLAINS, and editor of *Good Words:* is he the IGNORANT or the CONSCIOUS accomplice of the Dean of Canterbury? Is it in GROSS IGNORANCE, or in the ENLIGHTENMENT of KNOWLEDGE that he propagates the trash and slander of the Dean of Canterbury, and with *Good Words* mingles poison into the food of his readers?

And these two men are "masters and TEACHERS in Israel!"

Unless they can plead invincible or excusable ignorance, from whatever unsinful cause it may arise, God will demand of them the souls they have poisoned; before His throne they will have to give an account of their talents, will have to answer for every IDLE WORD EVEN they have uttered or propagated; and mercy on the Dean of Canterbury,—and most seriously do I mean it—if, as he says, "the whole tenet of purgatorial pains is an idle fiction," and that, as our Saviour tells us, "nothing defiled is to enter heaven." But if he does not believe in hell either, then we can understand why he rates us for "drawing false and unwarranted distinction between sins venial and sins deadly," for drawing a great distinction between an untruth told by an infant and the falsehood deliberately propagated by a Dean or Minister; between the stealing of a penny by a child and the sacrilegious stealing—in refined language confiscation, annexation, dissolution—of temples, sacred vessels, monasteries and convents, dedicated and consecrated to the service of the living God. The law of England, and even barbarous nations, make distinctions in offences and crimes, but not the Dean of Canterbury !

It is from men and authorities such as these that Sir John Herschel,—whose name, unfortunately, I have hitherto spelled with a double l,—Sir David Brewster and others of high moral

character and rectitude, have their knowledge of the Catholic
Church, and upon it base their slurs against her in matters of
morals and religion as well as science; but I trust that henceforth,
SOME at least, if not many noble hearts, will seek the right and
true CATHOLIC sources of information, which it is neither troublesome nor difficult to find.   Otherwise, no Dean of Canterbury,
not even one of Her Majesty's chaplains, will save them, no more
than "Laplace (see page 203) will save us from destruction, if,
—like a thief over night—it is thus to come upon us."

In treating of this delicate subject, as a matter of duty and
necessity, I am not aware of having overstepped the limits of
propriety, in any way mis-stated or exaggerated facts, drawn unwarranted or illegitimate conclusions, or wronged even the Dean
of Canterbury in the just indignation he has aroused, much less
any of the other writers who engaged my attention; and though
I have freely expressed myself as to the learning, knowledge,
enlightenment, and supposed intellectual superiority outside the
Church, according to my own humble unassisted truly private
judgment, yet I hope I know how to respect myself by otherwise
giving honor and respect to whom honor and respect are due,
whatever garb he wear, to whatever creed he may belong.

Would to God, that as a beginning at least, religious strife and
aspersion were for ever discarded from science which, above every
other pursuit, should bind men in harmony together.

If all astronomical works were as free from objection on this
ground as Mr. Hind's *Introduction* and *Solar System*—and I do not
see why ALL should not be so—I should have been spared the pain
of noticing the UNJUST reproaches against us, and saved a little
for printing and binding too.

We, on our side, shall never be found the aggressors unless
science be turned on Christianity itself; and a writer in the *Times*
will find that ours is not " a Church in which Biblical criticism

is impossible, and the discoveries of science must be ignored."
For 'the sake of science, private and public morality, the true
welfare of nations, and much more for eternal life, WE ARE
AFRAID of science being abused, the same as we are afraid of
seeing children, the inexperienced and the wicked, handling
dangerous weapons and sharp-edged tools.

Gladly, then, returning from these topics, and from the effects of
comets upon certain minds to their effect upon the sun, the varying
influence of our luminary upon the earth according to the intensity of his light, and the changes in his fiery envelope, it seems
to me most probable, that, if the upper surface of thunderclouds,
as noticed page 97, presents to us ridges and protuberances rising
upwards to great altitudes ; and if the aurora borealis (page 84.)
sends up lofty cones and streaks of light, such protuberances and
mountainous elevations, in conformity with actual observations,
should, at times, more or less happen or be produced on the
superior surface of the solar photosphere, particularly round the
equator* where the electric element is most active, manifesting
itself in the zodiacal light and other appearances in the envelope
of our luminary ; but, whether in that state the sun differently
influences the climate and productive condition of the earth,
remains still to be found out.

That great outbursts of light,† and extraordinary changes of
spots, more or less affect the magnetic needle, seems to be an
established fact ; and as the zodiacal light, the result of a modified continuance of such outbursts and protuberances, like a faint

---

* None have been observed at the poles.

† One of these outbursts of light,—and most extraordinary it must have
been,—was observed by Messrs. Carrington & Hodgson, on the 1st September,
1859, as noticed in the *Monthly Notices* of the Royal Astronomical Society,
of 11th November, 1859 ; it lasted but five minutes and disappeared instantaneously, yet, "the magnetic instruments at Kew were simultaneously
disturbed to a great extent."
    It is of vast importance, that, according to Mr. Carrington, " the magnetic
storm experienced here, subsequent accounts should have established to have

luminous cone, is said to extend beyond the orbits of Mercury and Venus, and even to rise nearly to the orbit of the earth, it is more than probable, that the sun at the time, or during the period of its appearance, should differently affect the electric condition and consequent fecundity of our planet.

X.

REFUTATION OF THE OLD THEORY OF THE ROTATION OF THE EARTH, THE PARALLELISM OF THE EARTH'S AXIS AND THE CAUSE OF THE SEASONS; ANOMALIES OF THE THEORY.—REVIEW OF THE VARIOUS SELF-ACTING FORCES.—THE FAMOUS DISCOVERY OF THE PRIMITIVE IMPULSE; NEWTON AND LAPLACE.—THE GYROSCOPE.—IMPORTANT EFFECT OF THE NEW THEORY ON THE CALCULATION OF ECLIPSES; THEIR INFLUENCE ON MAN.—MIRACLES; EFFECT OF UNBELIEF.

ACCORDING to the old theory, the earth, like all the planets, has arbitrarily been assumed to rotate round its UPRIGHT axis, giving to the said axis an inclination of about twenty-three degrees from the perpendicular to the plane of its orbit, and to the other planets more or less; and this inclination is what constitutes the so-called obliquity of the ecliptic.

The earth, therefore, like a spinning-top on its upright axis, in the most arbitrary manner, is made to dance forward, sideways, backward, sideways, and forward again in a circle round the sun, with its upper end, or north pole, always pointing to the same

been as considerable in the southern as in the northern hemisphere; that, though the contemporary occurrence might deserve noting, he would not have it supposed "—(why not?) "that he even leans towards hastily connecting them: One swallow does not make summer."

But the "contemporary occurrence" confirms the interior electric connection of the earth from pole to pole, as proved already page 85, by the contemporary occurrence of polar lights in the opposite hemispheres.

constellation, to the same part of the heavens, whilst it constantly rotates at angles, and at one time even in exactly the opposite direction, with the line of pro- or retro-gression, except in one instance, where rotation coincides with its path. The theory reminds me of the boy coming too late to school on a fine slippery winter morning. When asked the reason of his being behind time, he said, that when he made one step forward, he slipped two backward; and on being required to explain how by that means he came to school at all, he replied, that he began to go home. Thus the earth is made to rotate one way, and to progress the other in its stooping position.

There might be still some reason for this, if the south pole, the pivot of the earth's rotation, were resting and moving forward on a solid or hard substance, like the spinning-top on the table or floor. The friction of the pivot of the spinning-top is with the wood or the ground upon which it dances, and, of course, much greater than the friction of the body itself with the air in which it turns round. If it were the same with the earth, that the pivot, the point of its south pole, were resting and spinning round upon a solid body or foundation; that the south pole were attracted by and pointing to a body, to a centre of gravitation underneath, the same as the pivot of the spinning-top, as a balloon twisted or turned round its perpendicular axis in the air, and whose lower part, or the car, is pointing to and attracted by the centre of the earth as the fulcrum of gravitation of the whole body, there would be no anomaly in the old theory; but the earth moves within a fluid which everywhere presses equally upon it; friction, therefore, can take place on that part of the surface only which rotates or turns over; and this part being the equatorial belt of the earth, that part through which, from its centre to the CENTRE of the sun, passes the line of attraction or gravitation,—not through south or north pole,—it follows: that our globe rotates one way, and progresses in another, and this all the more so, when we consider that the earth is said to stoop forward, that its north pole is inclined twenty-three-and-a-half degrees to the plane of its path, and on which plane there is no support or friction.

If the earth and the balloon twist or rotate round their RESP : PERPENDICULAR axis : why does the lower pole, the south pole of the earth, point into blank space, whilst the lower pole of the balloon always points to the CENTRE OF THE EARTH? why does the lower pole of the earth not equally point to the CENTRE OF THE SUN? Whence this anomaly if the old theory be correct?

But more. If the theory of gravitation be true, the revolution of the earth round the sun becomes· still more arbitrary and inexplicable.

The earth, ASSUMED to be projected by the Creator in a straight line, is said to be drawn out of it by the attraction of the sun, and these two forces, projectile and attractive, draw and keep it in its orbit.

Now if this were true, the earth would move round the sun as a bird would fly round the earth, head always foremost, and in advance of the body, because the same forces which are said to make the earth move in a circular or elliptic orbit round the sun, would also influence the position of the body of the earth itself, and throw it into a diagonal direction between the said two forces ; the direction or inclination of the earth, as it was flung into space by the Almighty, would always coincide with the line or direction of its path or orbit, whether straight, elliptic, or circular, whereas, to sustain the fact of its north pole always pointing to the polar star, in the same direction of the heavens, to make the position of the body of the earth in its course round the sun remain always parallel to itself, to maintain the parallelism of the earth, as it is called, it is on its passage made to go six months in advance with the north, and, by a gradual change, unexplained and inex-

Q

plicable, six months in advance with the south pole, or, which is the same, six months foremost with the head, and six months foremost with the tail, including the sideward motion, or pro- and retro-gression, right and left, and all this without the assistance of any given third power to prevent the diagonal direction of the position of the body of the earth WHICH OUGHT TO RESULT from the projectile and attracting forces.

By what law, except the most arbitrary, however ingenious the invention, should otherwise the earth at one time present head, and at another, tail, and at others again front or side, to the sun? Bare assumption is NO LAW, and NO CAUSE is given by the old theorists why the body of the earth, under the influence of the said two forces, should preserve its parallelism; the thing is impossible upon their own showing: but the object is, to account for the seasons, which, without this arbitrary adoption, THEY CANNOT DO. And because they CANNOT do so, because THEY cannot give any other reason as the cause of the seasons: therefore, they say, the parallelism and inclination of the earth's axis must be true. " The simplicity with which " thereby " the seasons are explained . . . . . may certainly be adduced as a strong presumptive proof of the correctness of the principles already advanced ; for on no other RATIONAL SUPPOSITION with respect to the relations of the earth and sun, can these, and other as well-known phenomena be accounted for."

Thus Mr. Hind in his *Introduction* p. 13, where he may well call his " relation of the earth and sun " a SUPPOSITION; for there is not only not a shadow of physical, or even sensible, reason to prove it, but is moreover an unfinished, most illogical, conclusion of the effect of the composition of two forces, to which, in fact, the parallelism of the earth's axis is directly opposed. And yet this is called a RATIONAL supposition, beside which there CAN BE NO OTHER, to account for the seasons, though, with due deference

to superior learning, I invite comparison between this supposition without foundation, and my own theory as unfolded p. 187, and based upon physical and very simple REASONS.

If a straight line of railway of twenty or thirty miles in length were changed into a circle or an ellipse, and carried round London, so that end met end, the engine would always be in advance of the train, whether it travelled on the line when straight or when changed into a circle.

The same with the earth. If not attracted by the sun, it is said, it would have travelled in a straight line to the polar star regions, stooping with its north pole, its hoary head, as it were, in advance; but since that line was changed into a curve round the sun, why should the north pole not be in advance on the curved line, the same as it was on the straight one? Why should, on one side of the sun, the feet, the south pole, be in advance? Is there a third force and where, keeping it by the hair, attracting it to the polar star regions, making the train once turn on a pivot on the line round London, so as to keep the engine always working in the direction of the north or the polar star, to make it preserve its parallelism whilst it completes the circuit of the city?

"It is sufficient that the axis (of the earth) has been once directed to a certain point of the heavens in order to remain in that direction, though the earth has an annual motion."

Thus, in *The Earth and its Mechanism*, Henry Worms cuts short this ticklish question ; better, no doubt, than if he had tried to explain it and failed like so many others. Galileo, at least had the advantage of perceiving the anomaly, though the astronomers of our own advanced age do not seem to do so; for, in the words of Humboldt, *Cosmos*, p. 177, "in his celebrated *Dialogo* he ascribes the constant parallel direction of the earth's axis to a centre of magnetic attraction existing in space."

It is contended by some, that the composition of the two forces

Q 2

plicable, six months in advance with the south pole, or, which is the same, six months foremost with the head, and six months foremost with the tail, including the sideward motion, or pro- and retro-gression, right and left, and all this without the assistance of any given third power to prevent the diagonal direction of the position of the body of the earth WHICH OUGHT TO RESULT from the projectile and attracting forces.

By what law, except the most arbitrary, however ingenious the invention, should otherwise the earth at one time present head, and at another, tail, and at others again front or side, to the sun? Bare assumption is NO LAW, and NO CAUSE is given by the old theorists why the body of the earth, under the influence of the said two forces, should preserve its parallelism; the thing is impossible upon their own showing: but the object is, to account for the seasons, which, without this arbitrary adoption, THEY CANNOT DO. And because they CANNOT do so, because THEY cannot give any other reason as the cause of the seasons : therefore, they say, the parallelism and inclination of the earth's axis must be true. " The simplicity with which " thereby " the seasons are explained . . . . . may certainly be adduced as a strong presumptive proof of the correctness of the principles already advanced ; for on no other RATIONAL SUPPOSITION with respect to the relations of the earth and sun, can these, and other as well-known phenomena be accounted for."

Thus Mr. Hind in his *Introduction* p. 13, where he may well call his " relation of the earth and sun " a SUPPOSITION; for there is not only not a shadow of physical, or even sensible, reason to prove it, but is moreover an unfinished, most illogical, conclusion of the effect of the composition of two forces, to which, in fact, the parallelism of the earth's axis is directly opposed. And yet this is called a RATIONAL supposition, beside which there CAN BE NO OTHER, to account for the seasons, though, with due deference

to superior learning, I invite comparison between this supposition without foundation, and my own theory as unfolded p. 187, and based upon physical and very simple REASONS.

If a straight line of railway of twenty or thirty miles in length were changed into a circle or an ellipse, and carried round London, so that end met end, the engine would always be in advance of the train, whether it travelled on the line when straight or when changed into a circle.

The same with the earth. If not attracted by the sun, it is said, it would have travelled in a straight line to the polar star regions, stooping with its north pole, its hoary head, as it were, in advance; but since that line was changed into a curve round the sun, why should the north pole not be in advance on the curved line, the same as it was on the straight one? Why should, on one side of the sun, the feet, the south pole, be in advance? Is there a third force and where, keeping it by the hair, attracting it to the polar star regions, making the train once turn on a pivot on the line round London, so as to keep the engine always working in the direction of the north or the polar star, to make it preserve its parallelism whilst it completes the circuit of the city?

"It is sufficient that the axis (of the earth) has been once directed to a certain point of the heavens in order to remain in that direction, though the earth has an annual motion."

Thus, in *The Earth and its Mechanism*, Henry Worms cuts short this ticklish question; better, no doubt, than if he had tried to explain it and failed like so many others. Galileo, at least had the advantage of perceiving the anomaly, though the astronomers of our own advanced age do not seem to do so; for, in the words of Humboldt, *Cosmos*, p. 177, " in his celebrated *Dialogo* he ascribes the constant parallel direction of the earth's axis to a centre of magnetic attraction existing in space."

It is contended by some, that the composition of the two forces

plicable, six months in advance with the south pole, or, which is the same, six months foremost with the head, and six months foremost with the tail, including the sideward motion, or pro- and retro-gression, right and left, and all this without the assistance of any given third power to prevent the diagonal direction of the position of the body of the earth WHICH OUGHT TO RESULT from the projectile and attracting forces.

By what law, except the most arbitrary, however ingenious the invention, should otherwise the earth at one time present head, and at another, tail, and at others again front or side, to the sun? Bare assumption is NO LAW, and NO CAUSE is given by the old theorists why the body of the earth, under the influence of the said two forces, should preserve its parallelism; the thing is impossible upon their own showing: but the object is, to account for the seasons, which, without this arbitrary adoption, THEY CANNOT DO. And because they CANNOT do so, because THEY cannot give any other reason as the cause of the seasons : therefore, they say, the parallelism and inclination of the earth's axis must be true. " The simplicity with which " thereby " the seasons are explained . . . . . may certainly be adduced as a strong presumptive proof of the correctness of the principles already advanced ; for on no other RATIONAL SUPPOSITION with respect to the relations of the earth and sun, can these, and other as well-known phenomena be accounted for."

Thus Mr. Hind in his *Introduction* p. 13, where he may well call his " relation of the earth and sun " a SUPPOSITION; for there is not only not a shadow of physical, or even sensible, reason to prove it, but is moreover an unfinished, most illogical, conclusion of the effect of the composition of two forces, to which, in fact, the parallelism of the earth's axis is directly opposed. And yet this is called a RATIONAL supposition, beside which there CAN BE NO OTHER, to account for the seasons, though, with due deference

to superior learning, I invite comparison between this supposition without foundation, and my own theory as unfolded p. 187, and based upon physical and very simple REASONS.

If a straight line of railway of twenty or thirty miles in length were changed into a circle or an ellipse, and carried round London, so that end met end, the engine would always be in advance of the train, whether it travelled on the line when straight or when changed into a circle.

The same with the earth. If not attracted by the sun, it is said, it would have travelled in a straight line to the polar star regions, stooping with its north pole, its hoary head, as it were, in advance; but since that line was changed into a curve round the sun, why should the north pole not be in advance on the curved line, the same as it was on the straight one? Why should, on one side of the sun, the feet, the south pole, be in advance? Is there a third force and where, keeping it by the hair, attracting it to the polar star regions, making the train once turn on a pivot on the line round London, so as to keep the engine always working in the direction of the north or the polar star, to make it preserve its parallelism whilst it completes the circuit of the city?

"It is sufficient that the axis (of the earth) has been once directed to a certain point of the heavens in order to remain in that direction, though the earth has an annual motion."

Thus, in *The Earth and its Mechanism*, Henry Worms cuts short this ticklish question; better, no doubt, than if he had tried to explain it and failed like so many others. Galileo, at least had the advantage of perceiving the anomaly, though the astronomers of our own advanced age do not seem to do so; for, in the words of Humboldt, *Cosmos*, p. 177, " in his celebrated *Dialogo* he ascribes the constant parallel direction of the earth's axis to a centre of magnetic attraction existing in space."

It is contended by some, that the composition of the two forces

Q 2

plicable, six months in advance with the south pole, or, which is the same, six months foremost with the head, and six months foremost with the tail, including the sideward motion, or pro- and retro-gression, right and left, and all this without the assistance of any given third power to prevent the diagonal direction of the position of the body of the earth WHICH OUGHT TO RESULT from the projectile and attracting forces.

By what law, except the most arbitrary, however ingenious the invention, should otherwise the earth at one time present head, and at another, tail, and at others again front or side, to the sun? Bare assumption is NO LAW, and NO CAUSE is given by the old theorists why the body of the earth, under the influence of the said two forces, should preserve its parallelism; the thing is impossible upon their own showing: but the object is, to account for the seasons, which, without this arbitrary adoption, THEY CANNOT DO. And because they CANNOT do so, because THEY cannot give any other reason as the cause of the seasons : therefore, they say, the parallelism and inclination of the earth's axis must be true. " The simplicity with which " thereby " the seasons are explained . . . . . may certainly be adduced as a strong presumptive proof of the correctness of the principles already advanced ; for on no other RATIONAL SUPPOSITION with respect to the relations of the earth and sun, can these, and other as well-known phenomena be accounted for."

Thus Mr. Hind in his *Introduction* p. 13, where he may well call his " relation of the earth and sun " a SUPPOSITION; for there is not only not a shadow of physical, or even sensible, reason to prove it, but is moreover an unfinished, most illogical, conclusion of the effect of the composition of two forces, to which, in fact, the parallelism of the earth's axis is directly opposed. And yet this is called a RATIONAL supposition, beside which there CAN BE NO OTHER, to account for the seasons, though, with due deference

to superior learning, I invite comparison between this supposition without foundation, and my own theory as unfolded p. 187, and based upon physical and very simple REASONS.

If a straight line of railway of twenty or thirty miles in length were changed into a circle or an ellipse, and carried round London, so that end met end, the engine would always be in advance of the train, whether it travelled on the line when straight or when changed into a circle.

The same with the earth. If not attracted by the sun, it is said, it would have travelled in a straight line to the polar star regions, stooping with its north pole, its hoary head, as it were, in advance; but since that line was changed into a curve round the sun, why should the north pole not be in advance on the curved line, the same as it was on the straight one? Why should, on one side of the sun, the feet, the south pole, be in advance? Is there a third force and where, keeping it by the hair, attracting it to the polar star regions, making the train once turn on a pivot on the line round London, so as to keep the engine always working in the direction of the north or the polar star, to make it preserve its parallelism whilst it completes the circuit of the city?

"It is sufficient that the axis (of the earth) has been once directed to a certain point of the heavens in order to remain in that direction, though the earth has an annual motion."

Thus, in *The Earth and its Mechanism*, Henry Worms cuts short this ticklish question; better, no doubt, than if he had tried to explain it and failed like so many others. Galileo, at least had the advantage of perceiving the anomaly, though the astronomers of our own advanced age do not seem to do so; for, in the words of Humboldt, *Cosmos*, p. 177, " in his celebrated *Dialogo* he ascribes the constant parallel direction of the earth's axis to a centre of magnetic attraction existing in space."

It is contended by some, that the composition of the two forces

plicable, six months in advance with the south pole, or, which is the same, six months foremost with the head, and six months foremost with the tail, including the sideward motion, or pro- and retro-gression, right and left, and all this without the assistance of any given third power to prevent the diagonal direction of the position of the body of the earth WHICH OUGHT TO RESULT from the projectile and attracting forces.

By what law, except the most arbitrary, however ingenious the invention, should otherwise the earth at one time present head, and at another, tail, and at others again front or side, to the sun? Bare assumption is NO LAW, and NO CAUSE is given by the old theorists why the body of the earth, under the influence of the said two forces, should preserve its parallelism; the thing is impossible upon their own showing: but the object is, to account for the seasons, which, without this arbitrary adoption, THEY CANNOT DO. And because they CANNOT do so, because THEY cannot give any other reason as the cause of the seasons: therefore, they say, the parallelism and inclination of the earth's axis must be true. "The simplicity with which" thereby "the seasons are explained . . . . . . may certainly be adduced as a strong presumptive proof of the correctness of the principles already advanced; for on no other RATIONAL SUPPOSITION with respect to the relations of the earth and sun, can these, and other as well-known phenomena be accounted for."

Thus Mr. Hind in his *Introduction* p. 13, where he may well call his "relation of the earth and sun" a SUPPOSITION; for there is not only not a shadow of physical, or even sensible, reason to prove it, but is moreover an unfinished, most illogical, conclusion of the effect of the composition of two forces, to which, in fact, the parallelism of the earth's axis is directly opposed. And yet this is called a RATIONAL supposition, beside which there CAN BE NO OTHER, to account for the seasons, though, with due deference

to superior learning, I invite comparison between this supposition without foundation, and my own theory as unfolded p. 187, and based upon physical and very simple REASONS.

If a straight line of railway of twenty or thirty miles in length were changed into a circle or an ellipse, and carried round London, so that end met end, the engine would always be in advance of the train, whether it travelled on the line when straight or when changed into a circle.

The same with the earth. If not attracted by the sun, it is said, it would have travelled in a straight line to the polar star regions, stooping with its north pole, its hoary head, as it were, in advance; but since that line was changed into a curve round the sun, why should the north pole not be in advance on the curved line, the same as it was on the straight one? Why should, on one side of the sun, the feet, the south pole, be in advance? Is there a third force and where, keeping it by the hair, attracting it to the polar star regions, making the train once turn on a pivot on the line round London, so as to keep the engine always working in the direction of the north or the polar star, to make it preserve its parallelism whilst it completes the circuit of the city?

"It is sufficient that the axis (of the earth) has been once directed to a certain point of the heavens in order to remain in that direction, though the earth has an annual motion."

Thus, in *The Earth and its Mechanism*, Henry Worms cuts short this ticklish question; better, no doubt, than if he had tried to explain it and failed like so many others. Galileo, at least had the advantage of perceiving the anomaly, though the astronomers of our own advanced age do not seem to do so; for, in the words of Humboldt, *Cosmos*, p. 177, " in his celebrated *Dialogo* he ascribes the constant parallel direction of the earth's axis to a centre of magnetic attraction existing in space."

It is contended by some, that the composition of the two forces

Q 2

plicable, six months in advance with the south pole, or, which is the same, six months foremost with the head, and six months foremost with the tail, including the sideward motion, or pro- and retro-gression, right and left, and all this without the assistance of any given third power to prevent the diagonal direction of the position of the body of the earth WHICH OUGHT TO RESULT from the projectile and attracting forces.

By what law, except the most arbitrary, however ingenious the invention, should otherwise the earth at one time present head, and at another, tail, and at others again front or side, to the sun? Bare assumption is NO LAW, and NO CAUSE is given by the old theorists why the body of the earth, under the influence of the said two forces, should preserve its parallelism; the thing is impossible upon their own showing: but the object is, to account for the seasons, which, without this arbitrary adoption, THEY CANNOT DO. And because they CANNOT do so, because THEY cannot give any other reason as the cause of the seasons : therefore, they say, the parallelism and inclination of the earth's axis must be true. "The simplicity with which" thereby "the seasons are explained . . . . . may certainly be adduced as a strong presumptive proof of the correctness of the principles already advanced ; for on no other RATIONAL SUPPOSITION with respect to the relations of the earth and sun, can these, and other as well-known phenomena be accounted for."

Thus Mr. Hind in his *Introduction* p. 13, where he may well call his "relation of the earth and sun" a SUPPOSITION; for there is not only not a shadow of physical, or even sensible, reason to prove it, but is moreover an unfinished, most illogical, conclusion of the effect of the composition of two forces, to which, in fact, the parallelism of the earth's axis is directly opposed. And yet this is called a RATIONAL supposition, beside which there CAN BE NO OTHER, to account for the seasons, though, with due deference

to superior learning, I invite comparison between this supposition without foundation, and my own theory as unfolded p. 187, and based upon physical and very simple REASONS.

If a straight line of railway of twenty or thirty miles in length were changed into a circle or an ellipse, and carried round London, so that end met end, the engine would always be in advance of the train, whether it travelled on the line when straight or when changed into a circle.

The same with the earth. If not attracted by the sun, it is said, it would have travelled in a straight line to the polar star regions, stooping with its north pole, its hoary head, as it were, in advance; but since that line was changed into a curve round the sun, why should the north pole not be in advance on the curved line, the same as it was on the straight one? Why should, on one side of the sun, the feet, the south pole, be in advance? Is there a third force and where, keeping it by the hair, attracting it to the polar star regions, making the train once turn on a pivot on the line round London, so as to keep the engine always working in the direction of the north or the polar star, to make it preserve its parallelism whilst it completes the circuit of the city?

"It is sufficient that the axis (of the earth) has been once directed to a certain point of the heavens in order to remain in that direction, though the earth has an annual motion."

Thus, in *The Earth and its Mechanism*, Henry Worms cuts short this ticklish question; better, no doubt, than if he had tried to explain it and failed like so many others. Galileo, at least had the advantage of perceiving the anomaly, though the astronomers of our own advanced age do not seem to do so; for, in the words of Humboldt, *Cosmos*, p. 177, " in his celebrated *Dialogo* he ascribes the constant parallel direction of the earth's axis to a centre of magnetic attraction existing in space."

It is contended by some, that the composition of the two forces

plicable, six months in advance with the south pole, or, which is the same, six months foremost with the head, and six months foremost with the tail, including the sideward motion, or pro- and retro-gression, right and left, and all this without the assistance of any given third power to prevent the diagonal direction of the position of the body of the earth WHICH OUGHT TO RESULT from the projectile and attracting forces.

By what law, except the most arbitrary, however ingenious the invention, should otherwise the earth at one time present head, and at another, tail, and at others again front or side, to the sun? Bare assumption is NO LAW, and NO CAUSE is given by the old theorists why the body of the earth, under the influence of the said two forces, should preserve its parallelism; the thing is impossible upon their own showing: but the object is, to account for the seasons, which, without this arbitrary adoption, THEY CANNOT DO. And because they CANNOT do so, because THEY cannot give any other reason as the cause of the seasons: therefore, they say, the parallelism and inclination of the earth's axis must be true. " The simplicity with which " thereby " the seasons are explained . . . . . . may certainly be adduced as a strong presumptive proof of the correctness of the principles already advanced ; for on no other RATIONAL SUPPOSITION with respect to the relations of the earth and sun, can these, and other as well-known phenomena be accounted for."

Thus Mr. Hind in his *Introduction* p. 13, where he may well call his " relation of the earth and sun " a SUPPOSITION ; for there is not only not a shadow of physical, or even sensible, reason to prove it, but is moreover an unfinished, most illogical, conclusion of the effect of the composition of two forces, to which, in fact, the parallelism of the earth's axis is directly opposed. And yet this is called a RATIONAL supposition, beside which there CAN BE NO OTHER, to account for the seasons, though, with due deference

to superior learning, I invite comparison between this supposition without foundation, and my own theory as unfolded p. 187, and based upon physical and very simple REASONS.

If a straight line of railway of twenty or thirty miles in length were changed into a circle or an ellipse, and carried round London, so that end met end, the engine would always be in advance of the train, whether it travelled on the line when straight or when changed into a circle.

The same with the earth. If not attracted by the sun, it is said, it would have travelled in a straight line to the polar star regions, stooping with its north pole, its hoary head, as it were, in advance; but since that line was changed into a curve round the sun, why should the north pole not be in advance on the curved line, the same as it was on the straight one? Why should, on one side of the sun, the feet, the south pole, be in advance? Is there a third force and where, keeping it by the hair, attracting it to the polar star regions, making the train once turn on a pivot on the line round London, so as to keep the engine always working in the direction of the north or the polar star, to make it preserve its parallelism whilst it completes the circuit of the city?

"It is sufficient that the axis (of the earth) has been once directed to a certain point of the heavens in order to remain in that direction, though the earth has an annual motion."

Thus, in *The Earth and its Mechanism*, Henry Worms cuts short this ticklish question; better, no doubt, than if he had tried to explain it and failed like so many others. Galileo, at least had the advantage of perceiving the anomaly, though the astronomers of our own advanced age do not seem to do so; for, in the words of Humboldt, *Cosmos*, p. 177, " in his celebrated *Dialogo* he ascribes the constant parallel direction of the earth's axis to a centre of magnetic attraction existing in space."

It is contended by some, that the composition of the two forces

plicable, six months in advance with the south pole, or, which is
the same, six months foremost with the head, and six months
foremost with the tail, including the sideward motion, or pro- and
retro-gression, right and left, and all this without the assistance
of any given third power to prevent the diagonal direction of the
position of the body of the earth WHICH OUGHT TO RESULT from
the projectile and attracting forces.

By what law, except the most arbitrary, however ingenious the
invention, should otherwise the earth at one time present head,
and at another, tail, and at others again front or side, to the sun?
Bare assumption is NO LAW, and NO CAUSE is given by the old
theorists why the body of the earth, under the influence of the
said two forces, should preserve its parallelism; the thing is impossible upon their own showing: but the object is, to account for
the seasons, which, without this arbitrary adoption, THEY CANNOT
DO. And because they CANNOT do so, because THEY cannot give
any other reason as the cause of the seasons : therefore, they say,
the parallelism and inclination of the earth's axis must be true.
" The simplicity with which " thereby " the seasons are explained
. . . . . may certainly be adduced as a strong presumptive
proof of the correctness of the principles already advanced ; for
on no other RATIONAL SUPPOSITION with respect to the relations of
the earth and sun, can these, and other as well-known phenomena
be accounted for."

Thus Mr. Hind in his *Introduction* p. 13, where he may well
call his " relation of the earth and sun " a SUPPOSITION; for
there is not only not a shadow of physical, or even sensible, reason
to prove it, but is moreover an unfinished, most illogical, conclusion
of the effect of the composition of two forces, to which, in fact,
the parallelism of the earth's axis is directly opposed. And yet
this is called a RATIONAL supposition, beside which there CAN BE
NO OTHER, to account for the seasons, though, with due deference

to superior learning, I invite comparison between this supposition without foundation, and my own theory as unfolded p. 187, and based upon physical and very simple REASONS.

If a straight line of railway of twenty or thirty miles in length were changed into a circle or an ellipse, and carried round London, so that end met end, the engine would always be in advance of the train, whether it travelled on the line when straight or when changed into a circle.

The same with the earth. If not attracted by the sun, it is said, it would have travelled in a straight line to the polar star regions, stooping with its north pole, its hoary head, as it were, in advance; but since that line was changed into a curve round the sun, why should the north pole not be in advance on the curved line, the same as it was on the straight one? Why should, on one side of the sun, the feet, the south pole, be in advance? Is there a third force and where, keeping it by the hair, attracting it to the polar star regions, making the train once turn on a pivot on the line round London, so as to keep the engine always working in the direction of the north or the polar star, to make it preserve its parallelism whilst it completes the circuit of the city?

"It is sufficient that the axis (of the earth) has been once directed to a certain point of the heavens in order to remain in that direction, though the earth has an annual motion."

Thus, in *The Earth and its Mechanism*, Henry Worms cuts short this ticklish question; better, no doubt, than if he had tried to explain it and failed like so many others. Galileo, at least had the advantage of perceiving the anomaly, though the astronomers of our own advanced age do not seem to do so; for, in the words of Humboldt, *Cosmos*, p. 177, " in his celebrated *Dialogo* he ascribes the constant parallel direction of the earth's axis to a centre of magnetic attraction existing in space."

It is contended by some, that the composition of the two forces

plicable, six months in advance with the south pole, or, which is the same, six months foremost with the head, and six months foremost with the tail, including the sideward motion, or pro- and retro-gression, right and left, and all this without the assistance of any given third power to prevent the diagonal direction of the position of the body of the earth WHICH OUGHT TO RESULT from the projectile and attracting forces.

By what law, except the most arbitrary, however ingenious the invention, should otherwise the earth at one time present head, and at another, tail, and at others again front or side, to the sun? Bare assumption is NO LAW, and NO CAUSE is given by the old theorists why the body of the earth, under the influence of the said two forces, should preserve its parallelism; the thing is impossible upon their own showing: but the object is, to account for the seasons, which, without this arbitrary adoption, THEY CANNOT DO.   And because they CANNOT do so, because THEY cannot give any other reason as the cause of the seasons : therefore, they say, the parallelism and inclination of the earth's axis must be true. " The simplicity with which " thereby " the seasons are explained . . . . . . may certainly be adduced as a strong presumptive proof of the correctness of the principles already advanced ; for on no other RATIONAL SUPPOSITION with respect to the relations of the earth and sun, can these, and other as well-known phenomena be accounted for."

Thus Mr. Hind in his *Introduction* p. 13, where he may well call his " relation of the earth and sun " a SUPPOSITION; for there is not only not a shadow of physical, or even sensible, reason to prove it, but is moreover an unfinished, most illogical, conclusion of the effect of the composition of two forces, to which, in fact, the parallelism of the earth's axis is directly opposed. And yet this is called a RATIONAL supposition, beside which there CAN BE NO OTHER, to account for the seasons, though, with due deference

to superior learning, I invite comparison between this supposition without foundation, and my own theory as unfolded p. 187, and based upon physical and very simple REASONS.

If a straight line of railway of twenty or thirty miles in length were changed into a circle or an ellipse, and carried round London, so that end met end, the engine would always be in advance of the train, whether it travelled on the line when straight or when changed into a circle.

The same with the earth. If not attracted by the sun, it is said, it would have travelled in a straight line to the polar star regions, stooping with its north pole, its hoary head, as it were, in advance; but since that line was changed into a curve round the sun, why should the north pole not be in advance on the curved line, the same as it was on the straight one? Why should, on one side of the sun, the feet, the south pole, be in advance? Is there a third force and where, keeping it by the hair, attracting it to the polar star regions, making the train once turn on a pivot on the line round London, so as to keep the engine always working in the direction of the north or the polar star, to make it preserve its parallelism whilst it completes the circuit of the city?

"It is sufficient that the axis (of the earth) has been once directed to a certain point of the heavens in order to remain in that direction, though the earth has an annual motion."

Thus, in *The Earth and its Mechanism*, Henry Worms cuts short this ticklish question; better, no doubt, than if he had tried to explain it and failed like so many others. Galileo, at least had the advantage of perceiving the anomaly, though the astronomers of our own advanced age do not seem to do so; for, in the words of Humboldt, *Cosmos*, p. 177, "in his celebrated *Dialogo* he ascribes the constant parallel direction of the earth's axis to a centre of magnetic attraction existing in space."

It is contended by some, that the composition of the two forces

plicable, six months in advance with the south pole, or, which is the same, six months foremost with the head, and six months foremost with the tail, including the sideward motion, or pro- and retro-gression, right and left, and all this without the assistance of any given third power to prevent the diagonal direction of the position of the body of the earth WHICH OUGHT TO RESULT from the projectile and attracting forces.

By what law, except the most arbitrary, however ingenious the invention, should otherwise the earth at one time present head, and at another, tail, and at others again front or side, to the sun? Bare assumption is NO LAW, and NO CAUSE is given by the old theorists why the body of the earth, under the influence of the said two forces, should preserve its parallelism; the thing is impossible upon their own showing: but the object is, to account for the seasons, which, without this arbitrary adoption, THEY CANNOT DO. And because they CANNOT do so, because THEY cannot give any other reason as the cause of the seasons: therefore, they say, the parallelism and inclination of the earth's axis must be true. "The simplicity with which" thereby "the seasons are explained . . . . . . may certainly be adduced as a strong presumptive proof of the correctness of the principles already advanced; for on no other RATIONAL SUPPOSITION with respect to the relations of the earth and sun, can these, and other as well-known phenomena be accounted for."

Thus Mr. Hind in his *Introduction* p. 13, where he may well call his "relation of the earth and sun" a SUPPOSITION; for there is not only not a shadow of physical, or even sensible, reason to prove it, but is moreover an unfinished, most illogical, conclusion of the effect of the composition of two forces, to which, in fact, the parallelism of the earth's axis is directly opposed. And yet this is called a RATIONAL supposition, beside which there CAN BE NO OTHER, to account for the seasons, though, with due deference

to superior learning, I invite comparison between this supposition without foundation, and my own theory as unfolded p. 187, and based upon physical and very simple REASONS.

If a straight line of railway of twenty or thirty miles in length were changed into a circle or an ellipse, and carried round London, so that end met end, the engine would always be in advance of the train, whether it travelled on the line when straight or when changed into a circle.

The same with the earth. If not attracted by the sun, it is said, it would have travelled in a straight line to the polar star regions, stooping with its north pole, its hoary head, as it were, in advance; but since that line was changed into a curve round the sun, why should the north pole not be in advance on the curved line, the same as it was on the straight one? Why should, on one side of the sun, the feet, the south pole, be in advance? Is there a third force and where, keeping it by the hair, attracting it to the polar star regions, making the train once turn on a pivot on the line round London, so as to keep the engine always working in the direction of the north or the polar star, to make it preserve its parallelism whilst it completes the circuit of the city?

"It is sufficient that the axis (of the earth) has been once directed to a certain point of the heavens in order to remain in that direction, though the earth has an annual motion."

Thus, in *The Earth and its Mechanism*, Henry Worms cuts short this ticklish question; better, no doubt, than if he had tried to explain it and failed like so many others. Galileo, at least had the advantage of perceiving the anomaly, though the astronomers of our own advanced age do not seem to do so; for, in the words of Humboldt, *Cosmos*, p. 177, "in his celebrated *Dialogo* he ascribes the constant parallel direction of the earth's axis to a centre of magnetic attraction existing in space."

It is contended by some, that the composition of the two forces

plicable, six months in advance with the south pole, or, which is the same, six months foremost with the head, and six months foremost with the tail, including the sideward motion, or pro- and retro-gression, right and left, and all this without the assistance of any given third power to prevent the diagonal direction of the position of the body of the earth WHICH OUGHT TO RESULT from the projectile and attracting forces.

By what law, except the most arbitrary, however ingenious the invention, should otherwise the earth at one time present head, and at another, tail, and at others again front or side, to the sun? Bare assumption is NO LAW, and NO CAUSE is given by the old theorists why the body of the earth, under the influence of the said two forces, should preserve its parallelism; the thing is impossible upon their own showing: but the object is, to account for the seasons, which, without this arbitrary adoption, THEY CANNOT DO. And because they CANNOT do so, because THEY cannot give any other reason as the cause of the seasons : therefore, they say, the parallelism and inclination of the earth's axis must be true. "The simplicity with which" thereby "the seasons are explained . . . . . may certainly be adduced as a strong presumptive proof of the correctness of the principles already advanced ; for on no other RATIONAL SUPPOSITION with respect to the relations of the earth and sun, can these, and other as well-known phenomena be accounted for."

Thus Mr. Hind in his *Introduction* p. 13, where he may well call his "relation of the earth and sun" a SUPPOSITION; for there is not only not a shadow of physical, or even sensible, reason to prove it, but is moreover an unfinished, most illogical, conclusion of the effect of the composition of two forces, to which, in fact, the parallelism of the earth's axis is directly opposed. And yet this is called a RATIONAL supposition, beside which there CAN BE NO OTHER, to account for the seasons, though, with due deference

to superior learning, I invite comparison between this supposition without foundation, and my own theory as unfolded p. 187, and based upon physical and very simple REASONS.

If a straight line of railway of twenty or thirty miles in length were changed into a circle or an ellipse, and carried round London, so that end met end, the engine would always be in advance of the train, whether it travelled on the line when straight or when changed into a circle.

The same with the earth. If not attracted by the sun, it is said, it would have travelled in a straight line to the polar star regions, stooping with its north pole, its hoary head, as it were, in advance; but since that line was changed into a curve round the sun, why should the north pole not be in advance on the curved line, the same as it was on the straight one? Why should, on one side of the sun, the feet, the south pole, be in advance? Is there a third force and where, keeping it by the hair, attracting it to the polar star regions, making the train once turn on a pivot on the line round London, so as to keep the engine always working in the direction of the north or the polar star, to make it preserve its parallelism whilst it completes the circuit of the city?

"It is sufficient that the axis (of the earth) has been once directed to a certain point of the heavens in order to remain in that direction, though the earth has an annual motion."

Thus, in *The Earth and its Mechanism*, Henry Worms cuts short this ticklish question; better, no doubt, than if he had tried to explain it and failed like so many others. Galileo, at least had the advantage of perceiving the anomaly, though the astronomers of our own advanced age do not seem to do so; for, in the words of Humboldt, *Cosmos*, p. 177, "in his celebrated *Dialogo* he ascribes the constant parallel direction of the earth's axis to a centre of magnetic attraction existing in space."

It is contended by some, that the composition of the two forces

Q 2

plicable, six months in advance with the south pole, or, which is the same, six months foremost with the head, and six months foremost with the tail, including the sideward motion, or pro- and retro-gression, right and left, and all this without the assistance of any given third power to prevent the diagonal direction of the position of the body of the earth WHICH OUGHT TO RESULT from the projectile and attracting forces.

By what law, except the most arbitrary, however ingenious the invention, should otherwise the earth at one time present head, and at another, tail, and at others again front or side, to the sun? Bare assumption is NO LAW, and NO CAUSE is given by the old theorists why the body of the earth, under the influence of the said two forces, should preserve its parallelism; the thing is impossible upon their own showing: but the object is, to account for the seasons, which, without this arbitrary adoption, THEY CANNOT DO. And because they CANNOT do so, because THEY cannot give any other reason as the cause of the seasons : therefore, they say, the parallelism and inclination of the earth's axis must be true. " The simplicity with which " thereby " the seasons are explained . . . . . may certainly be adduced as a strong presumptive proof of the correctness of the principles already advanced ; for on no other RATIONAL SUPPOSITION with respect to the relations of the earth and sun, can these, and other as well-known phenomena be accounted for."

Thus Mr. Hind in his *Introduction* p. 13, where he may well call his " relation of the earth and sun " a SUPPOSITION; for there is not only not a shadow of physical, or even sensible, reason to prove it, but is moreover an unfinished, most illogical, conclusion of the effect of the composition of two forces, to which, in fact, the parallelism of the earth's axis is directly opposed. And yet this is called a RATIONAL supposition, beside which there CAN BE NO OTHER, to account for the seasons, though, with due deference

to superior learning, I invite comparison between this supposition without foundation, and my own theory as unfolded p. 187, and based upon physical and very simple REASONS.

If a straight line of railway of twenty or thirty miles in length were changed into a circle or an ellipse, and carried round London, so that end met end, the engine would always be in advance of the train, whether it travelled on the line when straight or when changed into a circle.

The same with the earth. If not attracted by the sun, it is said, it would have travelled in a straight line to the polar star regions, stooping with its north pole, its hoary head, as it were, in advance; but since that line was changed into a curve round the sun, why should the north pole not be in advance on the curved line, the same as it was on the straight one? Why should, on one side of the sun, the feet, the south pole, be in advance? Is there a third force and where, keeping it by the hair, attracting it to the polar star regions, making the train once turn on a pivot on the line round London, so as to keep the engine always working in the direction of the north or the polar star, to make it preserve its parallelism whilst it completes the circuit of the city?

"It is sufficient that the axis (of the earth) has been once directed to a certain point of the heavens in order to remain in that direction, though the earth has an annual motion."

Thus, in *The Earth and its Mechanism*, Henry Worms cuts short this ticklish question ; better, no doubt, than if he had tried to explain it and failed like so many others. Galileo, at least had the advantage of perceiving the anomaly, though the astronomers of our own advanced age do not seem to do so ; for, in the words of Humboldt, *Cosmos*, p. 177, " in his celebrated *Dialogo* he ascribes the constant parallel direction of the earth's axis to a centre of magnetic attraction existing in space."

It is contended by some, that the composition of the two forces

plicable, six months in advance with the south pole, or, which is the same, six months foremost with the head, and six months foremost with the tail, including the sideward motion, or pro- and retro-gression, right and left, and all this without the assistance of any given third power to prevent the diagonal direction of the position of the body of the earth WHICH OUGHT TO RESULT from the projectile and attracting forces.

By what law, except the most arbitrary, however ingenious the invention, should otherwise the earth at one time present head, and at another, tail, and at others again front or side, to the sun? Bare assumption is NO LAW, and NO CAUSE is given by the old theorists why the body of the earth, under the influence of the said two forces, should preserve its parallelism; the thing is impossible upon their own showing: but the object is, to account for the seasons, which, without this arbitrary adoption, THEY CANNOT DO. And because they CANNOT do so, because THEY cannot give any other reason as the cause of the seasons : therefore, they say, the parallelism and inclination of the earth's axis must be true. " The simplicity with which " thereby " the seasons are explained . . . . . may certainly be adduced as a strong presumptive proof of the correctness of the principles already advanced ; for on no other RATIONAL SUPPOSITION with respect to the relations of the earth and sun, can these, and other as well-known phenomena be accounted for."

Thus Mr. Hind in his *Introduction* p. 13, where he may well call his " relation of the earth and sun " a SUPPOSITION; for there is not only not a shadow of physical, or even sensible, reason to prove it, but is moreover an unfinished, most illogical, conclusion of the effect of the composition of two forces, to which, in fact, the parallelism of the earth's axis is directly opposed. And yet this is called a RATIONAL supposition, beside which there CAN BE NO OTHER, to account for the seasons, though, with due deference

to superior learning, I invite comparison between this supposition without foundation, and my own theory as unfolded p. 187, and based upon physical and very simple REASONS.

If a straight line of railway of twenty or thirty miles in length were changed into a circle or an ellipse, and carried round London, so that end met end, the engine would always be in advance of the train, whether it travelled on the line when straight or when changed into a circle.

The same with the earth. If not attracted by the sun, it is said, it would have travelled in a straight line to the polar star regions, stooping with its north pole, its hoary head, as it were, in advance; but since that line was changed into a curve round the sun, why should the north pole not be in advance on the curved line, the same as it was on the straight one? Why should, on one side of the sun, the feet, the south pole, be in advance? Is there a third force and where, keeping it by the hair, attracting it to the polar star regions, making the train once turn on a pivot on the line round London, so as to keep the engine always working in the direction of the north or the polar star, to make it preserve its parallelism whilst it completes the circuit of the city?

"It is sufficient that the axis (of the earth) has been once directed to a certain point of the heavens in order to remain in that direction, though the earth has an annual motion."

Thus, in *The Earth and its Mechanism*, Henry Worms cuts short this ticklish question; better, no doubt, than if he had tried to explain it and failed like so many others. Galileo, at least had the advantage of perceiving the anomaly, though the astronomers of our own advanced age do not seem to do so; for, in the words of Humboldt, *Cosmos*, p. 177, " in his celebrated *Dialogo* he ascribes the constant parallel direction of the earth's axis to a centre of magnetic attraction existing in space."

It is contended by some, that the composition of the two forces

plicable, six months in advance with the south pole, or, which is the same, six months foremost with the head, and six months foremost with the tail, including the sideward motion, or pro- and retro-gression, right and left, and all this without the assistance of any given third power to prevent the diagonal direction of the position of the body of the earth WHICH OUGHT TO RESULT from the projectile and attracting forces.

By what law, except the most arbitrary, however ingenious the invention, should otherwise the earth at one time present head, and at another, tail, and at others again front or side, to the sun? Bare assumption is NO LAW, and NO CAUSE is given by the old theorists why the body of the earth, under the influence of the said two forces, should preserve its parallelism; the thing is impossible upon their own showing: but the object is, to account for the seasons, which, without this arbitrary adoption, THEY CANNOT DO. And because they CANNOT do so, because THEY cannot give any other reason as the cause of the seasons: therefore, they say, the parallelism and inclination of the earth's axis must be true. " The simplicity with which " thereby " the seasons are explained . . . . . may certainly be adduced as a strong presumptive proof of the correctness of the principles already advanced; for on no other RATIONAL SUPPOSITION with respect to the relations of the earth and sun, can these, and other as well-known phenomena be accounted for."

Thus Mr. Hind in his *Introduction* p. 13, where he may well call his " relation of the earth and sun " a SUPPOSITION; for there is not only not a shadow of physical, or even sensible, reason to prove it, but is moreover an unfinished, most illogical, conclusion of the effect of the composition of two forces, to which, in fact, the parallelism of the earth's axis is directly opposed. And yet this is called a RATIONAL supposition, beside which there CAN BE NO OTHER, to account for the seasons, though, with due deference

to superior learning, I invite comparison between this supposition without foundation, and my own theory as unfolded p. 187, and based upon physical and very simple REASONS.

If a straight line of railway of twenty or thirty miles in length were changed into a circle or an ellipse, and carried round London, so that end met end, the engine would always be in advance of the train, whether it travelled on the line when straight or when changed into a circle.

The same with the earth. If not attracted by the sun, it is said, it would have travelled in a straight line to the polar star regions, stooping with its north pole, its hoary head, as it were, in advance; but since that line was changed into a curve round the sun, why should the north pole not be in advance on the curved line, the same as it was on the straight one ?   Why should, on one side of the sun, the feet, the south pole, be in advance ?   Is there a third force and where, keeping it by the hair, attracting it to the polar star regions, making the train once turn on a pivot on the line round London, so as to keep the engine always working in the direction of the north or the polar star, to make it preserve its parallelism whilst it completes the circuit of the city ?

"It is sufficient that the axis (of the earth) has been once directed to a certain point of the heavens in order to remain in that direction, though the earth has an annual motion."

Thus, in *The Earth and its Mechanism*, Henry Worms cuts short this ticklish question ; better, no doubt, than if he had tried to explain it and failed like so many others. Galileo, at least had the advantage of perceiving the anomaly, though the astronomers of our own advanced age do not seem to do so ; for, in the words of Humboldt, *Cosmos*, p. 177, " in his celebrated *Dialogo* he ascribes the constant parallel direction of the earth's axis to a centre of magnetic attraction existing in space."

It is contended by some, that the composition of the two forces

plicable, six months in advance with the south pole, or, which is the same, six months foremost with the head, and six months foremost with the tail, including the sideward motion, or pro- and retro-gression, right and left, and all this without the assistance of any given third power to prevent the diagonal direction of the position of the body of the earth WHICH OUGHT TO RESULT from the projectile and attracting forces.

By what law, except the most arbitrary, however ingenious the invention, should otherwise the earth at one time present head, and at another, tail, and at others again front or side, to the sun? Bare assumption is NO LAW, and NO CAUSE is given by the old theorists why the body of the earth, under the influence of the said two forces, should preserve its parallelism; the thing is impossible upon their own showing: but the object is, to account for the seasons, which, without this arbitrary adoption, THEY CANNOT DO. And because they CANNOT do so, because THEY cannot give any other reason as the cause of the seasons : therefore, they say, the parallelism and inclination of the earth's axis must be true. "The simplicity with which" thereby "the seasons are explained . . . . . may certainly be adduced as a strong presumptive proof of the correctness of the principles already advanced ; for on no other RATIONAL SUPPOSITION with respect to the relations of the earth and sun, can these, and other as well-known phenomena be accounted for."

Thus Mr. Hind in his *Introduction* p. 13, where he may well call his "relation of the earth and sun" a SUPPOSITION ; for there is not only not a shadow of physical, or even sensible, reason to prove it, but is moreover an unfinished, most illogical, conclusion of the effect of the composition of two forces, to which, in fact, the parallelism of the earth's axis is directly opposed. And yet this is called a RATIONAL supposition, beside which there CAN BE NO OTHER, to account for the seasons, though, with due deference

to superior learning, I invite comparison between this supposition without foundation, and my own theory as unfolded p. 187, and based upon physical and very simple REASONS.

If a straight line of railway of twenty or thirty miles in length were changed into a circle or an ellipse, and carried round London, so that end met end, the engine would always be in advance of the train, whether it travelled on the line when straight or when changed into a circle.

The same with the earth. If not attracted by the sun, it is said, it would have travelled in a straight line to the polar star regions, stooping with its north pole, its hoary head, as it were, in advance; but since that line was changed into a curve round the sun, why should the north pole not be in advance on the curved line, the same as it was on the straight one? Why should, on one side of the sun, the feet, the south pole, be in advance? Is there a third force and where, keeping it by the hair, attracting it to the polar star regions, making the train once turn on a pivot on the line round London, so as to keep the engine always working in the direction of the north or the polar star, to make it preserve its parallelism whilst it completes the circuit of the city?

"It is sufficient that the axis (of the earth) has been once directed to a certain point of the heavens in order to remain in that direction, though the earth has an annual motion."

Thus, in *The Earth and its Mechanism*, Henry Worms cuts short this ticklish question; better, no doubt, than if he had tried to explain it and failed like so many others. Galileo, at least had the advantage of perceiving the anomaly, though the astronomers of our own advanced age do not seem to do so; for, in the words of Humboldt, *Cosmos*, p. 177, "in his celebrated *Dialogo* he ascribes the constant parallel direction of the earth's axis to a centre of magnetic attraction existing in space."

It is contended by some, that the composition of the two forces

plicable, six months in advance with the south pole, or, which is the same, six months foremost with the head, and six months foremost with the tail, including the sideward motion, or pro- and retro-gression, right and left, and all this without the assistance of any given third power to prevent the diagonal direction of the position of the body of the earth WHICH OUGHT TO RESULT from the projectile and attracting forces.

By what law, except the most arbitrary, however ingenious the invention, should otherwise the earth at one time present head, and at another, tail, and at others again front or side, to the sun? Bare assumption is NO LAW, and NO CAUSE is given by the old theorists why the body of the earth, under the influence of the said two forces, should preserve its parallelism; the thing is impossible upon their own showing: but the object is, to account for the seasons, which, without this arbitrary adoption, THEY CANNOT DO. And because they CANNOT do so, because THEY cannot give any other reason as the cause of the seasons : therefore, they say, the parallelism and inclination of the earth's axis must be true. " The simplicity with which " thereby " the seasons are explained . . . . . may certainly be adduced as a strong presumptive proof of the correctness of the principles already advanced ; for on no other RATIONAL SUPPOSITION with respect to the relations of the earth and sun, can these, and other as well-known phenomena be accounted for."

Thus Mr. Hind in his *Introduction* p. 13, where he may well call his " relation of the earth and sun " a SUPPOSITION; for there is not only not a shadow of physical, or even sensible, reason to prove it, but is moreover an unfinished, most illogical, conclusion of the effect of the composition of two forces, to which, in fact, the parallelism of the earth's axis is directly opposed. And yet this is called a RATIONAL supposition, beside which there CAN BE NO OTHER, to account for the seasons, though, with due deference

to superior learning, I invite comparison between this supposition without foundation, and my own theory as unfolded p. 187, and based upon physical and very simple REASONS.

If a straight line of railway of twenty or thirty miles in length were changed into a circle or an ellipse, and carried round London, so that end met end, the engine would always be in advance of the train, whether it travelled on the line when straight or when changed into a circle.

The same with the earth. If not attracted by the sun, it is said, it would have travelled in a straight line to the polar star regions, stooping with its north pole, its hoary head, as it were, in advance; but since that line was changed into a curve round the sun, why should the north pole not be in advance on the curved line, the same as it was on the straight one? Why should, on one side of the sun, the feet, the south pole, be in advance? Is there a third force and where, keeping it by the hair, attracting it to the polar star regions, making the train once turn on a pivot on the line round London, so as to keep the engine always working in the direction of the north or the polar star, to make it preserve its parallelism whilst it completes the circuit of the city?

"It is sufficient that the axis (of the earth) has been once directed to a certain point of the heavens in order to remain in that direction, though the earth has an annual motion."

Thus, in *The Earth and its Mechanism*, Henry Worms cuts short this ticklish question; better, no doubt, than if he had tried to explain it and failed like so many others. Galileo, at least had the advantage of perceiving the anomaly, though the astronomers of our own advanced age do not seem to do so; for, in the words of Humboldt, *Cosmos*, p. 177, "in his celebrated *Dialogo* he ascribes the constant parallel direction of the earth's axis to a centre of magnetic attraction existing in space."

It is contended by some, that the composition of the two forces

plicable, six months in advance with the south pole, or, which is the same, six months foremost with the head, and six months foremost with the tail, including the sideward motion, or pro- and retro-gression, right and left, and all this without the assistance of any given third power to prevent the diagonal direction of the position of the body of the earth WHICH OUGHT TO RESULT from the projectile and attracting forces.

By what law, except the most arbitrary, however ingenious the invention, should otherwise the earth at one time present head, and at another, tail, and at others again front or side, to the sun? Bare assumption is NO LAW, and NO CAUSE is given by the old theorists why the body of the earth, under the influence of the said two forces, should preserve its parallelism; the thing is impossible upon their own showing: but the object is, to account for the seasons, which, without this arbitrary adoption, THEY CANNOT DO. And because they CANNOT do so, because THEY cannot give any other reason as the cause of the seasons : therefore, they say, the parallelism and inclination of the earth's axis must be true. " The simplicity with which " thereby " the seasons are explained . . . . . may certainly be adduced as a strong presumptive proof of the correctness of the principles already advanced ; for on no other RATIONAL SUPPOSITION with respect to the relations of the earth and sun, can these, and other as well-known phenomena be accounted for."

Thus Mr. Hind in his *Introduction* p. 13, where he may well call his "relation of the earth and sun " a SUPPOSITION; for there is not only not a shadow of physical, or even sensible, reason to prove it, but is moreover an unfinished, most illogical, conclusion of the effect of the composition of two forces, to which, in fact, the parallelism of the earth's axis is directly opposed. And yet this is called a RATIONAL supposition, beside which there CAN BE NO OTHER, to account for the seasons, though, with due deference

to superior learning, I invite comparison between this supposition without foundation, and my own theory as unfolded p. 187, and based upon physical and very simple REASONS.

If a straight line of railway of twenty or thirty miles in length were changed into a circle or an ellipse, and carried round London, so that end met end, the engine would always be in advance of the train, whether it travelled on the line when straight or when changed into a circle.

The same with the earth. If not attracted by the sun, it is said, it would have travelled in a straight line to the polar star regions, stooping with its north pole, its hoary head, as it were, in advance; but since that line was changed into a curve round the sun, why should the north pole not be in advance on the curved line, the same as it was on the straight one? Why should, on one side of the sun, the feet, the south pole, be in advance? Is there a third force and where, keeping it by the hair, attracting it to the polar star regions, making the train once turn on a pivot on the line round London, so as to keep the engine always working in the direction of the north or the polar star, to make it preserve its parallelism whilst it completes the circuit of the city?

"It is sufficient that the axis (of the earth) has been once directed to a certain point of the heavens in order to remain in that direction, though the earth has an annual motion."

Thus, in *The Earth and its Mechanism*, Henry Worms cuts short this ticklish question; better, no doubt, than if he had tried to explain it and failed like so many others. Galileo, at least had the advantage of perceiving the anomaly, though the astronomers of our own advanced age do not seem to do so; for, in the words of Humboldt, *Cosmos*, p. 177, "in his celebrated *Dialogo* he ascribes the constant parallel direction of the earth's axis to a centre of magnetic attraction existing in space."

It is contended by some, that the composition of the two forces

Q 2

plicable, six months in advance with the south pole, or, which is the same, six months foremost with the head, and six months foremost with the tail, including the sideward motion, or pro- and retro-gression, right and left, and all this without the assistance of any given third power to prevent the diagonal direction of the position of the body of the earth WHICH OUGHT TO RESULT from the projectile and attracting forces.

By what law, except the most arbitrary, however ingenious the invention, should otherwise the earth at one time present head, and at another, tail, and at others again front or side, to the sun? Bare assumption is NO LAW, and NO CAUSE is given by the old theorists why the body of the earth, under the influence of the said two forces, should preserve its parallelism; the thing is impossible upon their own showing: but the object is, to account for the seasons, which, without this arbitrary adoption, THEY CANNOT DO. And because they CANNOT do so, because THEY cannot give any other reason as the cause of the seasons : therefore, they say, the parallelism and inclination of the earth's axis must be true. "The simplicity with which" thereby "the seasons are explained . . . . . may certainly be adduced as a strong presumptive proof of the correctness of the principles already advanced ; for on no other RATIONAL SUPPOSITION with respect to the relations of the earth and sun, can these, and other as well-known phenomena be accounted for."

Thus Mr. Hind in his *Introduction* p. 13, where he may well call his "relation of the earth and sun" a SUPPOSITION; for there is not only not a shadow of physical, or even sensible, reason to prove it, but is moreover an unfinished, most illogical, conclusion of the effect of the composition of two forces, to which, in fact, the parallelism of the earth's axis is directly opposed. And yet this is called a RATIONAL supposition, beside which there CAN BE NO OTHER, to account for the seasons, though, with due deference

to superior learning, I invite comparison between this supposition without foundation, and my own theory as unfolded p. 187, and based upon physical and very simple REASONS.

If a straight line of railway of twenty or thirty miles in length were changed into a circle or an ellipse, and carried round London, so that end met end, the engine would always be in advance of the train, whether it travelled on the line when straight or when changed into a circle.

The same with the earth. If not attracted by the sun, it is said, it would have travelled in a straight line to the polar star regions, stooping with its north pole, its hoary head, as it were, in advance; but since that line was changed into a curve round the sun, why should the north pole not be in advance on the curved line, the same as it was on the straight one? Why should, on one side of the sun, the feet, the south pole, be in advance? Is there a third force and where, keeping it by the hair, attracting it to the polar star regions, making the train once turn on a pivot on the line round London, so as to keep the engine always working in the direction of the north or the polar star, to make it preserve its parallelism whilst it completes the circuit of the city?

"It is sufficient that the axis (of the earth) has been once directed to a certain point of the heavens in order to remain in that direction, though the earth has an annual motion."

Thus, in *The Earth and its Mechanism*, Henry Worms cuts short this ticklish question; better, no doubt, than if he had tried to explain it and failed like so many others. Galileo, at least had the advantage of perceiving the anomaly, though the astronomers of our own advanced age do not seem to do so; for, in the words of Humboldt, *Cosmos*, p. 177, " in his celebrated *Dialogo* he ascribes the constant parallel direction of the earth's axis to a centre of magnetic attraction existing in space."

It is contended by some, that the composition of the two forces

plicable, six months in advance with the south pole, or, which is the same, six months foremost with the head, and six months foremost with the tail, including the sideward motion, or pro- and retro-gression, right and left, and all this without the assistance of any given third power to prevent the diagonal direction of the position of the body of the earth WHICH OUGHT TO RESULT from the projectile and attracting forces.

By what law, except the most arbitrary, however ingenious the invention, should otherwise the earth at one time present head, and at another, tail, and at others again front or side, to the sun? Bare assumption is NO LAW, and NO CAUSE is given by the old theorists why the body of the earth, under the influence of the said two forces, should preserve its parallelism; the thing is impossible upon their own showing: but the object is, to account for the seasons, which, without this arbitrary adoption, THEY CANNOT DO. And because they CANNOT do so, because THEY cannot give any other reason as the cause of the seasons : therefore, they say, the parallelism and inclination of the earth's axis must be true. "The simplicity with which " thereby "the seasons are explained . . . . . may certainly be adduced as a strong presumptive proof of the correctness of the principles already advanced ; for on no other RATIONAL SUPPOSITION with respect to the relations of the earth and sun, can these, and other as well-known phenomena be accounted for."

Thus Mr. Hind in his *Introduction* p. 13, where he may well call his "relation of the earth and sun " a SUPPOSITION; for there is not only not a shadow of physical, or even sensible, reason to prove it, but is moreover an unfinished, most illogical, conclusion of the effect of the composition of two forces, to which, in fact, the parallelism of the earth's axis is directly opposed. And yet this is called a RATIONAL supposition, beside which there CAN BE NO OTHER, to account for the seasons, though, with due deference

to superior learning, I invite comparison between this supposition without foundation, and my own theory as unfolded p. 187, and based upon physical and very simple REASONS.

If a straight line of railway of twenty or thirty miles in length were changed into a circle or an ellipse, and carried round London, so that end met end, the engine would always be in advance of the train, whether it travelled on the line when straight or when changed into a circle.

The same with the earth. If not attracted by the sun, it is said, it would have travelled in a straight line to the polar star regions, stooping with its north pole, its hoary head, as it were, in advance; but since that line was changed into a curve round the sun, why should the north pole not be in advance on the curved line, the same as it was on the straight one? Why should, on one side of the sun, the feet, the south pole, be in advance? Is there a third force and where, keeping it by the hair, attracting it to the polar star regions, making the train once turn on a pivot on the line round London, so as to keep the engine always working in the direction of the north or the polar star, to make it preserve its parallelism whilst it completes the circuit of the city?

"It is sufficient that the axis (of the earth) has been once directed to a certain point of the heavens in order to remain in that direction, though the earth has an annual motion."

Thus, in *The Earth and its Mechanism*, Henry Worms cuts short this ticklish question; better, no doubt, than if he had tried to explain it and failed like so many others. Galileo, at least had the advantage of perceiving the anomaly, though the astronomers of our own advanced age do not seem to do so; for, in the words of Humboldt, *Cosmos*, p. 177, "in his celebrated *Dialogo* he ascribes the constant parallel direction of the earth's axis to a centre of magnetic attraction existing in space."

It is contended by some, that the composition of the two forces

plicable, six months in advance with the south pole, or, which is the same, six months foremost with the head, and six months foremost with the tail, including the sideward motion, or pro- and retro-gression, right and left, and all this without the assistance of any given third power to prevent the diagonal direction of the position of the body of the earth WHICH OUGHT TO RESULT from the projectile and attracting forces.

By what law, except the most arbitrary, however ingenious the invention, should otherwise the earth at one time present head, and at another, tail, and at others again front or side, to the sun? Bare assumption is NO LAW, and NO CAUSE is given by the old theorists why the body of the earth, under the influence of the said two forces, should preserve its parallelism; the thing is impossible upon their own showing: but the object is, to account for the seasons, which, without this arbitrary adoption, THEY CANNOT DO. And because they CANNOT do so, because THEY cannot give any other reason as the cause of the seasons: therefore, they say, the parallelism and inclination of the earth's axis must be true. "The simplicity with which" thereby "the seasons are explained . . . . . . may certainly be adduced as a strong presumptive proof of the correctness of the principles already advanced ; for on no other RATIONAL SUPPOSITION with respect to the relations of the earth and sun, can these, and other as well-known phenomena be accounted for."

Thus Mr. Hind in his *Introduction* p. 13, where he may well call his "relation of the earth and sun" a SUPPOSITION; for there is not only not a shadow of physical, or even sensible, reason to prove it, but is moreover an unfinished, most illogical, conclusion of the effect of the composition of two forces, to which, in fact, the parallelism of the earth's axis is directly opposed. And yet this is called a RATIONAL supposition, beside which there CAN BE NO OTHER, to account for the seasons, though, with due deference

to superior learning, I invite comparison between this supposition without foundation, and my own theory as unfolded p. 187, and based upon physical and very simple REASONS.

If a straight line of railway of twenty or thirty miles in length were changed into a circle or an ellipse, and carried round London, so that end met end, the engine would always be in advance of the train, whether it travelled on the line when straight or when changed into a circle.

The same with the earth. If not attracted by the sun, it is said, it would have travelled in a straight line to the polar star regions, stooping with its north pole, its hoary head, as it were, in advance; but since that line was changed into a curve round the sun, why should the north pole not be in advance on the curved line, the same as it was on the straight one? Why should, on one side of the sun, the feet, the south pole, be in advance? Is there a third force and where, keeping it by the hair, attracting it to the polar star regions, making the train once turn on a pivot on the line round London, so as to keep the engine always working in the direction of the north or the polar star, to make it preserve its parallelism whilst it completes the circuit of the city?

"It is sufficient that the axis (of the earth) has been once directed to a certain point of the heavens in order to remain in that direction, though the earth has an annual motion."

Thus, in *The Earth and its Mechanism*, Henry Worms cuts short this ticklish question; better, no doubt, than if he had tried to explain it and failed like so many others. Galileo, at least had the advantage of perceiving the anomaly, though the astronomers of our own advanced age do not seem to do so; for, in the words of Humboldt, *Cosmos*, p. 177, " in his celebrated *Dialogo* he ascribes the constant parallel direction of the earth's axis to a centre of magnetic attraction existing in space."

It is contended by some, that the composition of the two forces

Q 2

plicable, six months in advance with the south pole, or, which is the same, six months foremost with the head, and six months foremost with the tail, including the sideward motion, or pro- and retro-gression, right and left, and all this without the assistance of any given third power to prevent the diagonal direction of the position of the body of the earth WHICH OUGHT TO RESULT from the projectile and attracting forces.

By what law, except the most arbitrary, however ingenious the invention, should otherwise the earth at one time present head, and at another, tail, and at others again front or side, to the sun? Bare assumption is NO LAW, and NO CAUSE is given by the old theorists why the body of the earth, under the influence of the said two forces, should preserve its parallelism; the thing is impossible upon their own showing: but the object is, to account for the seasons, which, without this arbitrary adoption, THEY CANNOT DO. And because they CANNOT do so, because THEY cannot give any other reason as the cause of the seasons : therefore, they say, the parallelism and inclination of the earth's axis must be true. " The simplicity with which " thereby " the seasons are explained . . . . . may certainly be adduced as a strong presumptive proof of the correctness of the principles already advanced ; for on no other RATIONAL SUPPOSITION with respect to the relations of the earth and sun, can these, and other as well-known phenomena be accounted for."

Thus Mr. Hind in his *Introduction* p. 13, where he may well call his " relation of the earth and sun " a SUPPOSITION; for there is not only not a shadow of physical, or even sensible, reason to prove it, but is moreover an unfinished, most illogical, conclusion of the effect of the composition of two forces, to which, in fact, the parallelism of the earth's axis is directly opposed. And yet this is called a RATIONAL supposition, beside which there CAN BE NO OTHER, to account for the seasons, though, with due deference

to superior learning, I invite comparison between this supposition without foundation, and my own theory as unfolded p. 187, and based upon physical and very simple REASONS.

If a straight line of railway of twenty or thirty miles in length were changed into a circle or an ellipse, and carried round London, so that end met end, the engine would always be in advance of the train, whether it travelled on the line when straight or when changed into a circle.

The same with the earth. If not attracted by the sun, it is said, it would have travelled in a straight line to the polar star regions, stooping with its north pole, its hoary head, as it were, in advance; but since that line was changed into a curve round the sun, why should the north pole not be in advance on the curved line, the same as it was on the straight one? Why should, on one side of the sun, the feet, the south pole, be in advance? Is there a third force and where, keeping it by the hair, attracting it to the polar star regions, making the train once turn on a pivot on the line round London, so as to keep the engine always working in the direction of the north or the polar star, to make it preserve its parallelism whilst it completes the circuit of the city?

"It is sufficient that the axis (of the earth) has been once directed to a certain point of the heavens in order to remain in that direction, though the earth has an annual motion."

Thus, in *The Earth and its Mechanism*, Henry Worms cuts short this ticklish question; better, no doubt, than if he had tried to explain it and failed like so many others. Galileo, at least had the advantage of perceiving the anomaly, though the astronomers of our own advanced age do not seem to do so; for, in the words of Humboldt, *Cosmos*, p. 177, " in his celebrated *Dialogo* he ascribes the constant parallel direction of the earth's axis to a centre of magnetic attraction existing in space."

It is contended by some, that the composition of the two forces

Q 2

plicable, six months in advance with the south pole, or, which is the same, six months foremost with the head, and six months foremost with the tail, including the sideward motion, or pro- and retro-gression, right and left, and all this without the assistance of any given third power to prevent the diagonal direction of the position of the body of the earth WHICH OUGHT TO RESULT from the projectile and attracting forces.

By what law, except the most arbitrary, however ingenious the invention, should otherwise the earth at one time present head, and at another, tail, and at others again front or side, to the sun? Bare assumption is NO LAW, and NO CAUSE is given by the old theorists why the body of the earth, under the influence of the said two forces, should preserve its parallelism; the thing is impossible upon their own showing: but the object is, to account for the seasons, which, without this arbitrary adoption, THEY CANNOT DO. And because they CANNOT do so, because THEY cannot give any other reason as the cause of the seasons : therefore, they say, the parallelism and inclination of the earth's axis must be true. "The simplicity with which" thereby "the seasons are explained . . . . . may certainly be adduced as a strong presumptive proof of the correctness of the principles already advanced ; for on no other RATIONAL SUPPOSITION with respect to the relations of the earth and sun, can these, and other as well-known phenomena be accounted for."

Thus Mr. Hind in his *Introduction* p. 13, where he may well call his "relation of the earth and sun" a SUPPOSITION; for there is not only not a shadow of physical, or even sensible, reason to prove it, but is moreover an unfinished, most illogical, conclusion of the effect of the composition of two forces, to which, in fact, the parallelism of the earth's axis is directly opposed. And yet this is called a RATIONAL supposition, beside which there CAN BE NO OTHER, to account for the seasons, though, with due deference

to superior learning, I invite comparison between this supposition without foundation, and my own theory as unfolded p. 187, and based upon physical and very simple REASONS.

If a straight line of railway of twenty or thirty miles in length were changed into a circle or an ellipse, and carried round London, so that end met end, the engine would always be in advance of the train, whether it travelled on the line when straight or when changed into a circle.

The same with the earth. If not attracted by the sun, it is said, it would have travelled in a straight line to the polar star regions, stooping with its north pole, its hoary head, as it were, in advance; but since that line was changed into a curve round the sun, why should the north pole not be in advance on the curved line, the same as it was on the straight one? Why should, on one side of the sun, the feet, the south pole, be in advance? Is there a third force and where, keeping it by the hair, attracting it to the polar star regions, making the train once turn on a pivot on the line round London, so as to keep the engine always working in the direction of the north or the polar star, to make it preserve its parallelism whilst it completes the circuit of the city?

"It is sufficient that the axis (of the earth) has been once directed to a certain point of the heavens in order to remain in that direction, though the earth has an annual motion."

Thus, in *The Earth and its Mechanism*, Henry Worms cuts short this ticklish question; better, no doubt, than if he had tried to explain it and failed like so many others. Galileo, at least had the advantage of perceiving the anomaly, though the astronomers of our own advanced age do not seem to do so; for, in the words of Humboldt, *Cosmos*, p. 177, " in his celebrated *Dialogo* he ascribes the constant parallel direction of the earth's axis to a centre of magnetic attraction existing in space."

It is contended by some, that the composition of the two forces

Q 2

plicable, six months in advance with the south pole, or, which is the same, six months foremost with the head, and six months foremost with the tail, including the sideward motion, or pro- and retro-gression, right and left, and all this without the assistance of any given third power to prevent the diagonal direction of the position of the body of the earth WHICH OUGHT TO RESULT from the projectile and attracting forces.

By what law, except the most arbitrary, however ingenious the invention, should otherwise the earth at one time present head, and at another, tail, and at others again front or side, to the sun? Bare assumption is NO LAW, and NO CAUSE is given by the old theorists why the body of the earth, under the influence of the said two forces, should preserve its parallelism; the thing is impossible upon their own showing: but the object is, to account for the seasons, which, without this arbitrary adoption, THEY CANNOT DO. And because they CANNOT do so, because THEY cannot give any other reason as the cause of the seasons : therefore, they say, the parallelism and inclination of the earth's axis must be true. " The simplicity with which " thereby " the seasons are explained . . . . . may certainly be adduced as a strong presumptive proof of the correctness of the principles already advanced ; for on no other RATIONAL SUPPOSITION with respect to the relations of the earth and sun, can these, and other as well-known phenomena be accounted for."

Thus Mr. Hind in his *Introduction* p. 13, where he may well call his " relation of the earth and sun " a SUPPOSITION; for there is not only not a shadow of physical, or even sensible, reason to prove it, but is moreover an unfinished, most illogical, conclusion of the effect of the composition of two forces, to which, in fact, the parallelism of the earth's axis is directly opposed. And yet this is called a RATIONAL supposition, beside which there CAN BE NO OTHER, to account for the seasons, though, with due deference

to superior learning, I invite comparison between this supposition without foundation, and my own theory as unfolded p. 187, and based upon physical and very simple REASONS.

If a straight line of railway of twenty or thirty miles in length were changed into a circle or an ellipse, and carried round London, so that end met end, the engine would always be in advance of the train, whether it travelled on the line when straight or when changed into a circle.

The same with the earth. If not attracted by the sun, it is said, it would have travelled in a straight line to the polar star regions, stooping with its north pole, its hoary head, as it were, in advance; but since that line was changed into a curve round the sun, why should the north pole not be in advance on the curved line, the same as it was on the straight one? Why should, on one side of the sun, the feet, the south pole, be in advance? Is there a third force and where, keeping it by the hair, attracting it to the polar star regions, making the train once turn on a pivot on the line round London, so as to keep the engine always working in the direction of the north or the polar star, to make it preserve its parallelism whilst it completes the circuit of the city?

"It is sufficient that the axis (of the earth) has been once directed to a certain point of the heavens in order to remain in that direction, though the earth has an annual motion."

Thus, in *The Earth and its Mechanism*, Henry Worms cuts short this ticklish question; better, no doubt, than if he had tried to explain it and failed like so many others. Galileo, at least had the advantage of perceiving the anomaly, though the astronomers of our own advanced age do not seem to do so; for, in the words of Humboldt, *Cosmos*, p. 177, " in his celebrated *Dialogo* he ascribes the constant parallel direction of the earth's axis to a centre of magnetic attraction existing in space."

It is contended by some, that the composition of the two forces

Q 2

plicable, six months in advance with the south pole, or, which is the same, six months foremost with the head, and six months foremost with the tail, including the sideward motion, or pro- and retro-gression, right and left, and all this without the assistance of any given third power to prevent the diagonal direction of the position of the body of the earth WHICH OUGHT TO RESULT from the projectile and attracting forces.

By what law, except the most arbitrary, however ingenious the invention, should otherwise the earth at one time present head, and at another, tail, and at others again front or side, to the sun? Bare assumption is NO LAW, and NO CAUSE is given by the old theorists why the body of the earth, under the influence of the said two forces, should preserve its parallelism; the thing is impossible upon their own showing: but the object is, to account for the seasons, which, without this arbitrary adoption, THEY CANNOT DO. And because they CANNOT do so, because THEY cannot give any other reason as the cause of the seasons : therefore, they say, the parallelism and inclination of the earth's axis must be true. " The simplicity with which " thereby " the seasons are explained . . . . . may certainly be adduced as a strong presumptive proof of the correctness of the principles already advanced ; for on no other RATIONAL SUPPOSITION with respect to the relations of the earth and sun, can these, and other as well-known phenomena be accounted for."

Thus Mr. Hind in his *Introduction* p. 13, where he may well call his " relation of the earth and sun " a SUPPOSITION ; for there is not only not a shadow of physical, or even sensible, reason to prove it, but is moreover an unfinished, most illogical, conclusion of the effect of the composition of two forces, to which, in fact, the parallelism of the earth's axis is directly opposed. And yet this is called a RATIONAL supposition, beside which there CAN BE NO OTHER, to account for the seasons, though, with due deference

to superior learning, I invite comparison between this supposition without foundation, and my own theory as unfolded p. 187, and based upon physical and very simple REASONS.

If a straight line of railway of twenty or thirty miles in length were changed into a circle or an ellipse, and carried round London, so that end met end, the engine would always be in advance of the train, whether it travelled on the line when straight or when changed into a circle.

The same with the earth. If not attracted by the sun, it is said, it would have travelled in a straight line to the polar star regions, stooping with its north pole, its hoary head, as it were, in advance; but since that line was changed into a curve round the sun, why should the north pole not be in advance on the curved line, the same as it was on the straight one? Why should, on one side of the sun, the feet, the south pole, be in advance? Is there a third force and where, keeping it by the hair, attracting it to the polar star regions, making the train once turn on a pivot on the line round London, so as to keep the engine always working in the direction of the north or the polar star, to make it preserve its parallelism whilst it completes the circuit of the city?

"It is sufficient that the axis (of the earth) has been once directed to a certain point of the heavens in order to remain in that direction, though the earth has an annual motion."

Thus, in *The Earth and its Mechanism*, Henry Worms cuts short this ticklish question; better, no doubt, than if he had tried to explain it and failed like so many others. Galileo, at least had the advantage of perceiving the anomaly, though the astronomers of our own advanced age do not seem to do so; for, in the words of Humboldt, *Cosmos*, p. 177, " in his celebrated *Dialogo* he ascribes the constant parallel direction of the earth's axis to a centre of magnetic attraction existing in space."

It is contended by some, that the composition of the two forces

Q 2

plicable, six months in advance with the south pole, or, which is the same, six months foremost with the head, and six months foremost with the tail, including the sideward motion, or pro- and retro-gression, right and left, and all this without the assistance of any given third power to prevent the diagonal direction of the position of the body of the earth WHICH OUGHT TO RESULT from the projectile and attracting forces.

By what law, except the most arbitrary, however ingenious the invention, should otherwise the earth at one time present head, and at another, tail, and at others again front or side, to the sun? Bare assumption is NO LAW, and NO CAUSE is given by the old theorists why the body of the earth, under the influence of the said two forces, should preserve its parallelism; the thing is impossible upon their own showing: but the object is, to account for the seasons, which, without this arbitrary adoption, THEY CANNOT DO. And because they CANNOT do so, because THEY cannot give any other reason as the cause of the seasons : therefore, they say, the parallelism and inclination of the earth's axis must be true. " The simplicity with which " thereby " the seasons are explained . . . . . may certainly be adduced as a strong presumptive proof of the correctness of the principles already advanced ; for on no other RATIONAL SUPPOSITION with respect to the relations of the earth and sun, can these, and other as well-known phenomena be accounted for."

Thus Mr. Hind in his *Introduction* p. 13, where he may well call his " relation of the earth and sun " a SUPPOSITION; for there is not only not a shadow of physical, or even sensible, reason to prove it, but is moreover an unfinished, most illogical, conclusion of the effect of the composition of two forces, to which, in fact, the parallelism of the earth's axis is directly opposed. And yet this is called a RATIONAL supposition, beside which there CAN BE NO OTHER, to account for the seasons, though, with due deference

to superior learning, I invite comparison between this supposition without foundation, and my own theory as unfolded p. 187, and based upon physical and very simple REASONS.

If a straight line of railway of twenty or thirty miles in length were changed into a circle or an ellipse, and carried round London, so that end met end, the engine would always be in advance of the train, whether it travelled on the line when straight or when changed into a circle.

The same with the earth. If not attracted by the sun, it is said, it would have travelled in a straight line to the polar star regions, stooping with its north pole, its hoary head, as it were, in advance; but since that line was changed into a curve round the sun, why should the north pole not be in advance on the curved line, the same as it was on the straight one? Why should, on one side of the sun, the feet, the south pole, be in advance? Is there a third force and where, keeping it by the hair, attracting it to the polar star regions, making the train once turn on a pivot on the line round London, so as to keep the engine always working in the direction of the north or the polar star, to make it preserve its parallelism whilst it completes the circuit of the city?

"It is sufficient that the axis (of the earth) has been once directed to a certain point of the heavens in order to remain in that direction, though the earth has an annual motion."

Thus, in *The Earth and its Mechanism*, Henry Worms cuts short this ticklish question ; better, no doubt, than if he had tried to explain it and failed like so many others. Galileo, at least had the advantage of perceiving the anomaly, though the astronomers of our own advanced age do not seem to do so ; for, in the words of Humboldt, *Cosmos*, p. 177, " in his celebrated *Dialogo* he ascribes the constant parallel direction of the earth's axis to a centre of magnetic attraction existing in space."

It is contended by some, that the composition of the two forces

plicable, six months in advance with the south pole, or, which is the same, six months foremost with the head, and six months foremost with the tail, including the sideward motion, or pro- and retro-gression, right and left, and all this without the assistance of any given third power to prevent the diagonal direction of the position of the body of the earth WHICH OUGHT TO RESULT from the projectile and attracting forces.

By what law, except the most arbitrary, however ingenious the invention, should otherwise the earth at one time present head, and at another, tail, and at others again front or side, to the sun? Bare assumption is NO LAW, and NO CAUSE is given by the old theorists why the body of the earth, under the influence of the said two forces, should preserve its parallelism; the thing is impossible upon their own showing: but the object is, to account for the seasons, which, without this arbitrary adoption, THEY CANNOT DO. And because they CANNOT do so, because THEY cannot give any other reason as the cause of the seasons : therefore, they say, the parallelism and inclination of the earth's axis must be true. " The simplicity with which " thereby " the seasons are explained . . . . . . may certainly be adduced as a strong presumptive proof of the correctness of the principles already advanced ; for on no other RATIONAL SUPPOSITION with respect to the relations of the earth and sun, can these, and other as well-known phenomena be accounted for."

Thus Mr. Hind in his *Introduction* p. 13, where he may well call his " relation of the earth and sun " a SUPPOSITION ; for there is not only not a shadow of physical, or even sensible, reason to prove it, but is moreover an unfinished, most illogical, conclusion of the effect of the composition of two forces, to which, in fact, the parallelism of the earth's axis is directly opposed. And yet this is called a RATIONAL supposition, beside which there CAN BE NO OTHER, to account for the seasons, though, with due deference

to superior learning, I invite comparison between this supposition without foundation, and my own theory as unfolded p. 187, and based upon physical and very simple REASONS.

If a straight line of railway of twenty or thirty miles in length were changed into a circle or an ellipse, and carried round London, so that end met end, the engine would always be in advance of the train, whether it travelled on the line when straight or when changed into a circle.

The same with the earth. If not attracted by the sun, it is said, it would have travelled in a straight line to the polar star regions, stooping with its north pole, its hoary head, as it were, in advance; but since that line was changed into a curve round the sun, why should the north pole not be in advance on the curved line, the same as it was on the straight one? Why should, on one side of the sun, the feet, the south pole, be in advance? Is there a third force and where, keeping it by the hair, attracting it to the polar star regions, making the train once turn on a pivot on the line round London, so as to keep the engine always working in the direction of the north or the polar star, to make it preserve its parallelism whilst it completes the circuit of the city?

"It is sufficient that the axis (of the earth) has been once directed to a certain point of the heavens in order to remain in that direction, though the earth has an annual motion."

Thus, in *The Earth and its Mechanism*, Henry Worms cuts short this ticklish question; better, no doubt, than if he had tried to explain it and failed like so many others. Galileo, at least had the advantage of perceiving the anomaly, though the astronomers of our own advanced age do not seem to do so; for, in the words of Humboldt, *Cosmos*, p. 177, " in his celebrated *Dialogo* he ascribes the constant parallel direction of the earth's axis to a centre of magnetic attraction existing in space."

It is contended by some, that the composition of the two forces

plicable, six months in advance with the south pole, or, which is the same, six months foremost with the head, and six months foremost with the tail, including the sideward motion, or pro- and retro-gression, right and left, and all this without the assistance of any given third power to prevent the diagonal direction of the position of the body of the earth WHICH OUGHT TO RESULT from the projectile and attracting forces.

By what law, except the most arbitrary, however ingenious the invention, should otherwise the earth at one time present head, and at another, tail, and at others again front or side, to the sun? Bare assumption is NO LAW, and NO CAUSE is given by the old theorists why the body of the earth, under the influence of the said two forces, should preserve its parallelism; the thing is impossible upon their own showing: but the object is, to account for the seasons, which, without this arbitrary adoption, THEY CANNOT DO. And because they CANNOT do so, because THEY cannot give any other reason as the cause of the seasons : therefore, they say, the parallelism and inclination of the earth's axis must be true. " The simplicity with which " thereby " the seasons are explained . . . . . may certainly be adduced as a strong presumptive proof of the correctness of the principles already advanced ; for on no other RATIONAL SUPPOSITION with respect to the relations of the earth and sun, can these, and other as well-known phenomena be accounted for."

Thus Mr. Hind in his *Introduction* p. 13, where he may well call his " relation of the earth and sun " a SUPPOSITION ; for there is not only not a shadow of physical, or even sensible, reason to prove it, but is moreover an unfinished, most illogical, conclusion of the effect of the composition of two forces, to which, in fact, the parallelism of the earth's axis is directly opposed. And yet this is called a RATIONAL supposition, beside which there CAN BE NO OTHER, to account for the seasons, though, with due deference

to superior learning, I invite comparison between this supposition without foundation, and my own theory as unfolded p. 187, and based upon physical and very simple REASONS.

If a straight line of railway of twenty or thirty miles in length were changed into a circle or an ellipse, and carried round London, so that end met end, the engine would always be in advance of the train, whether it travelled on the line when straight or when changed into a circle.

The same with the earth. If not attracted by the sun, it is said, it would have travelled in a straight line to the polar star regions, stooping with its north pole, its hoary head, as it were, in advance; but since that line was changed into a curve round the sun, why should the north pole not be in advance on the curved line, the same as it was on the straight one? Why should, on one side of the sun, the feet, the south pole, be in advance? Is there a third force and where, keeping it by the hair, attracting it to the polar star regions, making the train once turn on a pivot on the line round London, so as to keep the engine always working in the direction of the north or the polar star, to make it preserve its parallelism whilst it completes the circuit of the city?

"It is sufficient that the axis (of the earth) has been once directed to a certain point of the heavens in order to remain in that direction, though the earth has an annual motion."

Thus, in *The Earth and its Mechanism*, Henry Worms cuts short this ticklish question; better, no doubt, than if he had tried to explain it and failed like so many others. Galileo, at least had the advantage of perceiving the anomaly, though the astronomers of our own advanced age do not seem to do so; for, in the words of Humboldt, *Cosmos*, p. 177, " in his celebrated *Dialogo* he ascribes the constant parallel direction of the earth's axis to a centre of magnetic attraction existing in space."

It is contended by some, that the composition of the two forces

plicable, six months in advance with the south pole, or, which is the same, six months foremost with the head, and six months foremost with the tail, including the sideward motion, or pro- and retro-gression, right and left, and all this without the assistance of any given third power to prevent the diagonal direction of the position of the body of the earth WHICH OUGHT TO RESULT from the projectile and attracting forces.

By what law, except the most arbitrary, however ingenious the invention, should otherwise the earth at one time present head, and at another, tail, and at others again front or side, to the sun? Bare assumption is NO LAW, and NO CAUSE is given by the old theorists why the body of the earth, under the influence of the said two forces, should preserve its parallelism; the thing is impossible upon their own showing: but the object is, to account for the seasons, which, without this arbitrary adoption, THEY CANNOT DO. And because they CANNOT do so, because THEY cannot give any other reason as the cause of the seasons : therefore, they say, the parallelism and inclination of the earth's axis must be true. " The simplicity with which " thereby " the seasons are explained . . . . . may certainly be adduced as a strong presumptive proof of the correctness of the principles already advanced ; for on no other RATIONAL SUPPOSITION with respect to the relations of the earth and sun, can these, and other as well-known phenomena be accounted for."

Thus Mr. Hind in his *Introduction* p. 13, where he may well call his " relation of the earth and sun " a SUPPOSITION; for there is not only not a shadow of physical, or even sensible, reason to prove it, but is moreover an unfinished, most illogical, conclusion of the effect of the composition of two forces, to which, in fact, the parallelism of the earth's axis is directly opposed. And yet this is called a RATIONAL supposition, beside which there CAN BE NO OTHER, to account for the seasons, though, with due deference

to superior learning, I invite comparison between this supposition without foundation, and my own theory as unfolded p. 187, and based upon physical and very simple REASONS.

If a straight line of railway of twenty or thirty miles in length were changed into a circle or an ellipse, and carried round London, so that end met end, the engine would always be in advance of the train, whether it travelled on the line when straight or when changed into a circle.

The same with the earth. If not attracted by the sun, it is said, it would have travelled in a straight line to the polar star regions, stooping with its north pole, its hoary head, as it were, in advance; but since that line was changed into a curve round the sun, why should the north pole not be in advance on the curved line, the same as it was on the straight one? Why should, on one side of the sun, the feet, the south pole, be in advance? Is there a third force and where, keeping it by the hair, attracting it to the polar star regions, making the train once turn on a pivot on the line round London, so as to keep the engine always working in the direction of the north or the polar star, to make it preserve its parallelism whilst it completes the circuit of the city?

"It is sufficient that the axis (of the earth) has been once directed to a certain point of the heavens in order to remain in that direction, though the earth has an annual motion."

Thus, in *The Earth and its Mechanism*, Henry Worms cuts short this ticklish question ; better, no doubt, than if he had tried to explain it and failed like so many others. Galileo, at least had the advantage of perceiving the anomaly, though the astronomers of our own advanced age do not seem to do so ; for, in the words of Humboldt, *Cosmos*, p. 177, " in his celebrated *Dialogo* he ascribes the constant parallel direction of the earth's axis to a centre of magnetic attraction existing in space."

It is contended by some, that the composition of the two forces

which turn the straight part of the earth into a circular or elliptical one, does not influence the body of the earth itself, because these two forces act upon the centre, upon a point, and that we can well imagine in this way the earth to preserve its parallelism during its passage round the sun. True, we may imagine and believe such a thing, because to God everything is possible; but it is nevertheless a great fallacy, and DOES NOT CORRESPOND TO THE LAW LAID DOWN.

A point, a centre, is either nothing, or something; if it is nothing, no power or composition of powers can affect it; if it is something, it must be something substantial, something tangible; and if it be substantial or tangible, the composition of the two forces must affect it, whether the point is beyond imagination little, or beyond conception grand. To God, the whole creation is but a point in eternity; the suns, the planets, the moons: all are but enlarged points, and as such must be subject to the action of forces exercised upon them, not merely as regards their path, but also their corporal position or situation. If the CENTRES OF BODIES, IMAGINARY POINTS without substance, are alone affected by the forces of attraction and gravitation, it seems to me strange that the centre of gravitation of the planets should be laid exterior to the sun, and that that of the moon should be placed outside the centre of the earth. But let us exemplify the fallacy of the old theory.

A substantial point or ball being the same, whether large or small, let us suppose such a ball, B, within the grasp of the two reputed forces, the dotted line, p, in the annexed diagram representing the projectile force coming from the hand of the Almighty, and the dotted line, a, representing the attractive force emanating from the centre of the sun. In this position, B 1, the point of such ball would be directed to the top of the page. The projectile force, p, does not allow the ball any rest, and the attracting force from the centre of the sun, NOT FROM ANY SPOT EXTERIOR TO THIS CENTRE, does not relax its hold; and thus the body of the ball must retain its relative position and inclination to the sun. On its onward path, this relative position would have been preserved, the point of the ball would have changed its original direction from the top of the page to the right hand, B 2, of the same; and further on, the same two forces driving and drawing the object of contention with untiring and unyielding energy, the point of the ball, B 3, is directed to the bottom of the page, and so on, the same as the railway train would travel round a town, point or engine always in advance. Let us invest the ball, the heavenly bodies, with four cardinal points culminating in the centre; let the propelling and attracting powers lay hold of two such points: and will it be said that in the course of a revolution of the ball these forces shift or remove their hold from one cardinal point to the other, that the two forces, acting always in the same direction, will twist the ball, as between two fingers, in a contrary direction from that of its path? that the ball will be the sport as well as the slave of the two forces? That, as in figure 2 at the following page, the primitive impulse, or the tangential force, should at one time impinge upon the side of the south pole of the earth, driving the north pole in advance; that by degrees it should strike on the west of the earth, driving the east in advance; that three months later the said impulse or force should pounce upon the side of the north

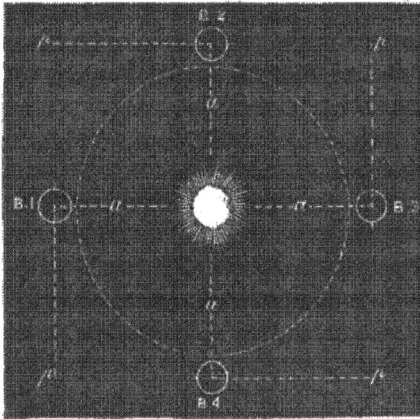

pole of the earth driving the south pole in advance; and that in
three months more it should gradually strike upon the east of the
earth, driving the west in advance? Or is it really possible still
to maintain, according to THE LAW of the composition of the two
forces in question, that a substantial point, or ball, the earth, the
train, moving in a circle on a plane, can, or could preserve
their parallelism under the influence of these two forces? Imagine
these bodies in the following position :—

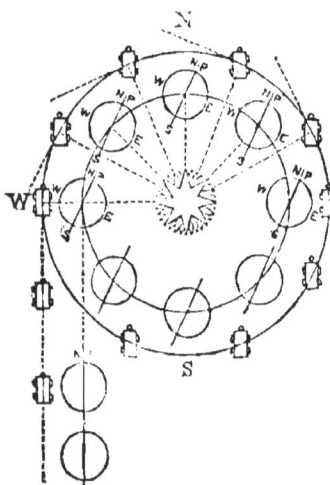

Fig. 2. As it is by ASSUMPTION.   Fig. 3. As it ought to be BY LAW.

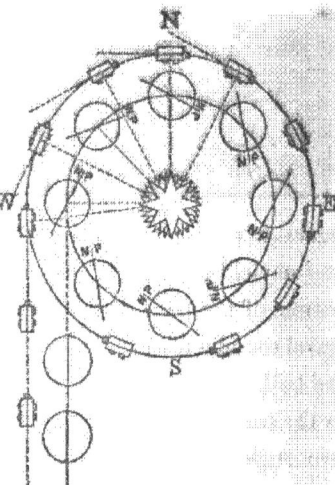

Propelled by the hand of the Almighty, in some imaginary
manner, or by electricity, steam, or otherwise, they come up in
straight lines, perhaps in a succession of bodies : what in the world,
except a third, unexplained power, a mystery, could make them,
as in fig. 2, preserve their original direction, their parallelism,
their north pole, N P, or foremost part, always pointing to the top
of the page whilst performing their journey round the centre of
their orbit? Whilst on the straight line engine FOLLOWS engine,
globe FOLLOWS globe, we find them on the northern curve SIDE BY
SIDE, the wheels of the engine rotating forward towards the polar

star regions, and yet the engine going SIDEWAYS on its plane ; on the eastern curve we find them again FOLLOWING one another, but BACK-WARDS, the engine-wheels positively going in the direction opposite to that of their rotation ; and on the southern curve we find them again SIDE BY SIDE, rotating at angles with the line of progression, and the globes rotating one way, and going in the opposite.

The planets, thus said to spin round the sun on an hori-zontal plane,—whether in a resisting medium or in an alleged vacant space, in this manner,—like the engines and globes above, on the theory of the day, move on a false bottom, on an airy foundation, on an imaginary plane, held in suspense by a fictitious lever, centered on a SHIFTING PHANTOM FULCRUM outside or exterior to the centre of the sun ; and the calculation of their weight against the weight of the sun, and the consequent cal-culation of their fulcrum, their point of gravitation as existing somewhere between the centre and the surface of the sun, is every day disturbed and falsified by the discovery of new planets, whilst it will make no difference in my theory, whatever number of planets may still be discovered, whatever their weight, size, and distance, because all find their point of rest in the solar centre.

In the old system there is no rest. How the suns were planted into their central positions we are not told ; but the planets were somehow or other LAUNCHED into space, and on they must go, carried by nothing, resting upon nothing, resisted by nothing, all at once laid hold of, attracted, by the sun, yet pushed away at a tangent by the primitive impulse attendant upon the launch ; if it were not for this impulse, the planets would have fallen into the sun, and if it were not for the attraction, they would have gone straight into eternity. By the same means ships on the sea, or balloons in the air, might be said to preserve their orbits, and ' neither go to the bottom, nor fly off into unknown regions, leaks or explosions, I suppose, excepted.

In the 19th century, the year of our Lord, 1864, the "primitive impulse"—with the parallelism of the earth's axis—like a deeply-rooted prejudice, is still haunting astronomers and keeping fast its hold on the minds of many.

In his *Introduction to Astronomy*, p. 21, with a clearness worthy of a better cause, Mr. Hind continues to detail and explain the motive power and helpless velocity communicated to the planet by the "Divine Hand," and its attraction by the sun; how at perihelion the primitive impulse, that famous discovery, overcomes gravitation, and how at aphelion attraction overcomes the primitive impulse; how in turn the one is getting exhausted whilst the other is gaining strength, how "the contending forces are so accurately adjusted, that neither can ever gain the ascendancy, and the planet must continue to describe its elliptic path through all time."

It may be a happy coincidence that at the singular rencontre of sun and planets the primitive impulse was equally matched by the attraction of the sun and *vice versa*, or we might now, carried along by the propelling launch, be travelling straightforward through an eternity of darkness, instead of being gently wafted round our glorious sun; but if ever I were to believe in this "hide-and-seek" theory, in this everlasting advance and retreat, meeting and greeting, starting and parting, in this most happy accommodation of two contending parties: I still could not understand, how the sun at all came to lay hold of the passing stranger, and that too at such a distance; and how, having overcome the first and greatest difficulty of all, that of drawing off from its path the planet in its rapid course, the attracting force of the sun should have stood still—as no magnet ever does—instead of increasing, and following up the advantage where no resistance offered, where no resisting medium stood in the way to draw it down to himself.

Then again, having a certain centrifugal force emanating in some mysterious manner from the sun by the side of, yet opposed to, attraction, acting upon the planets and trying to drive them off into space in a straight line, at a right angle with the primitive impulse; and again having a certain centripetal power of unfathomable origin, without parentage, in opposition to the centrifugal force and trying to drive them back again: the planets, by this means, like Mahomet's coffin, are kept in suspense between these two contending, each other neutralizing, forces, and thus preserved in their primitive positions.

Then there is the swinging force of the sun, holding, as it were, each planet at the end of a string, and thus, as in the gyroscope, * on the horizontal representation of the solar system,—preventing their fall into space, not having a spot to rest upon. This last force was taken into partnership with that famous discovery, the primitive impulse.

Mutual attraction and repulsion in equal degrees becomes neutralization, and a neutral state in physics means a state in which the balance of owners is equal—IT MEANS NOTHING; if there is no weight in a pair of scales, they are continually in balance.

Stripping, then, the old system of these neutralizing forces, NOTHING remains but the arbitrary primitive launch and disposition of the planets, the "one and the same force, modified only by distance from the sun, (*Herschel's Outlines*, 290), which retains ALL the planets in their orbits about it."

Instead of the planets floating and revolving upon their bellies and on their proper level within the atmosphere of the sun, towards whose centre they gravitate, the above law is so faithfully applied in the old system, "that if the earth were taken from its actual orbit, (*Herschel's Outlines*, 290), and launched anew in space, AT

---

* I have seen nothing better than the gyroscope to prove the exceeding fallacy of the old theory of the solar system, though it has been, and no doubt still is, considered the best evidence in its favour.

THE PLACE, in the direction, and with the velocity of ANY OF THE OTHER PLANETS, it would describe the very same orbit, and in the same period, WHICH THAT PLANET ACTUALLY DOES."

Thus the launch, differently applied to every planet, remains everywhere the same; the ever-ready forces accommodate themselves to every size, condition, constitution, and distance of the planet, keeping it in rank and file in whatever station or direction of the solar system it may have been pitched, though it would be all the same if it were neither pushed nor pulled, in which case, at least, there would be no perturbation.

Then there is the centrifugal force of the earth itself, for which NO CAUSE IS GIVEN, instead of an interior circumferential agent ; but the object is, to account for the rotation of the earth round its own axis.

Then we have a centripetal force of the earth, for which NO CAUSE is given, instead of the planetary attraction of our globe, which keeps hold of whatever belongs to its sphere; but the object is, that by the antagonism of the centripetal to the centrifugal force, both being equally matched, neutralizing each other, the latter does not cause the earth to fly into pieces, and the former its not shrinking into the size of a nutshell. The centrifugal force, nevertheless, by a number of inappropriate analogies, is said to have produced the protuberance of the earth at the equator, whilst, if the protuberance be really a matter of fact, it is simply and evidently the result of a more active life, and of that enormous oxidation or solidification of matter, which exists, constantly is, and for ages has been, going on there. If, on the other hand, our earth is kept together by this reputed centripetal force : the supposed planet of Olbers—which burst, and whose fragments we are told to behold in the asteroids—cannot certainly have been endowed with it ; or, the centripetal force must by degrees have escaped or worn out, and been overcome by its centrifugal antagonist.

This theory is stated and commented upon by Mr. Mackintosh in his *Electrical Theory of the Universe,* in the following manner. Speaking of Sir Isaac Newton's centrifugal force, he says : " This is the weak point of the Newtonian philosophy, for very little reflection will enable us to discern that the momentum or centrifugal FORCE, as it is absurdly denominated, is no force whatever, but an EFFECT derived from, and dependant upon a real force, whatever that force may be. The Creator, it is said, having formed the earth and moon, impelled them from his hand into space, somewhat after the manner that a man impels a cricket or skittle ball, by which PRIMITIVE IMPULSE, as it is technically termed, a momentum was created ; and space being a vacuum, it is said this momentum once created, must remain undiminished for ever, because a body once put in motion, must continue in motion, unless it meet with some resistance from another body to destroy that motion ; and further, under the influence of this momentum, as derived from the ' primitive impulse,' the earth or moving body, continually endeavours to proceed in a straight line, and would so move, and be thereby carried out of the solar system, were it not that she is continually deflected from this course, and turned towards the sun by the force of attraction. Again, as the earth revolves upon her axis, it is said, that at the time of her receiving the ' primitive impulse,' she was held, not exactly in the centre, but a little on one side, so that, when she received the great heave, she went off with a whirling motion, which she has retained ever since, just as a stick, held by one end, whirls, when thrown from the hand, with this difference, that the stick does not whirl for ever.   It will scarcely be believed that Laplace, ' LAPLACE LE GRAND,' actually entered into an elaborate calculation, with a view to determine at what particular point the Creator held the earth at the time of giving the grand push, and that after a most profound investigation, he arrived at the never-to be-for-

THE PLACE, in the direction, and with the velocity of ANY OF THE OTHER PLANETS, it would describe the very same orbit, and in the same period, WHICH THAT PLANÉT ACTUALLY DOES."

Thus the launch, differently applied to every planet, remains everywhere the same; the ever-ready forces accommodate themselves to every size, condition, constitution, and distance of the planet, keeping it in rank and file in whatever station or direction of the solar system it may have been pitched, though it would be all the same if it were neither pushed nor pulled, in which case, at least, there would be no perturbation.

Then there is the centrifugal force of the earth itself, for which NO CAUSE IS GIVEN, instead of an interior circumferential agent; but the object is, to account for the rotation of the earth round its own axis.

Then we have a centripetal force of the earth, for which NO CAUSE is given, instead of the planetary attraction of our globe, which keeps hold of whatever belongs to its sphere; but the object is, that by the antagonism of the centripetal to the centrifugal force, both being equally matched, neutralizing each other, the latter does not cause the earth to fly into pieces, and the former its not shrinking into the size of a nutshell. The centrifugal force, nevertheless, by a number of inappropriate analogies, is said to have produced the protuberance of the earth at the equator, whilst, if the protuberance be really a matter of fact, it is simply and evidently the result of a more active life, and of that enormous oxidation or solidification of matter, which exists, constantly is, and for ages has been, going on there. If, on the other hand, our earth is kept together by this reputed centripetal force: the supposed planet of Olbers—which burst, and whose fragments we are told to behold in the asteroids—cannot certainly have been endowed with it; or, the centripetal force must by degrees have escaped or worn out, and been overcome by its centrifugal antagonist.

This theory is stated and commented upon by Mr. Mackintosh in his *Electrical Theory of the Universe*, in the following manner. Speaking of Sir Isaac Newton's centrifugal force, he says : " This is the weak point of the Newtonian philosophy, for very little reflection will enable us to discern that the momentum or centrifugal FORCE, as it is absurdly denominated, is no force whatever, but an EFFECT derived from, and dependant upon a real force, whatever that force may be.   The Creator, it is said, having formed the earth and moon, impelled them from his hand into space, somewhat after the manner that a man impels a cricket or skittle ball, by which PRIMITIVE IMPULSE, as it is technically termed, a momentum was created ; and space being a vacuum, it is said this momentum once created, must remain undiminished for ever, because a body once put in motion, must continue in motion, unless it meet with some resistance from another body to destroy that motion ; and further, under the influence of this momentum, as derived from the ' primitive impulse,' the earth or moving body, continually endeavours to proceed in a straight line, and would so move, and be thereby carried out of the solar system, were it not that she is continually deflected from this course, and turned towards the sun by the force of attraction.   Again, as the earth revolves upon her axis, it is said, that at the time of her receiving the ' primitive impulse,' she was held, not exactly in the centre, but a little on one side, so that, when she received the great heave, she went off with a whirling motion, which she has retained ever since, just as a stick, held by one end, whirls, when thrown from the hand, with this difference, that the stick does not whirl for ever.   It will scarcely be believed that Laplace, ' LAPLACE LE GRAND,' actually entered into an elaborate calculation, with a view to determine at what particular point the Creator held the earth at the time of giving the grand push, and that after a most profound investigation, he arrived at the never-to be-for-

THE PLACE, in the direction, and with the velocity of ANY OF THE OTHER PLANETS, it would describe the very same orbit, and in the same period, WHICH THAT PLANET ACTUALLY DOES."

Thus the launch, differently applied to every planet, remains everywhere the same; the ever-ready forces accommodate themselves to every size, condition, constitution, and distance of the planet, keeping it in rank and file in whatever station or direction of the solar system it may have been pitched, though it would be all the same if it were neither pushed nor pulled, in which case, at least, there would be no perturbation.

Then there is the centrifugal force of the earth itself, for which NO CAUSE IS GIVEN, instead of an interior circumferential agent; but the object is, to account for the rotation of the earth round its own axis.

Then we have a centripetal force of the earth, for which NO CAUSE is given, instead of the planetary attraction of our globe, which keeps hold of whatever belongs to its sphere; but the object is, that by the antagonism of the centripetal to the centrifugal force, both being equally matched, neutralizing each other, the latter does not cause the earth to fly into pieces, and the former its not shrinking into the size of a nutshell. The centrifugal force, nevertheless, by a number of inappropriate analogies, is said to have produced the protuberance of the earth at the equator, whilst, if the protuberance be really a matter of fact, it is simply and evidently the result of a more active life, and of that enormous oxidation or solidification of matter, which exists, constantly is, and for ages has been, going on there. If, on the other hand, our earth is kept together by this reputed centripetal force: the supposed planet of Olbers—which burst, and whose fragments we are told to behold in the asteroids—cannot certainly have been endowed with it; or, the centripetal force must by degrees have escaped or worn out, and been overcome by its centrifugal antagonist.

This theory is stated and commented upon by Mr. Mackintosh in his *Electrical Theory of the Universe*, in the following manner. Speaking of Sir Isaac Newton's centrifugal force, he says : " This is the weak point of the Newtonian philosophy, for very little reflection will enable us to discern that the momentum or centrifugal FORCE, as it is absurdly denominated, is no force whatever, but an EFFECT derived from, and dependant upon a real force, whatever that force may be. The Creator, it is said, having formed the earth and moon, impelled them from his hand into space, somewhat after the manner that a man impels a cricket or skittle ball, by which PRIMITIVE IMPULSE, as it is technically termed, a momentum was created ; and space being a vacuum, it is said this momentum once created, must remain undiminished for ever, because a body once put in motion, must continue in motion, unless it meet with some resistance from another body to destroy that motion ; and further, under the influence of this momentum, as derived from the ' primitive impulse,' the earth or moving body, continually endeavours to proceed in a straight line, and would so move, and be thereby carried out of the solar system, were it not that she is continually deflected from this course, and turned towards the sun by the force of attraction. Again, as the earth revolves upon her axis, it is said, that at the time of her receiving the ' primitive impulse,' she was held, not exactly in the centre, but a little on one side, so that, when she received the great heave, she went off with a whirling motion, which she has retained ever since, just as a stick, held by one end, whirls, when thrown from the hand, with this difference, that the stick does not whirl for ever.   It will scarcely be believed that Laplace, ' LAPLACE LE GRAND,' actually entered into an elaborate calculation, with a view to determine at what particular point the Creator held the earth at the time of giving the grand push, and that after a most profound investigation, he arrived at the never-to be-for-

THE PLACE, in the direction, and with the velocity of ANY OF THE OTHER PLANETS, it would describe the very same orbit, and in the same period, WHICH THAT PLANET ACTUALLY DOES."

Thus the launch, differently applied to every planet, remains everywhere the same; the ever-ready forces accommodate themselves to every size, condition, constitution, and distance of the planet, keeping it in rank and file in whatever station or direction of the solar system it may have been pitched, though it would be all the same if it were neither pushed nor pulled, in which case, at least, there would be no perturbation.

Then there is the centrifugal force of the earth itself, for which NO CAUSE IS GIVEN, instead of an interior circumferential agent; but the object is, to account for the rotation of the earth round its own axis.

Then we have a centripetal force of the earth, for which NO CAUSE is given, instead of the planetary attraction of our globe, which keeps hold of whatever belongs to its sphere; but the object is, that by the antagonism of the centripetal to the centrifugal force, both being equally matched, neutralizing each other, the latter does not cause the earth to fly into pieces, and the former its not shrinking into the size of a nutshell. The centrifugal force, nevertheless, by a number of inappropriate analogies, is said to have produced the protuberance of the earth at the equator, whilst, if the protuberance be really a matter of fact, it is simply and evidently the result of a more active life, and of that enormous oxidation or solidification of matter, which exists, constantly is, and for ages has been, going on there. If, on the other hand, our earth is kept together by this reputed centripetal force: the supposed planet of Olbers—which burst, and whose fragments we are told to behold in the asteroids—cannot certainly have been endowed with it; or, the centripetal force must by degrees have escaped or worn out, and been overcome by its centrifugal antagonist.

This theory is stated and commented upon by Mr. Mackintosh in his *Electrical Theory of the Universe*, in the following manner. Speaking of Sir Isaac Newton's centrifugal force, he says : " This is the weak point of the Newtonian philosophy, for very little reflection will enable us to discern that the momentum or centrifugal FORCE, as it is absurdly denominated, is no force whatever, but an EFFECT derived from, and dependant upon a real force, whatever that force may be.   The Creator, it is said, having formed the earth and moon, impelled them from his hand into space, somewhat after the manner that a man impels a cricket or skittle ball, by which PRIMITIVE IMPULSE, as it is technically termed, a momentum was created ; and space being a vacuum, it is said this momentum once created, must remain undiminished for ever, because a body once put in motion, must continue in motion, unless it meet with some resistance from another body to destroy that motion ; and further, under the influence of this momentum, as derived from the ' primitive impulse,' the earth or moving body, continually endeavours to proceed in a straight line, and would so move, and be thereby carried out of the solar system, were it not that she is continually deflected from this course, and turned towards the sun by the force of attraction.   Again, as the earth revolves upon her axis, it is said, that at the time of her receiving the ' primitive impulse,' she was held, not exactly in the centre, but a little on one side, so that, when she received the great heave, she went off with a whirling motion, which she has retained ever since, just as a stick, held by one end, whirls, when thrown from the hand, with this difference, that the stick does not whirl for ever.   It will scarcely be believed that Laplace, ' LAPLACE LE GRAND,' actually entered into an elaborate calculation, with a view to determine at what particular point the Creator held the earth at the time of giving the grand push, and that after a most profound investigation, he arrived at the never-to be-for-

THE PLACE, in the direction, and with the velocity of ANY OF THE OTHER PLANETS, it would describe the very same orbit, and in the same period, WHICH THAT PLANET ACTUALLY DOES."

Thus the launch, differently applied to every planet, remains everywhere the same; the ever-ready forces accommodate themselves to every size, condition, constitution, and distance of the planet, keeping it in rank and file in whatever station or direction of the solar system it may have been pitched, though it would be all the same if it were neither pushed nor pulled, in which case, at least, there would be no perturbation.

Then there is the centrifugal force of the earth itself, for which NO CAUSE IS GIVEN, instead of an interior circumferential agent; but the object is, to account for the rotation of the earth round its own axis.

Then we have a centripetal force of the earth, for which NO CAUSE is given, instead of the planetary attraction of our globe, which keeps hold of whatever belongs to its sphere; but the object is, that by the antagonism of the centripetal to the centrifugal force, both being equally matched, neutralizing each other, the latter does not cause the earth to fly into pieces, and the former its not shrinking into the size of a nutshell. The centrifugal force, nevertheless, by a number of inappropriate analogies, is said to have produced the protuberance of the earth at the equator, whilst, if the protuberance be really a matter of fact, it is simply and evidently the result of a more active life, and of that enormous oxidation or solidification of matter, which exists, constantly is, and for ages has been, going on there. If, on the other hand, our earth is kept together by this reputed centripetal force: the supposed planet of Olbers—which burst, and whose fragments we are told to behold in the asteroids—cannot certainly have been endowed with it; or, the centripetal force must by degrees have escaped or worn out, and been overcome by its centrifugal antagonist.

This theory is stated and commented upon by Mr. Mackintosh in his *Electrical Theory of the Universe*, in the following manner. Speaking of Sir Isaac Newton's centrifugal force, he says : " This is the weak point of the Newtonian philosophy, for very little reflection will enable us to discern that the momentum or centrifugal FORCE, as it is absurdly denominated, is no force whatever, but an EFFECT derived from, and dependant upon a real force, whatever that force may be. The Creator, it is said, having formed the earth and moon, impelled them from his hand into space, somewhat after the manner that a man impels a cricket or skittle ball, by which PRIMITIVE IMPULSE, as it is technically termed, a momentum was created ; and space being a vacuum, it is said this momentum once created, must remain undiminished for ever, because a body once put in motion, must continue in motion, unless it meet with some resistance from another body to destroy that motion ; and further, under the influence of this momentum, as derived from the ' primitive impulse,' the earth or moving body, continually endeavours to proceed in a straight line, and would so move, and be thereby carried out of the solar system, were it not that she is continually deflected from this course, and turned towards the sun by the force of attraction. Again, as the earth revolves upon her axis, it is said, that at the time of her receiving the ' primitive impulse,' she was held, not exactly in the centre, but a little on one side, so that, when she received the great heave, she went off with a whirling motion, which she has retained ever since, just as a stick, held by one end, whirls, when thrown from the hand, with this difference, that the stick does not whirl for ever.   It will scarcely be believed that Laplace, ' LAPLACE LE GRAND,' actually entered into an elaborate calculation, with a view to determine at what particular point the Creator held the earth at the time of giving the grand push, and that after a most profound investigation, he arrived at the never-to be-for-

THE PLACE, in the direction, and with the velocity of ANY OF THE OTHER PLANETS, it would describe the very same orbit, and in the same period, WHICH THAT PLANET ACTUALLY DOES."

Thus the launch, differently applied to every planet, remains everywhere the same; the ever-ready forces accommodate themselves to every size, condition, constitution, and distance of the planet, keeping it in rank and file in whatever station or direction of the solar system it may have been pitched, though it would be all the same if it were neither pushed nor pulled, in which case, at least, there would be no perturbation.

Then there is the centrifugal force of the earth itself, for which NO CAUSE IS GIVEN, instead of an interior circumferential agent; but the object is, to account for the rotation of the earth round its own axis.

Then we have a centripetal force of the earth, for which NO CAUSE is given, instead of the planetary attraction of our globe, which keeps hold of whatever belongs to its sphere; but the object is, that by the antagonism of the centripetal to the centrifugal force, both being equally matched, neutralizing each other, the latter does not cause the earth to fly into pieces, and the former its not shrinking into the size of a nutshell. The centrifugal force, nevertheless, by a number of inappropriate analogies, is said to have produced the protuberance of the earth at the equator, whilst, if the protuberance be really a matter of fact, it is simply and evidently the result of a more active life, and of that enormous oxidation or solidification of matter, which exists, constantly is, and for ages has been, going on there. If, on the other hand, our earth is kept together by this reputed centripetal force: the supposed planet of Olbers—which burst, and whose fragments we are told to behold in the asteroids—cannot certainly have been endowed with it; or, the centripetal force must by degrees have escaped or worn out, and been overcome by its centrifugal antagonist.

This theory is stated and commented upon by Mr. Mackintosh in his *Electrical Theory of the Universe*, in the following manner. Speaking of Sir Isaac Newton's centrifugal force, he says: "This is the weak point of the Newtonian philosophy, for very little reflection will enable us to discern that the momentum or centrifugal FORCE, as it is absurdly denominated, is no force whatever, but an EFFECT derived from, and dependant upon a real force, whatever that force may be. The Creator, it is said, having formed the earth and moon, impelled them from his hand into space, somewhat after the manner that a man impels a cricket or skittle ball, by which PRIMITIVE IMPULSE, as it is technically termed, a momentum was created; and space being a vacuum, it is said this momentum once created, must remain undiminished for ever, because a body once put in motion, must continue in motion, unless it meet with some resistance from another body to destroy that motion; and further, under the influence of this momentum, as derived from the 'primitive impulse,' the earth or moving body, continually endeavours to proceed in a straight line, and would so move, and be thereby carried out of the solar system, were it not that she is continually deflected from this course, and turned towards the sun by the force of attraction. Again, as the earth revolves upon her axis, it is said, that at the time of her receiving the 'primitive impulse,' she was held, not exactly in the centre, but a little on one side, so that, when she received the great heave, she went off with a whirling motion, which she has retained ever since, just as a stick, held by one end, whirls, when thrown from the hand, with this difference, that the stick does not whirl for ever. It will scarcely be believed that Laplace, 'LAPLACE LE GRAND,' actually entered into an elaborate calculation, with a view to determine at what particular point the Creator held the earth at the time of giving the grand push, and that after a most profound investigation, he arrived at the never-to be-for-

THE PLACE, in the direction, and with the velocity of ANY OF THE OTHER PLANETS, it would describe the very same orbit, and in the same period, WHICH THAT PLANET ACTUALLY DOES."

Thus the launch, differently applied to every planet, remains everywhere the same; the ever-ready forces accommodate themselves to every size, condition, constitution, and distance of the planet, keeping it in rank and file in whatever station or direction of the solar system it may have been pitched, though it would be all the same if it were neither pushed nor pulled, in which case, at least, there would be no perturbation.

Then there is the centrifugal force of the earth itself, for which NO CAUSE IS GIVEN, instead of an interior circumferential agent; but the object is, to account for the rotation of the earth round its own axis.

Then we have a centripetal force of the earth, for which NO CAUSE is given, instead of the planetary attraction of our globe, which keeps hold of whatever belongs to its sphere; but the object is, that by the antagonism of the centripetal to the centrifugal force, both being equally matched, neutralizing each other, the latter does not cause the earth to fly into pieces, and the former its not shrinking into the size of a nutshell. The centrifugal force, nevertheless, by a number of inappropriate analogies, is said to have produced the protuberance of the earth at the equator, whilst, if the protuberance be really a matter of fact, it is simply and evidently the result of a more active life, and of that enormous oxidation or solidification of matter, which exists, constantly is, and for ages has been, going on there. If, on the other hand, our earth is kept together by this reputed centripetal force: the supposed planet of Olbers—which burst, and whose fragments we are told to behold in the asteroids—cannot certainly have been endowed with it; or, the centripetal force must by degrees have escaped or worn out, and been overcome by its centrifugal antagonist.

This theory is stated and commented upon by Mr. Mackintosh in his *Electrical Theory of the Universe*, in the following manner. Speaking of Sir Isaac Newton's centrifugal force, he says : " This is the weak point of the Newtonian philosophy, for very little reflection will enable us to discern that the momentum or centrifugal FORCE, as it is absurdly denominated, is no force whatever, but an EFFECT derived from, and dependant upon a real force, whatever that force may be. The Creator, it is said, having formed the earth and moon, impelled them from his hand into space, somewhat after the manner that a man impels a cricket or skittle ball, by which PRIMITIVE IMPULSE, as it is technically termed, a momentum was created ; and space being a vacuum, it is said this momentum once created, must remain undiminished for ever, because a body once put in motion, must continue in motion, unless it meet with some resistance from another body to destroy that motion ; and further, under the influence of this momentum, as derived from the ' primitive impulse,' the earth or moving body, continually endeavours to proceed in a straight line, and would so move, and be thereby carried out of the solar system, were it not that she is continually deflected from this course, and turned towards the sun by the force of attraction. Again, as the earth revolves upon her axis, it is said, that at the time of her receiving the 'primitive impulse,' she was held, not exactly in the centre, but a little on one side, so that, when she received the great heave, she went off with a whirling motion, which she has retained ever since, just as a stick, held by one end, whirls, when thrown from the hand, with this difference, that the stick does not whirl for ever.   It will scarcely be believed that Laplace, ' LAPLACE LE GRAND,' actually entered into an elaborate calculation, with a view to determine at what particular point the Creator held the earth at the time of giving the grand push, and that after a most profound investigation, he arrived at the never-to be-for-

THE PLACE, in the direction, and with the velocity of ANY OF THE OTHER PLANETS, it would describe the very same orbit, and in the same period, WHICH THAT PLANET ACTUALLY DOES."

Thus the launch, differently applied to every planet, remains everywhere the same; the ever-ready forces accommodate themselves to every size, condition, constitution, and distance of the planet, keeping it in rank and file in whatever station or direction of the solar system it may have been pitched, though it would be all the same if it were neither pushed nor pulled, in which case, at least, there would be no perturbation.

Then there is the centrifugal force of the earth itself, for which NO CAUSE IS GIVEN, instead of an interior circumferential agent; but the object is, to account for the rotation of the earth round its own axis.

Then we have a centripetal force of the earth, for which NO CAUSE is given, instead of the planetary attraction of our globe, which keeps hold of whatever belongs to its sphere; but the object is, that by the antagonism of the centripetal to the centrifugal force, both being equally matched, neutralizing each other, the latter does not cause the earth to fly into pieces, and the former its not shrinking into the size of a nutshell. The centrifugal force, nevertheless, by a number of inappropriate analogies, is said to have produced the protuberance of the earth at the equator, whilst, if the protuberance be really a matter of fact, it is simply and evidently the result of a more active life, and of that enormous oxidation or solidification of matter, which exists, constantly is, and for ages has been, going on there. If, on the other hand, our earth is kept together by this reputed centripetal force: the supposed planet of Olbers—which burst, and whose fragments we are told to behold in the asteroids—cannot certainly have been endowed with it; or, the centripetal force must by degrees have escaped or worn out, and been overcome by its centrifugal antagonist.

This theory is stated and commented upon by Mr. Mackintosh in his *Electrical Theory of the Universe*, in the following manner. Speaking of Sir Isaac Newton's centrifugal force, he says : " This is the weak point of the Newtonian philosophy, for very little reflection will enable us to discern that the momentum or centrifugal FORCE, as it is absurdly denominated, is no force whatever, but an EFFECT derived from, and dependant upon a real force, whatever that force may be. The Creator, it is said, having formed the earth and moon, impelled them from his hand into space, somewhat after the manner that a man impels a cricket or skittle ball, by which PRIMITIVE IMPULSE, as it is technically termed, a momentum was created ; and space being a vacuum, it is said this momentum once created, must remain undiminished for ever, because a body once put in motion, must continue in motion, unless it meet with some resistance from another body to destroy that motion ; and further, under the influence of this momentum, as derived from the ' primitive impulse,' the earth or moving body, continually endeavours to proceed in a straight line, and would so move, and be thereby carried out of the solar system, were it not that she is continually deflected from this course, and turned towards the sun by the force of attraction. Again, as the earth revolves upon her axis, it is said, that at the time of her receiving the ' primitive impulse,' she was held, not exactly in the centre, but a little on one side, so that, when she received the great heave, she went off with a whirling motion, which she has retained ever since, just as a stick, held by one end, whirls, when thrown from the hand, with this difference, that the stick does not whirl for ever.   It will scarcely be believed that Laplace, ' LAPLACE LE GRAND,' actually entered into an elaborate calculation, with a view to determine at what particular point the Creator held the earth at the time of giving the grand push, and that after a most profound investigation, he arrived at the never-to be-for-

THE PLACE, in the direction, and with the velocity of ANY OF THE OTHER PLANETS, it would describe the very same orbit, and in the same period, WHICH THAT PLANET ACTUALLY DOES."

Thus the launch, differently applied to every planet, remains everywhere the same; the ever-ready forces accommodate themselves to every size, condition, constitution, and distance of the planet, keeping it in rank and file in whatever station or direction of the solar system it may have been pitched, though it would be all the same if it were neither pushed nor pulled, in which case, at least, there would be no perturbation.

Then there is the centrifugal force of the earth itself, for which NO CAUSE IS GIVEN, instead of an interior circumferential agent; but the object is, to account for the rotation of the earth round its own axis.

Then we have a centripetal force of the earth, for which NO CAUSE is given, instead of the planetary attraction of our globe, which keeps hold of whatever belongs to its sphere; but the object is, that by the antagonism of the centripetal to the centrifugal force, both being equally matched, neutralizing each other, the latter does not cause the earth to fly into pieces, and the former its not shrinking into the size of a nutshell. The centrifugal force, nevertheless, by a number of inappropriate analogies, is said to have produced the protuberance of the earth at the equator, whilst, if the protuberance be really a matter of fact, it is simply and evidently the result of a more active life, and of that enormous oxidation or solidification of matter, which exists, constantly is, and for ages has been, going on there. If, on the other hand, our earth is kept together by this reputed centripetal force: the supposed planet of Olbers—which burst, and whose fragments we are told to behold in the asteroids—cannot certainly have been endowed with it; or, the centripetal force must by degrees have escaped or worn out, and been overcome by its centrifugal antagonist.

This theory is stated and commented upon by Mr. Mackintosh in his *Electrical Theory of the Universe*, in the following manner. Speaking of Sir Isaac Newton's centrifugal force, he says: " This is the weak point of the Newtonian philosophy, for very little reflection will enable us to discern that the momentum or centrifugal FORCE, as it is absurdly denominated, is no force whatever, but an EFFECT derived from, and dependant upon a real force, whatever that force may be. The Creator, it is said, having formed the earth and moon, impelled them from his hand into space, somewhat after the manner that a man impels a cricket or skittle ball, by which PRIMITIVE IMPULSE, as it is technically termed, a momentum was created ; and space being a vacuum, it is said this momentum once created, must remain undiminished for ever, because a body once put in motion, must continue in motion, unless it meet with some resistance from another body to destroy that motion ; and further, under the influence of this momentum, as derived from the ' primitive impulse,' the earth or moving body, continually endeavours to proceed in a straight line, and would so move, and be thereby carried out of the solar system, were it not that she is continually deflected from this course, and turned towards the sun by the force of attraction. Again, as the earth revolves upon her axis, it is said, that at the time of her receiving the ' primitive impulse,' she was held, not exactly in the centre, but a little on one side, so that, when she received the great heave, she went off with a whirling motion, which she has retained ever since, just as a stick, held by one end, whirls, when thrown from the hand, with this difference, that the stick does not whirl for ever. It will scarcely be believed that Laplace, ' LAPLACE LE GRAND,' actually entered into an elaborate calculation, with a view to determine at what particular point the Creator held the earth at the time of giving the grand push, and that after a most profound investigation, he arrived at the never-to be-for-

THE PLACE, in the direction, and with the velocity of ANY OF THE OTHER PLANETS, it would describe the very same orbit, and in the same period, WHICH THAT PLANET ACTUALLY DOES."

Thus the launch, differently applied to every planet, remains everywhere the same; the ever-ready forces accommodate themselves to every size, condition, constitution, and distance of the planet, keeping it in rank and file in whatever station or direction of the solar system it may have been pitched, though it would be all the same if it were neither pushed nor pulled, in which case, at least, there would be no perturbation.

Then there is the centrifugal force of the earth itself, for which NO CAUSE IS GIVEN, instead of an interior circumferential agent ; but the object is, to account for the rotation of the earth round its own axis.

Then we have a centripetal force of the earth, for which NO CAUSE is given, instead of the planetary attraction of our globe, which keeps hold of whatever belongs to its sphere; but the object is, that by the antagonism of the centripetal to the centrifugal force, both being equally matched, neutralizing each other, the latter does not cause the earth to fly into pieces, and the former its not shrinking into the size of a nutshell. The centrifugal force, nevertheless, by a number of inappropriate analogies, is said to have produced the protuberance of the earth at the equator, whilst, if the protuberance be really a matter of fact, it is simply and evidently the result of a more active life, and of that enormous oxidation or solidification of matter, which exists, constantly is, and for ages has been, going on there. If, on the other hand, our earth is kept together by this reputed centripetal force : the supposed planet of Olbers—which burst, and whose fragments we are told to behold in the asteroids—cannot certainly have been endowed with it ; or, the centripetal force must by degrees have escaped or worn out, and been overcome by its centrifugal antagonist.

This theory is stated and commented upon by Mr. Mackintosh in his *Electrical Theory of the Universe*, in the following manner. Speaking of Sir Isaac Newton's centrifugal force, he says : " This is the weak point of the Newtonian philosophy, for very little reflection will enable us to discern that the momentum or centrifugal FORCE, as it is absurdly denominated, is no force whatever, but an EFFECT derived from, and dependant upon a real force, whatever that force may be.   The Creator, it is said, having formed the earth and moon, impelled them from his hand into space, somewhat after the manner that a man impels a cricket or skittle ball, by which PRIMITIVE IMPULSE, as it is technically termed, a momentum was created ; and space being a vacuum, it is said this momentum once created, must remain undiminished for ever, because a body once put in motion, must continue in motion, unless it meet with some resistance from another body to destroy that motion ; and further, under the influence of this momentum, as derived from the ' primitive impulse,' the earth or moving body, continually endeavours to proceed in a straight line, and would so move, and be thereby carried out of the solar system, were it not that she is continually deflected from this course, and turned towards the sun by the force of attraction.   Again, as the earth revolves upon her axis, it is said, that at the time of her receiving the ' primitive impulse,' she was held, not exactly in the centre, but a little on one side, so that, when she received the great heave, she went off with a whirling motion, which she has retained ever since, just as a stick, held by one end, whirls, when thrown from the hand, with this difference, that the stick does not whirl for ever.   It will scarcely be believed that Laplace, ' LAPLACE LE GRAND,' actually entered into an elaborate calculation, with a view to determine at what particular point the Creator held the earth at the time of giving the grand push, and that after a most profound investigation, he arrived at the never-to be-for-

THE PLACE, in the direction, and with the velocity of ANY OF THE OTHER PLANETS, it would describe the very same orbit, and in the same period, WHICH THAT PLANET ACTUALLY DOES."

Thus the launch, differently applied to every planet, remains everywhere the same; the ever-ready forces accommodate themselves to every size, condition, constitution, and distance of the planet, keeping it in rank and file in whatever station or direction of the solar system it may have been pitched, though it would be all the same if it were neither pushed nor pulled, in which case, at least, there would be no perturbation.

Then there is the centrifugal force of the earth itself, for which NO CAUSE IS GIVEN, instead of an interior circumferential agent; but the object is, to account for the rotation of the earth round its own axis.

Then we have a centripetal force of the earth, for which NO CAUSE is given, instead of the planetary attraction of our globe, which keeps hold of whatever belongs to its sphere; but the object is, that by the antagonism of the centripetal to the centrifugal force, both being equally matched, neutralizing each other, the latter does not cause the earth to fly into pieces, and the former its not shrinking into the size of a nutshell. The centrifugal force, nevertheless, by a number of inappropriate analogies, is said to have produced the protuberance of the earth at the equator, whilst, if the protuberance be really a matter of fact, it is simply and evidently the result of a more active life, and of that enormous oxidation or solidification of matter, which exists, constantly is, and for ages has been, going on there. If, on the other hand, our earth is kept together by this reputed centripetal force: the supposed planet of Olbers—which burst, and whose fragments we are told to behold in the asteroids—cannot certainly have been endowed with it; or, the centripetal force must by degrees have escaped or worn out, and been overcome by its centrifugal antagonist.

This theory is stated and commented upon by Mr. Mackintosh in his *Electrical Theory of the Universe*, in the following manner. Speaking of Sir Isaac Newton's centrifugal force, he·says: "This is the weak point of the Newtonian philosophy, for very little reflection will enable us to discern that the momentum or centrifugal FORCE, as it is absurdly denominated, is no force whatever, but an EFFECT derived from, and dependant upon a real force, whatever that force may be. The Creator, it is said, having formed the earth and moon, impelled them from his hand into space, somewhat after the manner that a man impels a cricket or skittle ball, by which PRIMITIVE IMPULSE, as it is technically termed, a momentum was created ; and space being a vacuum, it is said this momentum once created, must remain undiminished for ever, because a body once put in motion, must continue in motion, unless it meet with some resistance from another body to destroy that motion ; and further, under the influence of this momentum, as derived from the 'primitive impulse,' the earth or moving body, continually endeavours to proceed in a straight line, and would so move, and be thereby carried out of the solar system, were it not that she is continually deflected from this course, and turned towards the sun by the force of attraction. Again, as the earth revolves upon her axis, it is said, that at the time of her receiving the 'primitive impulse,' she was held, not exactly in the centre, but a little on one side, so that, when she received the great heave, she went off with a whirling motion, which she has retained ever since, just as a stick, held by one end, whirls, when thrown from the hand, with this difference, that the stick does not whirl for ever. It will scarcely be believed that Laplace, 'LAPLACE LE GRAND,' actually entered into an elaborate calculation, with a view to determine at what particular point the Creator held the earth at the time of giving the grand push, and that after a most profound investigation, he arrived at the never-to be-for-

THE PLACE, in the direction, and with the velocity of ANY OF THE OTHER PLANETS, it would describe the very same orbit, and in the same period, WHICH THAT PLANET ACTUALLY DOES."

Thus the launch, differently applied to every planet, remains everywhere the same; the ever-ready forces accommodate themselves to every size, condition, constitution, and distance of the planet, keeping it in rank and file in whatever station or direction of the solar system it may have been pitched, though it would be all the same if it were neither pushed nor pulled, in which case, at least, there would be no perturbation.

Then there is the centrifugal force of the earth itself, for which NO CAUSE IS GIVEN, instead of an interior circumferential agent; but the object is, to account for the rotation of the earth round its own axis.

Then we have a centripetal force of the earth, for which NO CAUSE is given, instead of the planetary attraction of our globe, which keeps hold of whatever belongs to its sphere; but the object is, that by the antagonism of the centripetal to the centrifugal force, both being equally matched, neutralizing each other, the latter does not cause the earth to fly into pieces, and the former its not shrinking into the size of a nutshell. The centrifugal force, nevertheless, by a number of inappropriate analogies, is said to have produced the protuberance of the earth at the equator, whilst, if the protuberance be really a matter of fact, it is simply and evidently the result of a more active life, and of that enormous oxidation or solidification of matter, which exists, constantly is, and for ages has been, going on there. If, on the other hand, our earth is kept together by this reputed centripetal force: the supposed planet of Olbers—which burst, and whose fragments we are told to behold in the asteroids—cannot certainly have been endowed with it; or, the centripetal force must by degrees have escaped or worn out, and been overcome by its centrifugal antagonist.

This theory is stated and commented upon by Mr. Mackintosh in his *Electrical Theory of the Universe,* in the following manner. Speaking of Sir Isaac Newton's centrifugal force, he says : " This is the weak point of the Newtonian philosophy, for very little reflection will enable us to discern that the momentum or centrifugal FORCE, as it is absurdly denominated, is no force what-ever, but an EFFECT derived from, and dependant upon a real force, whatever that force may be. The Creator, it is said, having formed the earth and moon, impelled them from his hand into space, somewhat after the manner that a man impels a cricket or skittle ball, by which PRIMITIVE IMPULSE, as it is technically termed, a momentum was created ; and space being a vacuum, it is said this momentum once created, must remain undiminished for ever, because a body once put in motion, must continue in motion, unless it meet with some resistance from another body to destroy that motion ; and further, under the influence of this momentum, as derived from the 'primitive impulse,' the earth or moving body, continually endeavours to proceed in a straight line, and would so move, and be thereby carried out of the solar system, were it not that she is continually deflected from this course, and turned towards the sun by the force of attraction. Again, as the earth revolves upon her axis, it is said, that at the time of her receiving the 'primitive impulse,' she was held, not exactly in the centre, but a little on one side, so that, when she received the great heave, she went off with a whirling motion, which she has retained ever since, just as a stick, held by one end, whirls, when thrown from the hand, with this difference, that the stick does not whirl for ever.   It will scarcely be believed that Laplace, ' LAPLACE LE GRAND,' actually entered into an elaborate calculation, with a view to determine at what particular point the Creator held the earth at the time of giving the grand push, and that after a most profound investigation, he arrived at the never-to be-for-

THE PLACE, in the direction, and with the velocity of ANY OF THE
OTHER PLANETS, it would describe the very same orbit, and in the
same period, WHICH THAT PLANET ACTUALLY DOES."

Thus the launch, differently applied to every planet, remains
everywhere the same; the ever-ready forces accommodate themselves to every size, condition, constitution, and distance of the
planet, keeping it in rank and file in whatever station or direction
of the solar system it may have been pitched, though it would be
all the same if it were neither pushed nor pulled, in which case,
at least, there would be no perturbation.

Then there is the centrifugal force of the earth itself, for which
NO CAUSE IS GIVEN, instead of an interior circumferential agent ;
but the object is, to account for the rotation of the earth round its
own axis.

Then we have a centripetal force of the earth, for which NO
CAUSE is given, instead of the planetary attraction of our globe,
which keeps hold of whatever belongs to its sphere; but the object
is, that by the antagonism of the centripetal to the centrifugal
force, both being equally matched, neutralizing each other, the
latter does not cause the earth to fly into pieces, and the former its
not shrinking into the size of a nutshell. The centrifugal force,
nevertheless, by a number of inappropriate analogies, is said to
have produced the protuberance of the earth at the equator, whilst,
if the protuberance be really a matter of fact, it is simply and
evidently the result of a more active life, and of that enormous
oxidation or solidification of matter, which exists, constantly is,
and for ages has been, going on there. If, on the other hand, our
earth is kept together by this reputed centripetal force : the supposed planet of Olbers—which burst, and whose fragments we are
told to behold in the asteroids—cannot certainly have been endowed
with it ; or, the centripetal force must by degrees have escaped or
worn out, and been overcome by its centrifugal antagonist.

This theory is stated and commented upon by Mr. Mackintosh in his *Electrical Theory of the Universe*, in the following manner. Speaking of Sir Isaac Newton's centrifugal force, he says : " This is the weak point of the Newtonian philosophy, for very little reflection will enable us to discern that the momentum or centrifugal FORCE, as it is absurdly denominated, is no force whatever, but an EFFECT derived from, and dependant upon a real force, whatever that force may be. The Creator, it is said, having formed the earth and moon, impelled them from his hand into space, somewhat after the manner that a man impels a cricket or skittle ball, by which PRIMITIVE IMPULSE, as it is technically termed, a momentum was created ; and space being a vacuum, it is said this momentum once created, must remain undiminished for ever, because a body once put in motion, must continue in motion, unless it meet with some resistance from another body to destroy that motion ; and further, under the influence of this momentum, as derived from the ' primitive impulse,' the earth or moving body, continually endeavours to proceed in a straight line, and would so move, and be thereby carried out of the solar system, were it not that she is continually deflected from this course, and turned towards the sun by the force of attraction. Again, as the earth revolves upon her axis, it is said, that at the time of her receiving the ' primitive impulse,' she was held, not exactly in the centre, but a little on one side, so that, when she received the great heave, she went off with a whirling motion, which she has retained ever since, just as a stick, held by one end, whirls, when thrown from the hand, with this difference, that the stick does not whirl for ever. It will scarcely be believed that Laplace, ' LAPLACE LE GRAND,' actually entered into an elaborate calculation, with a view to determine at what particular point the Creator held the earth at the time of giving the grand push, and that after a most profound investigation, he arrived at the never-to be-for-

THE PLACE, in the direction, and with the velocity of ANY OF THE OTHER PLANETS, it would describe the very same orbit, and in the same period, WHICH THAT PLANET ACTUALLY DOES."

Thus the launch, differently applied to every planet, remains everywhere the same; the ever-ready forces accommodate themselves to every size, condition, constitution, and distance of the planet, keeping it in rank and file in whatever station or direction of the solar system it may have been pitched, though it would be all the same if it were neither pushed nor pulled, in which case, at least, there would be no perturbation.

Then there is the centrifugal force of the earth itself, for which NO CAUSE IS GIVEN, instead of an interior circumferential agent; but the object is, to account for the rotation of the earth round its own axis.

Then we have a centripetal force of the earth, for which NO CAUSE is given, instead of the planetary attraction of our globe, which keeps hold of whatever belongs to its sphere; but the object is, that by the antagonism of the centripetal to the centrifugal force, both being equally matched, neutralizing each other, the latter does not cause the earth to fly into pieces, and the former its not shrinking into the size of a nutshell. The centrifugal force, nevertheless, by a number of inappropriate analogies, is said to have produced the protuberance of the earth at the equator, whilst, if the protuberance be really a matter of fact, it is simply and evidently the result of a more active life, and of that enormous oxidation or solidification of matter, which exists, constantly is, and for ages has been, going on there. If, on the other hand, our earth is kept together by this reputed centripetal force: the supposed planet of Olbers—which burst, and whose fragments we are told to behold in the asteroids—cannot certainly have been endowed with it; or, the centripetal force must by degrees have escaped or worn out, and been overcome by its centrifugal antagonist.

This theory is stated and commented upon by Mr. Mackintosh in his *Electrical Theory of the Universe*, in the following manner. Speaking of Sir Isaac Newton's centrifugal force, he says : "This is the weak point of the Newtonian philosophy, for very little reflection will enable us to discern that the momentum or centrifugal FORCE, as it is absurdly denominated, is no force whatever, but an EFFECT derived from, and dependant upon a real force, whatever that force may be. The Creator, it is said, having formed the earth and moon, impelled them from his hand into space, somewhat after the manner that a man impels a cricket or skittle ball, by which PRIMITIVE IMPULSE, as it is technically termed, a momentum was created ; and space being a vacuum, it is said this momentum once created, must remain undiminished for ever, because a body once put in motion, must continue in motion, unless it meet with some resistance from another body to destroy that motion ; and further, under the influence of this momentum, as derived from the 'primitive impulse,' the earth or moving body, continually endeavours to proceed in a straight line, and would so move, and be thereby carried out of the solar system, were it not that she is continually deflected from this course, and turned towards the sun by the force of attraction. Again, as the earth revolves upon her axis, it is said, that at the time of her receiving the 'primitive impulse,' she was held, not exactly in the centre, but a little on one side, so that, when she received the great heave, she went off with a whirling motion, which she has retained ever since, just as a stick, held by one end, whirls, when thrown from the hand, with this difference, that the stick does not whirl for ever.   It will scarcely be believed that Laplace, ' LAPLACE LE GRAND,' actually entered into an elaborate calculation, with a view to determine at what particular point the Creator held the earth at the time of giving the grand push, and that after a most profound investigation, he arrived at the never-to be-for-

THE PLACE, in the direction, and with the velocity of ANY OF THE OTHER PLANETS, it would describe the very same orbit, and in the same period, WHICH THAT PLANET ACTUALLY DOES."

Thus the launch, differently applied to every planet, remains everywhere the same; the ever-ready forces accommodate themselves to every size, condition, constitution, and distance of the planet, keeping it in rank and file in whatever station or direction of the solar system it may have been pitched, though it would be all the same if it were neither pushed nor pulled, in which case, at least, there would be no perturbation.

Then there is the centrifugal force of the earth itself, for which NO CAUSE IS GIVEN, instead of an interior circumferential agent; but the object is, to account for the rotation of the earth round its own axis.

Then we have a centripetal force of the earth, for which NO CAUSE is given, instead of the planetary attraction of our globe, which keeps hold of whatever belongs to its sphere; but the object is, that by the antagonism of the centripetal to the centrifugal force, both being equally matched, neutralizing each other, the latter does not cause the earth to fly into pieces, and the former its not shrinking into the size of a nutshell. The centrifugal force, nevertheless, by a number of inappropriate analogies, is said to have produced the protuberance of the earth at the equator, whilst, if the protuberance be really a matter of fact, it is simply and evidently the result of a more active life, and of that enormous oxidation or solidification of matter, which exists, constantly is, and for ages has been, going on there. If, on the other hand, our earth is kept together by this reputed centripetal force: the supposed planet of Olbers—which burst, and whose fragments we are told to behold in the asteroids—cannot certainly have been endowed with it; or, the centripetal force must by degrees have escaped or worn out, and been overcome by its centrifugal antagonist.

This theory is stated and commented upon by Mr. Mackintosh in his *Electrical Theory of the Universe*, in the following manner. Speaking of Sir Isaac Newton's centrifugal force, he says : " This is the weak point of the Newtonian philosophy, for very little reflection will enable us to discern that the momentum or centrifugal FORCE, as it is absurdly denominated, is no force whatever, but an EFFECT derived from, and dependant upon a real force, whatever that force may be. The Creator, it is said, having formed the earth and moon, impelled them from his hand into space, somewhat after the manner that a man impels a cricket or skittle ball, by which PRIMITIVE IMPULSE, as it is technically termed, a momentum was created ; and space being a vacuum, it is said this momentum once created, must remain undiminished for ever, because a body once put in motion, must continue in motion, unless it meet with some resistance from another body to destroy that motion ; and further, under the influence of this momentum, as derived from the ' primitive impulse,' the earth or moving body, continually endeavours to proceed in a straight line, and would so move, and be thereby carried out of the solar system, were it not that she is continually deflected from this course, and turned towards the sun by the force of attraction. Again, as the earth revolves upon her axis, it is said, that at the time of her receiving the ' primitive impulse,' she was held, not exactly in the centre, but a little on one side, so that, when she received the great heave, she went off with a whirling motion, which she has retained ever since, just as a stick, held by one end, whirls, when thrown from the hand, with this difference, that the stick does not whirl for ever. It will scarcely be believed that Laplace, ' LAPLACE LE GRAND,' actually entered into an elaborate calculation, with a view to determine at what particular point the Creator held the earth at the time of giving the grand push, and that after a most profound investigation, he arrived at the never-to be-for-

THE PLACE, in the direction, and with the velocity of ANY OF THE OTHER PLANETS, it would describe the very same orbit, and in the same period, WHICH THAT PLANET ACTUALLY DOES."

Thus the launch, differently applied to every planet, remains everywhere the same; the ever-ready forces accommodate themselves to every size, condition, constitution, and distance of the planet, keeping it in rank and file in whatever station or direction of the solar system it may have been pitched, though it would be all the same if it were neither pushed nor pulled, in which case, at least, there would be no perturbation.

Then there is the centrifugal force of the earth itself, for which NO CAUSE IS GIVEN, instead of an interior circumferential agent; but the object is, to account for the rotation of the earth round its own axis.

Then we have a centripetal force of the earth, for which NO CAUSE is given, instead of the planetary attraction of our globe, which keeps hold of whatever belongs to its sphere; but the object is, that by the antagonism of the centripetal to the centrifugal force, both being equally matched, neutralizing each other, the latter does not cause the earth to fly into pieces, and the former its not shrinking into the size of a nutshell. The centrifugal force, nevertheless, by a number of inappropriate analogies, is said to have produced the protuberance of the earth at the equator, whilst, if the protuberance be really a matter of fact, it is simply and evidently the result of a more active life, and of that enormous oxidation or solidification of matter, which exists, constantly is, and for ages has been, going on there. If, on the other hand, our earth is kept together by this reputed centripetal force: the supposed planet of Olbers—which burst, and whose fragments we are told to behold in the asteroids—cannot certainly have been endowed with it; or, the centripetal force must by degrees have escaped or worn out, and been overcome by its centrifugal antagonist.

This theory is stated and commented upon by Mr. Mackintosh in his *Electrical Theory of the Universe*, in the following manner. Speaking of Sir Isaac Newton's centrifugal force, he says : " This is the weak point of the Newtonian philosophy, for very little reflection will enable us to discern that the momentum or centrifugal FORCE, as it is absurdly denominated, is no force whatever, but an EFFECT derived from, and dependant upon a real force, whatever that force may be. The Creator, it is said, having formed the earth and moon, impelled them from his hand into space, somewhat after the manner that a man impels a cricket or skittle ball, by which PRIMITIVE IMPULSE, as it is technically termed, a momentum was created ; and space being a vacuum, it is said this momentum once created, must remain undiminished for ever, because a body once put in motion, must continue in motion, unless it meet with some resistance from another body to destroy that motion ; and further, under the influence of this momentum, as derived from the ' primitive impulse,' the earth or moving body, continually endeavours to proceed in a straight line, and would so move, and be thereby carried out of the solar system, were it not that she is continually deflected from this course, and turned towards the sun by the force of attraction. Again, as the earth revolves upon her axis, it is said, that at the time of her receiving the 'primitive impulse,' she was held, not exactly in the centre, but a little on one side, so that, when she received the great heave, she went off with a whirling motion, which she has retained ever since, just as a stick, held by one end, whirls, when thrown from the hand, with this difference, that the stick does not whirl for ever.   It will scarcely be believed that Laplace, ' LAPLACE LE GRAND,' actually entered into an elaborate calculation, with a view to determine at what particular point the Creator held the earth at the time of giving the grand push, and that after a most profound investigation, he arrived at the never-to be-for-

THE PLACE, in the direction, and with the velocity of ANY OF THE OTHER PLANETS, it would describe the very same orbit, and in the same period, WHICH THAT PLANET ACTUALLY DOES."

Thus the launch, differently applied to every planet, remains everywhere the same ; the ever-ready forces accommodate themselves to every size, condition, constitution, and distance of the planet, keeping it in rank and file in whatever station or direction of the solar system it may have been pitched, though it would be all the same if it were neither pushed nor pulled, in which case, at least, there would be no perturbation.

Then there is the centrifugal force of the earth itself, for which NO CAUSE IS GIVEN, instead of an interior circumferential agent ; but the object is, to account for the rotation of the earth round its own axis.

Then we have a centripetal force of the earth, for which NO CAUSE is given, instead of the planetary attraction of our globe, which keeps hold of whatever belongs to its sphere; but the object is, that by the antagonism of the centripetal to the centrifugal force, both being equally matched, neutralizing each other, the latter does not cause the earth to fly into pieces, and the former its not shrinking into the size of a nutshell. The centrifugal force, nevertheless, by a number of inappropriate analogies, is said to have produced the protuberance of the earth at the equator, whilst, if the protuberance be really a matter of fact, it is simply and evidently the result of a more active life, and of that enormous oxidation or solidification of matter, which exists, constantly is, and for ages has been, going on there. If, on the other hand, our earth is kept together by this reputed centripetal force : the supposed planet of Olbers—which burst, and whose fragments we are told to behold in the asteroids—cannot certainly have been endowed with it ; or, the centripetal force must by degrees have escaped or worn out, and been overcome by its centrifugal antagonist.

This theory is stated and commented upon by Mr. Mackintosh in his *Electrical Theory of the Universe*, in the following manner. Speaking of Sir Isaac Newton's centrifugal force, he says : "This is the weak point of the Newtonian philosophy, for very little reflection will enable us to discern that the momentum or centrifugal FORCE, as it is absurdly denominated, is no force whatever, but an EFFECT derived from, and dependant upon a real force, whatever that force may be. The Creator, it is said, having formed the earth and moon, impelled them from his hand into space, somewhat after the manner that a man impels a cricket or skittle ball, by which PRIMITIVE IMPULSE, as it is technically termed, a momentum was created ; and space being a vacuum, it is said this momentum once created, must remain undiminished for ever, because a body once put in motion, must continue in motion, unless it meet with some resistance from another body to destroy that motion ; and further, under the influence of this momentum, as derived from the ' primitive impulse,' the earth or moving body, continually endeavours to proceed in a straight line, and would so move, and be thereby carried out of the solar system, were it not that she is continually deflected from this course, and turned towards the sun by the force of attraction. Again, as the earth revolves upon her axis, it is said, that at the time of her receiving the ' primitive impulse,' she was held, not exactly in the centre, but a little on one side, so that, when she received the great heave, she went off with a whirling motion, which she has retained ever since, just as a stick, held by one end, whirls, when thrown from the hand, with this difference, that the stick does not whirl for ever. It will scarcely be believed that Laplace, ' LAPLACE LE GRAND,' actually entered into an elaborate calculation, with a view to determine at what particular point the Creator held the earth at the time of giving the grand push, and that after a most profound investigation, he arrived at the never-to be-for-

THE PLACE, in the direction, and with the velocity of ANY OF THE OTHER PLANETS, it would describe the very same orbit, and in the same period, WHICH THAT PLANET ACTUALLY DOES."

Thus the launch, differently applied to every planet, remains everywhere the same; the ever-ready forces accommodate themselves to every size, condition, constitution, and distance of the planet, keeping it in rank and file in whatever station or direction of the solar system it may have been pitched, though it would be all the same if it were neither pushed nor pulled, in which case, at least, there would be no perturbation.

Then there is the centrifugal force of the earth itself, for which NO CAUSE IS GIVEN, instead of an interior circumferential agent; but the object is, to account for the rotation of the earth round its own axis.

Then we have a centripetal force of the earth, for which NO CAUSE is given, instead of the planetary attraction of our globe, which keeps hold of whatever belongs to its sphere; but the object is, that by the antagonism of the centripetal to the centrifugal force, both being equally matched, neutralizing each other, the latter does not cause the earth to fly into pieces, and the former its not shrinking into the size of a nutshell. The centrifugal force, nevertheless, by a number of inappropriate analogies, is said to have produced the protuberance of the earth at the equator, whilst, if the protuberance be really a matter of fact, it is simply and evidently the result of a more active life, and of that enormous oxidation or solidification of matter, which exists, constantly is, and for ages has been, going on there. If, on the other hand, our earth is kept together by this reputed centripetal force: the supposed planet of Olbers—which burst, and whose fragments we are told to behold in the asteroids—cannot certainly have been endowed with it; or, the centripetal force must by degrees have escaped or worn out, and been overcome by its centrifugal antagonist.

This theory is stated and commented upon by Mr. Mackintosh in his *Electrical Theory of the Universe*, in the following manner. Speaking of Sir Isaac Newton's centrifugal force, he says : " This is the weak point of the Newtonian philosophy, for very little reflection will enable us to discern that the momentum or centrifugal FORCE, as it is absurdly denominated, is no force whatever, but an EFFECT derived from, and dependant upon a real force, whatever that force may be. The Creator, it is said, having formed the earth and moon, impelled them from his hand into space, somewhat after the manner that a man impels a cricket or skittle ball, by which PRIMITIVE IMPULSE, as it is technically termed, a momentum was created ; and space being a vacuum, it is said this momentum once created, must remain undiminished for ever, because a body once put in motion, must continue in motion, unless it meet with some resistance from another body to destroy that motion ; and further, under the influence of this momentum, as derived from the ' primitive impulse,' the earth or moving body, continually endeavours to proceed in a straight line, and would so move, and be thereby carried out of the solar system, were it not that she is continually deflected from this course, and turned towards the sun by the force of attraction. Again, as the earth revolves upon her axis, it is said, that at the time of her receiving the ' primitive impulse,' she was held, not exactly in the centre, but a little on one side, so that, when she received the great heave, she went off with a whirling motion, which she has retained ever since, just as a stick, held by one end, whirls, when thrown from the hand, with this difference, that the stick does not whirl for ever.   It will scarcely be believed that Laplace, ' LAPLACE LE GRAND,' actually entered into an elaborate calculation, with a view to determine at what particular point the Creator held the earth at the time of giving the grand push, and that after a most profound investigation, he arrived at the never-to be-for-

THE PLACE, in the direction, and with the velocity of ANY OF THE OTHER PLANETS, it would describe the very same orbit, and in the same period, WHICH THAT PLANET ACTUALLY DOES."

Thus the launch, differently applied to every planet, remains everywhere the same; the ever-ready forces accommodate themselves to every size, condition, constitution, and distance of the planet, keeping it in rank and file in whatever station or direction of the solar system it may have been pitched, though it would be all the same if it were neither pushed nor pulled, in which case, at least, there would be no perturbation.

Then there is the centrifugal force of the earth itself, for which NO CAUSE IS GIVEN, instead of an interior circumferential agent; but the object is, to account for the rotation of the earth round its own axis.

Then we have a centripetal force of the earth, for which NO CAUSE is given, instead of the planetary attraction of our globe, which keeps hold of whatever belongs to its sphere; but the object is, that by the antagonism of the centripetal to the centrifugal force, both being equally matched, neutralizing each other, the latter does not cause the earth to fly into pieces, and the former its not shrinking into the size of a nutshell. The centrifugal force, nevertheless, by a number of inappropriate analogies, is said to have produced the protuberance of the earth at the equator, whilst, if the protuberance be really a matter of fact, it is simply and evidently the result of a more active life, and of that enormous oxidation or solidification of matter, which exists, constantly is, and for ages has been, going on there. If, on the other hand, our earth is kept together by this reputed centripetal force: the supposed planet of Olbers—which burst, and whose fragments we are told to behold in the asteroids—cannot certainly have been endowed with it; or, the centripetal force must by degrees have escaped or worn out, and been overcome by its centrifugal antagonist.

This theory is stated and commented upon by Mr. Mackintosh in his *Electrical Theory of the Universe*, in the following manner. Speaking of Sir Isaac Newton's centrifugal force, he says : " This is the weak point of the Newtonian philosophy, for very little reflection will enable us to discern that the momentum or centrifugal FORCE, as it is absurdly denominated, is no force whatever, but an EFFECT derived from, and dependant upon a real force, whatever that force may be.   The Creator, it is said, having formed the earth and moon, impelled them from his hand into space, somewhat after the manner that a man impels a cricket or skittle ball, by which PRIMITIVE IMPULSE, as it is technically termed, a momentum was created ; and space being a vacuum, it is said this momentum once created, must remain undiminished for ever, because a body once put in motion, must continue in motion, unless it meet with some resistance from another body to destroy that motion ; and further, under the influence of this momentum, as derived from the ' primitive impulse,' the earth or moving body, continually endeavours to proceed in a straight line, and would so move, and be thereby carried out of the solar system, were it not that she is continually deflected from this course, and turned towards the sun by the force of attraction.   Again, as the earth revolves upon her axis, it is said, that at the time of her receiving the ' primitive impulse,' she was held, not exactly in the centre, but a little on one side, so that, when she received the great heave, she went off with a whirling motion, which she has retained ever since, just as a stick, held by one end, whirls, when thrown from the hand, with this difference, that the stick does not whirl for ever.   It will scarcely be believed that Laplace, ' LAPLACE LE GRAND,' actually entered into an elaborate calculation, with a view to determine at what particular point the Creator held the earth at the time of giving the grand push, and that after a most profound investigation, he arrived at the never-to be-for-

THE PLACE, in the direction, and with the velocity of ANY OF THE OTHER PLANETS, it would describe the very same orbit, and in the same period, WHICH THAT PLANET ACTUALLY DOES."

Thus the launch, differently applied to every planet, remains everywhere the same; the ever-ready forces accommodate themselves to every size, condition, constitution, and distance of the planet, keeping it in rank and file in whatever station or direction of the solar system it may have been pitched, though it would be all the same if it were neither pushed nor pulled, in which case, at least, there would be no perturbation.

Then there is the centrifugal force of the earth itself, for which NO CAUSE IS GIVEN, instead of an interior circumferential agent; but the object is, to account for the rotation of the earth round its own axis.

Then we have a centripetal force of the earth, for which NO CAUSE is given, instead of the planetary attraction of our globe, which keeps hold of whatever belongs to its sphere; but the object is, that by the antagonism of the centripetal to the centrifugal force, both being equally matched, neutralizing each other, the latter does not cause the earth to fly into pieces, and the former its not shrinking into the size of a nutshell. The centrifugal force, nevertheless, by a number of inappropriate analogies, is said to have produced the protuberance of the earth at the equator, whilst, if the protuberance be really a matter of fact, it is simply and evidently the result of a more active life, and of that enormous oxidation or solidification of matter, which exists, constantly is, and for ages has been, going on there. If, on the other hand, our earth is kept together by this reputed centripetal force: the supposed planet of Olbers—which burst, and whose fragments we are told to behold in the asteroids—cannot certainly have been endowed with it; or, the centripetal force must by degrees have escaped or worn out, and been overcome by its centrifugal antagonist.

This theory is stated and commented upon by Mr. Mackintosh in his *Electrical Theory of the Universe*, in the following manner. Speaking of Sir Isaac Newton's centrifugal force, he says : " This is the weak point of the Newtonian philosophy, for very little reflection will enable us to discern that the momentum or centrifugal FORCE, as it is absurdly denominated, is no force whatever, but an EFFECT derived from, and dependant upon a real force, whatever that force may be. The Creator, it is said, having formed the earth and moon, impelled them from his hand into space, somewhat after the manner that a man impels a cricket or skittle ball, by which PRIMITIVE IMPULSE, as it is technically termed, a momentum was created ; and space being a vacuum, it is said this momentum once created, must remain undiminished for ever, because a body once put in motion, must continue in motion, unless it meet with some resistance from another body to destroy that motion ; and further, under the influence of this momentum, as derived from the ' primitive impulse,' the earth or moving body, continually endeavours to proceed in a straight line, and would so move, and be thereby carried out of the solar system, were it not that she is continually deflected from this course, and turned towards the sun by the force of attraction. Again, as the earth revolves upon her axis, it is said, that at the time of her receiving the ' primitive impulse,' she was held, not exactly in the centre, but a little on one side, so that, when she received the great heave, she went off with a whirling motion, which she has retained ever since, just as a stick, held by one end, whirls, when thrown from the hand, with this difference, that the stick does not whirl for ever. It will scarcely be believed that Laplace, ' LAPLACE LE GRAND,' actually entered into an elaborate calculation, with a view to determine at what particular point the Creator held the earth at the time of giving the grand push, and that after a most profound investigation, he arrived at the never-to be-for-

THE PLACE, in the direction, and with the velocity of ANY OF THE OTHER PLANETS, it would describe the very same orbit, and in the same period, WHICH THAT PLANET ACTUALLY DOES."

Thus the launch, differently applied to every planet, remains everywhere the same; the ever-ready forces accommodate themselves to every size, condition, constitution, and distance of the planet, keeping it in rank and file in whatever station or direction of the solar system it may have been pitched, though it would be all the same if it were neither pushed nor pulled, in which case, at least, there would be no perturbation.

Then there is the centrifugal force of the earth itself, for which NO CAUSE IS GIVEN, instead of an interior circumferential agent; but the object is, to account for the rotation of the earth round its own axis.

Then we have a centripetal force of the earth, for which NO CAUSE is given, instead of the planetary attraction of our globe, which keeps hold of whatever belongs to its sphere; but the object is, that by the antagonism of the centripetal to the centrifugal force, both being equally matched, neutralizing each other, the latter does not cause the earth to fly into pieces, and the former its not shrinking into the size of a nutshell. The centrifugal force, nevertheless, by a number of inappropriate analogies, is said to have produced the protuberance of the earth at the equator, whilst, if the protuberance be really a matter of fact, it is simply and evidently the result of a more active life, and of that enormous oxidation or solidification of matter, which exists, constantly is, and for ages has been, going on there. If, on the other hand, our earth is kept together by this reputed centripetal force: the supposed planet of Olbers—which burst, and whose fragments we are told to behold in the asteroids—cannot certainly have been endowed with it; or, the centripetal force must by degrees have escaped or worn out, and been overcome by its centrifugal antagonist.

This theory is stated and commented upon by Mr. Mackintosh in his *Electrical Theory of the Universe*, in the following manner. Speaking of Sir Isaac Newton's centrifugal force, he·says : " This is the weak point of the Newtonian philosophy, for very little reflection will enable us to discern that the momentum or centrifugal FORCE, as it is absurdly denominated, is no force what-ever, but an EFFECT derived from, and dependant upon a real force, whatever that force may be. The Creator, it is said, having formed the earth and moon, impelled them from his hand into space, somewhat after the manner that a man impels a cricket or skittle ball, by which PRIMITIVE IMPULSE, as it is technically termed, a momentum was created ; and space being a vacuum, it is said this momentum once created, must remain undiminished for ever, because a body once put in motion, must continue in motion, unless it meet with some resistance from another body to destroy that motion ; and further, under the influence of this momentum, as derived from the ' primitive impulse,' the earth or moving body, continually endeavours to proceed in a straight line, and would so move, and be thereby carried out of the solar system, were it not that she is continually deflected from this course, and turned towards the sun by the force of attraction.   Again, as the earth revolves upon her axis, it is said, that at the time of her receiving the ' primitive impulse,' she was held, not exactly in the centre, but a little on one side, so that, when she received the great heave, she went off with a whirling motion, which she has retained ever since, just as a stick, held by one end, whirls, when thrown from the hand, with this difference, that the stick does not whirl for ever.   It will scarcely be believed that Laplace, ' LAPLACE LE GRAND,' actually entered into an elaborate calculation, with a view to determine at what particular point the Creator held the earth at the time of giving the grand push, and that after a most profound investigation, he arrived at the never-to be-for-

THE PLACE, in the direction, and with the velocity of ANY OF THE OTHER PLANETS, it would describe the very same orbit, and in the same period, WHICH THAT PLANET ACTUALLY DOES."

Thus the launch, differently applied to every planet, remains everywhere the same ; the ever-ready forces accommodate themselves to every size, condition, constitution, and distance of the planet, keeping it in rank and file in whatever station or direction of the solar system it may have been pitched, though it would be all the same if it were neither pushed nor pulled, in which case, at least, there would be no perturbation.

Then there is the centrifugal force of the earth itself, for which NO CAUSE IS GIVEN, instead of an interior circumferential agent ; but the object is, to account for the rotation of the earth round its own axis.

Then we have a centripetal force of the earth, for which NO CAUSE is given, instead of the planetary attraction of our globe, which keeps hold of whatever belongs to its sphere; but the object is, that by the antagonism of the centripetal to the centrifugal force, both being equally matched, neutralizing each other, the latter does not cause the earth to fly into pieces, and the former its not shrinking into the size of a nutshell. The centrifugal force, nevertheless, by a number of inappropriate analogies, is said to have produced the protuberance of the earth at the equator, whilst, if the protuberance be really a matter of fact, it is simply and evidently the result of a more active life, and of that enormous oxidation or solidification of matter, which exists, constantly is, and for ages has been, going on there. If, on the other hand, our earth is kept together by this reputed centripetal force : the supposed planet of Olbers—which burst, and whose fragments we are told to behold in the asteroids—cannot certainly have been endowed with it ; or, the centripetal force must by degrees have escaped or worn out, and been overcome by its centrifugal antagonist.

This theory is stated and commented upon by Mr. Mackintosh in his *Electrical Theory of the Universe*, in the following manner. Speaking of Sir Isaac Newton's centrifugal force, he says : " This is the weak point of the Newtonian philosophy, for very little reflection will enable us to discern that the momentum or centrifugal FORCE, as it is absurdly denominated, is no force whatever, but an EFFECT derived from, and dependant upon a real force, whatever that force may be. The Creator, it is said, having formed the earth and moon, impelled them from his hand into space, somewhat after the manner that a man impels a cricket or skittle ball, by which PRIMITIVE IMPULSE, as it is technically termed, a momentum was created ; and space being a vacuum, it is said this momentum once created, must remain undiminished for ever, because a body once put in motion, must continue in motion, unless it meet with some resistance from another body to destroy that motion ; and further, under the influence of this momentum, as derived from the ' primitive impulse,' the earth or moving body, continually endeavours to proceed in a straight line, and would so move, and be thereby carried out of the solar system, were it not that she is continually deflected from this course, and turned towards the sun by the force of attraction. Again, as the earth revolves upon her axis, it is said, that at the time of her receiving the ' primitive impulse,' she was held, not exactly in the centre, but a little on one side, so that, when she received the great heave, she went off with a whirling motion, which she has retained ever since, just as a stick, held by one end, whirls, when thrown from the hand, with this difference, that the stick does not whirl for ever.   It will scarcely be believed that Laplace, ' LAPLACE LE GRAND,' actually entered into an elaborate calculation, with a view to determine at what particular point the Creator held the earth at the time of giving the grand push, and that after a most profound investigation, he arrived at the never-to be-for-

THE PLACE, in the direction, and with the velocity of ANY OF THE OTHER PLANETS, it would describe the very same orbit, and in the same period, WHICH THAT PLANET ACTUALLY DOES."

Thus the launch, differently applied to every planet, remains everywhere the same; the ever-ready forces accommodate themselves to every size, condition, constitution, and distance of the planet, keeping it in rank and file in whatever station or direction of the solar system it may have been pitched, though it would be all the same if it were neither pushed nor pulled, in which case, at least, there would be no perturbation.

Then there is the centrifugal force of the earth itself, for which NO CAUSE IS GIVEN, instead of an interior circumferential agent; but the object is, to account for the rotation of the earth round its own axis.

Then we have a centripetal force of the earth, for which NO CAUSE is given, instead of the planetary attraction of our globe, which keeps hold of whatever belongs to its sphere; but the object is, that by the antagonism of the centripetal to the centrifugal force, both being equally matched, neutralizing each other, the latter does not cause the earth to fly into pieces, and the former its not shrinking into the size of a nutshell. The centrifugal force, nevertheless, by a number of inappropriate analogies, is said to have produced the protuberance of the earth at the equator, whilst, if the protuberance be really a matter of fact, it is simply and evidently the result of a more active life, and of that enormous oxidation or solidification of matter, which exists, constantly is, and for ages has been, going on there. If, on the other hand, our earth is kept together by this reputed centripetal force: the supposed planet of Olbers—which burst, and whose fragments we are told to behold in the asteroids—cannot certainly have been endowed with it; or, the centripetal force must by degrees have escaped or worn out, and been overcome by its centrifugal antagonist.

This theory is stated and commented upon by Mr. Mackintosh in his *Electrical Theory of the Universe*, in the following manner. Speaking of Sir Isaac Newton's centrifugal force, he says : " This is the weak point of the Newtonian philosophy, for very little reflection will enable us to discern that the momentum or centrifugal FORCE, as it is absurdly denominated, is no force whatever, but an EFFECT derived from, and dependant upon a real force, whatever that force may be. The Creator, it is said, having formed the earth and moon, impelled them from his hand into space, somewhat after the manner that a man impels a cricket or skittle ball, by which PRIMITIVE IMPULSE, as it is technically termed, a momentum was created ; and space being a vacuum, it is said this momentum once created, must remain undiminished for ever, because a body once put in motion, must continue in motion, unless it meet with some resistance from another body to destroy that motion ; and further, under the influence of this momentum, as derived from the 'primitive impulse,' the earth or moving body, continually endeavours to proceed in a straight line, and would so move, and be thereby carried out of the solar system, were it not that she is continually deflected from this course, and turned towards the sun by the force of attraction. Again, as the earth revolves upon her axis, it is said, that at the time of her receiving the 'primitive impulse,' she was held, not exactly in the centre, but a little on one side, so that, when she received the great heave, she went off with a whirling motion, which she has retained ever since, just as a stick, held by one end, whirls, when thrown from the hand, with this difference, that the stick does not whirl for ever.   It will scarcely be believed that Laplace, ' LAPLACE LE GRAND,' actually entered into an elaborate calculation, with a view to determine at what particular point the Creator held the earth at the time of giving the grand push, and that after a most profound investigation, he arrived at the never-to be-for-

THE PLACE, in the direction, and with the velocity of ANY OF THE OTHER PLANETS, it would describe the very same orbit, and in the same period, WHICH THAT PLANET ACTUALLY DOES."

Thus the launch, differently applied to every planet, remains everywhere the same; the ever-ready forces accommodate themselves to every size, condition, constitution, and distance of the planet, keeping it in rank and file in whatever station or direction of the solar system it may have been pitched, though it would be all the same if it were neither pushed nor pulled, in which case, at least, there would be no perturbation.

Then there is the centrifugal force of the earth itself, for which NO CAUSE IS GIVEN, instead of an interior circumferential agent; but the object is, to account for the rotation of the earth round its own axis.

Then we have a centripetal force of the earth, for which NO CAUSE is given, instead of the planetary attraction of our globe, which keeps hold of whatever belongs to its sphere; but the object is, that by the antagonism of the centripetal to the centrifugal force, both being equally matched, neutralizing each other, the latter does not cause the earth to fly into pieces, and the former its not shrinking into the size of a nutshell. The centrifugal force, nevertheless, by a number of inappropriate analogies, is said to have produced the protuberance of the earth at the equator, whilst, if the protuberance be really a matter of fact, it is simply and evidently the result of a more active life, and of that enormous oxidation or solidification of matter, which exists, constantly is, and for ages has been, going on there. If, on the other hand, our earth is kept together by this reputed centripetal force: the supposed planet of Olbers—which burst, and whose fragments we are told to behold in the asteroids—cannot certainly have been endowed with it; or, the centripetal force must by degrees have escaped or worn out, and been overcome by its centrifugal antagonist.

This theory is stated and commented upon by Mr. Mackintosh in his *Electrical Theory of the Universe*, in the following manner. Speaking of Sir Isaac Newton's centrifugal force, he says : " This is the weak point of the Newtonian philosophy, for very little reflection will enable us to discern that the momentum or centrifugal FORCE, as it is absurdly denominated, is no force whatever, but an EFFECT derived from, and dependant upon a real force, whatever that force may be.   The Creator, it is said, having formed the earth and moon, impelled them from his hand into space, somewhat after the manner that a man impels a cricket or skittle ball, by which PRIMITIVE IMPULSE, as it is technically termed, a momentum was created ; and space being a vacuum, it is said this momentum once created, must remain undiminished for ever, because a body once put in motion, must continue in motion, unless it meet with some resistance from another body to destroy that motion ; and further, under the influence of this momentum, as derived from the ' primitive impulse,' the earth or moving body, continually endeavours to proceed in a straight line, and would so move, and be thereby carried out of the solar system, were it not that she is continually deflected from this course, and turned towards the sun by the force of attraction.   Again, as the earth revolves upon her axis, it is said, that at the time of her receiving the ' primitive impulse,' she was held, not exactly in the centre, but a little on one side, so that, when she received the great heave, she went off with a whirling motion, which she has retained ever since, just as a stick, held by one end, whirls, when thrown from the hand, with this difference, that the stick does not whirl for ever.   It will scarcely be believed that Laplace, ' LAPLACE LE GRAND,' actually entered into an elaborate calculation, with a view to determine at what particular point the Creator held the earth at the time of giving the grand push, and that after a most profound investigation, he arrived at the never-to be-for-

THE PLACE, in the direction, and with the velocity of ANY OF THE OTHER PLANETS, it would describe the very same orbit, and in the same period, WHICH THAT PLANET ACTUALLY DOES."

Thus the launch, differently applied to every planet, remains everywhere the same; the ever-ready forces accommodate themselves to every size, condition, constitution, and distance of the planet, keeping it in rank and file in whatever station or direction of the solar system it may have been pitched, though it would be all the same if it were neither pushed nor pulled, in which case, at least, there would be no perturbation.

Then there is the centrifugal force of the earth itself, for which NO CAUSE IS GIVEN, instead of an interior circumferential agent ; but the object is, to account for the rotation of the earth round its own axis.

Then we have a centripetal force of the earth, for which NO CAUSE is given, instead of the planetary attraction of our globe, which keeps hold of whatever belongs to its sphere; but the object is, that by the antagonism of the centripetal to the centrifugal force, both being equally matched, neutralizing each other, the latter does not cause the earth to fly into pieces, and the former its not shrinking into the size of a nutshell. The centrifugal force, nevertheless, by a number of inappropriate analogies, is said to have produced the protuberance of the earth at the equator, whilst, if the protuberance be really a matter of fact, it is simply and evidently the result of a more active life, and of that enormous oxidation or solidification of matter, which exists, constantly is, and for ages has been, going on there. If, on the other hand, our earth is kept together by this reputed centripetal force : the supposed planet of Olbers—which burst, and whose fragments we are told to behold in the asteroids—cannot certainly have been endowed with it ; or, the centripetal force must by degrees have escaped or worn out, and been overcome by its centrifugal antagonist.

This theory is stated and commented upon by Mr. Mackintosh in his *Electrical Theory of the Universe*, in the following manner. Speaking of Sir Isaac Newton's centrifugal force, he says: " This is the weak point of the Newtonian philosophy, for very little reflection will enable us to discern that the momentum or centrifugal FORCE, as it is absurdly denominated, is no force whatever, but an EFFECT derived from, and dependant upon a real force, whatever that force may be. The Creator, it is said, having formed the earth and moon, impelled them from his hand into space, somewhat after the manner that a man impels a cricket or skittle ball, by which PRIMITIVE IMPULSE, as it is technically termed, a momentum was created ; and space being a vacuum, it is said this momentum once created, must remain undiminished for ever, because a body once put in motion, must continue in motion, unless it meet with some resistance from another body to destroy that motion ; and further, under the influence of this momentum, as derived from the ' primitive impulse,' the earth or moving body, continually endeavours to proceed in a straight line, and would so move, and be thereby carried out of the solar system, were it not that she is continually deflected from this course, and turned towards the sun by the force of attraction. Again, as the earth revolves upon her axis, it is said, that at the time of her receiving the 'primitive impulse,' she was held, not exactly in the centre, but a little on one side, so that, when she received the great heave, she went off with a whirling motion, which she has retained ever since, just as a stick, held by one end, whirls, when thrown from the hand, with this difference, that the stick does not whirl for ever.   It will scarcely be believed that Laplace, ' LAPLACE LE GRAND,' actually entered into an elaborate calculation, with a view to determine at what particular point the Creator held the earth at the time of giving the grand push, and that after a most profound investigation, he arrived at the never-to be-for-

THE PLACE, in the direction, and with the velocity of ANY OF THE OTHER PLANETS, it would describe the very same orbit, and in the same period, WHICH THAT PLANET ACTUALLY DOES."

Thus the launch, differently applied to every planet, remains everywhere the same; the ever-ready forces accommodate themselves to every size, condition, constitution, and distance of the planet, keeping it in rank and file in whatever station or direction of the solar system it may have been pitched, though it would be all the same if it were neither pushed nor pulled, in which case, at least, there would be no perturbation.

Then there is the centrifugal force of the earth itself, for which NO CAUSE IS GIVEN, instead of an interior circumferential agent; but the object is, to account for the rotation of the earth round its own axis.

Then we have a centripetal force of the earth, for which NO CAUSE is given, instead of the planetary attraction of our globe, which keeps hold of whatever belongs to its sphere; but the object is, that by the antagonism of the centripetal to the centrifugal force, both being equally matched, neutralizing each other, the latter does not cause the earth to fly into pieces, and the former its not shrinking into the size of a nutshell. The centrifugal force, nevertheless, by a number of inappropriate analogies, is said to have produced the protuberance of the earth at the equator, whilst, if the protuberance be really a matter of fact, it is simply and evidently the result of a more active life, and of that enormous oxidation or solidification of matter, which exists, constantly is, and for ages has been, going on there. If, on the other hand, our earth is kept together by this reputed centripetal force: the supposed planet of Olbers—which burst, and whose fragments we are told to behold in the asteroids—cannot certainly have been endowed with it; or, the centripetal force must by degrees have escaped or worn out, and been overcome by its centrifugal antagonist.

This theory is stated and commented upon by Mr. Mackintosh in his *Electrical Theory of the Universe*, in the following manner. Speaking of Sir Isaac Newton's centrifugal force, he says : " This is the weak point of the Newtonian philosophy, for very little reflection will enable us to discern that the momentum or centrifugal FORCE, as it is absurdly denominated, is no force whatever, but an EFFECT derived from, and dependant upon a real force, whatever that force may be.   The Creator, it is said, having formed the earth and moon, impelled them from his hand into space, somewhat after the manner that a man impels a cricket or skittle ball, by which PRIMITIVE IMPULSE, as it is technically termed, a momentum was created ; and space being a vacuum, it is said this momentum once created, must remain undiminished for ever, because a body once put in motion, must continue in motion, unless it meet with some resistance from another body to destroy that motion ; and further, under the influence of this momentum, as derived from the ' primitive impulse,' the earth or moving body, continually endeavours to proceed in a straight line, and would so move, and be thereby carried out of the solar system, were it not that she is continually deflected from this course, and turned towards the sun by the force of attraction.   Again, as the earth revolves upon her axis, it is said, that at the time of her receiving the ' primitive impulse,' she was held, not exactly in the centre, but a little on one side, so that, when she received the great heave, she went off with a whirling motion, which she has retained ever since, just as a stick, held by one end, whirls, when thrown from the hand, with this difference, that the stick does not whirl for ever.   It will scarcely be believed that Laplace, ' LAPLACE LE GRAND,' actually entered into an elaborate calculation, with a view to determine at what particular point the Creator held the earth at the time of giving the grand push, and that after a most profound investigation, he arrived at the never-to be-for-

THE PLACE, in the direction, and with the velocity of ANY OF THE OTHER PLANETS, it would describe the very same orbit, and in the same period, WHICH THAT PLANET ACTUALLY DOES."

Thus the launch, differently applied to every planet, remains everywhere the same; the ever-ready forces accommodate themselves to every size, condition, constitution, and distance of the planet, keeping it in rank and file in whatever station or direction of the solar system it may have been pitched, though it would be all the same if it were neither pushed nor pulled, in which case, at least, there would be no perturbation.

Then there is the centrifugal force of the earth itself, for which NO CAUSE IS GIVEN, instead of an interior circumferential agent; but the object is, to account for the rotation of the earth round its own axis.

Then we have a centripetal force of the earth, for which NO CAUSE is given, instead of the planetary attraction of our globe, which keeps hold of whatever belongs to its sphere; but the object is, that by the antagonism of the centripetal to the centrifugal force, both being equally matched, neutralizing each other, the latter does not cause the earth to fly into pieces, and the former its not shrinking into the size of a nutshell. The centrifugal force, nevertheless, by a number of inappropriate analogies, is said to have produced the protuberance of the earth at the equator, whilst, if the protuberance be really a matter of fact, it is simply and evidently the result of a more active life, and of that enormous oxidation or solidification of matter, which exists, constantly is, and for ages has been, going on there. If, on the other hand, our earth is kept together by this reputed centripetal force: the supposed planet of Olbers—which burst, and whose fragments we are told to behold in the asteroids—cannot certainly have been endowed with it; or, the centripetal force must by degrees have escaped or worn out, and been overcome by its centrifugal antagonist.

This theory is stated and commented upon by Mr. Mackintosh in his *Electrical Theory of the Universe*, in the following manner. Speaking of Sir Isaac Newton's centrifugal force, he says: "This is the weak point of the Newtonian philosophy, for very little reflection will enable us to discern that the momentum or centrifugal FORCE, as it is absurdly denominated, is no force whatever, but an EFFECT derived from, and dependant upon a real force, whatever that force may be. The Creator, it is said, having formed the earth and moon, impelled them from his hand into space, somewhat after the manner that a man impels a cricket or skittle ball, by which PRIMITIVE IMPULSE, as it is technically termed, a momentum was created; and space being a vacuum, it is said this momentum once created, must remain undiminished for ever, because a body once put in motion, must continue in motion, unless it meet with some resistance from another body to destroy that motion; and further, under the influence of this momentum, as derived from the 'primitive impulse,' the earth or moving body, continually endeavours to proceed in a straight line, and would so move, and be thereby carried out of the solar system, were it not that she is continually deflected from this course, and turned towards the sun by the force of attraction. Again, as the earth revolves upon her axis, it is said, that at the time of her receiving the 'primitive impulse,' she was held, not exactly in the centre, but a little on one side, so that, when she received the great heave, she went off with a whirling motion, which she has retained ever since, just as a stick, held by one end, whirls, when thrown from the hand, with this difference, that the stick does not whirl for ever. It will scarcely be believed that Laplace, 'LAPLACE LE GRAND,' actually entered into an elaborate calculation, with a view to determine at what particular point the Creator held the earth at the time of giving the grand push, and that after a most profound investigation, he arrived at the never-to be-for-

THE PLACE, in the direction, and with the velocity of ANY OF THE OTHER PLANETS, it would describe the very same orbit, and in the same period, WHICH THAT PLANET ACTUALLY DOES."

Thus the launch, differently applied to every planet, remains everywhere the same; the ever-ready forces accommodate themselves to every size, condition, constitution, and distance of the planet, keeping it in rank and file in whatever station or direction of the solar system it may have been pitched, though it would be all the same if it were neither pushed nor pulled, in which case, at least, there would be no perturbation.

Then there is the centrifugal force of the earth itself, for which NO CAUSE IS GIVEN, instead of an interior circumferential agent; but the object is, to account for the rotation of the earth round its own axis.

Then we have a centripetal force of the earth, for which NO CAUSE is given, instead of the planetary attraction of our globe, which keeps hold of whatever belongs to its sphere; but the object is, that by the antagonism of the centripetal to the centrifugal force, both being equally matched, neutralizing each other, the latter does not cause the earth to fly into pieces, and the former its not shrinking into the size of a nutshell. The centrifugal force, nevertheless, by a number of inappropriate analogies, is said to have produced the protuberance of the earth at the equator, whilst, if the protuberance be really a matter of fact, it is simply and evidently the result of a more active life, and of that enormous oxidation or solidification of matter, which exists, constantly is, and for ages has been, going on there. If, on the other hand, our earth is kept together by this reputed centripetal force: the supposed planet of Olbers—which burst, and whose fragments we are told to behold in the asteroids—cannot certainly have been endowed with it; or, the centripetal force must by degrees have escaped or worn out, and been overcome by its centrifugal antagonist.

This theory is stated and commented upon by Mr. Mackintosh in his *Electrical Theory of the Universe*, in the following manner. Speaking of Sir Isaac Newton's centrifugal force, he says : " This is the weak point of the Newtonian philosophy, for very little reflection will enable us to discern that the momentum or centrifugal FORCE, as it is absurdly denominated, is no force whatever, but an EFFECT derived from, and dependant upon a real force, whatever that force may be. The Creator, it is said, having formed the earth and moon, impelled them from his hand into space, somewhat after the manner that a man impels a cricket or skittle ball, by which PRIMITIVE IMPULSE, as it is technically termed, a momentum was created ; and space being a vacuum, it is said this momentum once created, must remain undiminished for ever, because a body once put in motion, must continue in motion, unless it meet with some resistance from another body to destroy that motion ; and further, under the influence of this momentum, as derived from the ' primitive impulse,' the earth or moving body, continually endeavours to proceed in a straight line, and would so move, and be thereby carried out of the solar system, were it not that she is continually deflected from this course, and turned towards the sun by the force of attraction. Again, as the earth revolves upon her axis, it is said, that at the time of her receiving the ' primitive impulse,' she was held, not exactly in the centre, but a little on one side, so that, when she received the great heave, she went off with a whirling motion, which she has retained ever since, just as a stick, held by one end, whirls, when thrown from the hand, with this difference, that the stick does not whirl for ever. It will scarcely be believed that Laplace, ' LAPLACE LE GRAND,' actually entered into an elaborate calculation, with a view to determine at what particular point the Creator held the earth at the time of giving the grand push, and that after a most profound investigation, he arrived at the never-to be-for-

THE PLACE, in the direction, and with the velocity of ANY OF THE OTHER PLANETS, it would describe the very same orbit, and in the same period, WHICH THAT PLANET ACTUALLY DOES."

Thus the launch, differently applied to every planet, remains everywhere the same; the ever-ready forces accommodate themselves to every size, condition, constitution, and distance of the planet, keeping it in rank and file in whatever station or direction of the solar system it may have been pitched, though it would be all the same if it were neither pushed nor pulled, in which case, at least, there would be no perturbation.

Then there is the centrifugal force of the earth itself, for which NO CAUSE IS GIVEN, instead of an interior circumferential agent; but the object is, to account for the rotation of the earth round its own axis.

Then we have a centripetal force of the earth, for which NO CAUSE is given, instead of the planetary attraction of our globe, which keeps hold of whatever belongs to its sphere; but the object is, that by the antagonism of the centripetal to the centrifugal force, both being equally matched, neutralizing each other, the latter does not cause the earth to fly into pieces, and the former its not shrinking into the size of a nutshell. The centrifugal force, nevertheless, by a number of inappropriate analogies, is said to have produced the protuberance of the earth at the equator, whilst, if the protuberance be really a matter of fact, it is simply and evidently the result of a more active life, and of that enormous oxidation or solidification of matter, which exists, constantly is, and for ages has been, going on there. If, on the other hand, our earth is kept together by this reputed centripetal force: the supposed planet of Olbers—which burst, and whose fragments we are told to behold in the asteroids—cannot certainly have been endowed with it; or, the centripetal force must by degrees have escaped or worn out, and been overcome by its centrifugal antagonist.

This theory is stated and commented upon by Mr. Mackintosh in his *Electrical Theory of the Universe*, in the following manner. Speaking of Sir Isaac Newton's centrifugal force, he says: " This is the weak point of the Newtonian philosophy, for very little reflection will enable us to discern that the momentum or centrifugal FORCE, as it is absurdly denominated, is no force whatever, but an EFFECT derived from, and dependant upon a real force, whatever that force may be.   The Creator, it is said, having formed the earth and moon, impelled them from his hand into space, somewhat after the manner that a man impels a cricket or skittle ball, by which PRIMITIVE IMPULSE, as it is technically termed, a momentum was created ; and space being a vacuum, it is said this momentum once created, must remain undiminished for ever, because a body once put in motion, must continue in motion, unless it meet with some resistance from another body to destroy that motion ; and further, under the influence of this momentum, as derived from the ' primitive impulse,' the earth or moving body, continually endeavours to proceed in a straight line, and would so move, and be thereby carried out of the solar system, were it not that she is continually deflected from this course, and turned towards the sun by the force of attraction.   Again, as the earth revolves upon her axis, it is said, that at the time of her receiving the ' primitive impulse,' she was held, not exactly in the centre, but a little on one side, so that, when she received the great heave, she went off with a whirling motion, which she has retained ever since, just as a stick, held by one end, whirls, when thrown from the hand, with this difference, that the stick does not whirl for ever.   It will scarcely be believed that Laplace, ' LAPLACE LE GRAND,' actually entered into an elaborate calculation, with a view to determine at what particular point the Creator held the earth at the time of giving the grand push, and that after a most profound investigation, he arrived at the never-to be-for-

THE PLACE, in the direction, and with the velocity of ANY OF THE OTHER PLANETS, it would describe the very same orbit, and in the same period, WHICH THAT PLANET ACTUALLY DOES."

Thus the launch, differently applied to every planet, remains everywhere the same; the ever-ready forces accommodate themselves to every size, condition, constitution, and distance of the planet, keeping it in rank and file in whatever station or direction of the solar system it may have been pitched, though it would be all the same if it were neither pushed nor pulled, in which case, at least, there would be no perturbation.

Then there is the centrifugal force of the earth itself, for which NO CAUSE IS GIVEN, instead of an interior circumferential agent; but the object is, to account for the rotation of the earth round its own axis.

Then we have a centripetal force of the earth, for which NO CAUSE is given, instead of the planetary attraction of our globe, which keeps hold of whatever belongs to its sphere; but the object is, that by the antagonism of the centripetal to the centrifugal force, both being equally matched, neutralizing each other, the latter does not cause the earth to fly into pieces, and the former its not shrinking into the size of a nutshell. The centrifugal force, nevertheless, by a number of inappropriate analogies, is said to have produced the protuberance of the earth at the equator, whilst, if the protuberance be really a matter of fact, it is simply and evidently the result of a more active life, and of that enormous oxidation or solidification of matter, which exists, constantly is, and for ages has been, going on there. If, on the other hand, our earth is kept together by this reputed centripetal force: the supposed planet of Olbers—which burst, and whose fragments we are told to behold in the asteroids—cannot certainly have been endowed with it; or, the centripetal force must by degrees have escaped or worn out, and been overcome by its centrifugal antagonist.

This theory is stated and commented upon by Mr. Mackintosh in his *Electrical Theory of the Universe*, in the following manner. Speaking of Sir Isaac Newton's centrifugal force, he says : " This is the weak point of the Newtonian philosophy, for very little reflection will enable us to discern that the momentum or centrifugal FORCE, as it is absurdly denominated, is no force whatever, but an EFFECT derived from, and dependant upon a real force, whatever that force may be. The Creator, it is said, having formed the earth and moon, impelled them from his hand into space, somewhat after the manner that a man impels a cricket or skittle ball, by which PRIMITIVE IMPULSE, as it is technically termed, a momentum was created ; and space being a vacuum, it is said this momentum once created, must remain undiminished for ever, because a body once put in motion, must continue in motion, unless it meet with some resistance from another body to destroy that motion ; and further, under the influence of this momentum, as derived from the ' primitive impulse,' the earth or moving body, continually endeavours to proceed in a straight line, and would so move, and be thereby carried out of the solar system, were it not that she is continually deflected from this course, and turned towards the sun by the force of attraction. Again, as the earth revolves upon her axis, it is said, that at the time of her receiving the ' primitive impulse,' she was held, not exactly in the centre, but a little on one side, so that, when she received the great heave, she went off with a whirling motion, which she has retained ever since, just as a stick, held by one end, whirls, when thrown from the hand, with this difference, that the stick does not whirl for ever.   It will scarcely be believed that Laplace, ' LAPLACE LE GRAND,' actually entered into an elaborate calculation, with a view to determine at what particular point the Creator held the earth at the time of giving the grand push, and that after a most profound investigation, he arrived at the never-to be-for-

THE PLACE, in the direction, and with the velocity of ANY OF THE OTHER PLANETS, it would describe the very same orbit, and in the same period, WHICH THAT PLANET ACTUALLY DOES."

Thus the launch, differently applied to every planet, remains everywhere the same; the ever-ready forces accommodate themselves to every size, condition, constitution, and distance of the planet, keeping it in rank and file in whatever station or direction of the solar system it may have been pitched, though it would be all the same if it were neither pushed nor pulled, in which case, at least, there would be no perturbation.

Then there is the centrifugal force of the earth itself, for which NO CAUSE IS GIVEN, instead of an interior circumferential agent; but the object is, to account for the rotation of the earth round its own axis.

Then we have a centripetal force of the earth, for which NO CAUSE is given, instead of the planetary attraction of our globe, which keeps hold of whatever belongs to its sphere; but the object is, that by the antagonism of the centripetal to the centrifugal force, both being equally matched, neutralizing each other, the latter does not cause the earth to fly into pieces, and the former its not shrinking into the size of a nutshell. The centrifugal force, nevertheless, by a number of inappropriate analogies, is said to have produced the protuberance of the earth at the equator, whilst, if the protuberance be really a matter of fact, it is simply and evidently the result of a more active life, and of that enormous oxidation or solidification of matter, which exists, constantly is, and for ages has been, going on there. If, on the other hand, our earth is kept together by this reputed centripetal force: the supposed planet of Olbers—which burst, and whose fragments we are told to behold in the asteroids—cannot certainly have been endowed with it; or, the centripetal force must by degrees have escaped or worn out, and been overcome by its centrifugal antagonist.

This theory is stated and commented upon by Mr. Mackintosh in his *Electrical Theory of the Universe*, in the following manner. Speaking of Sir Isaac Newton's centrifugal force, he says : " This is the weak point of the Newtonian philosophy, for very little reflection will enable us to discern that the momentum or centrifugal FORCE, as it is absurdly denominated, is no force whatever, but an EFFECT derived from, and dependant upon a real force, whatever that force may be. The Creator, it is said, having formed the earth and moon, impelled them from his hand into space, somewhat after the manner that a man impels a cricket or skittle ball, by which PRIMITIVE IMPULSE, as it is technically termed, a momentum was created ; and space being a vacuum, it is said this momentum once created, must remain undiminished for ever, because a body once put in motion, must continue in motion, unless it meet with some resistance from another body to destroy that motion ; and further, under the influence of this momentum, as derived from the ' primitive impulse,' the earth or moving body, continually endeavours to proceed in a straight line, and would so move, and be thereby carried out of the solar system, were it not that she is continually deflected from this course, and turned towards the sun by the force of attraction. Again, as the earth revolves upon her axis, it is said, that at the time of her receiving the ' primitive impulse,' she was held, not exactly in the centre, but a little on one side, so that, when she received the great heave, she went off with a whirling motion, which she has retained ever since, just as a stick, held by one end, whirls, when thrown from the hand, with this difference, that the stick does not whirl for ever.   It will scarcely be believed that Laplace, ' LAPLACE LE GRAND,' actually entered into an elaborate calculation, with a view to determine at what particular point the Creator held the earth at the time of giving the grand push, and that after a most profound investigation, he arrived at the never-to be-for-

THE PLACE, in the direction, and with the velocity of ANY OF THE OTHER PLANETS, it would describe the very same orbit, and in the same period, WHICH THAT PLANET ACTUALLY DOES."

Thus the launch, differently applied to every planet, remains everywhere the same; the ever-ready forces accommodate themselves to every size, condition, constitution, and distance of the planet, keeping it in rank and file in whatever station or direction of the solar system it may have been pitched, though it would be all the same if it were neither pushed nor pulled, in which case, at least, there would be no perturbation.

Then there is the centrifugal force of the earth itself, for which NO CAUSE IS GIVEN, instead of an interior circumferential agent ; but the object is, to account for the rotation of the earth round its own axis.

Then we have a centripetal force of the earth, for which NO CAUSE is given, instead of the planetary attraction of our globe, which keeps hold of whatever belongs to its sphere; but the object is, that by the antagonism of the centripetal to the centrifugal force, both being equally matched, neutralizing each other, the latter does not cause the earth to fly into pieces, and the former its not shrinking into the size of a nutshell. The centrifugal force, nevertheless, by a number of inappropriate analogies, is said to have produced the protuberance of the earth at the equator, whilst, if the protuberance be really a matter of fact, it is simply and evidently the result of a more active life, and of that enormous oxidation or solidification of matter, which exists, constantly is, and for ages has been, going on there. If, on the other hand, our earth is kept together by this reputed centripetal force : the supposed planet of Olbers—which burst, and whose fragments we are told to behold in the asteroids—cannot certainly have been endowed with it ; or, the centripetal force must by degrees have escaped or worn out, and been overcome by its centrifugal antagonist.

This theory is stated and commented upon by Mr. Mackintosh in his *Electrical Theory of the Universe*, in the following manner. Speaking of Sir Isaac Newton's centrifugal force, he says: "This is the weak point of the Newtonian philosophy, for very little reflection will enable us to discern that the momentum or centrifugal FORCE, as it is absurdly denominated, is no force whatever, but an EFFECT derived from, and dependant upon a real force, whatever that force may be. The Creator, it is said, having formed the earth and moon, impelled them from his hand into space, somewhat after the manner that a man impels a cricket or skittle ball, by which PRIMITIVE IMPULSE, as it is technically termed, a momentum was created; and space being a vacuum, it is said this momentum once created, must remain undiminished for ever, because a body once put in motion, must continue in motion, unless it meet with some resistance from another body to destroy that motion; and further, under the influence of this momentum, as derived from the 'primitive impulse,' the earth or moving body, continually endeavours to proceed in a straight line, and would so move, and be thereby carried out of the solar system, were it not that she is continually deflected from this course, and turned towards the sun by the force of attraction. Again, as the earth revolves upon her axis, it is said, that at the time of her receiving the 'primitive impulse,' she was held, not exactly in the centre, but a little on one side, so that, when she received the great heave, she went off with a whirling motion, which she has retained ever since, just as a stick, held by one end, whirls, when thrown from the hand, with this difference, that the stick does not whirl for ever.   It will scarcely be believed that Laplace, 'LAPLACE LE GRAND,' actually entered into an elaborate calculation, with a view to determine at what particular point the Creator held the earth at the time of giving the grand push, and that after a most profound investigation, he arrived at the never-to be-for-

THE PLACE, in the direction, and with the velocity of ANY OF THE OTHER PLANETS, it would describe the very same orbit, and in the same period, WHICH THAT PLANET ACTUALLY DOES."

Thus the launch, differently applied to every planet, remains everywhere the same; the ever-ready forces accommodate themselves to every size, condition, constitution, and distance of the planet, keeping it in rank and file in whatever station or direction of the solar system it may have been pitched, though it would be all the same if it were neither pushed nor pulled, in which case, at least, there would be no perturbation.

Then there is the centrifugal force of the earth itself, for which NO CAUSE IS GIVEN, instead of an interior circumferential agent; but the object is, to account for the rotation of the earth round its own axis.

Then we have a centripetal force of the earth, for which NO CAUSE is given, instead of the planetary attraction of our globe, which keeps hold of whatever belongs to its sphere; but the object is, that by the antagonism of the centripetal to the centrifugal force, both being equally matched, neutralizing each other, the latter does not cause the earth to fly into pieces, and the former its not shrinking into the size of a nutshell. The centrifugal force, nevertheless, by a number of inappropriate analogies, is said to have produced the protuberance of the earth at the equator, whilst, if the protuberance be really a matter of fact, it is simply and evidently the result of a more active life, and of that enormous oxidation or solidification of matter, which exists, constantly is, and for ages has been, going on there. If, on the other hand, our earth is kept together by this reputed centripetal force: the supposed planet of Olbers—which burst, and whose fragments we are told to behold in the asteroids—cannot certainly have been endowed with it; or, the centripetal force must by degrees have escaped or worn out, and been overcome by its centrifugal antagonist.

This theory is stated and commented upon by Mr. Mackintosh in his *Electrical Theory of the Universe*, in the following manner. Speaking of Sir Isaac Newton's centrifugal force, he·says : " This is the weak point of the Newtonian philosophy, for very little reflection will enable us to discern that the momentum or centrifugal FORCE, as it is absurdly denominated, is no force whatever, but an EFFECT derived from, and dependant upon a real force, whatever that force may be.   The Creator, it is said, having formed the earth and moon, impelled them from his hand into space, somewhat after the manner that a man impels a cricket or skittle ball, by which PRIMITIVE IMPULSE, as it is technically termed, a momentum was created ; and space being a vacuum, it is said this momentum once created, must remain undiminished for ever, because a body once put in motion, must continue in motion, unless it meet with some resistance from another body to destroy that motion ; and further, under the influence of this momentum, as derived from the ' primitive impulse,' the earth or moving body, continually endeavours to proceed in a straight line, and would so move, and be thereby carried out of the solar system, were it not that she is continually deflected from this course, and turned towards the sun by the force of attraction.   Again, as the earth revolves upon her axis, it is said, that at the time of her receiving the ' primitive impulse,' she was held, not exactly in the centre, but a little on one side, so that, when she received the great heave, she went off with a whirling motion, which she has retained ever since, just as a stick, held by one end, whirls, when thrown from the hand, with this difference, that the stick does not whirl for ever.   It will scarcely be believed that Laplace, ' LAPLACE LE GRAND,' actually entered into an elaborate calculation, with a view to determine at what particular point the Creator held the earth at the time of giving the grand push, and that after a most profound investigation, he arrived at the never-to be-for-

gotten conclusion, that when the ' primitive impulse ' was imparted, the earth was held exactly twenty-five miles from the centre, and hence, quoth Laplace, the earth revolves upon her axis once in twenty-four hours. If she had been held a little nearer to the centre, our days would have been longer, and if a little farther off, she would have revolved with greater velocity, and our days would have been shorter. . . . . . . These assumptions comprehend the very essence of what is called the Newtonian philosophy, and the world has been gravely assured that every proposition has been rigidly demonstrated."

With singular infelicity the gyroscope has been made use of, not only to prove this absurdity, but even to improve upon it.and assign to the Creator also the spot at which He must have stood when He "hurled the earth into space."

I am sure the gyroscopians cannot have deeply reflected upon their favourite illustration, otherwise they would have seen this, and above all, that the instrument represents nothing else but the stone in the sling, with this difference, that in the one instance, the "primitive impulse," that famous discovery, proceeds from the person wielding the string, and in the other it is imparted by a "Prime Mover," standing outside and independent of the object to be set in motion.

The gyroscope, moreover, represents the earth rotating upon an HORIZONTAL axis by the side of the sun, instead of spinning perpendicularly around him; and therefore it is also an inappropriate illustration.

However, the turning-point of the theory is the "primitive impulse," that famous discovery,—and its "centrifugal assistant;" but, not resting on any solid basis, Newton was obliged to assume, the world has unhesitatingly believed him, and Sir John Herschel still echoes it in *Good Words* that the heavenly bodies move in VACANT SPACE! That there is a VACANT CREATION! And yet

many of those who believe in this vacant space, cannot believe in vacant space outside the ball of creation, cannot believe that the exterior of creation is filled by God alone from eternity to eternity. But, allow a substance ever so fine to pervade the universe, and THE FOUNDATION OF THE OLD THEORY IS REMOVED, AND THE WHOLE EDIFICE CRUMBLES TO PIECES. It is the giant-cripple of universal gravitation, limping upon the crutch of empty space, driven about by an unparalleled primitive impulse of the imagination, and, as yet, kept from tumbling, by the kind concurrence of, in this instance, real astronomical forces. For, that an invisible ether fills all creation is now not only believed by the many as a matter of course, but has also been demonstrated by eminent men.

After this, the old theory of the rotation of the earth round its own axis, and in its orbit round the sun, must strike every one still more as arbitrary, forced, unnatural and untenable, whilst my own propositions could not be more simple and reasonable, though, perhaps, not without some flaw or other. And if on the old theory, certain astronomical calculations have proved tolerably correct, it is because my new theory very little disturbs the present arbitrary adoption of the inclination of the ecliptic, the position of the earth in its annual course round the sun; for, if the axis of the earth, at the time of the equinox, were taken in the exact perpendicular instead of the incline of $23\frac{1}{2}$ degrees to the plane of its orbit, the whole theory, in this respect, would almost exactly coincide with my own, according to which, the oscillation of the earth in its rotary course round the sun, in a most simple manner, explains the so-called nutation of the earth's axis, the actual and only cause of the seasons at the same time, whilst more evidently still it explains the circle which the north pole of the earth, in the course of the earth round the sun, describes round the polar star; and the performing of this circle round the polar star is a NECESSARY CONSEQUENCE of the path and motion of the earth, the

gotten conclusion, that when the ' primitive impulse' was imparted, the earth was held exactly twenty-five miles from the centre, and hence, quoth Laplace, the earth revolves upon her axis once in twenty-four hours. If she had been held a little nearer to the centre, our days would have been longer, and if a little farther off, she would have revolved with greater velocity, and our days would have been shorter. . . . . . . These assumptions comprehend the very essence of what is called the Newtonian philosophy, and the world has been gravely assured that every proposition has been rigidly demonstrated."

With singular infelicity the gyroscope has been made use of, not only to prove this absurdity, but even to improve upon it and assign to the Creator also the spot at which He must have stood when He "hurled the earth into space."

I am sure the gyroscopians cannot have deeply reflected upon their favourite illustration, otherwise they would have seen this, and above all, that the instrument represents nothing else but the stone in the sling, with this difference, that in the one instance, the "primitive impulse," that famous discovery, proceeds from the person wielding the string, and in the other it is imparted by a "Prime Mover," standing outside and independent of the object to be set in motion.

The gyroscope, moreover, represents the earth rotating upon an HORIZONTAL axis by the side of the sun, instead of spinning perpendicularly around him; and therefore it is also an inappropriate illustration.

However, the turning-point of the theory is the "primitive impulse," that famous discovery,—and its "centrifugal assistant;" but, not resting on any solid basis, Newton was obliged to assume, the world has unhesitatingly believed him, and Sir John Herschel still echoes it in *Good Words* that the heavenly bodies move in VACANT SPACE! That there is a VACANT CREATION! And yet

many of those who believe in this vacant space, cannot believe in vacant space outside the ball of creation, cannot believe that the exterior of creation is filled by God alone from eternity to eternity. But, allow a substance ever so fine to pervade the universe, and THE FOUNDATION OF THE OLD THEORY IS REMOVED, AND THE WHOLE EDIFICE CRUMBLES TO PIECES. It is the giant-cripple of universal gravitation, limping upon the crutch of empty space, driven about by an unparalleled primitive impulse of the imagination, and, as yet, kept from tumbling, by the kind concurrence of, in this instance, real astronomical forces. For, that an invisible ether fills all creation is now not only believed by the many as a matter of course, but has also been demonstrated by eminent men.

After this, the old theory of the rotation of the earth round its own axis, and in its orbit round the sun, must strike every one still more as arbitrary, forced, unnatural and untenable, whilst my own propositions could not be more simple and reasonable, though, perhaps, not without some flaw or other. And if on the old theory, certain astronomical calculations have proved tolerably correct, it is because my new theory very little disturbs the present arbitrary adoption of the inclination of the ecliptic, the position of the earth in its annual course round the sun; for, if the axis of the earth, at the time of the equinox, were taken in the exact perpendicular instead of the incline of $23\frac{1}{2}$ degrees to the plane of its orbit, the whole theory, in this respect, would almost exactly coincide with my own, according to which, the oscillation of the earth in its rotary course round the sun, in a most simple manner, explains the so-called nutation of the earth's axis, the actual and only cause of the seasons at the same time, whilst more evidently still it explains the circle which the north pole of the earth, in the course of the earth round the sun, describes round the polar star; and the performing of this circle round the polar star is a NECESSARY CONSEQUENCE of the path and motion of the earth, the

gotten conclusion, that when the ' primitive impulse ' was imparted, the earth was held exactly twenty-five miles from the centre, and hence, quoth Laplace, the earth revolves upon her axis once in twenty-four hours. If she had been held a little nearer to the centre, our days would have been longer, and if a little farther off, she would have revolved with greater velocity, and our days would have been shorter. . . . . . . These assumptions comprehend the very essence of what is called the Newtonian philosophy, and the world has been gravely assured that every proposition has been rigidly demonstrated."

With singular infelicity the gyroscope has been made use of, not only to prove this absurdity, but even to improve upon it and assign to the Creator also the spot at which He must have stood when He " hurled the earth into space."

I am sure the gyroscopians cannot have deeply reflected upon their favourite illustration, otherwise they would have seen this, and above all, that the instrument represents nothing else but the stone in the sling, with this difference, that in the one instance, the " primitive impulse," that famous discovery, proceeds from the person wielding the string, and in the other it is imparted by a " Prime Mover," standing outside and independent of the object to be set in motion.

The gyroscope, moreover, represents the earth rotating upon an HORIZONTAL axis by the side of the sun, instead of spinning perpendicularly around him; and therefore it is also an inappropriate illustration.

However, the turning-point of the theory is the " primitive impulse," that famous discovery,—and its " centrifugal assistant;" but, not resting on any solid basis, Newton was obliged to assume, the world has unhesitatingly believed him, and Sir John Herschel still echoes it in *Good Words* that the heavenly bodies move in VACANT SPACE ! That there is a VACANT CREATION ! And yet

many of those who believe in this vacant space, cannot believe in vacant space outside the ball of creation, cannot believe that the exterior of creation is filled by God alone from eternity to eternity. But, allow a substance ever so fine to pervade the universe, and THE FOUNDATION OF THE OLD THEORY IS REMOVED, AND THE WHOLE EDIFICE CRUMBLES TO PIECES. It is the giant-cripple of universal gravitation, limping upon the crutch of empty space, driven about by an unparalleled primitive impulse of the imagination, and, as yet, kept from tumbling, by the kind concurrence of, in this instance, real astronomical forces. For, that an invisible ether fills all creation is now not only believed by the many as a matter of course, but has also been demonstrated by eminent men.

After this, the old theory of the rotation of the earth round its own axis, and in its orbit round the sun, must strike every one still more as arbitrary, forced, unnatural and untenable, whilst my own propositions could not be more simple and reasonable, though, perhaps, not without some flaw or other. And if on the old theory, certain astronomical calculations have proved tolerably correct, it is because my new theory very little disturbs the present arbitrary adoption of the inclination of the ecliptic, the position of the earth in its annual course round the sun; for, if the axis of the earth, at the time of the equinox, were taken in the exact perpendicular instead of the incline of 23½ degrees to the plane of its orbit, the whole theory, in this respect, would almost exactly coincide with my own, according to which, the oscillation of the earth in its rotary course round the sun, in a most simple manner, explains the so-called nutation of the earth's axis, the actual and only cause of the seasons at the same time, whilst more evidently still it explains the circle which the north pole of the earth, in the course of the earth round the sun, describes round the polar star; and the performing of this circle round the polar star is a NECESSARY CONSEQUENCE of the path and motion of the earth, the

gotten conclusion, that when the ' primitive impulse' was imparted, the earth was held exactly twenty-five miles from the centre, and hence, quoth Laplace, the earth revolves upon her axis once in twenty-four hours. If she had been held a little nearer to the centre, our days would have been longer, and if a little farther off, she would have revolved with greater velocity, and our days would have been shorter. . . . . . . These assumptions comprehend the very essence of what is called the Newtonian philosophy, and the world has been gravely assured that every proposition has been rigidly demonstrated."

With singular infelicity the gyroscope has been made use of, not only to prove this absurdity, but even to improve upon it.and assign to the Creator also the spot at which He must have stood when He " hurled the earth into space."

I am sure the gyroscopians cannot have deeply reflected upon their favourite illustration, otherwise they would have seen this, and above all, that the instrument represents nothing else but the stone in the sling, with this difference, that in the one instance, the " primitive impulse," that famous discovery, proceeds from the person wielding the string, and in the other it is imparted by a " Prime Mover," standing outside and independent of the object to be set in motion.

The gyroscope, moreover, represents the earth rotating upon an HORIZONTAL axis by the side of the sun, instead of spinning perpendicularly around him; and therefore it is also an inappropriate illustration.

However, the turning-point of the theory is the " primitive impulse," that famous discovery,—and its " centrifugal assistant;" but, not resting on any solid basis, Newton was obliged to assume, the world has unhesitatingly believed him, and Sir John Herschel still echoes it in *Good Words* that the heavenly bodies move in VACANT SPACE! That there is a VACANT CREATION! And yet

many of those who believe in this vacant space, cannot believe in vacant space outside the ball of creation, cannot believe that the exterior of creation is filled by God alone from eternity to eternity. But, allow a substance ever so fine to pervade the universe, and THE FOUNDATION OF THE OLD THEORY IS REMOVED, AND THE WHOLE EDIFICE CRUMBLES TO PIECES. It is the giant-cripple of universal gravitation, limping upon the crutch of empty space, driven about by an unparalleled primitive impulse of the imagination, and, as yet, kept from tumbling, by the kind concurrence of, in this instance, real astronomical forces. For, that an invisible ether fills all creation is now not only believed by the many as a matter of course, but has also been demonstrated by eminent men.

After this, the old theory of the rotation of the earth round its own axis, and in its orbit round the sun, must strike every one still more as arbitrary, forced, unnatural and untenable, whilst my own propositions could not be more simple and reasonable, though, perhaps, not without some flaw or other. And if on the old theory, certain astronomical calculations have proved tolerably correct, it is because my new theory very little disturbs the present arbitrary adoption of the inclination of the ecliptic, the position of the earth in its annual course round the sun; for, if the axis of the earth, at the time of the equinox, were taken in the exact perpendicular instead of the incline of $23\frac{1}{2}$ degrees to the plane of its orbit, the whole theory, in this respect, would almost exactly coincide with my own, according to which, the oscillation of the earth in its rotary course round the sun, in a most simple manner, explains the so-called nutation of the earth's axis, the actual and only cause of the seasons at the same time, whilst more evidently still it explains the circle which the north pole of the earth, in the course of the earth round the sun, describes round the polar star; and the performing of this circle round the polar star is a NECESSARY CONSEQUENCE of the path and motion of the earth, the

gotten conclusion, that when the ' primitive impulse ' was imparted, the earth was held exactly twenty-five miles from the centre, and hence, quoth Laplace, the earth revolves upon her axis once in twenty-four hours. If she had been held a little nearer to the centre, our days would have been longer, and if a little farther off, she would have revolved with greater velocity, and our days would have been shorter. . . . . . . These assumptions comprehend the very essence of what is called the Newtonian philosophy, and the world has been gravely assured that every proposition has been rigidly demonstrated."

With singular infelicity the gyroscope has been made use of, not only to prove this absurdity, but even to improve upon it and assign to the Creator also the spot at which He must have stood when He "hurled the earth into space."

I am sure the gyroscopians cannot have deeply reflected upon their favourite illustration, otherwise they would have seen this, and above all, that the instrument represents nothing else but the stone in the sling, with this difference, that in the one instance, the " primitive impulse," that famous discovery, proceeds from the person wielding the string, and in the other it is imparted by a " Prime Mover," standing outside and independent of the object to be set in motion.

The gyroscope, moreover, represents the earth rotating upon an HORIZONTAL axis by the side of the sun, instead of spinning perpendicularly around him ; and therefore it is also an inappropriate illustration.

However, the turning-point of the theory is the "primitive impulse," that famous discovery,—and its "centrifugal assistant ;" but, not resting on any solid basis, Newton was obliged to assume, the world has unhesitatingly believed him, and Sir John Herschel still echoes it in *Good Words* that the heavenly bodies move in VACANT SPACE ! That there is a VACANT CREATION ! And yet

many of those who believe in this vacant space, cannot believe in vacant space outside the ball of creation, cannot believe that the exterior of creation is filled by God alone from eternity to eternity. But, allow a substance ever so fine to pervade the universe, and THE FOUNDATION OF THE OLD THEORY IS REMOVED, AND THE WHOLE EDIFICE CRUMBLES TO PIECES. It is the giant-cripple of universal gravitation, limping upon the crutch of empty space, driven about by an unparalleled primitive impulse of the imagination, and, as yet, kept from tumbling, by the kind concurrence of, in this instance, real astronomical forces. For, that an invisible ether fills all creation is now not only believed by the many as a matter of course, but has also been demonstrated by eminent men.

After this, the old theory of the rotation of the earth round its own axis, and in its orbit round the sun, must strike every one still more as arbitrary, forced, unnatural and untenable, whilst my own propositions could not be more simple and reasonable, though, perhaps, not without some flaw or other. And if on the old theory, certain astronomical calculations have proved tolerably correct, it is because my new theory very little disturbs the present arbitrary adoption of the inclination of the ecliptic, the position of the earth in its annual course round the sun; for, if the axis of the earth, at the time of the equinox, were taken in the exact perpendicular instead of the incline of 23½ degrees to the plane of its orbit, the whole theory, in this respect, would almost exactly coincide with my own, according to which, the oscillation of the earth in its rotary course round the sun, in a most simple manner, explains the so-called nutation of the earth's axis, the actual and only cause of the seasons at the same time, whilst more evidently still it explains the circle which the north pole of the earth, in the course of the earth round the sun, describes round the polar star; and the performing of this circle round the polar star is a NECESSARY CONSEQUENCE of the path and motion of the earth, the

gotten conclusion, that when the ' primitive impulse ' was imparted, the earth was held exactly twenty-five miles from the centre, and hence, quoth Laplace, the earth revolves upon her axis once in twenty-four hours. If she had been held a little nearer to the centre, our days would have been longer, and if a little farther off, she would have revolved with greater velocity, and our days would have been shorter. . . . . . . These assumptions comprehend the very essence of what is called the Newtonian philosophy, and the world has been gravely assured that every proposition has been rigidly demonstrated."

With singular infelicity the gyroscope has been made use of, not only to prove this absurdity, but even to improve upon it. and assign to the Creator also the spot at which He must have stood when He " hurled the earth into space."

I am sure the gyroscopians cannot have deeply reflected upon their favourite illustration, otherwise they would have seen this, and above all, that the instrument represents nothing else but the stone in the sling, with this difference, that in the one instance, the " primitive impulse," that famous discovery, proceeds from the person wielding the string, and in the other it is imparted by a " Prime Mover," standing outside and independent of the object to be set in motion.

The gyroscope, moreover, represents the earth rotating upon an HORIZONTAL axis by the side of the sun, instead of spinning perpendicularly around him ; and therefore it is also an inappropriate illustration.

However, the turning-point of the theory is the " primitive impulse," that famous discovery,—and its " centrifugal assistant ;" but, not resting on any solid basis, Newton was obliged to assume, the world has unhesitatingly believed him, and Sir John Herschel still echoes it in *Good Words* that the heavenly bodies move in VACANT SPACE ! That there is a VACANT CREATION ! And yet

many of those who believe in this vacant space, cannot believe in vacant space outside the ball of creation, cannot believe that the exterior of creation is filled by God alone from eternity to eternity. But, allow a substance ever so fine to pervade the universe, and THE FOUNDATION OF THE OLD THEORY IS REMOVED, AND THE WHOLE EDIFICE CRUMBLES TO PIECES.   It is the giant-cripple of universal gravitation, limping upon the crutch of empty space, driven about by an unparalleled primitive impulse of the imagination, and, as yet, kept from tumbling, by the kind concurrence of, in this instance, real astronomical forces.   For, that an invisible ether fills all creation is now not only believed by the many as a matter of course, but has also been demonstrated by eminent men.

After this, the old theory of the rotation of the earth round its own axis, and in its orbit round the sun, must strike every one still more as arbitrary, forced, unnatural and untenable, whilst my own propositions could not be more simple and reasonable, though, perhaps, not without some flaw or other. And if on the old theory, certain astronomical calculations have proved tolerably correct, it is because my new theory very little disturbs the present arbitrary adoption of the inclination of the ecliptic, the position of the earth in its annual course round the sun; for, if the axis of the earth, at the time of the equinox, were taken in the exact perpendicular instead of the incline of $23\frac{1}{2}$ degrees to the plane of its orbit, the whole theory, in this respect, would almost exactly coincide with my own, according to which, the oscillation of the earth in its rotary course round the sun, in a most simple manner, explains the so-called nutation of the earth's axis, the actual and only cause of the seasons at the same time, whilst more evidently still it explains the circle which the north pole of the earth, in the course of the earth round the sun, describes round the polar star; and the performing of this circle round the polar star is a NECESSARY CONSEQUENCE of the path and motion of the earth, the

236 THE NEW THEORY OF THE SOLAR SYSTEM.

gotten conclusion, that when the ' primitive impulse' was imparted, the earth was held exactly twenty-five miles from the centre, and hence, quoth Laplace, the earth revolves upon her axis once in twenty-four hours. If she had been held a little nearer to the centre, our days would have been longer, and if a little farther off, she would have revolved with greater velocity, and our days would have been shorter. . . . . . . These assumptions comprehend the very essence of what is called the Newtonian philosophy, and the world has been gravely assured that every proposition has been rigidly demonstrated."

With singular infelicity the gyroscope has been made use of, not only to prove this absurdity, but even to improve upon it and assign to the Creator also the spot at which He must have stood when He "hurled the earth into space."

I am sure the gyroscopians cannot have deeply reflected upon their favourite illustration, otherwise they would have seen this, and above all, that the instrument represents nothing else but the stone in the sling, with this difference, that in the one instance, the "primitive impulse," that famous discovery, proceeds from the person wielding the string, and in the other it is imparted by a "Prime Mover," standing outside and independent of the object to be set in motion.

The gyroscope, moreover, represents the earth rotating upon an HORIZONTAL axis by the side of the sun, instead of spinning perpendicularly around him; and therefore it is also an inappropriate illustration.

However, the turning-point of the theory is the "primitive impulse," that famous discovery,—and its "centrifugal assistant;" but, not resting on any solid basis, Newton was obliged to assume, the world has unhesitatingly believed him, and Sir John Herschel still echoes it in *Good Words* that the heavenly bodies move in VACANT SPACE! That there is a VACANT CREATION! And yet

many of those who believe in this vacant space, cannot believe in vacant space outside the ball of creation, cannot believe that the exterior of creation is filled by God alone from eternity to eternity. But, allow a substance ever so fine to pervade the universe, and THE FOUNDATION OF THE OLD THEORY IS REMOVED, AND THE WHOLE EDIFICE CRUMBLES TO PIECES.   It is the giant-cripple of universal gravitation, limping upon the crutch of empty space, driven about by an unparalleled primitive impulse of the imagination, and, as yet, kept from tumbling, by the kind concurrence of, in this instance, real astronomical forces.   For, that an invisible ether fills all creation is now not only believed by the many as a matter of course, but has also been demonstrated by eminent men.

After this, the old theory of the rotation of the earth round its own axis, and in its orbit round the sun, must strike every one still more as arbitrary, forced, unnatural and untenable, whilst my own propositions could not be more simple and reasonable, though, perhaps, not without some flaw or other. And if on the old theory, certain astronomical calculations have proved tolerably correct, it is because my new theory very little disturbs the present arbitrary adoption of the inclination of the ecliptic, the position of the earth in its annual course round the sun; for, if the axis of the earth, at the time of the equinox, were taken in the exact perpendicular instead of the incline of $23\frac{1}{2}$ degrees to the plane of its orbit, the whole theory, in this respect, would almost exactly coincide with my own, according to which, the oscillation of the earth in its rotary course round the sun, in a most simple manner, explains the so-called nutation of the earth's axis, the actual and only cause of the seasons at the same time, whilst more evidently still it explains the circle which the north pole of the earth, in the course of the earth round the sun, describes round the polar star; and the performing of this circle round the polar star is a NECESSARY CONSEQUENCE of the path and motion of the earth, the

gotten conclusion, that when the 'primitive impulse' was imparted, the earth was held exactly twenty-five miles from the centre, and hence, quoth Laplace, the earth revolves upon her axis once in twenty-four hours. If she had been held a little nearer to the centre, our days would have been longer, and if a little farther off, she would have revolved with greater velocity, and our days would have been shorter. . . . . . . These assumptions comprehend the very essence of what is called the Newtonian philosophy, and the world has been gravely assured that every proposition has been rigidly demonstrated."

With singular infelicity the gyroscope has been made use of, not only to prove this absurdity, but even to improve upon it and assign to the Creator also the spot at which He must have stood when He "hurled the earth into space."

I am sure the gyroscopians cannot have deeply reflected upon their favourite illustration, otherwise they would have seen this, and above all, that the instrument represents nothing else but the stone in the sling, with this difference, that in the one instance, the "primitive impulse," that famous discovery, proceeds from the person wielding the string, and in the other it is imparted by a "Prime Mover," standing outside and independent of the object to be set in motion.

The gyroscope, moreover, represents the earth rotating upon an HORIZONTAL axis by the side of the sun, instead of spinning perpendicularly around him; and therefore it is also an inappropriate illustration.

However, the turning-point of the theory is the "primitive impulse," that famous discovery,—and its "centrifugal assistant;" but, not resting on any solid basis, Newton was obliged to assume, the world has unhesitatingly believed him, and Sir John Herschel still echoes it in *Good Words* that the heavenly bodies move in VACANT SPACE! That there is a VACANT CREATION! And yet

many of those who believe in this vacant space, cannot believe in vacant space outside the ball of creation, cannot believe that the exterior of creation is filled by God alone from eternity to eternity. But, allow a substance ever so fine to pervade the universe, and THE FOUNDATION OF THE OLD THEORY IS REMOVED, AND THE WHOLE EDIFICE CRUMBLES TO PIECES.   It is the giant-cripple of universal gravitation, limping upon the crutch of empty space, driven about by an unparalleled primitive impulse of the imagination, and, as yet, kept from tumbling, by the kind concurrence of, in this instance, real astronomical forces.   For, that an invisible ether fills all creation is now not only believed by the many as a matter of course, but has also been demonstrated by eminent men.

After this, the old theory of the rotation of the earth round its own axis, and in its orbit round the sun, must strike every one still more as arbitrary, forced, unnatural and untenable, whilst my own propositions could not be more simple and reasonable, though, perhaps, not without some flaw or other. And if on the old theory, certain astronomical calculations have proved tolerably correct, it is because my new theory very little disturbs the present arbitrary adoption of the inclination of the ecliptic, the position of the earth in its annual course round the sun; for, if the axis of the earth, at the time of the equinox, were taken in the exact perpendicular instead of the incline of 23½ degrees to the plane of its orbit, the whole theory, in this respect, would almost exactly coincide with my own, according to which, the oscillation of the earth in its rotary course round the sun, in a most simple manner, explains the so-called nutation of the earth's axis, the actual and only cause of the seasons at the same time, whilst more evidently still it explains the circle which the north pole of the earth, in the course of the earth round the sun, describes round the polar star; and the performing of this circle round the polar star is a NECESSARY CONSEQUENCE of the path and motion of the earth, the

gotten conclusion, that when the ' primitive impulse ' was imparted, the earth was held exactly twenty-five miles from the centre, and hence, quoth Laplace, the earth revolves upon her axis once in twenty-four hours. If she had been held a little nearer to the centre, our days would have been longer, and if a little farther off, she would have revolved with greater velocity, and our days would have been shorter. . . . . . . These assumptions comprehend the very essence of what is called the Newtonian philosophy, and the world has been gravely assured that every proposition has been rigidly demonstrated."

With singular infelicity the gyroscope has been made use of, not only to prove this absurdity, but even to improve upon it and assign to the Creator also the spot at which He must have stood when He "hurled the earth into space."

I am sure the gyroscopians cannot have deeply reflected upon their favourite illustration, otherwise they would have seen this, and above all, that the instrument represents nothing else but the stone in the sling, with this difference, that in the one instance, the " primitive impulse," that famous discovery, proceeds from the person wielding the string, and in the other it is imparted by a " Prime Mover," standing outside and independent of the object to be set in motion.

The gyroscope, moreover, represents the earth rotating upon an HORIZONTAL axis by the side of the sun, instead of spinning perpendicularly around him ; and therefore it is also an inappropriate illustration.

However, the turning-point of the theory is the "primitive impulse," that famous discovery,—and its "centrifugal assistant;" but, not resting on any solid basis, Newton was obliged to assume, the world has unhesitatingly believed him, and Sir John Herschel still echoes it in *Good Words* that the heavenly bodies move in VACANT SPACE! That there is a VACANT CREATION! And yet

many of those who believe in this vacant space, cannot believe in
vacant space outside the ball of creation, cannot believe that the
exterior of creation is filled by God alone from eternity to eternity.
But, allow a substance ever so fine to pervade the universe, and
THE FOUNDATION OF THE OLD THEORY IS REMOVED, AND THE WHOLE
EDIFICE CRUMBLES TO PIECES.   It is the giant-cripple of universal
gravitation, limping upon the crutch of empty space, driven about
by an unparalleled primitive impulse of the imagination, and, as
yet, kept from tumbling, by the kind concurrence of, in this instance,
real astronomical forces.   For, that an invisible ether fills all creation is now not only believed by the many as a matter of course,
but has also been demonstrated by eminent men.

After this, the old theory of the rotation of the earth round its
own axis, and in its orbit round the sun, must strike every one still
more as arbitrary, forced, unnatural and untenable, whilst my own
propositions could not be more simple and reasonable, though,
perhaps, not without some flaw or other. And if on the old theory,
certain astronomical calculations have proved tolerably correct,
it is because my new theory very little disturbs the present arbitrary adoption of the inclination of the ecliptic, the position of the
earth in its annual course round the sun; for, if the axis of the
earth, at the time of the equinox, were taken in the exact perpendicular instead of the incline of 23½ degrees to the plane of its
orbit, the whole theory, in this respect, would almost exactly
coincide with my own, according to which, the oscillation of the
earth in its rotary course round the sun, in a most simple manner,
explains the so-called nutation of the earth's axis, the actual and
only cause of the seasons at the same time, whilst more evidently
still it explains the circle which the north pole of the earth, in the
course of the earth round the sun, describes round the polar star;
and the performing of this circle round the polar star is a NECES-
SARY CONSEQUENCE of the path and motion of the earth, the

gotten conclusion, that when the ' primitive impulse' was imparted, the earth was held exactly twenty-five miles from the centre, and hence, quoth Laplace, the earth revolves upon her axis once in twenty-four hours. If she had been held a little nearer to the centre, our days would have been longer, and if a little farther off, she would have revolved with greater velocity, and our days would have been shorter. . . . . . . These assumptions comprehend the very essence of what is called the Newtonian philosophy, and the world has been gravely assured that every proposition has been rigidly demonstrated."

With singular infelicity the gyroscope has been made use of, not only to prove this absurdity, but even to improve upon it and assign to the Creator also the spot at which He must have stood when He "hurled the earth into space."

I am sure the gyroscopians cannot have deeply reflected upon their favourite illustration, otherwise they would have seen this, and above all, that the instrument represents nothing else but the stone in the sling, with this difference, that in the one instance, the "primitive impulse," that famous discovery, proceeds from the person wielding the string, and in the other it is imparted by a "Prime Mover," standing outside and independent of the object to be set in motion.

The gyroscope, moreover, represents the earth rotating upon an HORIZONTAL axis by the side of the sun, instead of spinning perpendicularly around him; and therefore it is also an inappropriate illustration.

However, the turning-point of the theory is the "primitive impulse," that famous discovery,—and its "centrifugal assistant;" but, not resting on any solid basis, Newton was obliged to assume, the world has unhesitatingly believed him, and Sir John Herschel still echoes it in *Good Words* that the heavenly bodies move in VACANT SPACE! That there is a VACANT CREATION! And yet

many of those who believe in this vacant space, cannot believe in vacant space outside the ball of creation, cannot believe that the exterior of creation is filled by God alone from eternity to eternity. But, allow a substance ever so fine to pervade the universe, and THE FOUNDATION OF THE OLD THEORY IS REMOVED, AND THE WHOLE EDIFICE CRUMBLES TO PIECES.   It is the giant-cripple of universal gravitation, limping upon the crutch of empty space, driven about by an unparalleled primitive impulse of the imagination, and, as yet, kept from tumbling, by the kind concurrence of, in this instance, real astronomical forces.   For, that an invisible ether fills all creation is now not only believed by the many as a matter of course, but has also been demonstrated by eminent men.

After this, the old theory of the rotation of the earth round its own axis, and in its orbit round the sun, must strike every one still more as arbitrary, forced, unnatural and untenable, whilst my own propositions could not be more simple and reasonable, though, perhaps, not without some flaw or other. And if on the old theory, certain astronomical calculations have proved tolerably correct, it is because my new theory very little disturbs the present arbitrary adoption of the inclination of the ecliptic, the position of the earth in its annual course round the sun; for, if the axis of the earth, at the time of the equinox, were taken in the exact perpendicular instead of the incline of 23½ degrees to the plane of its orbit, the whole theory, in this respect, would almost exactly coincide with my own, according to which, the oscillation of the earth in its rotary course round the sun, in a most simple manner, explains the so-called nutation of the earth's axis, the actual and only cause of the seasons at the same time, whilst more evidently still it explains the circle which the north pole of the earth, in the course of the earth round the sun, describes round the polar star; and the performing of this circle round the polar star is a NECESSARY CONSEQUENCE of the path and motion of the earth, the

gotten conclusion, that when the ' primitive impulse ' was imparted, the earth was held exactly twenty-five miles from the centre, and hence, quoth Laplace, the earth revolves upon her axis once in twenty-four hours. If she had been held a little nearer to the centre, our days would have been longer, and if a little farther off, she would have revolved with greater velocity, and our days would have been shorter. . . . . . . These assumptions comprehend the very essence of what is called the Newtonian philosophy, and the world has been gravely assured that every proposition has been rigidly demonstrated."

With singular infelicity the gyroscope has been made use of, not only to prove this absurdity, but even to improve upon it and assign to the Creator also the spot at which He must have stood when He "hurled the earth into space."

I am sure the gyroscopians cannot have deeply reflected upon their favourite illustration, otherwise they would have seen this, and above all, that the instrument represents nothing else but the stone in the sling, with this difference, that in the one instance, the "primitive impulse," that famous discovery, proceeds from the person wielding the string, and in the other it is imparted by a "Prime Mover," standing outside and independent of the object to be set in motion.

The gyroscope, moreover, represents the earth rotating upon an HORIZONTAL axis by the side of the sun, instead of spinning perpendicularly around him ; and therefore it is also an inappropriate illustration.

However, the turning-point of the theory is the "primitive impulse," that famous discovery,—and its "centrifugal assistant ;" but, not resting on any solid basis, Newton was obliged to assume, the world has unhesitatingly believed him, and Sir John Herschel still echoes it in *Good Words* that the heavenly bodies move in VACANT SPACE ! That there is a VACANT CREATION ! And yet

many of those who believe in this vacant space, cannot believe in vacant space outside the ball of creation, cannot believe that the exterior of creation is filled by God alone from eternity to eternity. But, allow a substance ever so fine to pervade the universe, and THE FOUNDATION OF THE OLD THEORY IS REMOVED, AND THE WHOLE EDIFICE CRUMBLES TO PIECES.   It is the giant-cripple of universal gravitation, limping upon the crutch of empty space, driven about by an unparalleled primitive impulse of the imagination, and, as yet, kept from tumbling, by the kind concurrence of, in this instance, real astronomical forces.   For, that an invisible ether fills all creation is now not only believed by the many as a matter of course, but has also been demonstrated by eminent men.

After this, the old theory of the rotation of the earth round its own axis, and in its orbit round the sun, must strike every one still more as arbitrary, forced, unnatural and untenable, whilst my own propositions could not be more simple and reasonable, though, perhaps, not without some flaw or other. And if on the old theory, certain astronomical calculations have proved tolerably correct, it is because my new theory very little disturbs the present arbitrary adoption of the inclination of the ecliptic, the position of the earth in its annual course round the sun ; for, if the axis of the earth, at the time of the equinox, were taken in the exact perpendicular instead of the incline of 23½ degrees to the plane of its orbit, the whole theory, in this respect, would almost exactly coincide with my own, according to which, the oscillation of the earth in its rotary course round the sun, in a most simple manner, explains the so-called nutation of the earth's axis, the actual and only cause of the seasons at the same time, whilst more evidently still it explains the circle which the north pole of the earth, in the course of the earth round the sun, describes round the polar star ; and the performing of this circle round the polar star is a NECESSARY CONSEQUENCE of the path and motion of the earth, the

gotten conclusion, that when the 'primitive impulse' was imparted, the earth was held exactly twenty-five miles from the centre, and hence, quoth Laplace, the earth revolves upon her axis once in twenty-four hours. If she had been held a little nearer to the centre, our days would have been longer, and if a little farther off, she would have revolved with greater velocity, and our days would have been shorter. . . . . . . These assumptions comprehend the very essence of what is called the Newtonian philosophy, and the world has been gravely assured that every proposition has been rigidly demonstrated."

With singular infelicity the gyroscope has been made use of, not only to prove this absurdity, but even to improve upon it and assign to the Creator also the spot at which He must have stood when He "hurled the earth into space."

I am sure the gyroscopians cannot have deeply reflected upon their favourite illustration, otherwise they would have seen this, and above all, that the instrument represents nothing else but the stone in the sling, with this difference, that in the one instance, the "primitive impulse," that famous discovery, proceeds from the person wielding the string, and in the other it is imparted by a "Prime Mover," standing outside and independent of the object to be set in motion.

The gyroscope, moreover, represents the earth rotating upon an HORIZONTAL axis by the side of the sun, instead of spinning perpendicularly around him; and therefore it is also an inappropriate illustration.

However, the turning-point of the theory is the "primitive impulse," that famous discovery,—and its "centrifugal assistant;" but, not resting on any solid basis, Newton was obliged to assume, the world has unhesitatingly believed him, and Sir John Herschel still echoes it in *Good Words* that the heavenly bodies move in VACANT SPACE! That there is a VACANT CREATION! And yet

many of those who believe in this vacant space, cannot believe in vacant space outside the ball of creation, cannot believe that the exterior of creation is filled by God alone from eternity to eternity. But, allow a substance ever so fine to pervade the universe, and THE FOUNDATION OF THE OLD THEORY IS REMOVED, AND THE WHOLE EDIFICE CRUMBLES TO PIECES. It is the giant-cripple of universal gravitation, limping upon the crutch of empty space, driven about by an unparalleled primitive impulse of the imagination, and, as yet, kept from tumbling, by the kind concurrence of, in this instance, real astronomical forces. For, that an invisible ether fills all creation is now not only believed by the many as a matter of course, but has also been demonstrated by eminent men.

After this, the old theory of the rotation of the earth round its own axis, and in its orbit round the sun, must strike every one still more as arbitrary, forced, unnatural and untenable, whilst my own propositions could not be more simple and reasonable, though, perhaps, not without some flaw or other. And if on the old theory, certain astronomical calculations have proved tolerably correct, it is because my new theory very little disturbs the present arbitrary adoption of the inclination of the ecliptic, the position of the earth in its annual course round the sun; for, if the axis of the earth, at the time of the equinox, were taken in the exact perpendicular instead of the incline of $23\frac{1}{2}$ degrees to the plane of its orbit, the whole theory, in this respect, would almost exactly coincide with my own, according to which, the oscillation of the earth in its rotary course round the sun, in a most simple manner, explains the so-called nutation of the earth's axis, the actual and only cause of the seasons at the same time, whilst more evidently still it explains the circle which the north pole of the earth, in the course of the earth round the sun, describes round the polar star; and the performing of this circle round the polar star is a NECESSARY CONSEQUENCE of the path and motion of the earth, the

gotten conclusion, that when the ' primitive impulse ' was imparted, the earth was held exactly twenty-five miles from the centre, and hence, quoth Laplace, the earth revolves upon her axis once in twenty-four hours. If she had been held a little nearer to the centre, our days would have been longer, and if a little farther off, she would have revolved with greater velocity, and our days would have been shorter. . . . . . . These assumptions comprehend the very essence of what is called the Newtonian philosophy, and the world has been gravely assured that every proposition has been rigidly demonstrated."

With singular infelicity the gyroscope has been made use of, not only to prove this absurdity, but even to improve upon it and assign to the Creator also the spot at which He must have stood when He "hurled the earth into space."

I am sure the gyroscopians cannot have deeply reflected upon their favourite illustration, otherwise they would have seen this, and above all, that the instrument represents nothing else but the stone in the sling, with this difference, that in the one instance, the "primitive impulse," that famous discovery, proceeds from the person wielding the string, and in the other it is imparted by a "Prime Mover," standing outside and independent of the object to be set in motion.

The gyroscope, moreover, represents the earth rotating upon an HORIZONTAL axis by the side of the sun, instead of spinning perpendicularly around him ; and therefore it is also an inappropriate illustration.

However, the turning-point of the theory is the "primitive impulse," that famous discovery,—and its "centrifugal assistant ;" but, not resting on any solid basis, Newton was obliged to assume, the world has unhesitatingly believed him, and Sir John Herschel still echoes it in *Good Words* that the heavenly bodies move in VACANT SPACE ! That there is a VACANT CREATION ! And yet

many of those who believe in this vacant space, cannot believe in vacant space outside the ball of creation, cannot believe that the exterior of creation is filled by God alone from eternity to eternity. But, allow a substance ever so fine to pervade the universe, and THE FOUNDATION OF THE OLD THEORY IS REMOVED, AND THE WHOLE EDIFICE CRUMBLES TO PIECES.   It is the giant-cripple of universal gravitation, limping upon the crutch of empty space, driven about by an unparalleled primitive impulse of the imagination, and, as yet, kept from tumbling, by the kind concurrence of, in this instance, real astronomical forces.   For, that an invisible ether fills all creation is now not only believed by the many as a matter of course, but has also been demonstrated by eminent men.

After this, the old theory of the rotation of the earth round its own axis, and in its orbit round the sun, must strike every one still more as arbitrary, forced, unnatural and untenable, whilst my own propositions could not be more simple and reasonable, though, perhaps, not without some flaw or other. And if on the old theory, certain astronomical calculations have proved tolerably correct, it is because my new theory very little disturbs the present arbitrary adoption of the inclination of the ecliptic, the position of the earth in its annual course round the sun; for, if the axis of the earth, at the time of the equinox, were taken in the exact perpendicular instead of the incline of 23½ degrees to the plane of its orbit, the whole theory, in this respect, would almost exactly coincide with my own, according to which, the oscillation of the earth in its rotary course round the sun, in a most simple manner, explains the so-called nutation of the earth's axis, the actual and only cause of the seasons at the same time, whilst more evidently still it explains the circle which the north pole of the earth, in the course of the earth round the sun, describes round the polar star; and the performing of this circle round the polar star is a NECESSARY CONSEQUENCE of the path and motion of the earth, the

gotten conclusion, that when the ' primitive impulse' was imparted, the earth was held exactly twenty-five miles from the centre, and hence, quoth Laplace, the earth revolves upon her axis once in twenty-four hours. If she had been held a little nearer to the centre, our days would have been longer, and if a little farther off, she would have revolved with greater velocity, and our days would have been shorter. . . . . . . These assumptions comprehend the very essence of what is called the Newtonian philosophy, and the world has been gravely assured that every proposition has been rigidly demonstrated."

With singular infelicity the gyroscope has been made use of, not only to prove this absurdity, but even to improve upon it and assign to the Creator also the spot at which He must have stood when He "hurled the earth into space."

I am sure the gyroscopians cannot have deeply reflected upon their favourite illustration, otherwise they would have seen this, and above all, that the instrument represents nothing else but the stone in the sling, with this difference, that in the one instance, the "primitive impulse," that famous discovery, proceeds from the person wielding the string, and in the other it is imparted by a "Prime Mover," standing outside and independent of the object to be set in motion.

The gyroscope, moreover, represents the earth rotating upon an HORIZONTAL axis by the side of the sun, instead of spinning perpendicularly around him; and therefore it is also an inappropriate illustration.

However, the turning-point of the theory is the "primitive impulse," that famous discovery,—and its "centrifugal assistant;" but, not resting on any solid basis, Newton was obliged to assume, the world has unhesitatingly believed him, and Sir John Herschel still echoes it in *Good Words* that the heavenly bodies move in VACANT SPACE! That there is a VACANT CREATION! And yet

many of those who believe in this vacant space, cannot believe in vacant space outside the ball of creation, cannot believe that the exterior of creation is filled by God alone from eternity to eternity. But, allow a substance ever so fine to pervade the universe, and THE FOUNDATION OF THE OLD THEORY IS REMOVED, AND THE WHOLE EDIFICE CRUMBLES TO PIECES. It is the giant-cripple of universal gravitation, limping upon the crutch of empty space, driven about by an unparalleled primitive impulse of the imagination, and, as yet, kept from tumbling, by the kind concurrence of, in this instance, real astronomical forces. For, that an invisible ether fills all creation is now not only believed by the many as a matter of course, but has also been demonstrated by eminent men.

After this, the old theory of the rotation of the earth round its own axis, and in its orbit round the sun, must strike every one still more as arbitrary, forced, unnatural and untenable, whilst my own propositions could not be more simple and reasonable, though, perhaps, not without some flaw or other. And if on the old theory, certain astronomical calculations have proved tolerably correct, it is because my new theory very little disturbs the present arbitrary adoption of the inclination of the ecliptic, the position of the earth in its annual course round the sun; for, if the axis of the earth, at the time of the equinox, were taken in the exact perpendicular instead of the incline of $23\frac{1}{2}$ degrees to the plane of its orbit, the whole theory, in this respect, would almost exactly coincide with my own, according to which, the oscillation of the earth in its rotary course round the sun, in a most simple manner, explains the so-called nutation of the earth's axis, the actual and only cause of the seasons at the same time, whilst more evidently still it explains the circle which the north pole of the earth, in the course of the earth round the sun, describes round the polar star; and the performing of this circle round the polar star is a NECESSARY CONSEQUENCE of the path and motion of the earth, the

gotten conclusion, that when the ' primitive impulse ' was imparted, the earth was held exactly twenty-five miles from the centre, and hence, quoth Laplace, the earth revolves upon her axis once in twenty-four hours. If she had been held a little nearer to the centre, our days would have been longer, and if a little farther off, she would have revolved with greater velocity, and our days would have been shorter. . . . . . . These assumptions comprehend the very essence of what is called the Newtonian philosophy, and the world has been gravely assured that every proposition has been rigidly demonstrated."

With singular infelicity the gyroscope has been made use of, not only to prove this absurdity, but even to improve upon it.and assign to the Creator also the spot at which He must have stood when He " hurled the earth into space."

I am sure the gyroscopians cannot have deeply reflected upon their favourite illustration, otherwise they would have seen this, and above all, that the instrument represents nothing else but the stone in the sling, with this difference, that in the one instance, the " primitive impulse," that famous discovery, proceeds from the person wielding the string, and in the other it is imparted by a " Prime Mover," standing outside and independent of the object to be set in motion.

The gyroscope, moreover, represents the earth rotating upon an HORIZONTAL axis by the side of the sun, instead of spinning perpendicularly around him; and therefore it is also an inappropriate illustration.

However, the turning-point of the theory is the " primitive impulse," that famous discovery,—and its " centrifugal assistant ;" but, not resting on any solid basis, Newton was obliged to assume, the world has unhesitatingly believed him, and Sir John Herschel still echoes it in *Good Words* that the heavenly bodies move in VACANT SPACE ! That there is a VACANT CREATION ! And yet

many of those who believe in this vacant space, cannot believe in vacant space outside the ball of creation, cannot believe that the exterior of creation is filled by God alone from eternity to eternity. But, allow a substance ever so fine to pervade the universe, and THE FOUNDATION OF THE OLD THEORY IS REMOVED, AND THE WHOLE EDIFICE CRUMBLES TO PIECES. It is the giant-cripple of universal gravitation, limping upon the crutch of empty space, driven about by an unparalleled primitive impulse of the imagination, and, as yet, kept from tumbling, by the kind concurrence of, in this instance, real astronomical forces. For, that an invisible ether fills all creation is now not only believed by the many as a matter of course, but has also been demonstrated by eminent men.

After this, the old theory of the rotation of the earth round its own axis, and in its orbit round the sun, must strike every one still more as arbitrary, forced, unnatural and untenable, whilst my own propositions could not be more simple and reasonable, though, perhaps, not without some flaw or other. And if on the old theory, certain astronomical calculations have proved tolerably correct, it is because my new theory very little disturbs the present arbitrary adoption of the inclination of the ecliptic, the position of the earth in its annual course round the sun; for, if the axis of the earth, at the time of the equinox, were taken in the exact perpendicular instead of the incline of 23½ degrees to the plane of its orbit, the whole theory, in this respect, would almost exactly coincide with my own, according to which, the oscillation of the earth in its rotary course round the sun, in a most simple manner, explains the so-called nutation of the earth's axis, the actual and only cause of the seasons at the same time, whilst more evidently still it explains the circle which the north pole of the earth, in the course of the earth round the sun, describes round the polar star; and the performing of this circle round the polar star is a NECESSARY CONSEQUENCE of the path and motion of the earth, the

gotten conclusion, that when the ' primitive impulse ' was imparted, the earth was held exactly twenty-five miles from the centre, and hence, quoth Laplace, the earth revolves upon her axis once in twenty-four hours. If she had been held a little nearer to the centre, our days would have been longer, and if a little farther off, she would have revolved with greater velocity, and our days would have been shorter. . . . . . . These assumptions comprehend the very essence of what is called the Newtonian philosophy, and the world has been gravely assured that every proposition has been rigidly demonstrated."

With singular infelicity the gyroscope has been made use of, not only to prove this absurdity, but even to improve upon it and assign to the Creator also the spot at which He must have stood when He "hurled the earth into space."

I am sure the gyroscopians cannot have deeply reflected upon their favourite illustration, otherwise they would have seen this, and above all, that the instrument represents nothing else but the stone in the sling, with this difference, that in the one instance, the " primitive impulse," that famous discovery, proceeds from the person wielding the string, and in the other it is imparted by a " Prime Mover," standing outside and independent of the object to be set in motion.

The gyroscope, moreover, represents the earth rotating upon an HORIZONTAL axis by the side of the sun, instead of spinning perpendicularly around him ; and therefore it is also an inappropriate illustration.

However, the turning-point of the theory is the "primitive impulse," that famous discovery,—and its "centrifugal assistant;" but, not resting on any solid basis, Newton was obliged to assume, the world has unhesitatingly believed him, and Sir John Herschel still echoes it in *Good Words* that the heavenly bodies move in VACANT SPACE !   That there is a VACANT CREATION !   And yet

many of those who believe in this vacant space, cannot believe in vacant space outside the ball of creation, cannot believe that the exterior of creation is filled by God alone from eternity to eternity. But, allow a substance ever so fine to pervade the universe, and THE FOUNDATION OF THE OLD THEORY IS REMOVED, AND THE WHOLE EDIFICE CRUMBLES TO PIECES. It is the giant-cripple of universal gravitation, limping upon the crutch of empty space, driven about by an unparalleled primitive impulse of the imagination, and, as yet, kept from tumbling, by the kind concurrence of, in this instance, real astronomical forces. For, that an invisible ether fills all creation is now not only believed by the many as a matter of course, but has also been demonstrated by eminent men.

After this, the old theory of the rotation of the earth round its own axis, and in its orbit round the sun, must strike every one still more as arbitrary, forced, unnatural and untenable, whilst my own propositions could not be more simple and reasonable, though, perhaps, not without some flaw or other. And if on the old theory, certain astronomical calculations have proved tolerably correct, it is because my new theory very little disturbs the present arbitrary adoption of the inclination of the ecliptic, the position of the earth in its annual course round the sun; for, if the axis of the earth, at the time of the equinox, were taken in the exact perpendicular instead of the incline of 23½ degrees to the plane of its orbit, the whole theory, in this respect, would almost exactly coincide with my own, according to which, the oscillation of the earth in its rotary course round the sun, in a most simple manner, explains the so-called nutation of the earth's axis, the actual and only cause of the seasons at the same time, whilst more evidently still it explains the circle which the north pole of the earth, in the course of the earth round the sun, describes round the polar star; and the performing of this circle round the polar star is a NECESSARY CONSEQUENCE of the path and motion of the earth, the

gotten conclusion, that when the ' primitive impulse ' was imparted, the earth was held exactly twenty-five miles from the centre, and hence, quoth Laplace, the earth revolves upon her axis once in twenty-four hours. If she had been held a little nearer to the centre, our days would have been longer, and if a little farther off, she would have revolved with greater velocity, and our days would have been shorter. . . . . . . These assumptions comprehend the very essence of what is called the Newtonian philosophy, and the world has been gravely assured that every proposition has been rigidly demonstrated."

With singular infelicity the gyroscope has been made use of, not only to prove this absurdity, but even to improve upon it and assign to the Creator also the spot at which He must have stood when He "hurled the earth into space."

I am sure the gyroscopians cannot have deeply reflected upon their favourite illustration, otherwise they would have seen this, and above all, that the instrument represents nothing else but the stone in the sling, with this difference, that in the one instance, the " primitive impulse," that famous discovery, proceeds from the person wielding the string, and in the other it is imparted by a " Prime Mover," standing outside and independent of the object to be set in motion.

The gyroscope, moreover, represents the earth rotating upon an HORIZONTAL axis by the side of the sun, instead of spinning perpendicularly around him; and therefore it is also an inappropriate illustration.

However, the turning-point of the theory is the "primitive impulse," that famous discovery,—and its "centrifugal assistant;" but, not resting on any solid basis, Newton was obliged to assume, the world has unhesitatingly believed him, and Sir John Herschel still echoes it in *Good Words* that the heavenly bodies move in VACANT SPACE! That there is a VACANT CREATION! And yet

many of those who believe in this vacant space, cannot believe in vacant space outside the ball of creation, cannot believe that the exterior of creation is filled by God alone from eternity to eternity. But, allow a substance ever so fine to pervade the universe, and THE FOUNDATION OF THE OLD THEORY IS REMOVED, AND THE WHOLE EDIFICE CRUMBLES TO PIECES. It is the giant-cripple of universal gravitation, limping upon the crutch of empty space, driven about by an unparalleled primitive impulse of the imagination, and, as yet, kept from tumbling, by the kind concurrence of, in this instance, real astronomical forces. For, that an invisible ether fills all creation is now not only believed by the many as a matter of course, but has also been demonstrated by eminent men.

After this, the old theory of the rotation of the earth round its own axis, and in its orbit round the sun, must strike every one still more as arbitrary, forced, unnatural and untenable, whilst my own propositions could not be more simple and reasonable, though, perhaps, not without some flaw or other. And if on the old theory, certain astronomical calculations have proved tolerably correct, it is because my new theory very little disturbs the present arbitrary adoption of the inclination of the ecliptic, the position of the earth in its annual course round the sun; for, if the axis of the earth, at the time of the equinox, were taken in the exact perpendicular instead of the incline of 23½ degrees to the plane of its orbit, the whole theory, in this respect, would almost exactly coincide with my own, according to which, the oscillation of the earth in its rotary course round the sun, in a most simple manner, explains the so-called nutation of the earth's axis, the actual and only cause of the seasons at the same time, whilst more evidently still it explains the circle which the north pole of the earth, in the course of the earth round the sun, describes round the polar star; and the performing of this circle round the polar star is a NECESSARY CONSEQUENCE of the path and motion of the earth, the

gotten conclusion, that when the ' primitive impulse ' was imparted, the earth was held exactly twenty-five miles from the centre, and hence, quoth Laplace, the earth revolves upon her axis once in twenty-four hours. If she had been held a little nearer to the centre, our days would have been longer, and if a little farther off, she would have revolved with greater velocity, and our days would have been shorter. . . . . . . These assumptions comprehend the very essence of what is called the Newtonian philosophy, and the world has been gravely assured that every proposition has been rigidly demonstrated."

With singular infelicity the gyroscope has been made use of, not only to prove this absurdity, but even to improve upon it and assign to the Creator also the spot at which He must have stood when He "hurled the earth into space."

I am sure the gyroscopians cannot have deeply reflected upon their favourite illustration, otherwise they would have seen this, and above all, that the instrument represents nothing else but the stone in the sling, with this difference, that in the one instance, the " primitive impulse," that famous discovery, proceeds from the person wielding the string, and in the other it is imparted by a " Prime Mover," standing outside and independent of the object to be set in motion.

The gyroscope, moreover, represents the earth rotating upon an HORIZONTAL axis by the side of the sun, instead of spinning perpendicularly around him ; and therefore it is also an inappropriate illustration.

However, the turning-point of the theory is the " primitive impulse," that famous discovery,—and its " centrifugal assistant ;" but, not resting on any solid basis, Newton was obliged to assume, the world has unhesitatingly believed him, and Sir John Herschel still echoes it in *Good Words* that the heavenly bodies move in VACANT SPACE ! That there is a VACANT CREATION ! And yet

many of those who believe in this vacant space, cannot believe in vacant space outside the ball of creation, cannot believe that the exterior of creation is filled by God alone from eternity to eternity. But, allow a substance ever so fine to pervade the universe, and THE FOUNDATION OF THE OLD THEORY IS REMOVED, AND THE WHOLE EDIFICE CRUMBLES TO PIECES.   It is the giant-cripple of universal gravitation, limping upon the crutch of empty space, driven about by an unparalleled primitive impulse of the imagination, and, as yet, kept from tumbling, by the kind concurrence of, in this instance, real astronomical forces.   For, that an invisible ether fills all creation is now not only believed by the many as a matter of course, but has also been demonstrated by eminent men.

After this, the old theory of the rotation of the earth round its own axis, and in its orbit round the sun, must strike every one still more as arbitrary, forced, unnatural and untenable, whilst my own propositions could not be more simple and reasonable, though, perhaps, not without some flaw or other. And if on the old theory, certain astronomical calculations have proved tolerably correct, it is because my new theory very little disturbs the present arbitrary adoption of the inclination of the ecliptic, the position of the earth in its annual course round the sun; for, if the axis of the earth, at the time of the equinox, were taken in the exact perpendicular instead of the incline of $23\frac{1}{2}$ degrees to the plane of its orbit, the whole theory, in this respect, would almost exactly coincide with my own, according to which, the oscillation of the earth in its rotary course round the sun, in a most simple manner, explains the so-called nutation of the earth's axis, the actual and only cause of the seasons at the same time, whilst more evidently still it explains the circle which the north pole of the earth, in the course of the earth round the sun, describes round the polar star ; and the performing of this circle round the polar star is a NECESSARY CONSEQUENCE of the path and motion of the earth, the

gotten conclusion, that when the 'primitive impulse' was imparted, the earth was held exactly twenty-five miles from the centre, and hence, quoth Laplace, the earth revolves upon her axis once in twenty-four hours. If she had been held a little nearer to the centre, our days would have been longer, and if a little farther off, she would have revolved with greater velocity, and our days would have been shorter. . . . . . . These assumptions compre- hend the very essence of what is called the Newtonian philosophy, and the world has been gravely assured that every proposition has been rigidly demonstrated."

With singular infelicity the gyroscope has been made use of, not only to prove this absurdity, but even to improve upon it and assign to the Creator also the spot at which He must have stood when He "hurled the earth into space."

I am sure the gyroscopians cannot have deeply reflected upon their favourite illustration, otherwise they would have seen this, and above all, that the instrument represents nothing else but the stone in the sling, with this difference, that in the one instance, the "primitive impulse," that famous discovery, proceeds from the person wielding the string, and in the other it is imparted by a "Prime Mover," standing outside and independent of the object to be set in motion.

The gyroscope, moreover, represents the earth rotating upon an HORIZONTAL axis by the side of the sun, instead of spinning perpendicularly around him; and therefore it is also an inappro- priate illustration.

However, the turning-point of the theory is the "primitive impulse," that famous discovery,—and its "centrifugal assistant;" but, not resting on any solid basis, Newton was obliged to assume, the world has unhesitatingly believed him, and Sir John Herschel still echoes it in *Good Words* that the heavenly bodies move in VACANT SPACE! That there is a VACANT CREATION! And yet

many of those who believe in this vacant space, cannot believe in vacant space outside the ball of creation, cannot believe that the exterior of creation is filled by God alone from eternity to eternity. But, allow a substance ever so fine to pervade the universe, and THE FOUNDATION OF THE OLD THEORY IS REMOVED, AND THE WHOLE EDIFICE CRUMBLES TO PIECES. It is the giant-cripple of universal gravitation, limping upon the crutch of empty space, driven about by an unparalleled primitive impulse of the imagination, and, as yet, kept from tumbling, by the kind concurrence of, in this instance, real astronomical forces. For, that an invisible ether fills all creation is now not only believed by the many as a matter of course, but has also been demonstrated by eminent men.

After this, the old theory of the rotation of the earth round its own axis, and in its orbit round the sun, must strike every one still more as arbitrary, forced, unnatural and untenable, whilst my own propositions could not be more simple and reasonable, though, perhaps, not without some flaw or other. And if on the old theory, certain astronomical calculations have proved tolerably correct, it is because my new theory very little disturbs the present arbitrary adoption of the inclination of the ecliptic, the position of the earth in its annual course round the sun; for, if the axis of the earth, at the time of the equinox, were taken in the exact perpendicular instead of the incline of $23\frac{1}{2}$ degrees to the plane of its orbit, the whole theory, in this respect, would almost exactly coincide with my own, according to which, the oscillation of the earth in its rotary course round the sun, in a most simple manner, explains the so-called nutation of the earth's axis, the actual and only cause of the seasons at the same time, whilst more evidently still it explains the circle which the north pole of the earth, in the course of the earth round the sun, describes round the polar star; and the performing of this circle round the polar star is a NECESSARY CONSEQUENCE of the path and motion of the earth, the

gotten conclusion, that when the ' primitive impulse ' was imparted, the earth was held exactly twenty-five miles from the centre, and hence, quoth Laplace, the earth revolves upon her axis once in twenty-four hours. If she had been held a little nearer to the centre, our days would have been longer, and if a little farther off, she would have revolved with greater velocity, and our days would have been shorter. . . . . . . These assumptions comprehend the very essence of what is called the Newtonian philosophy, and the world has been gravely assured that every proposition has been rigidly demonstrated."

With singular infelicity the gyroscope has been made use of, not only to prove this absurdity, but even to improve upon it. and assign to the Creator also the spot at which He must have stood when He "hurled the earth into space."

I am sure the gyroscopians cannot have deeply reflected upon their favourite illustration, otherwise they would have seen this, and above all, that the instrument represents nothing else but the stone in the sling, with this difference, that in the one instance, the "primitive impulse," that famous discovery, proceeds from the person wielding the string, and in the other it is imparted by a "Prime Mover," standing outside and independent of the object to be set in motion.

The gyroscope, moreover, represents the earth rotating upon an HORIZONTAL axis by the side of the sun, instead of spinning perpendicularly around him; and therefore it is also an inappropriate illustration.

However, the turning-point of the theory is the "primitive impulse," that famous discovery,—and its "centrifugal assistant;" but, not resting on any solid basis, Newton was obliged to assume, the world has unhesitatingly believed him, and Sir John Herschel still echoes it in *Good Words* that the heavenly bodies move in VACANT SPACE! That there is a VACANT CREATION! And yet

many of those who believe in this vacant space, cannot believe in vacant space outside the ball of creation, cannot believe that the exterior of creation is filled by God alone from eternity to eternity. But, allow a substance ever so fine to pervade the universe, and THE FOUNDATION OF THE OLD THEORY IS REMOVED, AND THE WHOLE EDIFICE CRUMBLES TO PIECES. It is the giant-cripple of universal gravitation, limping upon the crutch of empty space, driven about by an unparalleled primitive impulse of the imagination, and, as yet, kept from tumbling, by the kind concurrence of, in this instance, real astronomical forces. For, that an invisible ether fills all creation is now not only believed by the many as a matter of course, but has also been demonstrated by eminent men.

After this, the old theory of the rotation of the earth round its own axis, and in its orbit round the sun, must strike every one still more as arbitrary, forced, unnatural and untenable, whilst my own propositions could not be more simple and reasonable, though, perhaps, not without some flaw or other. And if on the old theory, certain astronomical calculations have proved tolerably correct, it is because my new theory very little disturbs the present arbitrary adoption of the inclination of the ecliptic, the position of the earth in its annual course round the sun; for, if the axis of the earth, at the time of the equinox, were taken in the exact perpendicular instead of the incline of 23½ degrees to the plane of its orbit, the whole theory, in this respect, would almost exactly coincide with my own, according to which, the oscillation of the earth in its rotary course round the sun, in a most simple manner, explains the so-called nutation of the earth's axis, the actual and only cause of the seasons at the same time, whilst more evidently still it explains the circle which the north pole of the earth, in the course of the earth round the sun, describes round the polar star; and the performing of this circle round the polar star is a NECESSARY CONSEQUENCE of the path and motion of the earth, the

gotten conclusion, that when the 'primitive impulse' was imparted, the earth was held exactly twenty-five miles from the centre, and hence, quoth Laplace, the earth revolves upon her axis once in twenty-four hours. If she had been held a little nearer to the centre, our days would have been longer, and if a little farther off, she would have revolved with greater velocity, and our days would have been shorter. . . . . . . These assumptions comprehend the very essence of what is called the Newtonian philosophy, and the world has been gravely assured that every proposition has been rigidly demonstrated."

With singular infelicity the gyroscope has been made use of, not only to prove this absurdity, but even to improve upon it and assign to the Creator also the spot at which He must have stood when He "hurled the earth into space."

I am sure the gyroscopians cannot have deeply reflected upon their favourite illustration, otherwise they would have seen this, and above all, that the instrument represents nothing else but the stone in the sling, with this difference, that in the one instance, the "primitive impulse," that famous discovery, proceeds from the person wielding the string, and in the other it is imparted by a "Prime Mover," standing outside and independent of the object to be set in motion.

The gyroscope, moreover, represents the earth rotating upon an HORIZONTAL axis by the side of the sun, instead of spinning perpendicularly around him; and therefore it is also an inappropriate illustration.

However, the turning-point of the theory is the "primitive impulse," that famous discovery,—and its "centrifugal assistant;" but, not resting on any solid basis, Newton was obliged to assume, the world has unhesitatingly believed him, and Sir John Herschel still echoes it in *Good Words* that the heavenly bodies move in VACANT SPACE! That there is a VACANT CREATION! And yet

many of those who believe in this vacant space, cannot believe in vacant space outside the ball of creation, cannot believe that the exterior of creation is filled by God alone from eternity to eternity. But, allow a substance ever so fine to pervade the universe, and THE FOUNDATION OF THE OLD THEORY IS REMOVED, AND THE WHOLE EDIFICE CRUMBLES TO PIECES. It is the giant-cripple of universal gravitation, limping upon the crutch of empty space, driven about by an unparalleled primitive impulse of the imagination, and, as yet, kept from tumbling, by the kind concurrence of, in this instance, real astronomical forces. For, that an invisible ether fills all creation is now not only believed by the many as a matter of course, but has also been demonstrated by eminent men.

After this, the old theory of the rotation of the earth round its own axis, and in its orbit round the sun, must strike every one still more as arbitrary, forced, unnatural and untenable, whilst my own propositions could not be more simple and reasonable, though, perhaps, not without some flaw or other. And if on the old theory, certain astronomical calculations have proved tolerably correct, it is because my new theory very little disturbs the present arbitrary adoption of the inclination of the ecliptic, the position of the earth in its annual course round the sun; for, if the axis of the earth, at the time of the equinox, were taken in the exact perpendicular instead of the incline of 23½ degrees to the plane of its orbit, the whole theory, in this respect, would almost exactly coincide with my own, according to which, the oscillation of the earth in its rotary course round the sun, in a most simple manner, explains the so-called nutation of the earth's axis, the actual and only cause of the seasons at the same time, whilst more evidently still it explains the circle which the north pole of the earth, in the course of the earth round the sun, describes round the polar star; and the performing of this circle round the polar star is a NECESSARY CONSEQUENCE of the path and motion of the earth, the

gotten conclusion, that when the ' primitive impulse ' was imparted, the earth was held exactly twenty-five miles from the centre, and hence, quoth Laplace, the earth revolves upon her axis once in twenty-four hours. If she had been held a little nearer to the centre, our days would have been longer, and if a little farther off, she would have revolved with greater velocity, and our days would have been shorter. . . . . . . These assumptions comprehend the very essence of what is called the Newtonian philosophy, and the world has been gravely assured that every proposition has been rigidly demonstrated."

With singular infelicity the gyroscope has been made use of, not only to prove this absurdity, but even to improve upon it and assign to the Creator also the spot at which He must have stood when He " hurled the earth into space."

I am sure the gyroscopians cannot have deeply reflected upon their favourite illustration, otherwise they would have seen this, and above all, that the instrument represents nothing else but the stone in the sling, with this difference, that in the one instance, the " primitive impulse," that famous discovery, proceeds from the person wielding the string, and in the other it is imparted by a " Prime Mover," standing outside and independent of the object to be set in motion.

The gyroscope, moreover, represents the earth rotating upon an HORIZONTAL axis by the side of the sun, instead of spinning perpendicularly around him ; and therefore it is also an inappropriate illustration.

However, the turning-point of the theory is the " primitive impulse," that famous discovery,—and its " centrifugal assistant ;" but, not resting on any solid basis, Newton was obliged to assume, the world has unhesitatingly believed him, and Sir John Herschel still echoes it in *Good Words* that the heavenly bodies move in VACANT SPACE ! That there is a VACANT CREATION ! And yet

many of those who believe in this vacant space, cannot believe in vacant space outside the ball of creation, cannot believe that the exterior of creation is filled by God alone from eternity to eternity. But, allow a substance ever so fine to pervade the universe, and THE FOUNDATION OF THE OLD THEORY IS REMOVED, AND THE WHOLE EDIFICE CRUMBLES TO PIECES. It is the giant-cripple of universal gravitation, limping upon the crutch of empty space, driven about by an unparalleled primitive impulse of the imagination, and, as yet, kept from tumbling, by the kind concurrence of, in this instance, real astronomical forces. For, that an invisible ether fills all creation is now not only believed by the many as a matter of course, but has also been demonstrated by eminent men.

After this, the old theory of the rotation of the earth round its own axis, and in its orbit round the sun, must strike every one still more as arbitrary, forced, unnatural and untenable, whilst my own propositions could not be more simple and reasonable, though, perhaps, not without some flaw or other. And if on the old theory, certain astronomical calculations have proved tolerably correct, it is because my new theory very little disturbs the present arbitrary adoption of the inclination of the ecliptic, the position of the earth in its annual course round the sun; for, if the axis of the earth, at the time of the equinox, were taken in the exact perpendicular instead of the incline of 23½ degrees to the plane of its orbit, the whole theory, in this respect, would almost exactly coincide with my own, according to which, the oscillation of the earth in its rotary course round the sun, in a most simple manner, explains the so-called nutation of the earth's axis, the actual and only cause of the seasons at the same time, whilst more evidently still it explains the circle which the north pole of the earth, in the course of the earth round the sun, describes round the polar star; and the performing of this circle round the polar star is a NECESSARY CONSEQUENCE of the path and motion of the earth, the

gotten conclusion, that when the 'primitive impulse' was imparted, the earth was held exactly twenty-five miles from the centre, and hence, quoth Laplace, the earth revolves upon her axis once in twenty-four hours. If she had been held a little nearer to the centre, our days would have been longer, and if a little farther off, she would have revolved with greater velocity, and our days would have been shorter. . . . . . . These assumptions comprehend the very essence of what is called the Newtonian philosophy, and the world has been gravely assured that every proposition has been rigidly demonstrated."

With singular infelicity the gyroscope has been made use of, not only to prove this absurdity, but even to improve upon it and assign to the Creator also the spot at which He must have stood when He "hurled the earth into space."

I am sure the gyroscopians cannot have deeply reflected upon their favourite illustration, otherwise they would have seen this, and above all, that the instrument represents nothing else but the stone in the sling, with this difference, that in the one instance, the "primitive impulse," that famous discovery, proceeds from the person wielding the string, and in the other it is imparted by a "Prime Mover," standing outside and independent of the object to be set in motion.

The gyroscope, moreover, represents the earth rotating upon an HORIZONTAL axis by the side of the sun, instead of spinning perpendicularly around him; and therefore it is also an inappropriate illustration.

However, the turning-point of the theory is the "primitive impulse," that famous discovery,—and its "centrifugal assistant;" but, not resting on any solid basis, Newton was obliged to assume, the world has unhesitatingly believed him, and Sir John Herschel still echoes it in *Good Words* that the heavenly bodies move in VACANT SPACE! That there is a VACANT CREATION! And yet

many of those who believe in this vacant space, cannot believe in vacant space outside the ball of creation, cannot believe that the exterior of creation is filled by God alone from eternity to eternity. But, allow a substance ever so fine to pervade the universe, and THE FOUNDATION OF THE OLD THEORY IS REMOVED, AND THE WHOLE EDIFICE CRUMBLES TO PIECES. It is the giant-cripple of universal gravitation, limping upon the crutch of empty space, driven about by an unparalleled primitive impulse of the imagination, and, as yet, kept from tumbling, by the kind concurrence of, in this instance, real astronomical forces. For, that an invisible ether fills all creation is now not only believed by the many as a matter of course, but has also been demonstrated by eminent men.

After this, the old theory of the rotation of the earth round its own axis, and in its orbit round the sun, must strike every one still more as arbitrary, forced, unnatural and untenable, whilst my own propositions could not be more simple and reasonable, though, perhaps, not without some flaw or other. And if on the old theory, certain astronomical calculations have proved tolerably correct, it is because my new theory very little disturbs the present arbitrary adoption of the inclination of the ecliptic, the position of the earth in its annual course round the sun; for, if the axis of the earth, at the time of the equinox, were taken in the exact perpendicular instead of the incline of $23\frac{1}{2}$ degrees to the plane of its orbit, the whole theory, in this respect, would almost exactly coincide with my own, according to which, the oscillation of the earth in its rotary course round the sun, in a most simple manner, explains the so-called nutation of the earth's axis, the actual and only cause of the seasons at the same time, whilst more evidently still it explains the circle which the north pole of the earth, in the course of the earth round the sun, describes round the polar star; and the performing of this circle round the polar star is a NECESSARY CONSEQUENCE of the path and motion of the earth, the

gotten conclusion, that when the ' primitive impulse ' was imparted, the earth was held exactly twenty-five miles from the centre, and hence, quoth Laplace, the earth revolves upon her axis once in twenty-four hours. If she had been held a little nearer to the centre, our days would have been longer, and if a little farther off, she would have revolved with greater velocity, and our days would have been shorter. . . . . . . These assumptions comprehend the very essence of what is called the Newtonian philosophy, and the world has been gravely assured that every proposition has been rigidly demonstrated."

With singular infelicity the gyroscope has been made use of, not only to prove this absurdity, but even to improve upon it.and assign to the Creator also the spot at which He must have stood when He "hurled the earth into space."

I am sure the gyroscopians cannot have deeply reflected upon their favourite illustration, otherwise they would have seen this, and above all, that the instrument represents nothing else but the stone in the sling, with this difference, that in the one instance, the " primitive impulse," that famous discovery, proceeds from the person wielding the string, and in the other it is imparted by a " Prime Mover," standing outside and independent of the object to be set in motion.

The gyroscope, moreover, represents the earth rotating upon an HORIZONTAL axis by the side of the sun, instead of spinning perpendicularly around him ; and therefore it is also an inappropriate illustration.

However, the turning-point of the theory is the " primitive impulse," that famous discovery,—and its " centrifugal assistant ;" but, not resting on any solid basis, Newton was obliged to assume, the world has unhesitatingly believed him, and Sir John Herschel still echoes it in *Good Words* that the heavenly bodies move in VACANT SPACE ! That there is a VACANT CREATION ! And yet

many of those who believe in this vacant space, cannot believe in vacant space outside the ball of creation, cannot believe that the exterior of creation is filled by God alone from eternity to eternity. But, allow a substance ever so fine to pervade the universe, and THE FOUNDATION OF THE OLD THEORY IS REMOVED, AND THE WHOLE EDIFICE CRUMBLES TO PIECES. It is the giant-cripple of universal gravitation, limping upon the crutch of empty space, driven about by an unparalleled primitive impulse of the imagination, and, as yet, kept from tumbling, by the kind concurrence of, in this instance, real astronomical forces. For, that an invisible ether fills all creation is now not only believed by the many as a matter of course, but has also been demonstrated by eminent men.

After this, the old theory of the rotation of the earth round its own axis, and in its orbit round the sun, must strike every one still more as arbitrary, forced, unnatural and untenable, whilst my own propositions could not be more simple and reasonable, though, perhaps, not without some flaw or other. And if on the old theory, certain astronomical calculations have proved tolerably correct, it is because my new theory very little disturbs the present arbitrary adoption of the inclination of the ecliptic, the position of the earth in its annual course round the sun; for, if the axis of the earth, at the time of the equinox, were taken in the exact perpendicular instead of the incline of $23\frac{1}{2}$ degrees to the plane of its orbit, the whole theory, in this respect, would almost exactly coincide with my own, according to which, the oscillation of the earth in its rotary course round the sun, in a most simple manner, explains the so-called nutation of the earth's axis, the actual and only cause of the seasons at the same time, whilst more evidently still it explains the circle which the north pole of the earth, in the course of the earth round the sun, describes round the polar star; and the performing of this circle round the polar star is a NECESSARY CONSEQUENCE of the path and motion of the earth, the

gotten conclusion, that when the 'primitive impulse' was imparted, the earth was held exactly twenty-five miles from the centre, and hence, quoth Laplace, the earth revolves upon her axis once in twenty-four hours. If she had been held a little nearer to the centre, our days would have been longer, and if a little farther off, she would have revolved with greater velocity, and our days would have been shorter. . . . . . . These assumptions comprehend the very essence of what is called the Newtonian philosophy, and the world has been gravely assured that every proposition has been rigidly demonstrated."

With singular infelicity the gyroscope has been made use of, not only to prove this absurdity, but even to improve upon it and assign to the Creator also the spot at which He must have stood when He "hurled the earth into space."

I am sure the gyroscopians cannot have deeply reflected upon their favourite illustration, otherwise they would have seen this, and above all, that the instrument represents nothing else but the stone in the sling, with this difference, that in the one instance, the " primitive impulse," that famous discovery, proceeds from the person wielding the string, and in the other it is imparted by a " Prime Mover," standing outside and independent of the object to be set in motion.

The gyroscope, moreover, represents the earth rotating upon an HORIZONTAL axis by the side of the sun, instead of spinning perpendicularly around him ; and therefore it is also an inappropriate illustration.

However, the turning-point of the theory is the "primitive impulse," that famous discovery,—and its "centrifugal assistant;" but, not resting on any solid basis, Newton was obliged to assume, the world has unhesitatingly believed him, and Sir John Herschel still echoes it in *Good Words* that the heavenly bodies move in VACANT SPACE ! That there is a VACANT CREATION ! And yet

many of those who believe in this vacant space, cannot believe in vacant space outside the ball of creation, cannot believe that the exterior of creation is filled by God alone from eternity to eternity. But, allow a substance ever so fine to pervade the universe, and THE FOUNDATION OF THE OLD THEORY IS REMOVED, AND THE WHOLE EDIFICE CRUMBLES TO PIECES. It is the giant-cripple of universal gravitation, limping upon the crutch of empty space, driven about by an unparalleled primitive impulse of the imagination, and, as yet, kept from tumbling, by the kind concurrence of, in this instance, real astronomical forces. For, that an invisible ether fills all creation is now not only believed by the many as a matter of course, but has also been demonstrated by eminent men.

After this, the old theory of the rotation of the earth round its own axis, and in its orbit round the sun, must strike every one still more as arbitrary, forced, unnatural and untenable, whilst my own propositions could not be more simple and reasonable, though, perhaps, not without some flaw or other. And if on the old theory, certain astronomical calculations have proved tolerably correct, it is because my new theory very little disturbs the present arbitrary adoption of the inclination of the ecliptic, the position of the earth in its annual course round the sun; for, if the axis of the earth, at the time of the equinox, were taken in the exact perpendicular instead of the incline of 23½ degrees to the plane of its orbit, the whole theory, in this respect, would almost exactly coincide with my own, according to which, the oscillation of the earth in its rotary course round the sun, in a most simple manner, explains the so-called nutation of the earth's axis, the actual and only cause of the seasons at the same time, whilst more evidently still it explains the circle which the north pole of the earth, in the course of the earth round the sun, describes round the polar star; and the performing of this circle round the polar star is a NECESSARY CONSEQUENCE of the path and motion of the earth, the

gotten conclusion, that when the ' primitive impulse ' was imparted, the earth was held exactly twenty-five miles from the centre, and hence, quoth Laplace, the earth revolves upon her axis once in twenty-four hours. If she had been held a little nearer to the centre, our days would have been longer, and if a little farther off, she would have revolved with greater velocity, and our days would have been shorter. . . . . . . These assumptions comprehend the very essence of what is called the Newtonian philosophy, and the world has been gravely assured that every proposition has been rigidly demonstrated."

With singular infelicity the gyroscope has been made use of, not only to prove this absurdity, but even to improve upon it.and assign to the Creator also the spot at which He must have stood when He "hurled the earth into space."

I am sure the gyroscopians cannot have deeply reflected upon their favourite illustration, otherwise they would have seen this, and above all, that the instrument represents nothing else but the stone in the sling, with this difference, that in the one instance, the " primitive impulse," that famous discovery, proceeds from the person wielding the string, and in the other it is imparted by a " Prime Mover," standing outside and independent of the object to be set in motion.

The gyroscope, moreover, represents the earth rotating upon an HORIZONTAL axis by the side of the sun, instead of spinning perpendicularly around him ; and therefore it is also an inappropriate illustration.

However, the turning-point of the theory is the "primitive impulse," that famous discovery,—and its "centrifugal assistant ;" but, not resting on any solid basis, Newton was obliged to assume, the world has unhesitatingly believed him, and Sir John Herschel still echoes it in *Good Words* that the heavenly bodies move in VACANT SPACE ! That there is a VACANT CREATION ! And yet

many of those who believe in this vacant space, cannot believe in vacant space outside the ball of creation, cannot believe that the exterior of creation is filled by God alone from eternity to eternity. But, allow a substance ever so fine to pervade the universe, and THE FOUNDATION OF THE OLD THEORY IS REMOVED, AND THE WHOLE EDIFICE CRUMBLES TO PIECES. It is the giant-cripple of universal gravitation, limping upon the crutch of empty space, driven about by an unparalleled primitive impulse of the imagination, and, as yet, kept from tumbling, by the kind concurrence of, in this instance, real astronomical forces. For, that an invisible ether fills all creation is now not only believed by the many as a matter of course, but has also been demonstrated by eminent men.

After this, the old theory of the rotation of the earth round its own axis, and in its orbit round the sun, must strike every one still more as arbitrary, forced, unnatural and untenable, whilst my own propositions could not be more simple and reasonable, though, perhaps, not without some flaw or other. And if on the old theory, certain astronomical calculations have proved tolerably correct, it is because my new theory very little disturbs the present arbitrary adoption of the inclination of the ecliptic, the position of the earth in its annual course round the sun; for, if the axis of the earth, at the time of the equinox, were taken in the exact perpendicular instead of the incline of $23\frac{1}{2}$ degrees to the plane of its orbit, the whole theory, in this respect, would almost exactly coincide with my own, according to which, the oscillation of the earth in its rotary course round the sun, in a most simple manner, explains the so-called nutation of the earth's axis, the actual and only cause of the seasons at the same time, whilst more evidently still it explains the circle which the north pole of the earth, in the course of the earth round the sun, describes round the polar star; and the performing of this circle round the polar star is a NECESSARY CONSEQUENCE of the path and motion of the earth, the

gotten conclusion, that when the ' primitive impulse ' was imparted, the earth was held exactly twenty-five miles from the centre, and hence, quoth Laplace, the earth revolves upon her axis once in twenty-four hours. If she had been held a little nearer to the centre, our days would have been longer, and if a little farther off, she would have revolved with greater velocity, and our days would have been shorter. . . . . . . These assumptions comprehend the very essence of what is called the Newtonian philosophy, and the world has been gravely assured that every proposition has been rigidly demonstrated."

With singular infelicity the gyroscope has been made use of, not only to prove this absurdity, but even to improve upon it and assign to the Creator also the spot at which He must have stood when He "hurled the earth into space."

I am sure the gyroscopians cannot have deeply reflected upon their favourite illustration, otherwise they would have seen this, and above all, that the instrument represents nothing else but the stone in the sling, with this difference, that in the one instance, the "primitive impulse," that famous discovery, proceeds from the person wielding the string, and in the other it is imparted by a "Prime Mover," standing outside and independent of the object to be set in motion.

The gyroscope, moreover, represents the earth rotating upon an HORIZONTAL axis by the side of the sun, instead of spinning perpendicularly around him; and therefore it is also an inappropriate illustration.

However, the turning-point of the theory is the "primitive impulse," that famous discovery,—and its "centrifugal assistant;" but, not resting on any solid basis, Newton was obliged to assume, the world has unhesitatingly believed him, and Sir John Herschel still echoes it in *Good Words* that the heavenly bodies move in VACANT SPACE! That there is a VACANT CREATION! And yet

many of those who believe in this vacant space, cannot believe in vacant space outside the ball of creation, cannot believe that the exterior of creation is filled by God alone from eternity to eternity. But, allow a substance ever so fine to pervade the universe, and THE FOUNDATION OF THE OLD THEORY IS REMOVED, AND THE WHOLE EDIFICE CRUMBLES TO PIECES. It is the giant-cripple of universal gravitation, limping upon the crutch of empty space, driven about by an unparalleled primitive impulse of the imagination, and, as yet, kept from tumbling, by the kind concurrence of, in this instance, real astronomical forces. For, that an invisible ether fills all creation is now not only believed by the many as a matter of course, but has also been demonstrated by eminent men.

After this, the old theory of the rotation of the earth round its own axis, and in its orbit round the sun, must strike every one still more as arbitrary, forced, unnatural and untenable, whilst my own propositions could not be more simple and reasonable, though, perhaps, not without some flaw or other. And if on the old theory, certain astronomical calculations have proved tolerably correct, it is because my new theory very little disturbs the present arbitrary adoption of the inclination of the ecliptic, the position of the earth in its annual course round the sun; for, if the axis of the earth, at the time of the equinox, were taken in the exact perpendicular instead of the incline of 23½ degrees to the plane of its orbit, the whole theory, in this respect, would almost exactly coincide with my own, according to which, the oscillation of the earth in its rotary course round the sun, in a most simple manner, explains the so-called nutation of the earth's axis, the actual and only cause of the seasons at the same time, whilst more evidently still it explains the circle which the north pole of the earth, in the course of the earth round the sun, describes round the polar star; and the performing of this circle round the polar star is a NECESSARY CONSEQUENCE of the path and motion of the earth, the

gotten conclusion, that when the ' primitive impulse ' was imparted, the earth was held exactly twenty-five miles from the centre, and hence, quoth Laplace, the earth revolves upon her axis once in twenty-four hours. If she had been held a little nearer to the centre, our days would have been longer, and if a little farther off, she would have revolved with greater velocity, and our days would have been shorter. . . . . . . These assumptions comprehend the very essence of what is called the Newtonian philosophy, and the world has been gravely assured that every proposition has been rigidly demonstrated."

With singular infelicity the gyroscope has been made use of, not only to prove this absurdity, but even to improve upon it, and assign to the Creator also the spot at which He must have stood when He "hurled the earth into space."

I am sure the gyroscopians cannot have deeply reflected upon their favourite illustration, otherwise they would have seen this, and above all, that the instrument represents nothing else but the stone in the sling, with this difference, that in the one instance, the "primitive impulse," that famous discovery, proceeds from the person wielding the string, and in the other it is imparted by a "Prime Mover," standing outside and independent of the object to be set in motion.

The gyroscope, moreover, represents the earth rotating upon an HORIZONTAL axis by the side of the sun, instead of spinning perpendicularly around him; and therefore it is also an inappropriate illustration.

However, the turning-point of the theory is the "primitive impulse," that famous discovery,—and its "centrifugal assistant;" but, not resting on any solid basis, Newton was obliged to assume, the world has unhesitatingly believed him, and Sir John Herschel still echoes it in *Good Words* that the heavenly bodies move in VACANT SPACE! That there is a VACANT CREATION! And yet

many of those who believe in this vacant space, cannot believe in vacant space outside the ball of creation, cannot believe that the exterior of creation is filled by God alone from eternity to eternity. But, allow a substance ever so fine to pervade the universe, and THE FOUNDATION OF THE OLD THEORY IS REMOVED, AND THE WHOLE EDIFICE CRUMBLES TO PIECES. It is the giant-cripple of universal gravitation, limping upon the crutch of empty space, driven about by an unparalleled primitive impulse of the imagination, and, as yet, kept from tumbling, by the kind concurrence of, in this instance, real astronomical forces. For, that an invisible ether fills all creation is now not only believed by the many as a matter of course, but has also been demonstrated by eminent men.

After this, the old theory of the rotation of the earth round its own axis, and in its orbit round the sun, must strike every one still more as arbitrary, forced, unnatural and untenable, whilst my own propositions could not be more simple and reasonable, though, perhaps, not without some flaw or other. And if on the old theory, certain astronomical calculations have proved tolerably correct, it is because my new theory very little disturbs the present arbitrary adoption of the inclination of the ecliptic, the position of the earth in its annual course round the sun; for, if the axis of the earth, at the time of the equinox, were taken in the exact perpendicular instead of the incline of $23\frac{1}{2}$ degrees to the plane of its orbit, the whole theory, in this respect, would almost exactly coincide with my own, according to which, the oscillation of the earth in its rotary course round the sun, in a most simple manner, explains the so-called nutation of the earth's axis, the actual and only cause of the seasons at the same time, whilst more evidently still it explains the circle which the north pole of the earth, in the course of the earth round the sun, describes round the polar star; and the performing of this circle round the polar star is a NECESSARY CONSEQUENCE of the path and motion of the earth, the

gotten conclusion, that when the ' primitive impulse ' was imparted, the earth was held exactly twenty-five miles from the centre, and hence, quoth Laplace, the earth revolves upon her axis once in twenty-four hours. If she had been held a little nearer to the centre, our days would have been longer, and if a little farther off, she would have revolved with greater velocity, and our days would have been shorter. . . . . . . These assumptions comprehend the very essence of what is called the Newtonian philosophy, and the world has been gravely assured that every proposition has been rigidly demonstrated."

With singular infelicity the gyroscope has been made use of, not only to prove this absurdity, but even to improve upon it.and assign to the Creator also the spot at which He must have stood when He " hurled the earth into space."

I am sure the gyroscopians cannot have deeply reflected upon their favourite illustration, otherwise they would have seen this, and above all, that the instrument represents nothing else but the stone in the sling, with this difference, that in the one instance, the " primitive impulse," that famous discovery, proceeds from the person wielding the string, and in the other it is imparted by a " Prime Mover," standing outside and independent of the object to be set in motion.

The gyroscope, moreover, represents the earth rotating upon an HORIZONTAL axis by the side of the sun, instead of spinning perpendicularly around him ; and therefore it is also an inappropriate illustration.

However, the turning-point of the theory is the " primitive impulse," that famous discovery,—and its " centrifugal assistant ;" but, not resting on any solid basis, Newton was obliged to assume, the world has unhesitatingly believed him, and Sir John Herschel still echoes it in *Good Words* that the heavenly bodies move in VACANT SPACE ! That there is a VACANT CREATION ! And yet

many of those who believe in this vacant space, cannot believe in vacant space outside the ball of creation, cannot believe that the exterior of creation is filled by God alone from eternity to eternity. But, allow a substance ever so fine to pervade the universe, and THE FOUNDATION OF THE OLD THEORY IS REMOVED, AND THE WHOLE EDIFICE CRUMBLES TO PIECES. It is the giant-cripple of universal gravitation, limping upon the crutch of empty space, driven about by an unparalleled primitive impulse of the imagination, and, as yet, kept from tumbling, by the kind concurrence of, in this instance, real astronomical forces. For, that an invisible ether fills all creation is now not only believed by the many as a matter of course, but has also been demonstrated by eminent men.

After this, the old theory of the rotation of the earth round its own axis, and in its orbit round the sun, must strike every one still more as arbitrary, forced, unnatural and untenable, whilst my own propositions could not be more simple and reasonable, though, perhaps, not without some flaw or other. And if on the old theory, certain astronomical calculations have proved tolerably correct, it is because my new theory very little disturbs the present arbitrary adoption of the inclination of the ecliptic, the position of the earth in its annual course round the sun; for, if the axis of the earth, at the time of the equinox, were taken in the exact perpendicular instead of the incline of 23½ degrees to the plane of its orbit, the whole theory, in this respect, would almost exactly coincide with my own, according to which, the oscillation of the earth in its rotary course round the sun, in a most simple manner, explains the so-called nutation of the earth's axis, the actual and only cause of the seasons at the same time, whilst more evidently still it explains the circle which the north pole of the earth, in the course of the earth round the sun, describes round the polar star; and the performing of this circle round the polar star is a NECESSARY CONSEQUENCE of the path and motion of the earth, the

gotten conclusion, that when the ' primitive impulse ' was imparted, the earth was held exactly twenty-five miles from the centre, and hence, quoth Laplace, the earth revolves upon her axis once in twenty-four hours. If she had been held a little nearer to the centre, our days would have been longer, and if a little farther off, she would have revolved with greater velocity, and our days would have been shorter. . . . . . . These assumptions compre-hend the very essence of what is called the Newtonian philosophy, and the world has been gravely assured that every proposition has been rigidly demonstrated."

With singular infelicity the gyroscope has been made use of, not only to prove this absurdity, but even to improve upon it and assign to the Creator also the spot at which He must have stood when He " hurled the earth into space."

I am sure the gyroscopians cannot have deeply reflected upon their favourite illustration, otherwise they would have seen this, and above all, that the instrument represents nothing else but the stone in the sling, with this difference, that in the one instance, the " primitive impulse," that famous discovery, proceeds from the person wielding the string, and in the other it is imparted by a " Prime Mover," standing outside and independent of the object to be set in motion.

The gyroscope, moreover, represents the earth rotating upon an HORIZONTAL axis by the side of the sun, instead of spinning perpendicularly around him ; and therefore it is also an inappro-priate illustration.

However, the turning-point of the theory is the " primitive impulse," that famous discovery,—and its " centrifugal assistant ;" but, not resting on any solid basis, Newton was obliged to assume, the world has unhesitatingly believed him, and Sir John Herschel still echoes it in *Good Words* that the heavenly bodies move in VACANT SPACE ! That there is a VACANT CREATION ! And yet

many of those who believe in this vacant space, cannot believe in vacant space outside the ball of creation, cannot believe that the exterior of creation is filled by God alone from eternity to eternity. But, allow a substance ever so fine to pervade the universe, and THE FOUNDATION OF THE OLD THEORY IS REMOVED, AND THE WHOLE EDIFICE CRUMBLES TO PIECES. It is the giant-cripple of universal gravitation, limping upon the crutch of empty space, driven about by an unparalleled primitive impulse of the imagination, and, as yet, kept from tumbling, by the kind concurrence of, in this instance, real astronomical forces. For, that an invisible ether fills all creation is now not only believed by the many as a matter of course, but has also been demonstrated by eminent men.

After this, the old theory of the rotation of the earth round its own axis, and in its orbit round the sun, must strike every one still more as arbitrary, forced, unnatural and untenable, whilst my own propositions could not be more simple and reasonable, though, perhaps, not without some flaw or other. And if on the old theory, certain astronomical calculations have proved tolerably correct, it is because my new theory very little disturbs the present arbitrary adoption of the inclination of the ecliptic, the position of the earth in its annual course round the sun; for, if the axis of the earth, at the time of the equinox, were taken in the exact perpendicular instead of the incline of $23\frac{1}{2}$ degrees to the plane of its orbit, the whole theory, in this respect, would almost exactly coincide with my own, according to which, the oscillation of the earth in its rotary course round the sun, in a most simple manner, explains the so-called nutation of the earth's axis, the actual and only cause of the seasons at the same time, whilst more evidently still it explains the circle which the north pole of the earth, in the course of the earth round the sun, describes round the polar star; and the performing of this circle round the polar star is a NECESSARY CONSEQUENCE of the path and motion of the earth, the

gotten conclusion, that when the ' primitive impulse' was imparted, the earth was held exactly twenty-five miles from the centre, and hence, quoth Laplace, the earth revolves upon her axis once in twenty-four hours. If she had been held a little nearer to the centre, our days would have been longer, and if a little farther off, she would have revolved with greater velocity, and our days would have been shorter. . . . . . . These assumptions comprehend the very essence of what is called the Newtonian philosophy, and the world has been gravely assured that every proposition has been rigidly demonstrated."

With singular infelicity the gyroscope has been made use of, not only to prove this absurdity, but even to improve upon it and assign to the Creator also the spot at which He must have stood when He "hurled the earth into space."

I am sure the gyroscopians cannot have deeply reflected upon their favourite illustration, otherwise they would have seen this, and above all, that the instrument represents nothing else but the stone in the sling, with this difference, that in the one instance, the " primitive impulse," that famous discovery, proceeds from the person wielding the string, and in the other it is imparted by a " Prime Mover," standing outside and independent of the object to be set in motion.

The gyroscope, moreover, represents the earth rotating upon an HORIZONTAL axis by the side of the sun, instead of spinning perpendicularly around him ; and therefore it is also an inappropriate illustration.

However, the turning-point of the theory is the "primitive impulse," that famous discovery,—and its "centrifugal assistant ;" but, not resting on any solid basis, Newton was obliged to assume, the world has unhesitatingly believed him, and Sir John Herschel still echoes it in *Good Words* that the heavenly bodies move in VACANT SPACE ! That there is a VACANT CREATION ! And yet

many of those who believe in this vacant space, cannot believe in vacant space outside the ball of creation, cannot believe that the exterior of creation is filled by God alone from eternity to eternity. But, allow a substance ever so fine to pervade the universe, and THE FOUNDATION OF THE OLD THEORY IS REMOVED, AND THE WHOLE EDIFICE CRUMBLES TO PIECES. It is the giant-cripple of universal gravitation, limping upon the crutch of empty space, driven about by an unparalleled primitive impulse of the imagination, and, as yet, kept from tumbling, by the kind concurrence of, in this instance, real astronomical forces. For, that an invisible ether fills all creation is now not only believed by the many as a matter of course, but has also been demonstrated by eminent men.

After this, the old theory of the rotation of the earth round its own axis, and in its orbit round the sun, must strike every one still more as arbitrary, forced, unnatural and untenable, whilst my own propositions could not be more simple and reasonable, though, perhaps, not without some flaw or other. And if on the old theory, certain astronomical calculations have proved tolerably correct, it is because my new theory very little disturbs the present arbitrary adoption of the inclination of the ecliptic, the position of the earth in its annual course round the sun; for, if the axis of the earth, at the time of the equinox, were taken in the exact perpendicular instead of the incline of 23½ degrees to the plane of its orbit, the whole theory, in this respect, would almost exactly coincide with my own, according to which, the oscillation of the earth in its rotary course round the sun, in a most simple manner, explains the so-called nutation of the earth's axis, the actual and only cause of the seasons at the same time, whilst more evidently still it explains the circle which the north pole of the earth, in the course of the earth round the sun, describes round the polar star; and the performing of this circle round the polar star is a NECESSARY CONSEQUENCE of the path and motion of the earth, the

gotten conclusion, that when the ' primitive impulse' was imparted, the earth was held exactly twenty-five miles from the centre, and hence, quoth Laplace, the earth revolves upon her axis once in twenty-four hours. If she had been held a little nearer to the centre, our days would have been longer, and if a little farther off, she would have revolved with greater velocity, and our days would have been shorter. . . . . . . These assumptions comprehend the very essence of what is called the Newtonian philosophy, and the world has been gravely assured that every proposition has been rigidly demonstrated."

With singular infelicity the gyroscope has been made use of, not only to prove this absurdity, but even to improve upon it and assign to the Creator also the spot at which He must have stood when He "hurled the earth into space."

I am sure the gyroscopians cannot have deeply reflected upon their favourite illustration, otherwise they would have seen this, and above all, that the instrument represents nothing else but the stone in the sling, with this difference, that in the one instance, the " primitive impulse," that famous discovery, proceeds from the person wielding the string, and in the other it is imparted by a " Prime Mover," standing outside and independent of the object to be set in motion.

The gyroscope, moreover, represents the earth rotating upon an HORIZONTAL axis by the side of the sun, instead of spinning perpendicularly around him; and therefore it is also an inappropriate illustration.

However, the turning-point of the theory is the " primitive impulse," that famous discovery,—and its " centrifugal assistant;" but, not resting on any solid basis, Newton was obliged to assume, the world has unhesitatingly believed him, and Sir John Herschel still echoes it in *Good Words* that the heavenly bodies move in VACANT SPACE! That there is a VACANT CREATION! And yet

many of those who believe in this vacant space, cannot believe in vacant space outside the ball of creation, cannot believe that the exterior of creation is filled by God alone from eternity to eternity. But, allow a substance ever so fine to pervade the universe, and THE FOUNDATION OF THE OLD THEORY IS REMOVED, AND THE WHOLE EDIFICE CRUMBLES TO PIECES. It is the giant-cripple of universal gravitation, limping upon the crutch of empty space, driven about by an unparalleled primitive impulse of the imagination, and, as yet, kept from tumbling, by the kind concurrence of, in this instance, real astronomical forces. For, that an invisible ether fills all creation is now not only believed by the many as a matter of course, but has also been demonstrated by eminent men.

After this, the old theory of the rotation of the earth round its own axis, and in its orbit round the sun, must strike every one still more as arbitrary, forced, unnatural and untenable, whilst my own propositions could not be more simple and reasonable, though, perhaps, not without some flaw or other. And if on the old theory, certain astronomical calculations have proved tolerably correct, it is because my new theory very little disturbs the present arbitrary adoption of the inclination of the ecliptic, the position of the earth in its annual course round the sun; for, if the axis of the earth, at the time of the equinox, were taken in the exact perpendicular instead of the incline of $23\frac{1}{2}$ degrees to the plane of its orbit, the whole theory, in this respect, would almost exactly coincide with my own, according to which, the oscillation of the earth in its rotary course round the sun, in a most simple manner, explains the so-called nutation of the earth's axis, the actual and only cause of the seasons at the same time, whilst more evidently still it explains the circle which the north pole of the earth, in the course of the earth round the sun, describes round the polar star; and the performing of this circle round the polar star is a NECES-SARY CONSEQUENCE of the path and motion of the earth, the

gotten conclusion, that when the 'primitive impulse' was imparted, the earth was held exactly twenty-five miles from the centre, and hence, quoth Laplace, the earth revolves upon her axis once in twenty-four hours. If she had been held a little nearer to the centre, our days would have been longer, and if a little farther off, she would have revolved with greater velocity, and our days would have been shorter. . . . . . . These assumptions comprehend the very essence of what is called the Newtonian philosophy, and the world has been gravely assured that every proposition has been rigidly demonstrated."

With singular infelicity the gyroscope has been made use of, not only to prove this absurdity, but even to improve upon it and assign to the Creator also the spot at which He must have stood when He "hurled the earth into space."

I am sure the gyroscopians cannot have deeply reflected upon their favourite illustration, otherwise they would have seen this, and above all, that the instrument represents nothing else but the stone in the sling, with this difference, that in the one instance, the "primitive impulse," that famous discovery, proceeds from the person wielding the string, and in the other it is imparted by a "Prime Mover," standing outside and independent of the object to be set in motion.

The gyroscope, moreover, represents the earth rotating upon an HORIZONTAL axis by the side of the sun, instead of spinning perpendicularly around him; and therefore it is also an inappropriate illustration.

However, the turning-point of the theory is the "primitive impulse," that famous discovery,—and its "centrifugal assistant;" but, not resting on any solid basis, Newton was obliged to assume, the world has unhesitatingly believed him, and Sir John Herschel still echoes it in *Good Words* that the heavenly bodies move in VACANT SPACE! That there is a VACANT CREATION! And yet

many of those who believe in this vacant space, cannot believe in vacant space outside the ball of creation, cannot believe that the exterior of creation is filled by God alone from eternity to eternity. But, allow a substance ever so fine to pervade the universe, and THE FOUNDATION OF THE OLD THEORY IS REMOVED, AND THE WHOLE EDIFICE CRUMBLES TO PIECES. It is the giant-cripple of universal gravitation, limping upon the crutch of empty space, driven about by an unparalleled primitive impulse of the imagination, and, as yet, kept from tumbling, by the kind concurrence of, in this instance, real astronomical forces. For, that an invisible ether fills all creation is now not only believed by the many as a matter of course, but has also been demonstrated by eminent men.

After this, the old theory of the rotation of the earth round its own axis, and in its orbit round the sun, must strike every one still more as arbitrary, forced, unnatural and untenable, whilst my own propositions could not be more simple and reasonable, though, perhaps, not without some flaw or other. And if on the old theory, certain astronomical calculations have proved tolerably correct, it is because my new theory very little disturbs the present arbitrary adoption of the inclination of the ecliptic, the position of the earth in its annual course round the sun; for, if the axis of the earth, at the time of the equinox, were taken in the exact perpendicular instead of the incline of 23½ degrees to the plane of its orbit, the whole theory, in this respect, would almost exactly coincide with my own, according to which, the oscillation of the earth in its rotary course round the sun, in a most simple manner, explains the so-called nutation of the earth's axis, the actual and only cause of the seasons at the same time, whilst more evidently still it explains the circle which the north pole of the earth, in the course of the earth round the sun, describes round the polar star; and the performing of this circle round the polar star is a NECESSARY CONSEQUENCE of the path and motion of the earth, the

gotten conclusion, that when the ' primitive impulse ' was imparted, the earth was held exactly twenty-five miles from the centre, and hence, quoth Laplace, the earth revolves upon her axis once in twenty-four hours. If she had been held a little nearer to the centre, our days would have been longer, and if a little farther off, she would have revolved with greater velocity, and our days would have been shorter. . . . . . . These assumptions comprehend the very essence of what is called the Newtonian philosophy, and the world has been gravely assured that every proposition has been rigidly demonstrated."

With singular infelicity the gyroscope has been made use of, not only to prove this absurdity, but even to improve upon it and assign to the Creator also the spot at which He must have stood when He " hurled the earth into space."

I am sure the gyroscopians cannot have deeply reflected upon their favourite illustration, otherwise they would have seen this, and above all, that the instrument represents nothing else but the stone in the sling, with this difference, that in the one instance, the " primitive impulse," that famous discovery, proceeds from the person wielding the string, and in the other it is imparted by a " Prime Mover," standing outside and independent of the object to be set in motion.

The gyroscope, moreover, represents the earth rotating upon an HORIZONTAL axis by the side of the sun, instead of spinning perpendicularly around him ; and therefore it is also an inappropriate illustration.

However, the turning-point of the theory is the " primitive impulse," that famous discovery,—and its " centrifugal assistant ;" but, not resting on any solid basis, Newton was obliged to assume, the world has unhesitatingly believed him, and Sir John Herschel still echoes it in *Good Words* that the heavenly bodies move in VACANT SPACE ! That there is a VACANT CREATION ! And yet

many of those who believe in this vacant space, cannot believe in vacant space outside the ball of creation, cannot believe that the exterior of creation is filled by God alone from eternity to eternity. But, allow a substance ever so fine to pervade the universe, and THE FOUNDATION OF THE OLD THEORY IS REMOVED, AND THE WHOLE EDIFICE CRUMBLES TO PIECES. It is the giant-cripple of universal gravitation, limping upon the crutch of empty space, driven about by an unparalleled primitive impulse of the imagination, and, as yet, kept from tumbling, by the kind concurrence of, in this instance, real astronomical forces. For, that an invisible ether fills all creation is now not only believed by the many as a matter of course, but has also been demonstrated by eminent men.

After this, the old theory of the rotation of the earth round its own axis, and in its orbit round the sun, must strike every one still more as arbitrary, forced, unnatural and untenable, whilst my own propositions could not be more simple and reasonable, though, perhaps, not without some flaw or other. And if on the old theory, certain astronomical calculations have proved tolerably correct, it is because my new theory very little disturbs the present arbitrary adoption of the inclination of the ecliptic, the position of the earth in its annual course round the sun; for, if the axis of the earth, at the time of the equinox, were taken in the exact perpendicular instead of the incline of $23\frac{1}{2}$ degrees to the plane of its orbit, the whole theory, in this respect, would almost exactly coincide with my own, according to which, the oscillation of the earth in its rotary course round the sun, in a most simple manner, explains the so-called nutation of the earth's axis, the actual and only cause of the seasons at the same time, whilst more evidently still it explains the circle which the north pole of the earth, in the course of the earth round the sun, describes round the polar star; and the performing of this circle round the polar star is a NECESSARY CONSEQUENCE of the path and motion of the earth, the

gotten conclusion, that when the ' primitive impulse' was imparted, the earth was held exactly twenty-five miles from the centre, and hence, quoth Laplace, the earth revolves upon her axis once in twenty-four hours. If she had been held a little nearer to the centre, our days would have been longer, and if a little farther off, she would have revolved with greater velocity, and our days would have been shorter. . . . . . . These assumptions comprehend the very essence of what is called the Newtonian philosophy, and the world has been gravely assured that every proposition has been rigidly demonstrated."

With singular infelicity the gyroscope has been made use of, not only to prove this absurdity, but even to improve upon it and assign to the Creator also the spot at which He must have stood when He "hurled the earth into space."

I am sure the gyroscopians cannot have deeply reflected upon their favourite illustration, otherwise they would have seen this, and above all, that the instrument represents nothing else but the stone in the sling, with this difference, that in the one instance, the "primitive impulse," that famous discovery, proceeds from the person wielding the string, and in the other it is imparted by a "Prime Mover," standing outside and independent of the object to be set in motion.

The gyroscope, moreover, represents the earth rotating upon an HORIZONTAL axis by the side of the sun, instead of spinning perpendicularly around him; and therefore it is also an inappropriate illustration.

However, the turning-point of the theory is the "primitive impulse," that famous discovery,—and its "centrifugal assistant;" but, not resting on any solid basis, Newton was obliged to assume, the world has unhesitatingly believed him, and Sir John Herschel still echoes it in *Good Words* that the heavenly bodies move in VACANT SPACE! That there is a VACANT CREATION! And yet

many of those who believe in this vacant space, cannot believe in vacant space outside the ball of creation, cannot believe that the exterior of creation is filled by God alone from eternity to eternity. But, allow a substance ever so fine to pervade the universe, and THE FOUNDATION OF THE OLD THEORY IS REMOVED, AND THE WHOLE EDIFICE CRUMBLES TO PIECES. It is the giant-cripple of universal gravitation, limping upon the crutch of empty space, driven about by an unparalleled primitive impulse of the imagination, and, as yet, kept from tumbling, by the kind concurrence of, in this instance, real astronomical forces. For, that an invisible ether fills all creation is now not only believed by the many as a matter of course, but has also been demonstrated by eminent men.

After this, the old theory of the rotation of the earth round its own axis, and in its orbit round the sun, must strike every one still more as arbitrary, forced, unnatural and untenable, whilst my own propositions could not be more simple and reasonable, though, perhaps, not without some flaw or other. And if on the old theory, certain astronomical calculations have proved tolerably correct, it is because my new theory very little disturbs the present arbitrary adoption of the inclination of the ecliptic, the position of the earth in its annual course round the sun; for, if the axis of the earth, at the time of the equinox, were taken in the exact perpendicular instead of the incline of 23½ degrees to the plane of its orbit, the whole theory, in this respect, would almost exactly coincide with my own, according to which, the oscillation of the earth in its rotary course round the sun, in a most simple manner, explains the so-called nutation of the earth's axis, the actual and only cause of the seasons at the same time, whilst more evidently still it explains the circle which the north pole of the earth, in the course of the earth round the sun, describes round the polar star; and the performing of this circle round the polar star is a NECESSARY CONSEQUENCE of the path and motion of the earth, the

same as its orbital course and oscillation brings us nearer to it at one time than at the other.

The slight difference in the axial position of the earth above stated as between the old and my new theory, accounting for the tolerable correctness of some, and the inaccuracy and absolute falsity of many, astronomical calculations, is of the greatest importance in the determination of eclipses both backward and forward in time.

Without entering into details I may say, that Von Gumpach, in his *Baby Worlds*, adduces the most positive evidence of the solar eclipses of Thales, Xerxes, Agathocles, and others, being HISTORICALLY true and correct; that the central part of the moon's shadow has been CALCULATED to have swept over portions of the earth over which, according to HISTORICAL records, it could not have passed, neither as to time or place; that, therefore, these and other historically-verified eclipses are not represented even in the latest tables of the sun and moon, nor the THEORETICALLY-CALCULATED eclipses any where found in the pages of history.

Owing to these serious discrepancies between historical and theoretical eclipses, Professor Airy, as Von Gumpach tells us, was induced to go so far as to substitute an obscuration of the moon for the solar eclipse of Xerxes HISTORICALLY CERTAIN. "Indeed," says the same author, "a comparison of the calculated and the observed times, connected with the various places of the total eclipse of July 18th, 1860, referred to different geographical positions, proves to evidence what LITTLE TRUST can be placed in them and in the theory on which they rest, even as applied to the present period." And yet astronomers "call upon our credulity to give them credit for learning; and upon our faith in science, to prefer ASTRONOMICAL THEORIES TO HISTORICAL FACTS."

It appears, according to *Baby Worlds*, that the miraculous retrogradation of the shadow on the sun-dial of Ahaz has been tried to be explained by the occurrence of a solar eclipse nowhere recorded, as if, notwithstanding the solar light, God could not make the shadow go back by other means at his command, whether by superseding the light of the sun, as by itself our earthly fires are subdued and superseded, or otherwise, or whether only to the sight of those concerned.

When in the French revolution the servants and children of the Church, of the God of Revelation, like wild beasts were hunted by the ministers, satellites and abettors of the Goddess of Reason, by the apostles of liberty, fraternity and equality, Abbé Colmar, of Strasbourg, later Bishop of Mayence, regardless of his life, was carrying the consolations of the Church in every direction.

One day he found himself so closely pursued that he had scarcely time to run into his sister's house to which he happened to be near. Hurriedly saying the "Memorare" of St. Bernard, recommending himself to the protection of the Mother of God, he entered the room where the family were at dinner, and with as much composure as possible sat down to table. The monsters in human form were at his heels. "Where is Abbé Colmar?" they cried. "If you want to know, look for him," was the sister's reply. They searched the whole house from top to bottom, but all in vain; they did not find him, and in fury quitted the house.

When they had left, one of the children, about four or five years of age, said to her mother: "My Dear Mamma, who was that beautiful Lady who covered uncle with her veil whilst the men were here?"—The INNOCENT CHILD ALONE had seen it, ALONE had been made the unimpeachable witness of God's power and protection.[*]

By the Holy Ghost, God made the Apostles, and numbers of

---

[*] "Out of the mouth of infants and sucklings Thou hast perfected praise, because of Thy enemies, that Thou mayest destroy the enemy and the avenger." *Psalm* viii. 3.

their LEGITIMATE successors—BUT NO OTHERS—speak in foreign tongues which they never knew before; and in the case of St. Vincent Ferrerius who, travelling through France, Italy, Germany, England, and into Ireland, everywhere preaching in his native Valencian dialect, yet—as in Genoa, where Greeks, Hungarians, Germans, Sardinians and others listened to him—He made all of them hear as if in their own tongue, all understood him alike, and moreover, the furthest as well as the nearest.

Thus, in the former instances, God manifested himself by the sense of seeing, and in the latter ones by the sense of hearing.

When we see these things, and more extraordinary phenomena still in the life of the Church, we may well believe the miracle wrought by Isaiah, though there is nothing wrong in trying—as the Church always does by a strict and searching process un-exampled—to find out the true character of an event we cannot fathom. In this way it is that she establishes whether there is imposition, the ordinary course of nature, or that which supersedes, goes beyond, the law of nature, a miracle.

It is inconsistent, however, and all the more reprehensible and culpable where there is a profession of faith, if Von Gumpach—though without intention I hope—makes the Holy Scripture a book of lies, by representing Isaiah the prophet as a most clumsy rogue and impostor, and Hezekiah, the king of Judah, with probably many witnesses, as nothing less than a short-sighted idiot.

In a more ASTRONOMICAL way, so VERY LIKELY, the star of the Magi is, by Von Gumpach and others, turned into a conjunction of Jupiter and Saturn !

As if He, who allows ships and other objects and pictures by refraction to appear high in the air, and Who sets the bow in the clouds, could not have at His command one of those wonderful temporary, or variable, stars, long since disappeared and at once bursting out again in light upon our vision, and from the stable of

Bethlehem reflect the lovely image of the Virgin-Mother and Child upon its radiant disc, direct to it the searching eyes of the three WISE kings longing and looking for the Star of Jacob, and by its bright connecting link between heaven and earth guiding them, as WE still are cheeringly guided, from the image to the living reality.

And are there no other stars, no meteors, no other creations, to be taken into the service of the Almighty? and for such a purpose?

When in all those cases where the supernatural law of God goes hand in hand with, or supersedes, the law of nature, learned men attempt to find scientific substitutes and think themselves so clever, they inevitably make fools of themselves and ridiculous to the REFLECTING CHRISTIAN mind; but when impositions and frauds are alleged and related of AUTHENTICATED miracles, they simply become absurd; and if true: these alleged deceptions both in object and duration, as in the case of the blood of St. Januarius, would be far greater miracles than the miracles themselves. And yet, in almost the very words of the above author: these anti-miracle-mongers, miracles themselves, "call upon our credulity to give them credit for learning," where in such matters there is neither knowledge nor common sense; "and upon our faith in science to prefer astronomical theories," and irrational and imbecile tales and explanations " TO HISTORICAL FACTS."

That these things should be so is in itself one of the greatest proofs how the law of grace and justice go hand-in-hand with the law of nature; for, persons speaking against the miraculous testimony of the Church instituted by Christ himself—not by vicious princes and parliaments, or abandoned monks or ecclesiastics,—have the "understanding darkened by the pride or prejudice which is within them," reason is eclipsed, and ENTIRELY FAILS in things divine; when sun and moon are obscured nothing is seen so distinctly.

R

their LEGITIMATE successors—BUT NO OTHERS—speak in foreign tongues which they never knew before; and in the case of St. Vincent Ferrerius who, travelling through France, Italy, Germany, England, and into Ireland, everywhere preaching in his native Valencian dialect, yet—as in Genoa, where Greeks, Hungarians, Germans, Sardinians and others listened to him—He made all of them hear as if in their own tongue, àll understood him alike, and moreover, the furthest as well as the nearest.

Thus, in the former instances, God manifested himself by the sense of seeing, and in the latter ones by the sense of hearing.

When we see these things, and more extraordinary phenomena still in the life of the Church, we may well believe the miracle wrought by Isaiah, though there is nothing wrong in trying—as the Church always does by a strict and searching process un-exampled—to find out the true character of an event we cannot fathom. In this way it is that she establishes whether there is imposition, the ordinary course of nature, or that which supersedes, goes beyond, the law of nature, a miracle.

It is inconsistent, however, and all the more reprehensible and culpable where there is a profession of faith, if Von Gumpach—though without intention I hope—makes the Holy Scripture a book of lies, by representing Isaiah the prophet as a most clumsy rogue and impostor, and Hezekiah, the king of Judah, with probably many witnesses, as nothing less than a short-sighted idiot.

In a more ASTRONOMICAL way, so VERY LIKELY, the star of the Magi is, by Von Gumpach and others, turned into a conjunction of Jupiter and Saturn !

As if He, who allows ships and other objects and pictures by refraction to appear high in the air, and Who sets the bow in the clouds, could not have at His command one of those wonderful temporary, or variable, stars, long since disappeared and at once bursting out again in light upon our vision, and from the stable of

Bethlehem reflect the lovely image of the Virgin-Mother and Child upon its radiant disc, direct to it the searching eyes of the three WISE kings longing and looking for the Star of Jacob, and by its bright connecting link between heaven and earth guiding them, as WE still are cheeringly guided, from the image to the living reality.

And are there no other stars, no meteors, no other creations, to be taken into the service of the Almighty? and for such a purpose?

When in all those cases where the supernatural law of God goes hand in hand with, or supersedes, the law of nature, learned men attempt to find scientific substitutes and think themselves so clever, they inevitably make fools of themselves and ridiculous to the REFLECTING CHRISTIAN mind; but when impositions and frauds are alleged and related of AUTHENTICATED miracles, they simply become absurd; and if true: these alleged deceptions both in object and duration, as in the case of the blood of St. Januarius, would be far greater miracles than the miracles themselves. And yet, in almost the very words of the above author: these anti-miracle-mongers, miracles themselves, "call upon our credulity to give them credit for learning," where in such matters there is neither knowledge nor common sense; "and upon our faith in science to prefer astronomical theories," and irrational and imbecile tales and explanations "TO HISTORICAL FACTS."

That these things should be so is in itself one of the greatest proofs how the law of grace and justice go hand-in-hand with the law of nature; for, persons speaking against the miraculous testimony of the Church instituted by Christ himself—not by vicious princes and parliaments, or abandoned monks or ecclesiastics,—have the "understanding darkened by the pride or prejudice which is within them," reason is eclipsed, and ENTIRELY FAILS in things divine; when sun and moon are obscured nothing is seen so distinctly.

R

their LEGITIMATE successors—BUT NO OTHERS—speak in foreign tongues which they never knew before ; and in the case of St. Vincent Ferrerius who, travelling through France, Italy, Germany, England, and into Ireland, everywhere preaching in his native Valencian dialect, yet—as in Genoa, where Greeks, Hungarians, Germans, Sardinians and others listened to him—He made all of them hear as if in their own tongue, àll understood him alike, and moreover, the furthest as well as the nearest.

Thus, in the former instances, God manifested himself by the sense of seeing, and in the latter ones by the sense of hearing.

When we see these things, and more extraordinary phenomena still in the life of the Church, we may well believe the miracle wrought by Isaiah, though there is nothing wrong in trying—as the Church always does by a strict and searching process un-exampled—to find out the true character of an event we cannot fathom. In this way it is that she establishes whether there is imposition, the ordinary course of nature, or that which supersedes, goes beyond, the law of nature, a miracle.

It is inconsistent, however, and all the more reprehensible and culpable where there is a profession of faith, if Von Gumpach—though without intention I hope—makes the Holy Scripture a book of lies, by representing Isaiah the prophet as a most clumsy rogue and impostor, and Hezekiah, the king of Judah, with probably many witnesses, as nothing less than a short-sighted idiot.

In a more ASTRONOMICAL way, SO VERY LIKELY, the star of the Magi is, by Von Gumpach and others, turned into a conjunction of Jupiter and Saturn !

As if He, who allows ships and other objects and pictures by refraction to appear high in the air, and Who sets the bow in the clouds, could not have at His command one of those wonderful temporary, or variable, stars, long since disappeared and at once bursting out again in light upon our vision, and from the stable of

Bethlehem reflect the lovely image of the Virgin-Mother and Child upon its radiant disc, direct to it the searching eyes of the three WISE kings longing and looking for the Star of Jacob, and by its bright connecting link between heaven and earth guiding them, as WE still are cheeringly guided, from the image to the living reality.

And are there no other stars, no meteors, no other creations, to be taken into the service of the Almighty ? and for such a purpose ?

When in all those cases where the supernatural law of God goes hand in hand with, or supersedes, the law of nature, learned men attempt to find scientific substitutes and think themselves so clever, they inevitably make fools of themselves and ridiculous to the REFLECTING CHRISTIAN mind; but when impositions and frauds are alleged and related of AUTHENTICATED miracles, they simply become absurd; and if true: these alleged deceptions both in object and duration, as in the case of the blood of St. Januarius, would be far greater miracles than the miracles themselves. And yet, in almost the very words of the above author: these anti-miracle-mongers, miracles themselves, "call upon our credulity to give them credit for learning," where in such matters there is neither knowledge nor common sense; "and upon our faith in science to prefer astronomical theories," and irrational and imbecile tales and explanations " TO HISTORICAL FACTS."

That these things should be so is in itself one of the greatest proofs how the law of grace and justice go hand-in-hand with the law of nature; for, persons speaking against the miraculous testimony of the Church instituted by Christ himself—not by vicious princes and parliaments, or abandoned monks or ecclesiastics,—have the " understanding darkened by the pride or prejudice which is within them," reason is eclipsed, and ENTIRELY FAILS in things divine; when sun and moon are obscured nothing is seen so distinctly.

R

their LEGITIMATE successors—BUT NO OTHERS—speak in foreign tongues which they never knew before; and in the case of St. Vincent Ferrerius who, travelling through France, Italy, Germany, England, and into Ireland, everywhere preaching in his native Valencian dialect, yet—as in Genoa, where Greeks, Hungarians, Germans, Sardinians and others listened to him—He made all of them hear as if in their own tongue, all understood him alike, and moreover, the furthest as well as the nearest.

Thus, in the former instances, God manifested himself by the sense of seeing, and in the latter ones by the sense of hearing.

When we see these things, and more extraordinary phenomena still in the life of the Church, we may well believe the miracle wrought by Isaiah, though there is nothing wrong in trying—as the Church always does by a strict and searching process un-exampled—to find out the true character of an event we cannot fathom. In this way it is that she establishes whether there is imposition, the ordinary course of nature, or that which supersedes, goes beyond, the law of nature, a miracle.

It is inconsistent, however, and all the more reprehensible and culpable where there is a profession of faith, if Von Gumpach—though without intention I hope—makes the Holy Scripture a book of lies, by representing Isaiah the prophet as a most clumsy rogue and impostor, and Hezekiah, the king of Judah, with probably many witnesses, as nothing less than a short-sighted idiot.

In a more ASTRONOMICAL way, so VERY LIKELY, the star of the Magi is, by Von Gumpach and others, turned into a conjunction of Jupiter and Saturn !

As if He, who allows ships and other objects and pictures by refraction to appear high in the air, and Who sets the bow in the clouds, could not have at His command one of those wonderful temporary, or variable, stars, long since disappeared and at once bursting out again in light upon our vision, and from the stable of

Bethlehem reflect the lovely image of the Virgin-Mother and Child upon its radiant disc, direct to it the searching eyes of the three WISE kings longing and looking for the Star of Jacob, and by its bright connecting link between heaven and earth guiding them, as WE still are cheeringly guided, from the image to the living reality.

And are there no other stars, no meteors, no other creations, to be taken into the service of the Almighty? and for such a purpose?

When in all those cases where the supernatural law of God goes hand in hand with, or supersedes, the law of nature, learned men attempt to find scientific substitutes and think themselves so clever, they inevitably make fools of themselves and ridiculous to the REFLECTING CHRISTIAN mind; but when impositions and frauds are alleged and related of AUTHENTICATED miracles, they simply become absurd; and if true: these alleged deceptions both in object and duration, as in the case of the blood of St. Januarius, would be far greater miracles than the miracles themselves. And yet, in almost the very words of the above author: these anti-miracle-mongers, miracles themselves, "call upon our credulity to give them credit for learning," where in such matters there is neither knowledge nor common sense; "and upon our faith in science to prefer astronomical theories," and irrational and imbecile tales and explanations "TO HISTORICAL FACTS."

That these things should be so is in itself one of the greatest proofs how the law of grace and justice go hand-in-hand with the law of nature; for, persons speaking against the miraculous testimony of the Church instituted by Christ himself—not by vicious princes and parliaments, or abandoned monks or ecclesiastics,—have the "understanding darkened by the pride or prejudice which is within them," reason is eclipsed, and ENTIRELY FAILS in things divine; when sun and moon are obscured nothing is seen so distinctly.

R

their LEGITIMATE successors—BUT NO OTHERS—speak in foreign
tongues which they never knew before; and in the case of St.
Vincent Ferrerius who, travelling through France, Italy, Germany,
England, and into Ireland, everywhere preaching in his native
Valencian dialect, yet—as in Genoa, where Greeks, Hungarians,
Germans, Sardinians and others listened to him—He made all of
them hear as if in their own tongue, all understood him alike, and
moreover, the furthest as well as the nearest.

Thus, in the former instances, God manifested himself by the
sense of seeing, and in the latter ones by the sense of hearing.

When we see these things, and more extraordinary phenomena
still in the life of the Church, we may well believe the miracle
wrought by Isaiah, though there is nothing wrong in trying—as
the Church always does by a strict and searching process un-
exampled—to find out the true character of an event we cannot
fathom. In this way it is that she establishes whether there is
imposition, the ordinary course of nature, or that which supersedes,
goes beyond, the law of nature, a miracle.

It is inconsistent, however, and all the more reprehensible and
culpable where there is a profession of faith, if Von Gumpach—
though without intention I hope—makes the Holy Scripture a
book of lies, by representing Isaiah the prophet as a most clumsy
rogue and impostor, and Hezekiah, the king of Judah, with probably
many witnesses, as nothing less than a short-sighted idiot.

In a more ASTRONOMICAL way, so VERY LIKELY, the star of the
Magi is, by Von Gumpach and others, turned into a conjunction
of Jupiter and Saturn !

As if He, who allows ships and other objects and pictures by
refraction to appear high in the air, and Who sets the bow in the
clouds, could not have at His command one of those wonderful
temporary, or variable, stars, long since disappeared and at once
bursting out again in light upon our vision, and from the stable of

Bethlehem reflect the lovely image of the Virgin-Mother and Child upon its radiant disc, direct to it the searching eyes of the three WISE kings longing and looking for the Star of Jacob, and by its bright connecting link between heaven and earth guiding them, as WE still are cheeringly guided, from the image to the living reality.

And are there no other stars, no meteors, no other creations, to be taken into the service of the Almighty ? and for such a purpose ?

When in all those cases where the supernatural law of God goes hand in hand with, or supersedes, the law of nature, learned men attempt to find scientific substitutes and think themselves so clever, they inevitably make fools of themselves and ridiculous to the REFLECTING CHRISTIAN mind; but when impositions and frauds are alleged and related of AUTHENTICATED miracles, they simply become absurd; and if true: these alleged deceptions both in object and duration, as in the case of the blood of St. Januarius, would be far greater miracles than the miracles themselves. And yet, in almost the very words of the above author: these anti-miracle-mongers, miracles themselves, "call upon our credulity to give them credit for learning," where in such matters there is neither knowledge nor common sense; "and upon our faith in science to prefer astronomical theories," and irrational and imbecile tales and explanations "TO HISTORICAL FACTS."

That these things should be so is in itself one of the greatest proofs how the law of grace and justice go hand-in-hand with the law of nature; for, persons speaking against the miraculous testimony of the Church instituted by Christ himself—not by vicious princes and parliaments, or abandoned monks or ecclesiastics,—have the "understanding darkened by the pride or prejudice which is within them," reason is eclipsed, and ENTIRELY FAILS in things divine; when sun and moon are obscured nothing is seen so distinctly.

R

their LEGITIMATE successors—BUT NO OTHERS—speak in foreign tongues which they never knew before ; and in the case of St. Vincent Ferrerius who, travelling through France, Italy, Germany, England, and into Ireland, everywhere preaching in his native Valencian dialect, yet—as in Genoa, where Greeks, Hungarians, Germans, Sardinians and others listened to him—He made all of them hear as if in their own tongue, all understood him alike, and moreover, the furthest as well as the nearest.

Thus, in the former instances, God manifested himself by the sense of seeing, and in the latter ones by the sense of hearing.

When we see these things, and more extraordinary phenomena still in the life of the Church, we may well believe the miracle wrought by Isaiah, though there is nothing wrong in trying—as the Church always does by a strict and searching process un-exampled—to find out the true character of an event we cannot fathom. In this way it is that she establishes whether there is imposition, the ordinary course of nature, or that which supersedes, goes beyond, the law of nature, a miracle.

It is inconsistent, however, and all the more reprehensible and culpable where there is a profession of faith, if Von Gumpach—though without intention I hope—makes the Holy Scripture a book of lies, by representing Isaiah the prophet as a most clumsy rogue and impostor, and Hezekiah, the king of Judah, with probably many witnesses, as nothing less than a short-sighted idiot.

In a more ASTRONOMICAL way, so VERY LIKELY, the star of the Magi is, by Von Gumpach and others, turned into a conjunction of Jupiter and Saturn !

As if He, who allows ships and other objects and pictures by refraction to appear high in the air, and Who sets the bow in the clouds, could not have at His command one of those wonderful temporary, or variable, stars, long since disappeared and at once bursting out again in light upon our vision, and from the stable of

Bethlehem reflect the lovely image of the Virgin-Mother and Child upon its radiant disc, direct to it the searching eyes of the three WISE kings longing and looking for the Star of Jacob, and by its bright connecting link between heaven and earth guiding them, as WE still are cheeringly guided, from the image to the living reality.

And are there no other stars, no meteors, no other creations, to be taken into the service of the Almighty ? and for such a purpose ?

When in all those cases where the supernatural law of God goes hand in hand with, or supersedes, the law of nature, learned men attempt to find scientific substitutes and think themselves so clever, they inevitably make fools of themselves and ridiculous to the REFLECTING CHRISTIAN mind; but when impositions and frauds are alleged and related of AUTHENTICATED miracles, they simply become absurd; and if true: these alleged deceptions both in object and duration, as in the case of the blood of St. Januarius, would be far greater miracles than the miracles themselves. And yet, in almost the very words of the above author: these anti-miracle-mongers, miracles themselves, " call upon our credulity to give them credit for learning," where in such matters there is neither knowledge nor common sense; "and upon our faith in science to prefer astronomical theories," and irrational and imbecile tales and explanations " TO HISTORICAL FACTS."

That these things should be so is in itself one of the greatest proofs how the law of grace and justice go hand-in-hand with the law of nature; for, persons speaking against the miraculous testimony of the Church instituted by Christ himself—not by vicious princes and parliaments, or abandoned monks or ecclesiastics,—have the " understanding darkened by the pride or prejudice which is within them," reason is eclipsed, and ENTIRELY FAILS in things divine; when sun and moon are obscured nothing is seen so distinctly.

their LEGITIMATE successors—BUT NO OTHERS—speak in foreign tongues which they never knew before; and in the case of St. Vincent Ferrerius who, travelling through France, Italy, Germany, England, and into Ireland, everywhere preaching in his native Valencian dialect, yet—as in Genoa, where Greeks, Hungarians, Germans, Sardinians and others listened to him—He made all of them hear as if in their own tongue, all understood him alike, and moreover, the furthest as well as the nearest.

Thus, in the former instances, God manifested himself by the sense of seeing, and in the latter ones by the sense of hearing.

When we see these things, and more extraordinary phenomena still in the life of the Church, we may well believe the miracle wrought by Isaiah, though there is nothing wrong in trying—as the Church always does by a strict and searching process unexampled—to find out the true character of an event we cannot fathom. In this way it is that she establishes whether there is imposition, the ordinary course of nature, or that which supersedes, goes beyond, the law of nature, a miracle.

It is inconsistent, however, and all the more reprehensible and culpable where there is a profession of faith, if Von Gumpach—though without intention I hope—makes the Holy Scripture a book of lies, by representing Isaiah the prophet as a most clumsy rogue and impostor, and Hezekiah, the king of Judah, with probably many witnesses, as nothing less than a short-sighted idiot.

In a more ASTRONOMICAL way, so VERY LIKELY, the star of the Magi is, by Von Gumpach and others, turned into a conjunction of Jupiter and Saturn !

As if He, who allows ships and other objects and pictures by refraction to appear high in the air, and Who sets the bow in the clouds, could not have at His command one of those wonderful temporary, or variable, stars, long since disappeared and at once bursting out again in light upon our vision, and from the stable of

Bethlehem reflect the lovely image of the Virgin-Mother and Child upon its radiant disc, direct to it the searching eyes of the three WISE kings longing and looking for the Star of Jacob, and by its bright connecting link between heaven and earth guiding them, as WE still are cheeringly guided, from the image to the living reality.

And are there no other stars, no meteors, no other creations, to be taken into the service of the Almighty ? and for such a purpose ?

When in all those cases where the supernatural law of God goes hand in hand with, or supersedes, the law of nature, learned men attempt to find scientific substitutes and think themselves so clever, they inevitably make fools of themselves and ridiculous to the REFLECTING CHRISTIAN mind; but when impositions and frauds are alleged and related of AUTHENTICATED miracles, they simply become absurd; and if true: these alleged deceptions both in object and duration, as in the case of the blood of St. Januarius, would be far greater miracles than the miracles themselves. And yet, in almost the very words of the above author: these anti-miracle-mongers, miracles themselves, "call upon our credulity to give them credit for learning," where in such matters there is neither knowledge nor common sense; "and upon our faith in science to prefer astronomical theories," and irrational and imbecile tales and explanations "TO HISTORICAL FACTS."

That these things should be so is in itself one of the greatest proofs how the law of grace and justice go hand-in-hand with the law of nature; for, persons speaking against the miraculous testimony of the Church instituted by Christ himself—not by vicious princes and parliaments, or abandoned monks or ecclesiastics,—have the "understanding darkened by the pride or prejudice which is within them," reason is eclipsed, and ENTIRELY FAILS in things divine; when sun and moon are obscured nothing is seen so distinctly.

R

their LEGITIMATE successors—BUT NO OTHERS—speak in foreign tongues which they never knew before; and in the case of St. Vincent Ferrerius who, travelling through France, Italy, Germany, England, and into Ireland, everywhere preaching in his native Valencian dialect, yet—as in Genoa, where Greeks, Hungarians, Germans, Sardinians and others listened to him—He made all of them hear as if in their own tongue, all understood him alike, and moreover, the furthest as well as the nearest.

Thus, in the former instances, God manifested himself by the sense of seeing, and in the latter ones by the sense of hearing.

When we see these things, and more extraordinary phenomena still in the life of the Church, we may well believe the miracle wrought by Isaiah, though there is nothing wrong in trying—as the Church always does by a strict and searching process un-exampled—to find out the true character of an event we cannot fathom. In this way it is that she establishes whether there is imposition, the ordinary course of nature, or that which supersedes, goes beyond, the law of nature, a miracle.

It is inconsistent, however, and all the more reprehensible and culpable where there is a profession of faith, if Von Gumpach—though without intention I hope—makes the Holy Scripture a book of lies, by representing Isaiah the prophet as a most clumsy rogue and impostor, and Hezekiah, the king of Judah, with probably many witnesses, as nothing less than a short-sighted idiot.

In a more ASTRONOMICAL way, so VERY LIKELY, the star of the Magi is, by Von Gumpach and others, turned into a conjunction of Jupiter and Saturn !

As if He, who allows ships and other objects and pictures by refraction to appear high in the air, and Who sets the bow in the clouds, could not have at His command one of those wonderful temporary, or variable, stars, long since disappeared and at once bursting out again in light upon our vision, and from the stable of

Bethlehem reflect the lovely image of the Virgin-Mother and Child upon its radiant disc, direct to it the searching eyes of the three WISE kings longing and looking for the Star of Jacob, and by its bright connecting link between heaven and earth guiding them, as WE still are cheeringly guided, from the image to the living reality.

And are there no other stars, no meteors, no other creations, to be taken into the service of the Almighty ? and for such a purpose ?

When in all those cases where the supernatural law of God goes hand in hand with, or supersedes, the law of nature, learned men attempt to find scientific substitutes and think themselves so clever, they inevitably make fools of themselves and ridiculous to the REFLECTING CHRISTIAN mind; but when impositions and frauds are alleged and related of AUTHENTICATED miracles, they simply become absurd; and if true: these alleged deceptions both in object and duration, as in the case of the blood of St. Januarius, would be far greater miracles than the miracles themselves. And yet, in almost the very words of the above author: these anti-miracle-mongers, miracles themselves, "call upon our credulity to give them credit for learning," where in such matters there is neither knowledge nor common sense; "and upon our faith in science to prefer astronomical theories," and irrational and imbecile tales and explanations "TO HISTORICAL FACTS."

That these things should be so is in itself one of the greatest proofs how the law of grace and justice go hand-in-hand with the law of nature; for, persons speaking against the miraculous testimony of the Church instituted by Christ himself—not by vicious princes and parliaments, or abandoned monks or eccle-siastics,—have the "understanding darkened by the pride or prejudice which is within them," reason is eclipsed, and ENTIRELY FAILS in things divine; when sun and moon are obscured nothing is seen so distinctly.

R

their LEGITIMATE successors—BUT NO OTHERS—speak in foreign tongues which they never knew before; and in the case of St. Vincent Ferrerius who, travelling through France, Italy, Germany, England, and into Ireland, everywhere preaching in his native Valencian dialect, yet—as in Genoa, where Greeks, Hungarians, Germans, Sardinians and others listened to him—He made all of them hear as if in their own tongue, àll understood him alike, and moreover, the furthest as well as the nearest.

Thus, in the former instances, God manifested himself by the sense of seeing, and in the latter ones by the sense of hearing.

When we see these things, and more extraordinary phenomena still in the life of the Church, we may well believe the miracle wrought by Isaiah, though there is nothing wrong in trying—as the Church always does by a strict and searching process un-exampled—to find out the true character of an event we cannot fathom. In this way it is that she establishes whether there is imposition, the ordinary course of nature, or that which supersedes, goes beyond, the law of nature, a miracle.

It is inconsistent, however, and all the more reprehensible and culpable where there is a profession of faith, if Von Gumpach—though without intention I hope—makes the Holy Scripture a book of lies, by representing Isaiah the prophet as a most clumsy rogue and impostor, and Hezekiah, the king of Judah, with probably many witnesses, as nothing less than a short-sighted idiot.

In a more ASTRONOMICAL way, so VERY LIKELY, the star of the Magi is, by Von Gumpach and others, turned into a conjunction of Jupiter and Saturn !

As if He, who allows ships and other objects and pictures by refraction to appear high in the air, and Who sets the bow in the clouds, could not have at His command one of those wonderful temporary, or variable, stars, long since disappeared and at once bursting out again in light upon our vision, and from the stable of

Bethlehem reflect the lovely image of the Virgin-Mother and Child upon its radiant disc, direct to it the searching eyes of the three WISE kings longing and looking for the Star of Jacob, and by its bright connecting link between heaven and earth guiding them, as WE still are cheeringly guided, from the image to the living reality.

And are there no other stars, no meteors, no other creations, to be taken into the service of the Almighty? and for such a purpose?

When in all those cases where the supernatural law of God goes hand in hand with, or supersedes, the law of nature, learned men attempt to find scientific substitutes and think themselves so clever, they inevitably make fools of themselves and ridiculous to the REFLECTING CHRISTIAN mind; but when impositions and frauds are alleged and related of AUTHENTICATED miracles, they simply become absurd; and if true: these alleged deceptions both in object and duration, as in the case of the blood of St. Januarius, would be far greater miracles than the miracles themselves. And yet, in almost the very words of the above author: these anti-miracle-mongers, miracles themselves, "call upon our credulity to give them credit for learning," where in such matters there is neither knowledge nor common sense; "and upon our faith in science to prefer astronomical theories," and irrational and imbecile tales and explanations " TO HISTORICAL FACTS."

That these things should be so is in itself one of the greatest proofs how the law of grace and justice go hand-in-hand with the law of nature; for, persons speaking against the miraculous testimony of the Church instituted by Christ himself—not by vicious princes and parliaments, or abandoned monks or ecclesiastics,—have the " understanding darkened by the pride or prejudice which is within them," reason is eclipsed, and ENTIRELY FAILS in things divine; when sun and moon are obscured nothing is seen so distinctly.

R

their LEGITIMATE successors—BUT NO OTHERS—speak in foreign tongues which they never knew before; and in the case of St. Vincent Ferrerius who, travelling through France, Italy, Germany, England, and into Ireland, everywhere preaching in his native Valencian dialect, yet—as in Genoa, where Greeks, Hungarians, Germans, Sardinians and others listened to him—He made all of them hear as if in their own tongue, all understood him alike, and moreover, the furthest as well as the nearest.

Thus, in the former instances, God manifested himself by the sense of seeing, and in the latter ones by the sense of hearing.

When we see these things, and more extraordinary phenomena still in the life of the Church, we may well believe the miracle wrought by Isaiah, though there is nothing wrong in trying—as the Church always does by a strict and searching process un-exampled—to find out the true character of an event we cannot fathom. In this way it is that she establishes whether there is imposition, the ordinary course of nature, or that which supersedes, goes beyond, the law of nature, a miracle.

It is inconsistent, however, and all the more reprehensible and culpable where there is a profession of faith, if Von Gumpach—though without intention I hope—makes the Holy Scripture a book of lies, by representing Isaiah the prophet as a most clumsy rogue and impostor, and Hezekiah, the king of Judah, with probably many witnesses, as nothing less than a short-sighted idiot.

In a more ASTRONOMICAL way, so VERY LIKELY, the star of the Magi is, by Von Gumpach and others, turned into a conjunction of Jupiter and Saturn !

As if He, who allows ships and other objects and pictures by refraction to appear high in the air, and Who sets the bow in the clouds, could not have at His command one of those wonderful temporary, or variable, stars, long since disappeared and at once bursting out again in light upon our vision, and from the stable of

Bethlehem reflect the lovely image of the Virgin-Mother and Child upon its radiant disc, direct to it the searching eyes of the three WISE kings longing and looking for the Star of Jacob, and by its bright connecting link between heaven and earth guiding them, as WE still are cheeringly guided, from the image to the living reality.

And are there no other stars, no meteors, no other creations, to be taken into the service of the Almighty ? and for such a purpose ?

When in all those cases where the supernatural law of God goes hand in hand with, or supersedes, the law of nature, learned men attempt to find scientific substitutes and think themselves so clever, they inevitably make fools of themselves and ridiculous to the REFLECTING CHRISTIAN mind ; but when impositions and frauds are alleged and related of AUTHENTICATED miracles, they simply become absurd ; and if true : these alleged deceptions both in object and duration, as in the case of the blood of St. Januarius, would be far greater miracles than the miracles themselves. And yet, in almost the very words of the above author : these anti-miracle-mongers, miracles themselves, " call upon our credulity to give them credit for learning," where in such matters there is neither knowledge nor common sense ; " and upon our faith in science to prefer astronomical theories," and irrational and imbecile tales and explanations " TO HISTORICAL FACTS."

That these things should be so is in itself one of the greatest proofs how the law of grace and justice go hand-in-hand with the law of nature ; for, persons speaking against the miraculous testimony of the Church instituted by Christ himself—not by vicious princes and parliaments, or abandoned monks or ecclesiastics,—have the " understanding darkened by the pride or prejudice which is within them," reason is eclipsed, and ENTIRELY FAILS in things divine ; when sun and moon are obscured nothing is seen so distinctly.

R

their LEGITIMATE successors—BUT NO OTHERS—speak in foreign
tongues which they never knew before; and in the case of St.
Vincent Ferrerius who, travelling through France, Italy, Germany,
England, and into Ireland, everywhere preaching in his native
Valencian dialect, yet—as in Genoa, where Greeks, Hungarians,
Germans, Sardinians and others listened to him—He made all of
them hear as if in their own tongue, àll understood him alike, and
moreover, the furthest as well as the nearest.

Thus, in the former instances, God manifested himself by the
sense of seeing, and in the latter ones by the sense of hearing.

When we see these things, and more extraordinary phenomena
still in the life of the Church, we may well believe the miracle
wrought by Isaiah, though there is nothing wrong in trying—as
the Church always does by a strict and searching process un-
exampled—to find out the true character of an event we cannot
fathom.   In this way it is that she establishes whether there is
imposition, the ordinary course of nature, or that which supersedes,
goes beyond, the law of nature, a miracle.

It is inconsistent, however, and all the more reprehensible and
culpable where there is a profession of faith, if Von Gumpach—
though without intention I hope—makes the Holy Scripture a
book of lies, by representing Isaiah the prophet as a most clumsy
rogue and impostor, and Hezekiah, the king of Judah, with probably
many witnesses, as nothing less than a short-sighted idiot.

In a more ASTRONOMICAL way, so VERY LIKELY, the star of the
Magi is, by Von Gumpach and others, turned into a conjunction
of Jupiter and Saturn !

As if He, who allows ships and other objects and pictures by
refraction to appear high in the air, and Who sets the bow in the
clouds, could not have at His command one of those wonderful
temporary, or variable, stars, long since disappeared and at once
bursting out again in light upon our vision, and from the stable of

Bethlehem reflect the lovely image of the Virgin-Mother and Child upon its radiant disc, direct to it the searching eyes of the three WISE kings longing and looking for the Star of Jacob, and by its bright connecting link between heaven and earth guiding them, as WE still are cheeringly guided, from the image to the living reality.

And are there no other stars, no meteors, no other creations, to be taken into the service of the Almighty ? and for such a purpose ?

When in all those cases where the supernatural law of God goes hand in hand with, or supersedes, the law of nature, learned men attempt to find scientific substitutes and think themselves so clever, they inevitably make fools of themselves and ridiculous to the REFLECTING CHRISTIAN mind; but when impositions and frauds are alleged and related of AUTHENTICATED miracles, they simply become absurd; and if true : these alleged deceptions both in object and duration, as in the case of the blood of St. Januarius, would be far greater miracles than the miracles themselves. And yet, in almost the very words of the above author : these anti-miracle-mongers, miracles themselves, "call upon our credulity to give them credit for learning," where in such matters there is neither knowledge nor common sense; "and upon our faith in science to prefer astronomical theories," and irrational and imbecile tales and explanations " TO HISTORICAL FACTS."

That these things should be so is in itself one of the greatest proofs how the law of grace and justice go hand-in-hand with the law of nature; for, persons speaking against the miraculous testimony of the Church instituted by Christ himself—not by vicious princes and parliaments, or abandoned monks or ecclesiastics,—have the " understanding darkened by the pride or prejudice which is within them," reason is eclipsed, and ENTIRELY FAILS in things divine ; when sun and moon are obscured nothing is seen so distinctly.

R

their LEGITIMATE successors—BUT NO OTHERS—speak in foreign tongues which they never knew before; and in the case of St. Vincent Ferrerius who, travelling through France, Italy, Germany, England, and into Ireland, everywhere preaching in his native Valencian dialect, yet—as in Genoa, where Greeks, Hungarians, Germans, Sardinians and others listened to him—He made all of them hear as if in their own tongue, àll understood him alike, and moreover, the furthest as well as the nearest.

Thus, in the former instances, God manifested himself by the sense of seeing, and in the latter ones by the sense of hearing.

When we see these things, and more extraordinary phenomena still in the life of the Church, we may well believe the miracle wrought by Isaiah, though there is nothing wrong in trying—as the Church always does by a strict and searching process un-exampled—to find out the true character of an event we cannot fathom. In this way it is that she establishes whether there is imposition, the ordinary course of nature, or that which supersedes, goes beyond, the law of nature, a miracle.

It is inconsistent, however, and all the more reprehensible and culpable where there is a profession of faith, if Von Gumpach—though without intention I hope—makes the Holy Scripture a book of lies, by representing Isaiah the prophet as a most clumsy rogue and impostor, and Hezekiah, the king of Judah, with probably many witnesses, as nothing less than a short-sighted idiot.

In a more ASTRONOMICAL way, so VERY LIKELY, the star of the Magi is, by Von Gumpach and others, turned into a conjunction of Jupiter and Saturn !

As if He, who allows ships and other objects and pictures by refraction to appear high in the air, and Who sets the bow in the clouds, could not have at His command one of those wonderful temporary, or variable, stars, long since disappeared and at once bursting out again in light upon our vision, and from the stable of

Bethlehem reflect the lovely image of the Virgin-Mother and Child upon its radiant disc, direct to it the searching eyes of the three WISE kings longing and looking for the Star of Jacob, and by its bright connecting link between heaven and earth guiding them, as WE still are cheeringly guided, from the image to the living reality.

And are there no other stars, no meteors, no other creations, to be taken into the service of the Almighty ? and for such a purpose ?

When in all those cases where the supernatural law of God goes hand in hand with, or supersedes, the law of nature, learned men attempt to find scientific substitutes and think themselves so clever, they inevitably make fools of themselves and ridiculous to the REFLECTING CHRISTIAN mind; but when impositions and frauds are alleged and related of AUTHENTICATED miracles, they simply become absurd; and if true: these alleged deceptions both in object and duration, as in the case of the blood of St. Januarius, would be far greater miracles than the miracles themselves. And yet, in almost the very words of the above author: these anti-miracle-mongers, miracles themselves, " call upon our credulity to give them credit for learning," where in such matters there is neither knowledge nor common sense; " and upon our faith in science to prefer astronomical theories," and irrational and imbecile tales and explanations " TO HISTORICAL FACTS."

That these things should be so is in itself one of the greatest proofs how the law of grace and justice go hand-in-hand with the law of nature; for, persons speaking against the miraculous testimony of the Church instituted by Christ himself—not by vicious princes and parliaments, or abandoned monks or ecclesiastics,—have the " understanding darkened by the pride or prejudice which is within them," reason is eclipsed, and ENTIRELY FAILS in things divine ; when sun and moon are obscured nothing is seen so distinctly.

their LEGITIMATE successors—BUT NO OTHERS—speak in foreign tongues which they never knew before; and in the case of St. Vincent Ferrerius who, travelling through France, Italy, Germany, England, and into Ireland, everywhere preaching in his native Valencian dialect, yet—as in Genoa, where Greeks, Hungarians, Germans, Sardinians and others listened to him—He made all of them hear as if in their own tongue, all understood him alike, and moreover, the furthest as well as the nearest.

Thus, in the former instances, God manifested himself by the sense of seeing, and in the latter ones by the sense of hearing.

When we see these things, and more extraordinary phenomena still in the life of the Church, we may well believe the miracle wrought by Isaiah, though there is nothing wrong in trying—as the Church always does by a strict and searching process un-exampled—to find out the true character of an event we cannot fathom. In this way it is that she establishes whether there is imposition, the ordinary course of nature, or that which supersedes, goes beyond, the law of nature, a miracle.

It is inconsistent, however, and all the more reprehensible and culpable where there is a profession of faith, if Von Gumpach—though without intention I hope—makes the Holy Scripture a book of lies, by representing Isaiah the prophet as a most clumsy rogue and impostor, and Hezekiah, the king of Judah, with probably many witnesses, as nothing less than a short-sighted idiot.

In a more ASTRONOMICAL way, so VERY LIKELY, the star of the Magi is, by Von Gumpach and others, turned into a conjunction of Jupiter and Saturn !

As if He, who allows ships and other objects and pictures by refraction to appear high in the air, and Who sets the bow in the clouds, could not have at His command one of those wonderful temporary, or variable, stars, long since disappeared and at once bursting out again in light upon our vision, and from the stable of

Bethlehem reflect the lovely image of the Virgin-Mother and Child upon its radiant disc, direct to it the searching eyes of the three WISE kings longing and looking for the Star of Jacob, and by its bright connecting link between heaven and earth guiding them, as WE still are cheeringly guided, from the image to the living reality.

And are there no other stars, no meteors, no other creations, to be taken into the service of the Almighty ? and for such a purpose ?

When in all those cases where the supernatural law of God goes hand in hand with, or supersedes, the law of nature, learned men attempt to find scientific substitutes and think themselves so clever, they inevitably make fools of themselves and ridiculous to the REFLECTING CHRISTIAN mind; but when impositions and frauds are alleged and related of AUTHENTICATED miracles, they simply become absurd; and if true: these alleged deceptions both in object and duration, as in the case of the blood of St. Januarius, would be far greater miracles than the miracles themselves. And yet, in almost the very words of the above author: these anti-miracle-mongers, miracles themselves, "call upon our credulity to give them credit for learning," where in such matters there is neither knowledge nor common sense; "and upon our faith in science to prefer astronomical theories," and irrational and imbecile tales and explanations " TO HISTORICAL FACTS."

That these things should be so is in itself one of the greatest proofs how the law of grace and justice go hand-in-hand with the law of nature; for, persons speaking against the miraculous testimony of the Church instituted by Christ himself—not by vicious princes and parliaments, or abandoned monks or ecclesiastics,—have the "understanding darkened by the pride or prejudice which is within them," reason is eclipsed, and ENTIRELY FAILS in things divine; when sun and moon are obscured nothing is seen so distinctly.

R

their LEGITIMATE successors—BUT NO OTHERS—speak in foreign tongues which they never knew before; and in the case of St. Vincent Ferrerius who, travelling through France, Italy, Germany, England, and into Ireland, everywhere preaching in his native Valencian dialect, yet—as in Genoa, where Greeks, Hungarians, Germans, Sardinians and others listened to him—He made all of them hear as if in their own tongue, all understood him alike, and moreover, the furthest as well as the nearest.

Thus, in the former instances, God manifested himself by the sense of seeing, and in the latter ones by the sense of hearing.

When we see these things, and more extraordinary phenomena still in the life of the Church, we may well believe the miracle wrought by Isaiah, though there is nothing wrong in trying—as the Church always does by a strict and searching process unexampled—to find out the true character of an event we cannot fathom. In this way it is that she establishes whether there is imposition, the ordinary course of nature, or that which supersedes, goes beyond, the law of nature, a miracle.

It is inconsistent, however, and all the more reprehensible and culpable where there is a profession of faith, if Von Gumpach—though without intention I hope—makes the Holy Scripture a book of lies, by representing Isaiah the prophet as a most clumsy rogue and impostor, and Hezekiah, the king of Judah, with probably many witnesses, as nothing less than a short-sighted idiot.

In a more ASTRONOMICAL way, so VERY LIKELY, the star of the Magi is, by Von Gumpach and others, turned into a conjunction of Jupiter and Saturn !

As if He, who allows ships and other objects and pictures by refraction to appear high in the air, and Who sets the bow in the clouds, could not have at His command one of those wonderful temporary, or variable, stars, long since disappeared and at once bursting out again in light upon our vision, and from the stable of

Bethlehem reflect the lovely image of the Virgin-Mother and Child upon its radiant disc, direct to it the searching eyes of the three WISE kings longing and looking for the Star of Jacob, and by its bright connecting link between heaven and earth guiding them, as WE still are cheeringly guided, from the image to the living reality.

And are there no other stars, no meteors, no other creations, to be taken into the service of the Almighty? and for such a purpose?

When in all those cases where the supernatural law of God goes hand in hand with, or supersedes, the law of nature, learned men attempt to find scientific substitutes and think themselves so clever, they inevitably make fools of themselves and ridiculous to the REFLECTING CHRISTIAN mind; but when impositions and frauds are alleged and related of AUTHENTICATED miracles, they simply become absurd; and if true: these alleged deceptions both in object and duration, as in the case of the blood of St. Januarius, would be far greater miracles than the miracles themselves. And yet, in almost the very words of the above author: these anti-miracle-mongers, miracles themselves, "call upon our credulity to give them credit for learning," where in such matters there is neither knowledge nor common sense; "and upon our faith in science to prefer astronomical theories," and irrational and imbecile tales and explanations "TO HISTORICAL FACTS."

That these things should be so is in itself one of the greatest proofs how the law of grace and justice go hand-in-hand with the law of nature; for, persons speaking against the miraculous testimony of the Church instituted by Christ himself—not by vicious princes and parliaments, or abandoned monks or ecclesiastics,—have the "understanding darkened by the pride or prejudice which is within them," reason is eclipsed, and ENTIRELY FAILS in things divine; when sun and moon are obscured nothing is seen so distinctly.

R

their LEGITIMATE successors—BUT NO OTHERS—speak in foreign tongues which they never knew before; and in the case of St. Vincent Ferrerius who, travelling through France, Italy, Germany, England, and into Ireland, everywhere preaching in his native Valencian dialect, yet—as in Genoa, where Greeks, Hungarians, Germans, Sardinians and others listened to him—He made all of them hear as if in their own tongue, àll understood him alike, and moreover, the furthest as well as the nearest.

Thus, in the former instances, God manifested himself by the sense of seeing, and in the latter ones by the sense of hearing.

When we see these things, and more extraordinary phenomena still in the life of the Church, we may well believe the miracle wrought by Isaiah, though there is nothing wrong in trying—as the Church always does by a strict and searching process un-exampled—to find out the true character of an event we cannot fathom. In this way it is that she establishes whether there is imposition, the ordinary course of nature, or that which supersedes, goes beyond, the law of nature, a miracle.

It is inconsistent, however, and all the more reprehensible and culpable where there is a profession of faith, if Von Gumpach—though without intention I hope—makes the Holy Scripture a book of lies, by representing Isaiah the prophet as a most clumsy rogue and impostor, and Hezekiah, the king of Judah, with probably many witnesses, as nothing less than a short-sighted idiot.

In a more ASTRONOMICAL way, so VERY LIKELY, the star of the Magi is, by Von Gumpach and others, turned into a conjunction of Jupiter and Saturn !

As if He, who allows ships and other objects and pictures by refraction to appear high in the air, and Who sets the bow in the clouds, could not have at His command one of those wonderful temporary, or variable, stars, long since disappeared and at once bursting out again in light upon our vision, and from the stable of

Bethlehem reflect the lovely image of the Virgin-Mother and Child upon its radiant disc, direct to it the searching eyes of the three WISE kings longing and looking for the Star of Jacob, and by its bright connecting link between heaven and earth guiding them, as WE still are cheeringly guided, from the image to the living reality.

And are there no other stars, no meteors, no other creations, to be taken into the service of the Almighty? and for such a purpose?

When in all those cases where the supernatural law of God goes hand in hand with, or supersedes, the law of nature, learned men attempt to find scientific substitutes and think themselves so clever, they inevitably make fools of themselves and ridiculous to the REFLECTING CHRISTIAN mind; but when impositions and frauds are alleged and related of AUTHENTICATED miracles, they simply become absurd; and if true: these alleged deceptions both in object and duration, as in the case of the blood of St. Januarius, would be far greater miracles than the miracles themselves. And yet, in almost the very words of the above author: these anti-miracle-mongers, miracles themselves, "call upon our credulity to give them credit for learning," where in such matters there is neither knowledge nor common sense; "and upon our faith in science to prefer astronomical theories," and irrational and imbecile tales and explanations "TO HISTORICAL FACTS."

That these things should be so is in itself one of the greatest proofs how the law of grace and justice go hand-in-hand with the law of nature; for, persons speaking against the miraculous testimony of the Church instituted by Christ himself—not by vicious princes and parliaments, or abandoned monks or ecclesiastics,—have the "understanding darkened by the pride or prejudice which is within them," reason is eclipsed, and ENTIRELY FAILS in things divine; when sun and moon are obscured nothing is seen so distinctly.

R

their LEGITIMATE successors—BUT NO OTHERS—speak in foreign tongues which they never knew before; and in the case of St. Vincent Ferrerius who, travelling through France, Italy, Germany, England, and into Ireland, everywhere preaching in his native Valencian dialect, yet—as in Genoa, where Greeks, Hungarians, Germans, Sardinians and others listened to him—He made all of them hear as if in their own tongue, all understood him alike, and moreover, the furthest as well as the nearest.

Thus, in the former instances, God manifested himself by the sense of seeing, and in the latter ones by the sense of hearing.

When we see these things, and more extraordinary phenomena still in the life of the Church, we may well believe the miracle wrought by Isaiah, though there is nothing wrong in trying—as the Church always does by a strict and searching process un-exampled—to find out the true character of an event we cannot fathom. In this way it is that she establishes whether there is imposition, the ordinary course of nature, or that which supersedes, goes beyond, the law of nature, a miracle.

It is inconsistent, however, and all the more reprehensible and culpable where there is a profession of faith, if Von Gumpach—though without intention I hope—makes the Holy Scripture a book of lies, by representing Isaiah the prophet as a most clumsy rogue and impostor, and Hezekiah, the king of Judah, with probably many witnesses, as nothing less than a short-sighted idiot.

In a more ASTRONOMICAL way, so VERY LIKELY, the star of the Magi is, by Von Gumpach and others, turned into a conjunction of Jupiter and Saturn !

As if He, who allows ships and other objects and pictures by refraction to appear high in the air, and Who sets the bow in the clouds, could not have at His command one of those wonderful temporary, or variable, stars, long since disappeared and at once bursting out again in light upon our vision, and from the stable of

Bethlehem reflect the lovely image of the Virgin-Mother and Child upon its radiant disc, direct to it the searching eyes of the three WISE kings longing and looking for the Star of Jacob, and by its bright connecting link between heaven and earth guiding them, as WE still are cheeringly guided, from the image to the living reality.

And are there no other stars, no meteors, no other creations, to be taken into the service of the Almighty? and for such a purpose?

When in all those cases where the supernatural law of God goes hand in hand with, or supersedes, the law of nature, learned men attempt to find scientific substitutes and think themselves so clever, they inevitably make fools of themselves and ridiculous to the REFLECTING CHRISTIAN mind; but when impositions and frauds are alleged and related of AUTHENTICATED miracles, they simply become absurd; and if true: these alleged deceptions both in object and duration, as in the case of the blood of St. Januarius, would be far greater miracles than the miracles themselves. And yet, in almost the very words of the above author: these anti-miracle-mongers, miracles themselves, "call upon our credulity to give them credit for learning," where in such matters there is neither knowledge nor common sense; "and upon our faith in science to prefer astronomical theories," and irrational and imbecile tales and explanations "TO HISTORICAL FACTS."

That these things should be so is in itself one of the greatest proofs how the law of grace and justice go hand-in-hand with the law of nature; for, persons speaking against the miraculous testimony of the Church instituted by Christ himself—not by vicious princes and parliaments, or abandoned monks or ecclesiastics,—have the "understanding darkened by the pride or prejudice which is within them," reason is eclipsed, and ENTIRELY FAILS in things divine; when sun and moon are obscured nothing is seen so distinctly.

their LEGITIMATE successors—BUT NO OTHERS—speak in foreign tongues which they never knew before; and in the case of St. Vincent Ferrerius who, travelling through France, Italy, Germany, England, and into Ireland, everywhere preaching in his native Valencian dialect, yet—as in Genoa, where Greeks, Hungarians, Germans, Sardinians and others listened to him—He made all of them hear as if in their own tongue, all understood him alike, and moreover, the furthest as well as the nearest.

Thus, in the former instances, God manifested himself by the sense of seeing, and in the latter ones by the sense of hearing.

When we see these things, and more extraordinary phenomena still in the life of the Church, we may well believe the miracle wrought by Isaiah, though there is nothing wrong in trying—as the Church always does by a strict and searching process unexampled—to find out the true character of an event we cannot fathom. In this way it is that she establishes whether there is imposition, the ordinary course of nature, or that which supersedes, goes beyond, the law of nature, a miracle.

It is inconsistent, however, and all the more reprehensible and culpable where there is a profession of faith, if Von Gumpach—though without intention I hope—makes the Holy Scripture a book of lies, by representing Isaiah the prophet as a most clumsy rogue and impostor, and Hezekiah, the king of Judah, with probably many witnesses, as nothing less than a short-sighted idiot.

In a more ASTRONOMICAL way, so VERY LIKELY, the star of the Magi is, by Von Gumpach and others, turned into a conjunction of Jupiter and Saturn !

As if He, who allows ships and other objects and pictures by refraction to appear high in the air, and Who sets the bow in the clouds, could not have at His command one of those wonderful temporary, or variable, stars, long since disappeared and at once bursting out again in light upon our vision, and from the stable of

Bethlehem reflect the lovely image of the Virgin-Mother and Child upon its radiant disc, direct to it the searching eyes of the three WISE kings longing and looking for the Star of Jacob, and by its bright connecting link between heaven and earth guiding them, as WE still are cheeringly guided, from the image to the living reality.

And are there no other stars, no meteors, no other creations, to be taken into the service of the Almighty ? and for such a purpose ?

When in all those cases where the supernatural law of God goes hand in hand with, or supersedes, the law of nature, learned men attempt to find scientific substitutes and think themselves so clever, they inevitably make fools of themselves and ridiculous to the REFLECTING CHRISTIAN mind ; but when impositions and frauds are alleged and related of AUTHENTICATED miracles, they simply become absurd ; and if true : these alleged deceptions both in object and duration, as in the case of the blood of St. Januarius, would be far greater miracles than the miracles themselves. And yet, in almost the very words of the above author : these anti-miracle-mongers, miracles themselves, "call upon our credulity to give them credit for learning," where in such matters there is neither knowledge nor common sense ; " and upon our faith in science to prefer astronomical theories," and irrational and imbecile tales and explanations " TO HISTORICAL FACTS."

That these things should be so is in itself one of the greatest proofs how the law of grace and justice go hand-in-hand with the law of nature ; for, persons speaking against the miraculous testimony of the Church instituted by Christ himself—not by vicious princes and parliaments, or abandoned monks or ecclesiastics,—have the " understanding darkened by the pride or prejudice which is within them," reason is eclipsed, and ENTIRELY FAILS in things divine ; when sun and moon are obscured nothing is seen so distinctly.

R

their LEGITIMATE successors—BUT NO OTHERS—speak in foreign tongues which they never knew before; and in the case of St. Vincent Ferrerius who, travelling through France, Italy, Germany, England, and into Ireland, everywhere preaching in his native Valencian dialect, yet—as in Genoa, where Greeks, Hungarians, Germans, Sardinians and others listened to him—He made all of them hear as if in their own tongue, all understood him alike, and moreover, the furthest as well as the nearest.

Thus, in the former instances, God manifested himself by the sense of seeing, and in the latter ones by the sense of hearing.

When we see these things, and more extraordinary phenomena still in the life of the Church, we may well believe the miracle wrought by Isaiah, though there is nothing wrong in trying—as the Church always does by a strict and searching process un-exampled—to find out the true character of an event we cannot fathom. In this way it is that she establishes whether there is imposition, the ordinary course of nature, or that which supersedes, goes beyond, the law of nature, a miracle.

It is inconsistent, however, and all the more reprehensible and culpable where there is a profession of faith, if Von Gumpach—though without intention I hope—makes the Holy Scripture a book of lies, by representing Isaiah the prophet as a most clumsy rogue and impostor, and Hezekiah, the king of Judah, with probably many witnesses, as nothing less than a short-sighted idiot.

In a more ASTRONOMICAL way, so VERY LIKELY, the star of the Magi is, by Von Gumpach and others, turned into a conjunction of Jupiter and Saturn !

As if He, who allows ships and other objects and pictures by refraction to appear high in the air, and Who sets the bow in the clouds, could not have at His command one of those wonderful temporary, or variable, stars, long since disappeared and at once bursting out again in light upon our vision, and from the stable of

Bethlehem reflect the lovely image of the Virgin-Mother and Child upon its radiant disc, direct to it the searching eyes of the three WISE kings longing and looking for the Star of Jacob, and by its bright connecting link between heaven and earth guiding them, as WE still are cheeringly guided, from the image to the living reality.

And are there no other stars, no meteors, no other creations, to be taken into the service of the Almighty ? and for such a purpose ?

When in all those cases where the supernatural law of God goes hand in hand with, or supersedes, the law of nature, learned men attempt to find scientific substitutes and think themselves so clever, they inevitably make fools of themselves and ridiculous to the REFLECTING CHRISTIAN mind ; but when impositions and frauds are alleged and related of AUTHENTICATED miracles, they simply become absurd ; and if true : these alleged deceptions both in object and duration, as in the case of the blood of St. Januarius, would be far greater miracles than the miracles themselves. And yet, in almost the very words of the above author : these anti-miracle-mongers, miracles themselves, "call upon our credulity to give them credit for learning," where in such matters there is neither knowledge nor common sense ; "and upon our faith in science to prefer astronomical theories," and irrational and imbecile tales and explanations "TO HISTORICAL FACTS."

That these things should be so is in itself one of the greatest proofs how the law of grace and justice go hand-in-hand with the law of nature ; for, persons speaking against the miraculous testimony of the Church instituted by Christ himself—not by vicious princes and parliaments, or abandoned monks or ecclesiastics,—have the "understanding darkened by the pride or prejudice which is within them," reason is eclipsed, and ENTIRELY FAILS in things divine ; when sun and moon are obscured nothing is seen so distinctly.

R

their LEGITIMATE successors—BUT NO OTHERS—speak in foreign tongues which they never knew before; and in the case of St. Vincent Ferrerius who, travelling through France, Italy, Germany, England, and into Ireland, everywhere preaching in his native Valencian dialect, yet—as in Genoa, where Greeks, Hungarians, Germans, Sardinians and others listened to him—He made all of them hear as if in their own tongue, all understood him alike, and moreover, the furthest as well as the nearest.

Thus, in the former instances, God manifested himself by the sense of seeing, and in the latter ones by the sense of hearing.

When we see these things, and more extraordinary phenomena still in the life of the Church, we may well believe the miracle wrought by Isaiah, though there is nothing wrong in trying—as the Church always does by a strict and searching process un-exampled—to find out the true character of an event we cannot fathom. In this way it is that she establishes whether there is imposition, the ordinary course of nature, or that which supersedes, goes beyond, the law of nature, a miracle.

It is inconsistent, however, and all the more reprehensible and culpable where there is a profession of faith, if Von Gumpach—though without intention I hope—makes the Holy Scripture a book of lies, by representing Isaiah the prophet as a most clumsy rogue and impostor, and Hezekiah, the king of Judah, with probably many witnesses, as nothing less than a short-sighted idiot.

In a more ASTRONOMICAL way, so VERY LIKELY, the star of the Magi is, by Von Gumpach and others, turned into a conjunction of Jupiter and Saturn !

As if He, who allows ships and other objects and pictures by refraction to appear high in the air, and Who sets the bow in the clouds, could not have at His command one of those wonderful temporary, or variable, stars, long since disappeared and at once bursting out again in light upon our vision, and from the stable of

Bethlehem reflect the lovely image of the Virgin-Mother and Child upon its radiant disc, direct to it the searching eyes of the three WISE kings longing and looking for the Star of Jacob, and by its bright connecting link between heaven and earth guiding them, as WE still are cheeringly guided, from the image to the living reality.

And are there no other stars, no meteors, no other creations, to be taken into the service of the Almighty ? and for such a purpose ?

When in all those cases where the supernatural law of God goes hand in hand with, or supersedes, the law of nature, learned men attempt to find scientific substitutes and think themselves so clever, they inevitably make fools of themselves and ridiculous to the REFLECTING CHRISTIAN mind; but when impositions and frauds are alleged and related of AUTHENTICATED miracles, they simply become absurd; and if true: these alleged deceptions both in object and duration, as in the case of the blood of St. Januarius, would be far greater miracles than the miracles themselves. And yet, in almost the very words of the above author: these anti-miracle-mongers, miracles themselves, "call upon our credulity to give them credit for learning," where in such matters there is neither knowledge nor common sense; "and upon our faith in science to prefer astronomical theories," and irrational and imbecile tales and explanations "TO HISTORICAL FACTS."

That these things should be so is in itself one of the greatest proofs how the law of grace and justice go hand-in-hand with the law of nature; for, persons speaking against the miraculous testimony of the Church instituted by Christ himself—not by vicious princes and parliaments, or abandoned monks or ecclesiastics,—have the "understanding darkened by the pride or prejudice which is within them," reason is eclipsed, and ENTIRELY FAILS in things divine; when sun and moon are obscured nothing is seen so distinctly.

R

their LEGITIMATE successors—BUT NO OTHERS—speak in foreign tongues which they never knew before ; and in the case of St. Vincent Ferrerius who, travelling through France, Italy, Germany, England, and into Ireland, everywhere preaching in his native Valencian dialect, yet—as in Genoa, where Greeks, Hungarians, Germans, Sardinians and others listened to him—He made all of them hear as if in their own tongue, àll understood him alike, and moreover, the furthest as well as the nearest.

Thus, in the former instances, God manifested himself by the sense of seeing, and in the latter ones by the sense of hearing.

When we see these things, and more extraordinary phenomena still in the life of the Church, we may well believe the miracle wrought by Isaiah, though there is nothing wrong in trying—as the Church always does by a strict and searching process un-exampled—to find out the true character of an event we cannot fathom.   In this way it is that she establishes whether there is imposition, the ordinary course of nature, or that which supersedes, goes beyond, the law of nature, a miracle.

It is inconsistent, however, and all the more reprehensible and culpable where there is a profession of faith, if Von Gumpach—though without intention I hope—makes the Holy Scripture a book of lies, by representing Isaiah the prophet as a most clumsy rogue and impostor, and Hezekiah, the king of Judah, with probably many witnesses, as nothing less than a short-sighted idiot.

In a more ASTRONOMICAL way, so VERY LIKELY, the star of the Magi is, by Von Gumpach and others, turned into a conjunction of Jupiter and Saturn !

As if He, who allows ships and other objects and pictures by refraction to appear high in the air, and Who sets the bow in the clouds, could not have at His command one of those wonderful temporary, or variable, stars, long since disappeared and at once bursting out again in light upon our vision, and from the stable of

Bethlehem reflect the lovely image of the Virgin-Mother and Child upon its radiant disc, direct to it the searching eyes of the three WISE kings longing and looking for the Star of Jacob, and by its bright connecting link between heaven and earth guiding them, as WE still are cheeringly guided, from the image to the living reality.

And are there no other stars, no meteors, no other creations, to be taken into the service of the Almighty ? and for such a purpose ?

When in all those cases where the supernatural law of God goes hand in hand with, or supersedes, the law of nature, learned men attempt to find scientific substitutes and think themselves so clever, they inevitably make fools of themselves and ridiculous to the REFLECTING CHRISTIAN mind; but when impositions and frauds are alleged and related of AUTHENTICATED miracles, they simply become absurd; and if true: these alleged deceptions both in object and duration, as in the case of the blood of St. Januarius, would be far greater miracles than the miracles themselves. And yet, in almost the very words of the above author: these anti-miracle-mongers, miracles themselves, "call upon our credulity to give them credit for learning," where in such matters there is neither knowledge nor common sense; "and upon our faith in science to prefer astronomical theories," and irrational and imbecile tales and explanations "TO HISTORICAL FACTS."

That these things should be so is in itself one of the greatest proofs how the law of grace and justice go hand-in-hand with the law of nature; for, persons speaking against the miraculous testimony of the Church instituted by Christ himself—not by vicious princes and parliaments, or abandoned monks or ecclesiastics,—have the "understanding darkened by the pride or prejudice which is within them," reason is eclipsed, and ENTIRELY FAILS in things divine; when sun and moon are obscured nothing is seen so distinctly.

R

their LEGITIMATE successors—BUT NO OTHERS—speak in foreign tongues which they never knew before; and in the case of St. Vincent Ferrerius who, travelling through France, Italy, Germany, England, and into Ireland, everywhere preaching in his native Valencian dialect, yet—as in Genoa, where Greeks, Hungarians, Germans, Sardinians and others listened to him—He made all of them hear as if in their own tongue, àll understood him alike, and moreover, the furthest as well as the nearest.

Thus, in the former instances, God manifested himself by the sense of seeing, and in the latter ones by the sense of hearing.

When we see these things, and more extraordinary phenomena still in the life of the Church, we may well believe the miracle wrought by Isaiah, though there is nothing wrong in trying—as the Church always does by a strict and searching process un-exampled—to find out the true character of an event we cannot fathom. In this way it is that she establishes whether there is imposition, the ordinary course of nature, or that which supersedes, goes beyond, the law of nature, a miracle.

It is inconsistent, however, and all the more reprehensible and culpable where there is a profession of faith, if Von Gumpach—though without intention I hope—makes the Holy Scripture a book of lies, by representing Isaiah the prophet as a most clumsy rogue and impostor, and Hezekiah, the king of Judah, with probably many witnesses, as nothing less than a short-sighted idiot.

In a more ASTRONOMICAL way, so VERY LIKELY, the star of the Magi is, by Von Gumpach and others, turned into a conjunction of Jupiter and Saturn !

As if He, who allows ships and other objects and pictures by refraction to appear high in the air, and Who sets the bow in the clouds, could not have at His command one of those wonderful temporary, or variable, stars, long since disappeared and at once bursting out again in light upon our vision, and from the stable of

Bethlehem reflect the lovely image of the Virgin-Mother and Child upon its radiant disc, direct to it the searching eyes of the three WISE kings longing and looking for the Star of Jacob, and by its bright connecting link between heaven and earth guiding them, as WE still are cheeringly guided, from the image to the living reality.

And are there no other stars, no meteors, no other creations, to be taken into the service of the Almighty ? and for such a purpose ?

When in all those cases where the supernatural law of God goes hand in hand with, or supersedes, the law of nature, learned men attempt to find scientific substitutes and think themselves so clever, they inevitably make fools of themselves and ridiculous to the REFLECTING CHRISTIAN mind; but when impositions and frauds are alleged and related of AUTHENTICATED miracles, they simply become absurd; and if true: these alleged deceptions both in object and duration, as in the case of the blood of St. Januarius, would be far greater miracles than the miracles themselves. And yet, in almost the very words of the above author: these anti-miracle-mongers, miracles themselves, "call upon our credulity to give them credit for learning," where in such matters there is neither knowledge nor common sense; "and upon our faith in science to prefer astronomical theories," and irrational and imbecile tales and explanations "TO HISTORICAL FACTS."

That these things should be so is in itself one of the greatest proofs how the law of grace and justice go hand-in-hand with the law of nature; for, persons speaking against the miraculous testimony of the Church instituted by Christ himself—not by vicious princes and parliaments, or abandoned monks or ecclesiastics,—have the "understanding darkened by the pride or prejudice which is within them," reason is eclipsed, and ENTIRELY FAILS in things divine; when sun and moon are obscured nothing is seen so distinctly.

R

their LEGITIMATE successors—BUT NO OTHERS—speak in foreign tongues which they never knew before; and in the case of St. Vincent Ferrerius who, travelling through France, Italy, Germany, England, and into Ireland, everywhere preaching in his native Valencian dialect, yet—as in Genoa, where Greeks, Hungarians, Germans, Sardinians and others listened to him—He made all of them hear as if in their own tongue, all understood him alike, and moreover, the furthest as well as the nearest.

Thus, in the former instances, God manifested himself by the sense of seeing, and in the latter ones by the sense of hearing.

When we see these things, and more extraordinary phenomena still in the life of the Church, we may well believe the miracle wrought by Isaiah, though there is nothing wrong in trying—as the Church always does by a strict and searching process un-exampled—to find out the true character of an event we cannot fathom. In this way it is that she establishes whether there is imposition, the ordinary course of nature, or that which supersedes, goes beyond, the law of nature, a miracle.

It is inconsistent, however, and all the more reprehensible and culpable where there is a profession of faith, if Von Gumpach—though without intention I hope—makes the Holy Scripture a book of lies, by representing Isaiah the prophet as a most clumsy rogue and impostor, and Hezekiah, the king of Judah, with probably many witnesses, as nothing less than a short-sighted idiot.

In a more ASTRONOMICAL way, so VERY LIKELY, the star of the Magi is, by Von Gumpach and others, turned into a conjunction of Jupiter and Saturn !

As if He, who allows ships and other objects and pictures by refraction to appear high in the air, and Who sets the bow in the clouds, could not have at His command one of those wonderful temporary, or variable, stars, long since disappeared and at once bursting out again in light upon our vision, and from the stable of

Bethlehem reflect the lovely image of the Virgin-Mother and Child upon its radiant disc, direct to it the searching eyes of the three WISE kings longing and looking for the Star of Jacob, and by its bright connecting link between heaven and earth guiding them, as WE still are cheeringly guided, from the image to the living reality.

And are there no other stars, no meteors, no other creations, to be taken into the service of the Almighty ? and for such a purpose ?

When in all those cases where the supernatural law of God goes hand in hand with, or supersedes, the law of nature, learned men attempt to find scientific substitutes and think themselves so clever, they inevitably make fools of themselves and ridiculous to the REFLECTING CHRISTIAN mind; but when impositions and frauds are alleged and related of AUTHENTICATED miracles, they simply become absurd; and if true: these alleged deceptions both in object and duration, as in the case of the blood of St. Januarius, would be far greater miracles than the miracles themselves. And yet, in almost the very words of the above author: these anti-miracle-mongers, miracles themselves, "call upon our credulity to give them credit for learning," where in such matters there is neither knowledge nor common sense; "and upon our faith in science to prefer astronomical theories," and irrational and imbecile tales and explanations "TO HISTORICAL FACTS."

That these things should be so is in itself one of the greatest proofs how the law of grace and justice go hand-in-hand with the law of nature; for, persons speaking against the miraculous testimony of the Church instituted by Christ himself—not by vicious princes and parliaments, or abandoned monks or ecclesiastics,—have the "understanding darkened by the pride or prejudice which is within them," reason is eclipsed, and ENTIRELY FAILS in things divine; when sun and moon are obscured nothing is seen so distinctly.

R

their LEGITIMATE successors—BUT NO OTHERS—speak in foreign tongues which they never knew before; and in the case of St. Vincent Ferrerius who, travelling through France, Italy, Germany, England, and into Ireland, everywhere preaching in his native Valencian dialect, yet—as in Genoa, where Greeks, Hungarians, Germans, Sardinians and others listened to him—He made all of them hear as if in their own tongue, all understood him alike, and moreover, the furthest as well as the nearest.

Thus, in the former instances, God manifested himself by the sense of seeing, and in the latter ones by the sense of hearing.

When we see these things, and more extraordinary phenomena still in the life of the Church, we may well believe the miracle wrought by Isaiah, though there is nothing wrong in trying—as the Church always does by a strict and searching process un-exampled—to find out the true character of an event we cannot fathom. In this way it is that she establishes whether there is imposition, the ordinary course of nature, or that which supersedes, goes beyond, the law of nature, a miracle.

It is inconsistent, however, and all the more reprehensible and culpable where there is a profession of faith, if Von Gumpach—though without intention I hope—makes the Holy Scripture a book of lies, by representing Isaiah the prophet as a most clumsy rogue and impostor, and Hezekiah, the king of Judah, with probably many witnesses, as nothing less than a short-sighted idiot.

In a more ASTRONOMICAL way, so VERY LIKELY, the star of the Magi is, by Von Gumpach and others, turned into a conjunction of Jupiter and Saturn !

As if He, who allows ships and other objects and pictures by refraction to appear high in the air, and Who sets the bow in the clouds, could not have at His command one of those wonderful temporary, or variable, stars, long since disappeared and at once bursting out again in light upon our vision, and from the stable of

Bethlehem reflect the lovely image of the Virgin-Mother and Child upon its radiant disc, direct to it the searching eyes of the three WISE kings longing and looking for the Star of Jacob, and by its bright connecting link between heaven and earth guiding them, as WE still are cheeringly guided, from the image to the living reality.

And are there no other stars, no meteors, no other creations, to be taken into the service of the Almighty ? and for such a purpose ?

When in all those cases where the supernatural law of God goes hand in hand with, or supersedes, the law of nature, learned men attempt to find scientific substitutes and think themselves so clever, they inevitably make fools of themselves and ridiculous to the REFLECTING CHRISTIAN mind; but when impositions and frauds are alleged and related of AUTHENTICATED miracles, they simply become absurd; and if true: these alleged deceptions both in object and duration, as in the case of the blood of St. Januarius, would be far greater miracles than the miracles themselves. And yet, in almost the very words of the above author: these anti-miracle-mongers, miracles themselves, "call upon our credulity to give them credit for learning," where in such matters there is neither knowledge nor common sense; "and upon our faith in science to prefer astronomical theories," and irrational and imbecile tales and explanations " TO HISTORICAL FACTS."

That these things should be so is in itself one of the greatest proofs how the law of grace and justice go hand-in-hand with the law of nature; for, persons speaking against the miraculous testimony of the Church instituted by Christ himself—not by vicious princes and parliaments, or abandoned monks or ecclesiastics,—have the "understanding darkened by the pride or prejudice which is within them," reason is eclipsed, and ENTIRELY FAILS in things divine; when sun and moon are obscured nothing is seen so distinctly.

R

their LEGITIMATE successors—BUT NO OTHERS—speak in foreign tongues which they never knew before; and in the case of St. Vincent Ferrerius who, travelling through France, Italy, Germany, England, and into Ireland, everywhere preaching in his native Valencian dialect, yet—as in Genoa, where Greeks, Hungarians, Germans, Sardinians and others listened to him—He made all of them hear as if in their own tongue, all understood him alike, and moreover, the furthest as well as the nearest.

Thus, in the former instances, God manifested himself by the sense of seeing, and in the latter ones by the sense of hearing.

When we see these things, and more extraordinary phenomena still in the life of the Church, we may well believe the miracle wrought by Isaiah, though there is nothing wrong in trying—as the Church always does by a strict and searching process un-exampled—to find out the true character of an event we cannot fathom. In this way it is that she establishes whether there is imposition, the ordinary course of nature, or that which supersedes, goes beyond, the law of nature, a miracle.

It is inconsistent, however, and all the more reprehensible and culpable where there is a profession of faith, if Von Gumpach—though without intention I hope—makes the Holy Scripture a book of lies, by representing Isaiah the prophet as a most clumsy rogue and impostor, and Hezekiah, the king of Judah, with probably many witnesses, as nothing less than a short-sighted idiot.

In a more ASTRONOMICAL way, so VERY LIKELY, the star of the Magi is, by Von Gumpach and others, turned into a conjunction of Jupiter and Saturn !

As if He, who allows ships and other objects and pictures by refraction to appear high in the air, and Who sets the bow in the clouds, could not have at His command one of those wonderful temporary, or variable, stars, long since disappeared and at once bursting out again in light upon our vision, and from the stable of

Bethlehem reflect the lovely image of the Virgin-Mother and Child upon its radiant disc, direct to it the searching eyes of the three WISE kings longing and looking for the Star of Jacob, and by its bright connecting link between heaven and earth guiding them, as WE still are cheeringly guided, from the image to the living reality.

And are there no other stars, no meteors, no other creations, to be taken into the service of the Almighty ? and for such a purpose ?

When in all those cases where the supernatural law of God goes hand in hand with, or supersedes, the law of nature, learned men attempt to find scientific substitutes and think themselves so clever, they inevitably make fools of themselves and ridiculous to the REFLECTING CHRISTIAN mind; but when impositions and frauds are alleged and related of AUTHENTICATED miracles, they simply become absurd; and if true: these alleged deceptions both in object and duration, as in the case of the blood of St. Januarius, would be far greater miracles than the miracles themselves. And yet, in almost the very words of the above author: these anti-miracle-mongers, miracles themselves, "call upon our credulity to give them credit for learning," where in such matters there is neither knowledge nor common sense; "and upon our faith in science to prefer astronomical theories," and irrational and imbecile tales and explanations " TO HISTORICAL FACTS."

That these things should be so is in itself one of the greatest proofs how the law of grace and justice go hand-in-hand with the law of nature; for, persons speaking against the miraculous testimony of the Church instituted by Christ himself—not by vicious princes and parliaments, or abandoned monks or ecclesiastics,—have the "understanding darkened by the pride or prejudice which is within them," reason is eclipsed, and ENTIRELY FAILS in things divine; when sun and moon are obscured nothing is seen so distinctly.

R

their LEGITIMATE successors—BUT NO OTHERS—speak in foreign
tongues which they never knew before; and in the case of St.
Vincent Ferrerius who, travelling through France, Italy, Germany,
England, and into Ireland, everywhere preaching in his native
Valencian dialect, yet—as in Genoa, where Greeks, Hungarians,
Germans, Sardinians and others listened to him—He made all of
them hear as if in their own tongue, all understood him alike, and
moreover, the furthest as well as the nearest.

Thus, in the former instances, God manifested himself by the
sense of seeing, and in the latter ones by the sense of hearing.

When we see these things, and more extraordinary phenomena
still in the life of the Church, we may well believe the miracle
wrought by Isaiah, though there is nothing wrong in trying—as
the Church always does by a strict and searching process un-
exampled—to find out the true character of an event we cannot
fathom. In this way it is that she establishes whether there is
imposition, the ordinary course of nature, or that which supersedes,
goes beyond, the law of nature, a miracle.

It is inconsistent, however, and all the more reprehensible and
culpable where there is a profession of faith, if Von Gumpach—
though without intention I hope—makes the Holy Scripture a
book of lies, by representing Isaiah the prophet as a most clumsy
rogue and impostor, and Hezekiah, the king of Judah, with probably
many witnesses, as nothing less than a short-sighted idiot.

In a more ASTRONOMICAL way, so VERY LIKELY, the star of the
Magi is, by Von Gumpach and others, turned into a conjunction
of Jupiter and Saturn !

As if He, who allows ships and other objects and pictures by
refraction to appear high in the air, and Who sets the bow in the
clouds, could not have at His command one of those wonderful
temporary, or variable, stars, long since disappeared and at once
bursting out again in light upon our vision, and from the stable of

Bethlehem reflect the lovely image of the Virgin-Mother and Child upon its radiant disc, direct to it the searching eyes of the three WISE kings longing and looking for the Star of Jacob, and by its bright connecting link between heaven and earth guiding them, as WE still are cheeringly guided, from the image to the living reality.

And are there no other stars, no meteors, no other creations, to be taken into the service of the Almighty? and for such a purpose?

When in all those cases where the supernatural law of God goes hand in hand with, or supersedes, the law of nature, learned men attempt to find scientific substitutes and think themselves so clever, they inevitably make fools of themselves and ridiculous to the REFLECTING CHRISTIAN mind; but when impositions and frauds are alleged and related of AUTHENTICATED miracles, they simply become absurd; and if true: these alleged deceptions both in object and duration, as in the case of the blood of St. Januarius, would be far greater miracles than the miracles themselves. And yet, in almost the very words of the above author: these anti-miracle-mongers, miracles themselves, "call upon our credulity to give them credit for learning," where in such matters there is neither knowledge nor common sense; "and upon our faith in science to prefer astronomical theories," and irrational and imbecile tales and explanations "TO HISTORICAL FACTS."

That these things should be so is in itself one of the greatest proofs how the law of grace and justice go hand-in-hand with the law of nature; for, persons speaking against the miraculous testimony of the Church instituted by Christ himself—not by vicious princes and parliaments, or abandoned monks or ecclesiastics,—have the "understanding darkened by the pride or prejudice which is within them," reason is eclipsed, and ENTIRELY FAILS in things divine; when sun and moon are obscured nothing is seen so distinctly.

R

To ourselves it belongs to look where and why our souls and senses are obscured, where lies the central path of the shadow ; HERE we may calculate to a nicety, though not without labour, and ourselves put an end to the eclipse ; but above all set right the axis, and if that be in its true position and inclination, all calculation will be easy and infallibly correct.

And so it is with the earth.

The parallelism of the earth's axis with the axis of the sun at the time of the equinoxes, and the subsequent oscillation of the earth in its course round the sun, will materially change the path of the moon's shadow over the earth, and bring within its borders portions hitherto excluded by the position which the old theory assigns to the axis of the earth, and pass by others, said to have been obscured. And I have no manner of doubt, that, by the adoption of my theory, the calculation of eclipses, both as to time and place, will reconcile the THEN more exact astronomical science with the positive record of historical fact.

XL.

PRECESSION OF THE EQUINOX.—SIR JOHN HERSCHEL ; THE EARTH A SPINNING-TOP.—NUTATION OF THE EARTH'S AXIS.—REMARKABLE THEOREMS AND CONCLUSIONS RESPECTING "MASSES" AND THE STABILITY OF THE SEASONS ; VARIATION OF THE ECLIPTIC.— FUTILITY OF THE CALCULATION OF MASSES ON THE GROUND OF UNIVERSAL-GRAVITATION.—CONFUSION OF THE OLD THEORY.— HOLLOWNESS OF SUPPOSED PERTURBATIONS.—HYPOTHETICAL CHANGE OF PLACE OF HEAVENLY BODIES.—PREVIOUS CALCULATIONS FALSI-FIED BY LATER DISCOVERIES.—ADAPTATION OF ELEMENTS AND CALCULATIONS.

MY theory of rotary progression accounts also for the precession of the equinox, which seems to be nothing but a slight annual retardation of the earth in its path round the sun, so that the earth does not perform a complete revolution round the sun during its two oscillations in the year; or, the apparent retardation proves, that the whole of the heavens are moving eastward round their common centre, our own solar system.

To an observer on a clear night the whole of the heavens seems to turn round a fixed axis, the one end of which is placed above our horizon in the vicinity of the polar star; and the constellations around the polar star, the constellations of the Great Bear, Dragon, Cepheus, Cassiopeja, Capella, and the Little Bear, of which group the polar star forms the centre, seem to perform a circular movement every day, without altering their respective positions to each other. This circular movement, however, is only apparent. The constellations in question are sufficiently distant from us beyond the northern part of the earth, so as to be always visible to us; they form a circle in the heavens corresponding to the circle of the earth. When thus the earth revolves round its own axis, the spectator passes through, or describes, a circle in twenty-four hours; and during this time the constellations ranged in a similar circle at an inclination above him 'will in turn appear in the zenith and rise and fall towards the horizon, the same as the sun on the day he does not set to the observer in the polar regions. The polar star, being the centre of the above constellations, and at a much greater distance from us than they themselves, will naturally describe but a small apparent circle during the rotation of the earth in twenty-four hours. Thus, the circle described by the rotation of the earth as well as that described by its revolution round the sun, makes it appear as if the vault of heaven were moving round a fixed axis having one of its ends near the polar star. But, as I have said before, it is more likely that the precession of the equinox indicates a universal and REAL movement of the heavens from west to east.

In the following diagram, facing the south pole of the sun and

Fig. 1

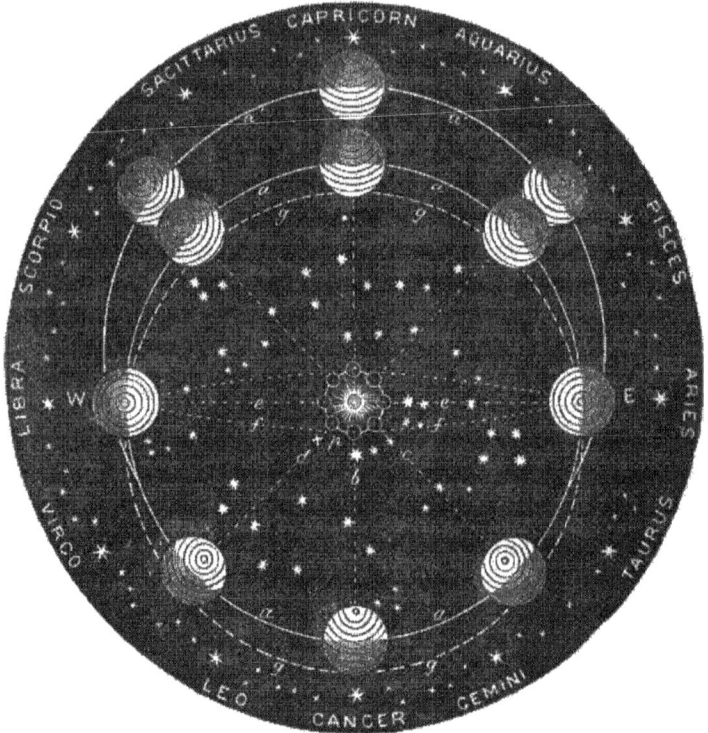

·of the earth, the circle *a* represents the path of the earth in which
it rotates ON ITS EQUATOR round itself every twenty-four hours,
performing its journey round the sun in a little more than 365 days,
both motions taking place from the left to the right, from west to
east.

In the relative position of the sun and the earth at the time of
the equinoxes at Libra and Aries, the prolongation into the distance
of the sphere of the universe of the horizontal and parallel axes
of the sun and the earth, the further end of the solar axis in space
would be to us the NORTH POLE OF THE UNIVERSE, as mentioned

already, p. 184, and remain unchanged to our view and calculations if the earth always retained the parallelism, the horizontality, of its axis with the axis of the sun as indicated; and the distance of this POLE OF THE UNIVERSE from where we are being so enormous that the circle of the path of the earth represented in the diagram vanishes altogether in the reality of creation, and falls in with the said pole of the universe ; that, as faintly illustrated by fig. 2 E E, at Libra and at Aries, 180 millions of miles apart, the annual parallax of the earth, the diameter of the orbit of the earth round the sun indicated by line *e e* in fig. 1, we see the same star or point in the heavens always in the zenith, right over head : the north pole of our earth in all its positions round the sun would CONSTANTLY POINT to that pole without any deviation.

On this supposition our days and nights and climate all over the earth would always be equal ; as, however, the days and nights become respectively longer and shorter, and the seasons are called forth by the oscillation of the earth : so that point in the north of the universe to which the north pole of our earth is directed, no longer remains the POLE OF THE UNIVERSE, the POLE OF THE PATH OR ORBIT of the earth, but changes, as in fig. 1, from the centre line *e e*, into the angle of lines *f f*, and becomes the POLE OF THE ECLIPTIC, the centre of THAT path or circle described by the oscillation of the POLES of the earth, the APPARENT path of the sun round the earth, when in summer we see him on this side, and in winter on the other side of the equator, in the southern hemisphere.

If, as in the diagram 1, the earth in its course round

the sun, from Libra to Capricorn, its south pole is repelled by the sun
and the north pole attracted, and going to the opposite extreme at
Cancer, the latter is repelled and the former attracted, the CENTRE
of the circle g g thus described by the NORTH POLE of the earth
will fall into the angle between lines f near the star b in the Little
Bear ; hence it is close to this star, therefore called the polar star,
that the north pole for a long time to come will constantly be
pointing all the year round, being the centre, the vanishing-point
of the circle described by it in the heavens, and round which point
the polar star itself, owing to its present position and proximity,
describes a small circle in the general apparent movement of the
circumpolar constellations round the pole of the ecliptic.

Thus we have the POLE OF THE ECLIPTIC as quite distinct from
the POLE OF THE UNIVERSE, the PATH of the earth itself as quite
distinct from the ecliptic or the APPARENT path of the sun produced
by the oscillation of the poles of the earth ; and in the opposite,
the southern hemisphere of the universe, the same will be the case
except that THERE the position of the POLE OF THE ECLIPTIC will
be determined by the direction of the SOUTH POLE of the earth.

As the time of noon is exactly ascertained by the stand of the
sun, so the time of the equinox, day and night being equal, was
always perfectly well known to the ancients, and the starting
point and anchor of their observations and calculations.   And thus
it had been noticed already by Hipparchus, 2000 years ago, in
comparing his observations with those of some astronomers about
150 years before him, that our earth did not maintain its relative
position with respect to the stars, that they had moved forward,
and that at the time of the equinox he saw a constellation different
from that seen by his predecessors, that Scorpio had advanced and
Libra taken its place, that the equinox formerly happening under
the sign of Scorpio, had slowly receded to under the sign of Libra,
that the equinox preceded the arrival of the earth at Scorpio.

And this precession has been steadily increasing, so that, instead of being under the sign of Libra at the time of Hipparchus, the equinox now takes place nearer towards the sign of Virgo.

By either of the two motions indicated, the pole of the ecliptic will undergo a gradual cycloidal change of place.

As in the diagram, we now see it close to the PRESENT polar star b; but by the forward movement of the fixed stars to the dotted line $d+$, or the retrogression of the equinox to dotted line d instead of its assumed position at Libra, corresponding to the recession of the earth from dotted line b to c, the star near c would be the polar star, the pole of the ecliptic would have gradually shifted from its present position to the spot ⊙ indicated by the dotted line c, and, as in fig. 1, after an immensity of time, performed a CIRCLE round the pole of the universe, if our system and earth be then what they are now. If, however, the path of our earth lies in an ellipse $a\,a'$ round the sun, then the pole of the ecliptic would have described an ELLIPSE round the pole of the universe, and our north pole itself trace a minute imperceptibly-elliptic circle in the heavens in its annual course round the sun.

If the hour and minute-hand on a dial-plate start together on their course at XII, it will take twelve revolutions of the minute, and one revolution of the hour-hand before they will again be in the same position and place from which they set out. And thus it has been calculated that the earth will have to perform 25,848 revolutions round the sun, whilst the ball of creation is ONCE going round on its own axis—such axis passing from the north pole of the universe through the centre of the sun to the south pole of the universe—until the same star or constellation appears again in the zenith of the earth which at present is seen at the time of the equinox.

The interval occupied by the earth in running its course from one vernal equinox to the other is called the tropical year, from the end of which it would further take 22 minutes and 10 seconds to reach the fixed star from under which it started twelve months before,—but gone in advance like the hour-hand,—and thus complete the "sideral year," which therefore, by so much, is longer than the tropical year.

The forward movement of the fixed stars, relative to the equinox of the earth, or the precession of the equinox relative to the fixed stars, has been observed also to take place with regard to the other planets of our system. The phenomenon of precession, therefore, can only be REASONABLY accounted for, as stated already, by the earth not completing a perfect orbit round the sun in one year, dating from one vernal equinox to the other; or, by the perfect completion of its orbit within the time given, whilst in the meanwhile the unfathomable sphere of creation is gently rolling forward into eternity, with and around our own solar system, with its glorious sun forming the centre, as the prolongation of its own axis north and south forms also the axis of created worlds; and thus the planets with their independent course and orbits are every year a little left behind the onward motion of the fixed stars.

It will be evident on referring to fig. 1, that by the gradual shifting of the equinox, and the oscillation of the earth, stars will gradually pass away from, and others rise to, our view; that with the change of direction of the north pole of the earth, as, for instance, the earth is receding in the heavens from its position at Capricorn to Libra, and further back to Cancer, the circumpolar constellations will be totally different from what they are now.

In fig. 1, I have followed the example of astronomers; they have adapted their theories to their calculations and their calculations to their theories; and so I have adapted my theory to my diagram, and my diagram to my theory. In the drawing I do not claim exactness of position, size, distance, or proportion, but in principle it represents faithfully the "simplicity of the creation;" and it will be perceived that, by my theory, the difficulties about the nutation of the earth's axis, its conical motion, its subordinate gyratory movements, ALTOGETHER VANISH, and that these subjects, with the precession of the equinox, have given a vast deal of trouble to scientific men, and involved them into the most trying reasoning, as we shall presently see.

According to the old theory, the precession of the equinox and the nutation of the earth's axis, so simply accounted for by my own theory, have been CALCULATED to be " due to the action of the sun and moon upon the PROTUBERANT MATTER of the earth's equator; and that a minute effect is due to the influence of the planet Venus."

"The precession of the equinox consists in a continual retrogradation of the node of the earth's equator on the ecliptic." Notwithstanding the theory of universal gravitation, every heavenly body affecting the other, "its cause must not be looked for IN THE PLANETS, THEIR ACTION—excepting Venus—is out of the question, and we must therefore look to the massive though distant sun, and to our near though minute neighbour, the moon, for its explanation. This will, accordingly, be found in their disturbing action on the redundant matter accumulated on the equator of the earth, by which its figure is rendered spheroidal, combined with the earth's rotation on its own axis. It is to the sagacity of Newton that we owe the discovery of this singular mode of action." (*Outlines*, 642).

Then, as far as I understand this "SAGACITY" and "SINGULAR MODE OF ACTION" indeed: "the molecules composing the redundant matter at the equator, have contrary dispositions"—how found out and by whom, unless by some astronomical Münchhausen, I do not know; but so it is SUPPOSED—"those who, owing to the disturbing sun and moon, tend to become more inclined, will counteract the effect of those which have at the SAME INSTANT a

contrary tendency. The excess on the one or the other side of these
contending molecules will change the inclination, and an average be
struck "—not at the end of the season, or the year, but—" at every
moment." Certainly an interesting and truly singular mode of
action

Then there are "retrograding molecules, reactionists, their
general tendency being to retrograde; a struggle again will take
place within the protuberant ring of the equator by the counteracting efforts of the pregressists, of the molecules contrarily
disposed."

By these means, changing the inclination in order to produce
this precession, the earth's axis is said to have "a conical motion,
like a child's peg-top, when it SPINS NOT UPRIGHT, or that amusing
toy the te-to-tum." This extremely slow conical motion, as
affecting the place of the pole of the equinoxial "is not uniform,
but compounded of one uniform, or nearly uniform, part, and
other smaller and subordinate periodical fluctuations, the former
giving rise to the phenomenon of precession; the latter to another
distinct phenomenon called nutation of the earth's axis, which is a
small gyratory movement."

This conical—not to say comical—motion, and subordinate
gyratory movement of the earth's axis,—no doubt important
discoveries,—are said to be the " necessary consequences of the
rotation of the earth, combined with its elliptical FIGURE, and the
UNEQUAL (yet permanently uniform) attraction on its POLAR and
EQUATORIAL regions. Accordingly, the monthly revolution of the
moon, and the annual motion of the sun, produce, each of them,
small NUTATIONS of the earth's axis."

"Thus we see that the two phenomena of nutation and
precession are intimately connected, or rather both of them
constituent parts of one and the same phenomenon."

The preceding attempts at definition of the equinox and the

nutation of the earth's axis, I have more or less LITERALLY, with a few elucidating additions, extracted from Sir John Herschel's *Outlines*, page 385 to 390; but, properly to estimate the : "It is to the sagacity of Newton that we owe the discovery of this singular mode of action," the whole chapter should be read in its entirety, as it is only by a rigourous analysis of every word, line and sentence, as done by Lieutenant R. J. Morrison, in his *Solar System as it is*, that the absurdity of the old theory and its explanation can be duly appreciated.

The rigid parallelism of the earth's axis in its orbit round the sun, no more than the theory of the tides, is found from observation to be true IN FACT. Therefore, the precession and nutation, the conical motion and subordinate gyratory movement of a body not spinning on a solid bottom, like the peg-top and the te-to-tum ; the restless and unruly molecules ; the sun and moon becoming disturbers of a body upon which their action is said also to be uniform and permanent, displaying at the SAME INSTANT a composing and a disturbing power, whilst the moon, in return, is said also to be subject to and kept in order by the earth and acted upon by the protuberant part of the equator ; the action of the planets, universal gravitation, for the nonce conveniently put aside, the influence of Venus excepted : All are had recourse to in this WONDERFULLY-SINGULAR mode of action, so SAGACIOUSLY discovered, calculations made and adapted accordingly—All to be swept away by the "simplicity of the creation," by the "oscillation of the earth whilst revolving upon its equator under the ecliptic, on her proper level, within the atmosphere, and round the equator of the sun," showing thereby, that "calculation goes for nothing against simplicity of a theory."

It is remarkable that Cardinal Cusa already (see *Cosmos* ii. note 403) ascribed a THREEFOLD movement to the earth. The old theory gives it an inclined rotary spinning-top progression, a nutation, a separate gyratory or conical motion, and further periodical fluctuations; but how can this adoption of FOUR movements, all so indifferently and poorly accounted for, be compared with the simple rotary progression of a body oscillating twice during its course round the sun, and for reasons so strikingly simple?

This oscillation of the earth, as well as of the other planets, this consequent obliquity of the earth and planets to the ecliptic, constitutes, as mentioned already page 187, the so-called "obliquity of the ecliptic;" and what is called "the diminution of the obliquity of the ecliptic," and the " variation of the plane of the ecliptic," is nothing else but a slight diminution of this oscillation IF IT REALLY DOES EXIST.

Whether in this way—as it is at the time of the equinox—the axis of the earth will ever come to be permanently horizontal to the horizontal axis of the sun, permanently horizontal to the ecliptic, no one knows but He who created, set in motion, and by His persevering will keeps in action, the magnificent clockwork of creation.

Sir John Herschel, however, respecting the variation of the ecliptic, says in his *Outlines*, page 385; ". . . . . the sum of all these products (of calculation) is actually very small; THEREFORE it must always remain so!" This is called a "remarkable theorem," and certainly no better appellation could have been applied to it; it is REMARKABLE IN LOGIC.

"Meanwhile there is no doubt that the plane of the ecliptic does actually vary by the ACTION OF ALL THE PLANETS, . . ." thus ". . . . . . this diminution of the obliquity of the ecliptic will not go beyond certain very moderate limits, after which—although in an immense period of ages, being a compound cycle resulting from the JOINT ACTION OF ALL THE PLANETS—it will again increase, and thus oscillate backward and forward about a mean position."

Here the obliquity of the ecliptic, the plane of the ecliptic, is oscillating; with me they are the planets which are oscillating. Whether, as said before, having once come to a comparative state of rest, to a state of equilibrium from oscillation, they would ever again resume their oscillating motion, God only knows; but, from the "remarkable theorem" above, astronomers have found out that "the diminution of the obliquity will not go on but again increase, etc." Remarkable conclusion!

"This REMARKABLE THEOREM, ALONE, then, would guarantee the stability of the orbits of the greater planets, but what has been shown of the tendency of EACH PLANET to work out a compensation on EVERY OTHER, it IS EVIDENT that the MINOR ONES are not excluded from this beneficial arrangement," though same page, § 642, THEY are dispensed from working out a compensation, "THEIR action upon the earth is out of the question;" the stability of THEIR orbits is only evident, not guaranteed!

THERE is some work cut out for the simple reader!

The theorem in question, so remarkable because so silly, has apparently been adopted by all astronomers, for, Mr. Hind, one of the most learned among them, in his *Introduction*, p. 15, endorses it without any reserve as to its consistency.

Applying it to the "permanency of the seasons," he says: "The inclination of the earth's axis to her annual path is subject at present to a very slow diminution . . . . . Supposing this could go on continually, the equator and ecliptic would at length coincide": there would, in the end, be no succession of seasons . . . "But it has been otherwise ordained: before the ecliptic can have approached the equator to the degree sufficient to produce any sensible alteration of climate upon the surface of our globe, its motion in that direction MUST cease," BECAUSE IT HAS BEEN SO ORDAINED, "and after becoming stationary for a time, it will begin to recede towards its present state" .   .

The seasons " must therefore succeed each other through all time ; and this ASTRONOMICAL FACT recalls to our recollection the promise of the Creator . . . . . . The permanency of the seasons is one of the most beautiful facts, which ASTRONOMY enables us to explain ! ! "

In vain I have pondered over the theorem, in vain gone through Sir John Herschel's definitions, in vain through the otherwise beautiful passages of Mr. Hind : but in neither have I found ANY LOGIC, ANY PHYSICAL, OR ASTRONOMICAL, REASON,—except the COMPOUND CYCLE OF UNIVERSAL GRAVITATION, the joint action of planets discovered and undiscovered, miscalculated and uncalculated—WHY the slow diminution of the inclination of the earth's axis at some future time MUST CEASE, why it should retrace its steps, why the diminution of the ecliptic will not go on, but again increase ?

In the theorem itself the logic is actually BAD ; " BECAUSE the product is small, THEREFORE it must remain so ; because the diminution of a something is slow, as, for instance, the estimation of the Newtonian theory, THEREFORE it will not continue, but again recede and stand as high as ever ! The system is actually but VERY SLOWLY dying from internal derangment ; THEREFORE it will not go beyond certain very moderate limits, but revive again, and thus oscillate backward and forward about a mean position between life and death ! " And this is called an astronomical FACT by which ASTRONOMY accounts for, and establishes, the permanency of the seasons through all time !

" By the researches of Legrange (of whose analitical conduct it is impossible here to give any idea), the following elegant theorem* has been demonstrated.

' IF THE MASS OF EVERY PLANET BE MULTIPLIED BY THE SQUARE ROOT OF THE MAJOR AXIS OF ITS ORBIT, AND THE PRODUCT BY THE

---

* The remarkable one of which a portion has been cited before.

SQUARE OF THE TANGENT OF ITS INCLINATION TO A FIXED PLANE,
THE SUM OF ALL THESE PRODUCTS WILL BE CONSTANTLY THE SAME
UNDER THE INFLUENCE OF THEIR MUTUAL ATTRACTION.'

But what is all this calculation against simplicity of theory?
What is it, moreover, "if the mass of every planet and satellite"
be incorrect, its calculation based upon false premises, its composition and constitution unknown and disregarded, and no doubt
many planets still undiscovered?

Thus, *Outlines*, § 759, "the mass of the moon is concluded, 1st
from the proportion of the lunar to the solar tide, as observed at
various stations, the effect being separated from each other by a
long series of observations of the relative heights of spring and
neap-tides which, we have seen (§ 752), depend upon the proportional influence of the two luminaries.   2ndly from the
phenomenon of nutation, which, being the result of the moon's
attraction alone, affords a means of calculating her mass, independent of any knowledge of the sun's.   Both methods agree in
assigning to our satellite a mass of about one seventy-fifth that
of the earth."

But what, I ask again, becomes of the CONCLUSION and CAL-
CULATION of the moon's mass, when based upon a theory of the
tides in which, as will be seen by and bye, there is not a particle,
not a shadow, of truth, which is contrary to absolutely ascertained
and established facts?   From want of a better reason it is CON-
CLUDED, that the moon LIFTS UP the water; and because this is
CONCLUDED, it is CONCLUDED again that her mass must be such
and such.

In the same way it is CONCLUDED, "that the phenomenon of
nutation is the result of the moon's attraction alone," that the
moon affects the north pole (why not also the south pole?) of the
earth by acting upon that protuberant stumbling-block at the
equator, thereby giving to the axis of the earth a gyratory

movement similar to what is witnessed in the peg-top, when TOWARDS THE END OF ITS CAREER, it SPINS NOT UPRIGHT; and from this extraordinary explanation of the phenomenon of nutation, being "the result of the moon's attraction alone" and of the consequent war of molecules; from this CONCLUSION the mass of the moon is CONCLUDED, and the said conclusion has offered a means of CALCULATING her mass, independent of any knowledge of the sun's attraction.

Thus, erroneous conclusions must naturally lead to erroneous calculations, and erroneous calculations must lead to erroneous conclusions.

From the perturbation of the WATERS AND OF THE AXIS OF THE EARTH by the moon, the moon's mass is concluded and calculated (§ 759); the perturbations of the MOONS of Jupiter have led to a knowledge of the proportion THEIR masses bear to their primary. (§ 758).　　　　　　　　　　　　　　　　　　　✦

Thus, a test totally different to that applied to the moon, is applied to the satellites of Jupiter for concluding and calculating THEIR masses; in one case, as it were, the mass of a balloon is calculated by its disturbing effect upon the earth, and in the other it is calculated by the disturbing effect of the earth upon the balloon, as if there were no atmosphere between to disturb the motion of balloon and moon, no solar atmosphere to affect the motions of planets and comets, apart from the diminished or increased influence of the sun upon two planets in conjunction, as noticed page 190, and apart still from the consideration, that a small mass, according to its constitution, may be infinitely more magnetic, attracting or repelling, disturbing, than a large mass differently composed. And who is there that will say, that there is no difference in the constitution of the heavenly bodies, and that, for instance, on account for the SMALLNESS of certain masses, THEIR ACTION is out of the question?

Irrespective of the exceptional influence of the sun upon two planets when in conjunction, of the motion and currents of the solar atmosphere, and perhaps also the interior condition of the heavenly bodies: it is nothing but a great though rooted fallacy, that "every planet produces an amount of perturbation in the motions of every other, apportioned to its mass, and to the degree of advantage or PURCHASE which its situation gives it over their movements." (§ 757, which should be read *in toto*).—Take away the atmosphere of the sun extending beyond all our planets to the limit of our system; take away the atmospheres of the planets extending beyond their satellites; suppose even—contrary to the teaching of the old school, contrary to the absurd idea of vacant space—that there is a resisting medium within our solar system, neutral, without gravitation or difference of density in any direction: what then becomes of the heavenly bodies and their masses?—Without recourse to the assistance of opposite, each other neutralizing, imaginary forces, having neither father nor mother, no existence or origin, save in the arbitrary suppositions and conclusions of astronomers, however learned: the whole theory of your system, all your calculations, with all their consequences, are reduced to nothing, all is naked, a mirage of long-fostered imagination; the moons would fall upon their primaries by which they are attracted, and the planets with their satellites would descend into the sun by whom they are attracted and towards whom all are gravitating; you would establish the airless, the exhausted, receiver, in which the feather and the lead descend to the bottom with equal velocity, you would ANNIHILATE your masses; or, in your scale of the heavens, your pound of cork would weigh heavier than the pound of lead; your primitive impulse even would not save you, for, the magnet, able to draw a swiftly passing object from its course, will not fail either, with ever-increasing intensity, to draw it also to itself.

As the risen balloon must return to the earth, so the heavenly bodies would descend to their centres of attraction, whatever curves they might have described, and however aided by the primitive push, that famous discovery.   They would, if possible, more surely still return again towards their respective centres of gravity than the stone flung from the hand of man ; for, he who flings or whirls the stone, is centered in and upon the centre of the earth, and any force, power, or impulse, emanating from his hand, is virtually emanating from the centre upon which he bases.   And in a similar manner the base of the planets is the solar centre, and the base of the moons is the centre of their primaries, whether, like balloons, or the flung-off stone, they rose from their bases, or were launched, or dropped, or created, into their respective positions.

Planets, therefore, are affected only by the sun, and moons only by the planets.   But as regards the position of the planets in the solar sphere, their masses go for nothing upon the old suppositions, conclusions and calculations.   For, as has been noticed already, " if the earth were taken from its actual orbit, and launched anew into space, at the place, in the direction, and with the velocity of any of the other planets, it would describe the very same orbit, and in the same period, which that planet actually does, a minute correction of the period only excepted, arising from the difference between the mass of the earth and that of the planet ; " so that " a MINUTE correction of the period only excepted," as far as the MASSES of the respective planets between each other are concerned, it would make no difference for Jupiter to move in the orbit of the earth, and the earth in that of Jupiter, whether Mercury were to supplant Neptune, and Neptune to revolve in the path of Mercury, whether the asteroids were to rise to the altitude of Saturn, and Saturn to descend to their present level, whether the child were substituted for the giant, or the giant put into the shoes of the child; the LAUNCH, that famous discovery; would make everything right, all would go on as it does now ; one planet would adapt itself to the level, depth, or elevation, of the other in the space of heaven on their exchange of places ; the goose would soar as high as the eagle, the ostrich trill at the elevation of the lark, and the proud peacock fly as fast as the gentle dove.

Evident as it is that, if launched with the same velocity, the cannon-ball and the rifle-bullet would go alike; that a horse, a cow and a donkey, careering along in a race, would perform the same distance in the same period of time if running with equal speed and in the same direction ; that a letter would travel as fast as a telegram if conveyed at the same rate as the electric spark; or, in a few words, that two or three different things would be equal if they were equal: still, it is as well to have it confirmed by so high an astronomical authority.

The accommodation of the planets to their new simulated positions, at first sight, seems MORE wonderful, but, on the old theory, likewise presents little difficulty. For, those self-acting forces would set all right; these fairies of the solar system, the invaluable co-operators of the launch and motors in astronomy would, in every instance, and particularly in the hands of an experienced mathematician, with facility adapt themselves to the various masses and circumstances of the bodies of our system in their translated stations ; " EVERY PLANET would still produce an amount of perturbation in the motion of EVERY OTHER, apportioned to its mass, and to the degree of advantage or purchase which its situation in the system gives it over their movements," and all would go on AS IT DOES NOW, as if no transposition had taken place; for, THE EARTH WOULD GO JUST THE SAME AS JUPITER if launched into that planet's orbit, and JUPITER WOULD GO JUST THE SAME AS THE EARTH if launched into its orbit.

From Sir John Herschel's reasoning it follows, as a rigid conse-

s 2

quence, that, "if all our planets and asteroids were taken from their actual orbits, and launched anew into space, at the place, in the direction, and with the velocity of any one of the planets, they would describe the very same orbit, and in the same period, which that planet,—whose orbit might have been chosen,—actually does, MINUTE CORRECTIONS of the periods only excepted, arising from the difference between their masses and that of the planet in question," ALL GOING JUST THE SAME AS THE BODY IN WHOSE ORBIT THEY WERE LAUNCHED WITH THE SAME VELOCITY; and thus, as I have said before, masses would be ANNIHILATED, would go FOR NOTHING, and our astronomers must agree ·with M. Von Gumpach, that the celestial bodies WEIGH NOTHING.

This is strictly correct, for, the launch, that famous discovery, and the attraction of the sun, would retain all of them in the self-same orbit, irrespective of everything, none would rise, none would fall, great or small ; and it is clear that, if the earth were to be placed in the orbit of Jupiter, and *vice versa*, and that with this exchange of place they were also to exchange their respective velocity and motion, "a minute correction of period only excepted," the result would be just the same, whether two, four, or the whole of our planets and asteroids, were re-launched with the same velocity of any one of the planets into that planet's orbit.

Therefore, the above proposition, like the whole Newtonian system, is based upon nothing but the bare ARBITRARY ASSUMPTION of the launch, upon a false theory of universal gravitation, or universal neutralization, and as a necessary consequence, a confused and contradictory theory of forces and· masses; and to speak of these as LAWS OF NATURE would be supremely ridiculous.

In the preceding simulated transposition of the planets, not only nothing has been said as to the effect which would be produced upon them by the sun, on the one being further removed from him, or the other brought nearer, which, if the theory of masses were worth anything, would be of the utmost consequence; but the supposition itself is confounded by its own author.

"EVERY PLANET would affect EVERY OTHER, etc.," as above; yet the EARTH WOULD GO JUST AS JUPITER if launched into that planet's orbit, THOUGH then so much further from the sun, and so much nearer to Saturn, Uranus and Neptune, than now, and THOUGH in our PRESENT state of the solar system, according to *Outlines*, § 642, "the immense distance of these (superior) planets from the earth puts THEIR action upon it out of the question;" and yet in opposi- . tion to this again it is said that "it is by means of the perturbations of the (inferior) planets that a knowledge of those planets is obtained WHICH HAVE NO SATELLITES."

If, then, the earth, HAVING a satellite, is not affected by the superior planets, how are Mars, Venus, and Mercury, and the asteroids to be affected by them and *vice versa*; and how are these last little ones to affect even one another? and how then are their masses to be calculated? And what, moreover, is the value of these calculations, based upon the contradictory theory of mutual attraction, perturbation, compensation, and the rest, when every day, as I have said once before, they are so palpably falsified and upset by the discovery of new planets?

On this head, listen to Lieut. Morrison, R.N., although an astrologer.

In his *Solar System as it is*, he says, page 197: "the reader will observe, that when this was written," (Sir John's disquisition on the subject,) "the great planet Neptune was unknown, and the discovery of some thirty minor planets, (now found to be as plenty as blackberries), had not yet been made. And it is worthy of note, that when Legrange gave out his NOTED THEOREM, the true mass of Jupiter had not been ascertained." (Has it now?) "Surely these great and important discoveries must in some way affect the "compensation theory": if things were 'all right' before, there must be a screw somewhere loose now!"

Never were astronomical calculations more confounded than by these facts.

This will appear still more so when we consider § 532 of Sir John Herschel's *Outlines*; for, he says: ". . . . . . the mass of a planet which is attended by one or more satellites can be as it were weighed against the sun, provided we have learnt from observation the dimensions of the orbits described by the planet about the sun, and the satellites about the planet, and also the periods, in which these orbits are respectively described. IT IS BY THIS METHOD that the masses of Jupiter, Saturn, Uranus, and Neptune have been ascertained," though in *Good Words*, no doubt referring to Jupiter and his famous visitor, he tells us that "they, the comets, have ACTUALLY informed us of the WEIGHT of one of the planets, which could not have been determined with any exactness if a comet had not on one occasion passed very near to it." And yet, see p. 33, Sir John says: "that not even the SATELLITES of Jupiter suffered any perceptible derangement in THEIR motions, owing to the smallness of the comet's mass, whilst, as above, "it ACTUALLY informed us of the weight of the PLANET," NOTWITHSTANDING the smallness of its mass.   What next!

Now it follows, according to the theorem of Legrange, viz., "If the mass of EVERY planet be multiplied . . . . under the influence of their MUTUAL attraction," that when Neptune and about fifty asteroids were NOT KNOWN TO EXIST, THEIR attraction could naturally not be thought of, much less calculated, in determining the masses and perturbations of the other KNOWN planets, as will be strikingly demonstrated by the discovery of Neptune itself, and its attempted substitution for the supposed disturber of Uranus.

Therefore, and apart from the failure established by the discovery of Neptune, calculations made in ignorance of undiscovered planets, based upon a MUTUAL attraction of the mass of EVERY planet,

MUST HAVE BEEN INCORRECT, and—SUPPOSING the theorem to be true—WOULD BE incorrect until ALL the planets, great and small of our system had been discovered, and until "we have learned from OBSERVATION the dimensions of the orbits described by as yet unknown planets about the sun, and by the unknown satellites about their planets, and also the period in which these orbits are respectively described."

Thus, as we have seen, reciprocal, imaginary perturbations, mutual attractions of the heavenly bodies of every description, are concluded, disregarded, or denied, when and where it suits; from these assumed reciprocal perturbations, masses are concluded and calculations accordingly MADE and ADAPTED; presumed facts are taken into the elements of calculation, and these preconceived facts will be sure to be the resultants; if, as in *Hamlet*, a camel be imagined in the astronomical cloud, a camel will be sure to appear, if a weasel be fancied, a weasel will be seen, and if a whale be taken into the account, something VERY LIKE A WHALE will come out again.

In this manner, as has been proved above, and as more particularly we shall see in the theory of the tides, calculations, in themselves difficult, can be, and as easily are, made to fit, as self-acting forces made to govern and correspond to, the harmony of heaven. Von Gumpach, in his *Baby Worlds*, is overwhelming in his evidence on these points.

Nevertheless, profound intellects may have found, or hit upon, formula concerning the masses of the bodies of our solar system, by which, in some instances, after months of calculation, reasonable, and to them satisfactory results, have been obtained, the same as getting from London-Babylon to Rome by way of Oxford or Cambridge, St. Petersburg or Constantinople. My new system, however, will lead to a short and direct road; for, in itself already, simplicity of theory goes beyond all suppositions, abstract notions and calculations, and is, and ever will be, the CHARACTERISTIC of TRUTH.

XII.

EQUATORIAL AND POLAR ATTRACTION; PENDULUM OBSERVATIONS; THE PROTUBERANCE AT THE EQUATOR.—ELLIPTICITY OF THE EARTH; STRANGE REASONING OF ASTRONOMERS.—ABSURDITY OF A CENTRAL SWINGING, OR SO-CALLED, CENTRIFUGAL FORCE.—THE FORMATION OF THE SOLAR SYSTEM AND LAPLACE.—DENSITY AND FIGURE OF THE EARTH.—THE TORSION-BALANCE.—THE ACCIDENTAL DISCOVERY OF NEPTUNE BY LE VERRIER.

THE earth is a spherical body, perfect in its interior constitution and organization, the same as the human body is perfect, for the accomplishment of certain ends.

That the earth is merely a huge lump of mixed heterogenous material, or its crust only a few hundred miles thick, and the interior in a molten fluid, or a watery liquid state, is contrary to anything created we know of.

As a perfectly organized body, more properly, perhaps, an organized spherical acting magnet, continually excited by the sun, we know it to have poles at which the magnetic needle is perpendicularly inclined, whilst at the equator it remains horizontal, without variation or declination. Hence also the greater attraction at the poles of every other mass, and its consequent greater weight there than at the equator; the 86,535 ascertained vibrations of a pendulum in London, against 86,400 vibrations of the same pendulum at the equator, where the longer vibrations, no doubt, are assisted by the rarety of the atmosphere, and impeded by its greater density towards the polar regions.

Hence also, aided by the solar heat upon the equatorial regions, the overflow of water, air, electricity, etc., towards the attracting poles, and their reflux in a colder and regenerated state towards the equator. By reason also of the greater and permanent heat of the sun at the equator, we there have the luxuriance of vegetation and animal life, and the gradual increase in bulk of the girdle of the earth.

By this constant heating of the equator by the sun, the electric life-blood of the earth is set and kept in motion, and this motion within the earth makes it turn round its own axis, like a kettle of water hung up by a cord before the fire, producing, once in twenty-four hours, a revolution of its body, upon and within the solar atmosphere. The rotating as well as progressive motion is equally participated in by every atom of which the earth is composed, even of water, air, and the superior oceans of matter; and this so much so, that at the equator, where the velocity of revolution is greatest, there are not only no tides, but contrarily, it is exactly there where there are the great calms of air and water, the standing still of these two great oceans, not even betraying the westerly course of their surface, though left a little behind by the solid part of the earth in its rotation.

This, unless the equilibrium of air and water be disturbed by temporary, periodical, or local causes, is also quite natural, because all that constitutes the earth is but ONE BODY, and equally—lead and feather—gravitates to the centre of the earth, and this in the case of the air so much so that it presses upon the surface of our globe with a weight equal to 15 lbs. upon the square inch, rooting, moreover, in the earth and water into which it penetrates and descends in its gravitation.

Hence, no motion of the earth, whether rotary or progressive, can have any influence upon its own figure; for, as far as THESE TWO CAUSES are concerned, the whole will go on and remain for ever AS IT WAS AND AS IT IS, because, as in the railway train, every atom of its parts, nay, every action of the passenger, objects dropt, a stone thrown, all partake of the motion of the train, the water in the tank even will remain as smooth and still as in a basin on the table in the room.

How beautifully simple this divine arrangement for the stability of the earth and all that is upon it—that ALL partakes of the same motion, ALL EQUALLY GRAVITATES to the same central nucleus, to the same heart of the earth, acted upon by the sun, making its electric blood to circulate, turn the earth on its axis to give us day and night, and to preponderate here, and preponderate there, for the production of the seasons and their attendant changes and blessings for the enjoyment of animated creation. These are the natural effects of a natural, interior, CENTRIFUGAL AGENT.

How, according to the old theory, the rotation of the earth is brought about, or whence it comes, I have not been able to find out; the rotation exists, and there the CAUSE of it ends, though not its effects, which we will now proceed to consider.

Sir John Herschel, in his *Outlines*, § 237, says : " the rotation of the earth GIVES RISE to the centrifugal force ; the centrifugal force PRODUCES an ellipticity (the protuberance at the equator) in the form of the earth itself; and this very ellipticity of form, etc."

According to this theory, a fly-wheel does not move by virtue of a centrifugal force, a force emanating from, having its seat in, the centre of the wheel, howsoever imparted ; but, because the fly-wheel rotates round its own centre, the rotation GIVES RISE TO the centrifugal force, which reasoning I consider both against logic and fact.

Then : " the effected centrifugal force effects the elliptical form of the earth, the protuberance at the equator."

How this effect can be produced in a body rotating round its own centre, free to move, unimpeded, unarrested by an unyielding frame or fixed axis, with its every atom gravitating to that centre, and every one of its particles equally partaking of the rotating and progressive motion of the whole body, being moreover equally pressed upon by the super-incumbent atmosphere, likewise gravitating to the common centre, I am at a loss to conceive.

However, to prove the ellipticity of the earth being produced by ITS OWN centrifugal force, Sir John has recourse to a most unfortunate comparison, viz. : " a pail of water, suspended by a cord, and made to SPIN ROUND, while the cord hangs perpendicularly." Upon this a vast amount of labour is spent to show, that " the centrifugal force"—CAUSED by the spinning pail— " generates a tendency in ALL the water to leave the axis, and press towards the circumference, where it is forced up the sides, &c."

Every atom of the body of the earth, like truth itself, always gravitates to its centre, to the common point of union, as we have seen before ; the whole structure requires no exterior force, no wall, no bond, foreign to its interior constitution, to keep it together. But the pail of water has no centre, no point of union, no more than error, towards which to gravitate. Hence, when spinning round, the water recedes from an apparent, an unreal, centre, and, like error, would fly in every direction, if the pail, the bondage of superior external force, did not hold it, though impotently rising against the wooden wall. Pail and water, moreover, are two distinct bodies with regard to each other, they have no common centre, save the centre of the earth. And upon such a puerile illustration so much ingenious argument has been bestowed, and the alleged cause of the elliptic form of the earth made to rest.

In the analogy cited, the paramount object is swung round within the pail, both bodies spinning round with unequal speed, by means of a force, a cause, exterior and foreign to each of them. In the case of a weight, or a stone, whirled round at the end of a string, the swinging proceeds from a real centre, and from the will of an animated being; but the swinging centre, the swinging person, is a body different from, exterior and foreign to, the object swung round; they are not one body like the earth with its every particle, its water, its air, the ship, the balloon, and the moon; all these the earth simply carries, but does not swing, and they partake of its own proper motion; there is no will or power in the earth to give them a swing, no mind to guide, no arm or hand to be employed for the purpose, nor are these at all necessary for making them move as they do; if the rotary motion of the earth by itself involved the swinging of bodies upon and above its solid parts,—as the outward effort of its centrifugal force is said to give a swing to the moon,—these bodies would make a revolution round the earth in a time equal to its own rotation round its own axis.

The same holds good with the sun and planets.

The sun carries the planets as the earth carries the water, the ship, the air, the balloon and the moon; they partake of the motion proper to his own body and sphere; but, there being no will, no power, no animate effort, and least of all an outward one, in the sun, in order to give the planets a swing,—as, experimentally, we may swing any object around ourselves,—so they lag behind his own time of rotation, as the water does on the earth, the ship on the flowing stream, the balloon in the air, and the moon in a still higher ocean.

With the old theory of a primitive impulse; the swinging, or centrifugal force, the outward effect said to be exercised by the sun upon the planets, becomes altogether superfluous.

To ascribe, then, a centrifugal force to the sun, the earth, and the other planets and their satellites, with an inner and outward effort, and effects resulting therefrom, as compared to, and illustrated by, a centrifugal machine or a fixed wheel ; or by an animate being using his arms or hand as an instrument, with the addition of a string, to whirl a stone or weight round about his head, and to make gravity act the part of a living intelligence and supply also both the office of arm and string, seems to me as ridiculous as if a learned man were to make a summersault and expect the moon to do the same.

It is, moreover, very unkind, that in opposition to the centrifugal force produced by Uranus and Neptune in their rotation from west to east, as well as in opposition to that famous discovery, the primitive impulse, their moons should swing around them in the contrary direction from east to west ; that comets, as if tantalizing the outward effort of the centrifugal force produced by the sun in his rotation from west to east, should swing around in and traverse the solar system in every direction ; and that both these facts should give the finishing stroke to the centrifugal force and primitive push theory.

The previous reasoning may be applied to Laplace's theory of the formation of the solar system, of rings being thrown off by the centrifugal force of the sun when rotating in its supposed original gaseous state; these rings breaking and contracting forming the planets ; and these again condensing and cooling somehow or other, and by their centrifugal force throwing up the equatorial portion, and in the words of Humboldt, *Cosmos*, pp. 350, 351, " showing, announcing, in the Mathematical figure of the Earth, in the existence of compression at the poles . . . . the primitive condition of fluidity of the rotating mass and "— though, or because, not confined in a pail,—" its solidification into the present form of the terrestrial spheroid !"

" The dimension of the earth and time of its rotation being known, it is easy thence to calculate the exact amount of the centrifugal force . . . by which the sea at the equator is lightened . . . . . . whilst at the poles, where no such force exists, the water may be considered specifically heavier." Sir John, § 229.

Dimension and time of rotation being known, it is, no doubt, easy to calculate the centrifugal force PRODUCED BY a mill or water-wheel, by a barrel turned by a squirrel or other animal, or by a billiard ball set in motion; if the rotation of these and similar objects PRODUCE a centrifugal force with an inner and outward effort or effect, it is, as said already, because they are fixed to, or moving on, or about, stationary bodies; they are not free to move without the application of exterior force, or a force residing within themselves as an integral part and property of their own substance; but the heavenly bodies are free to move, the moving force resides within and is part and parcel of their own substance, every atom partakes of the same motion uninfluenced by the centrifugal force produced by the whole; and if a centrifugal force be produced by the whole, it could only affect bodies exterior and foreign to itself.

To be sure, the earth and planets WERE set in motion by a force foreign to themselves, by that famous discovery, the primitive impulse, and thus far the analogy between them and the wheel and the rolling ball would hold good; but then the constituents of the celestial bodies gravitate to their own interior centres, and the terrestrial objects DO NOT.

However, waiving this also, the theorem of Sir John still falls to the ground, because he makes the heavenly bodies move in EMPTY SPACE, free in every direction, so that there the centrifugal force of the revolving bodies can much less influence their own exterior parts or molecules, particularly when in a gaseous state, than the billiard ball moving along unrestrained influences the particles that form its crust.

Hence, the centrifugal force produced by the heavenly bodies can throw up no water from their own surface, cannot LIGHTEN THE SEA, not even their atmosphere, though they were moving in vacant space, and of course much less in a resisting medium which, as we have seen, depresses the atmosphere of our earth and elongates into a tail of enormous length the hair of a comet.

It, moreover, becomes an interesting question, at what time and stage of their existence that famous discovery, the primitive launch, was applied to the planets when in the original form of rings they were, or had been, thrown off from the sun by his centrifugal force, and without the launch would have gone straight into space? when and at what stage of vaporosity or condensation the equatorial belt began to belge out and the poles to flatten —to give later so much trouble to calculators?—Laplace, who found no God in his *Celestial Mechanism*, has failed to give us the solution of his centrifugal off-shot.

Page 265 I have said,—and there is nothing new in it,—that by the heat of the sun the water at the equator is warmed or lightened to overflowing; but according to Sir John, as we have seen, it is " by the exact amount of the CENTRIFUGAL FORCE calculated in § 229, that the SEA at the equator is LIGHTENED," that the water recedes from the centre of the pail !

Why the land, or solid part of the earth, its molecules at the equator, are not also mentioned as being lightened by the same means, and "thereby, like the sea, rendered susceptible of being supported on a higher level, or more remote from the centre of the earth than at the poles " (same article), we are not told, though no doubt the protuberance of the earth at the equator, the hunchback of the old theory, is meant to include both LAND and sea, pail and water, and could not exist without the LIGHTENING of the molecules of the one as well as the other.

The SEA BEING THUS LIGHTENED outside the pail, "the centrifugal force produces an ellipticity in the form of the earth itself ; and this VERY ELLIPTICITY OF FORM modifies its power of attraction on bodies placed at its surface, and thus gives rise to the difference in question (§ 237) " namely : that a mass of matter in London, or nearer the pole still, weighing 100,000 lbs., exerts the same pressure on the ground, that 100,335 of THE SAME POUNDS transported to the equator would exert there.°

We have seen already that it is the polar magnetism of the earth, and not the elliptic form of our globe, that causes the difference ; that attraction, or gravity, does not depend upon the form, not upon the size, but upon the magnetic constitution of a body : I therefore need scarcely ask, which of the two is the most natural explanation of the phenomenon, of one pound weight exerting more pressure, and of a pendulum making more vibrations, at the poles than at the equator ?

My reasoning, moreover, will hold good, whatever be the form of the earth, particularly when we consider, that that form is still a problem. For, Humboldt remarks in his *Cosmos*, p. 159, that " pendulum experiments and measurements of degrees give such different results for different parts of the earth's surface, that it is impossible to assign any regular figure which shall satisfy all the results obtained by these methods."

However, to MAKE these methods agree, " the density of the earth " according to Von Gumpach, " was arbitrarily so fixed as to RECONCILE the results of pendulum observations with the results of geodetic measurements of meridian degrees, with the view of adapting both "—the one and the other resting on a false theory—" to the theory of the earth's figure, SUPPOSED to be flattened at the poles."

* How the lightening of the SEA at the equator can produce an ellipticity in the form of the earth, I cannot conceive ; but, at any rate, the proposition is quite at variance with the molecular war so beautifully described by Sir John Herschel, (see p. 249.)

According to the same authority, Professor Airy is said to state in positive terms : " Unless you KNOW what is the structure of the INTERIOR, you cannot say what the ellipticity of the earth will be."

Whether the above proceedings convey that KNOWLEDGE can be no riddle.

Nevertheless, with an impudent assurance of infallibility in astronomical science, see *Baby Worlds*, p. 2, Laplace observes, with reference to certain lunar inequalities: " It is a very remarkable thing, that an astronomer, without leaving his observatory, and simply by a comparison of his observations with the result of analysis, should have been able to determine the EXACT SIZE, and the EXACT AMOUNT, of the earth's polar depression, as well as ITS DISTANCE from the sun and moon ! "

When everything respecting "masses and density" is based upon error, how can truth come out of it? As a proof look ONLY at the error made in the weight of the sun and the earth, and consequently of all the other planets with their moons, and their distances from each other, as noticed p. 152, and the true value of all these "VERY REMARKABLE THINGS " will be strikingly evident ; and with Humboldt you will say " that it is impossible to assign any regular figure (to the earth) which shall satisfy all the results obtained by these methods " by the erroneous theories upon which they are founded.

" Though nearly, the degrees are not precisely equal. In all geographical longitudes the degrees of latitude are found to increase in length in going from the equator towards the poles. An increase in the length of a degree indicates a less amount of curvature. The earth's surface is therefore less curved, or less convex—that is, flatter—as we approach its poles on all sides from the equator. Its form then is elliptical, or oblate, and its polar diameter somewhat shorter than its equatorial." Sir John

T

Herschel's " Celestial Measurings and Weighings :" *Good Words,*
June, 1864.

Now I would ask whether this "increase in length of a degree,
or less amount of curvature of the earth's surface," would not·
equally, nay better, be obtained, by a swelling out of the polar
portions of the earth, by a protuberance of the poles of the earth
BEYOND a perfect sphere, rather than by a flattening of its poles
BELOW the perfect sphere ?

Besides this we must bear in mind, that the measurement of a
degree of the meridian has not yet been made at anything LIKE
NEAR the poles ; and thus we are still at liberty to take the old
orange, or the fresh lemon, side of the question. To myself it
seems more natural, more in harmony with my new theory of the
solar system, and more particularly corresponding with the polar-
magnetic character of the heavenly bodies, that the earth should
be of the form of a lemon, rather than of an orange.

Considering, however, that the earth is a REAL MAGNET, that
electricity with its attracting or magnetic property is the binding
element of all solid substances, and in another state also the
dissolving, the expanding element of matter ; that bodies upon the
earth in a certain electric condition as powerfully attract, as in
another they repel each other: it cannot be laid down as law,
that, according to Newton " every particle of matter attracts
every other particle, with a force directly proportioned to the mass
of the attracting particle, and inversely to the square of the distance
between them."

If this were true, all things created would either amalgamate
into one lump, or, the heavenly bodies being separately posted at
their different stations, but attracting each other—"as every particle
in nature attracts the other "—that attraction would keep all of
them in their mutual relative position, and no centrifugal force, still
less its outward effort, would be able to swing them off their stand.

Upon this proposition, nevertheless, very ingenious experiments have been made with the torsion balance to find out the degree of attraction between bodies of different size and material, and thence to DEDUCE THE MASS OF THE EARTH.

The bodies so operated with, were isolated, and electric influences removed, as far as human ingenuity could do it; but it must not be forgotten, that electricity is the INHERENT CEMENT of all solid substances, and that, therefore, solid bodies and particles can never be deprived of it or be made perfectly neutral relative to each other; and if the latter could be done, they would no longer attract each other. For this reason, though there may be some remote approximation to truth in the present calculation of the mass of the earth, something like calculating the weight of an elephant from the weight of a fly, ANOTHER TEST must be discovered before it can be calculated with certainty, and, as we have seen also, another test for calculating the masses of the sun, the planets and their satellites, instead of that of a mutual attraction and perturbation, and of an imaginary centrifugal force and swinging disposition, existing only in the artificial arrangements of man. And in these arrangements an isolation of particles and their mutual attraction bears no reference whatever to the attraction and mass of the earth, since all particles, however carefully electric influences may have been removed, are and must always be under the influence of the attraction of the earth itself, whose mass can as little be calculated by its attraction as the mass of a magnet by the weight or mass of the body which it attracts.

By means of another test, then, it will be found, that the masses of the celestial bodies bear a different proportion to each other, whilst those at present accepted MUST BE WRONG, being based upon laws and assumptions altogether fictitious and untenable, as PROVED ALSO SUFFICIENTLY by the discovery of Neptune and so many asteroids, and in 1863 by Le Verrier himself. For, " while studying some of the phenomena of gravitation, he found certain difficulties which he could explain only on the SUPPOSITION, that the sun is not so heavy by one-tenth as has been estimated, or that the earth is one-tenth heavier.   His own opinion was in favor of the latter conclusion."

This startling shift CANNOT BE TRUE EITHER, for, NOTHING CAN BE TRUE THAT IS BASED UPON THE ABSURD THEORY OF UNIVERSAL GRAVITATION. I cannot see, moreover, why THE SAME PROPORTION of one-tenth more or less for sun or earth should apply to two bodies, not only not equal, but so vastly different from each other in size and weight. Supposing, however, M. Le Verrier's inference and estimate to be true : what then again becomes of the hitherto adopted masses of the moon and of ALL THE OTHER HEAVENLY BODIES ; their own proper and their common centres of gravitation, their distances and even their sizes, when, in the words of Sir John Herschel, "the FIRST STEP towards ascertaining the real size of the sun is to determine its distance," and that this distance hitherto has been wrong ? what becomes of the theorems and experiments spoken of before, and of all the calculations with which astronomical works are crowded; of the 13,500,000,000,000,000,000,000,000lb. for instance, which Dr. Mädler found the earth to be weighing when he put her into the scale at Dorpat ?—Bewildering thoughts these for the mathematician !

We have seen before that "universal gravitation " is a delusion ; that APART from this delusion the BASES upon which the masses of the heavenly bodies are and have been calculated have no foundation whatever, and that THEREFORE these masses, with all the influences they are said to exercise, MUST BE incorrect. We have seen that, if these bases WERE correct, and that the law of universal gravitation WERE true, the masses of the heavenly bodies ALREADY KNOWN, could NEVERTHELESS NOT BE ascertained by calculation, on account of those STILL UNKNOWN, and that the masses of these unknown bodies could still less be calculated, as we know nothing about their number, position and orbits; except ALL the planets of our system, together with each one of their satellites, HAD BEEN DISCOVERED, neither universal gravitation nor the calculation of masses could be tested.   The present mode of proceeding is something like a mathematician trying to find out the weight and ages of children standing before him without putting them into the scale, and without evidence as to their time of birth, and pretending thereby to calculate correctly the weight and ages of their distant unseen and unknown parents.

But WE HAVE SEEN the discovery of planets, and moons, and asteroids, AFTER the heavenly bodies previously known HAD been weighed in the scale of universal gravitation, and the discovery found them wanting.   From the many corrections made by astronomers, and by the preceding evidence of Le Verrier himself, WE HAVE SEEN, and we DO SEE AS A MATTER OF FACT, that all the calculations made respecting the masses of the celestial bodies ARE, AND MUST BE, failures ; that, according to the latest edition of Hind's *Introduction*, p. 93, irrespective of the NEW WEIGHT of the sun and the earth, the mass of Mars is not exactly known, and, p. 95, not even the sizes of the minor planets, and what is more, the size of the sun, as seen at previous page ; that therefore, starting from the same wrong elements, and, of course, arriving at nearly one and the same erroneous result, both the calculations of this eminent astronomer and of Adams, respecting their PRESUMED disturber of Uranus CANNOT HAVE BEEN TRUE ; that, in further proof thereof, the CALCULATED body when looked for, was from two to nearly three times larger, its weight, according to Von Gumpach, "some quadrillions of pounds AVOIRDUPOIS" greater than Neptune, which, in its far inferior orbit and course round the sun, happened at the time to pass in the neighbourhood of the line of observation ; that, as an additional proof, the CALCULATED distance and orbit of the STILL UNDISCOVERED perturbator exceed by 500,000,000 miles the actual distance, and by 50 years the actual orbit, of the interloper—saving future corrections ;—that if the CALCULATED distance of the supposed planet had been nearer the mark, nay, if by a "singular coincidence," its distance and whereabouts had been calculated to the VERY SPOT where Neptune was found at the time, the calculation, notwithstanding, would HAVE BEEN ABSOLUTELY ILLUSIVE, its result a happy chance, being based upon a distance between the earth and the sun, and upon their relative weight, mass, and attraction, now ACKNOWLEDGED TO BE FALSE,—and falsehood begets no truth ; that, as further evidence of a great error "the eccentricity of Neptune is only ONE-TWELFTH of that of the theoretical planet ; its longitude of perihelion, in 1847, 47° instead of 285° or 299.°"—Von Gumpach.

We see, moreover, on account of the fallacy of the theory of masses, that the alleged perturbations or anomalies in the motion of Uranus which gave rise to the supposition of an ulterior planet and the calculation of his bodily constitution and place in the heavens, are quite unfounded, and neither exist nor ever existed, and that, therefore, the old planet COULD NEVER have been troubled by the new one, no more than by the one still non-existing. Besides this, though totally independent of the above, Neptune was twice observed in the year 1795 by Lalande, a French astronomer, and by others in 1846, all of whom mistook it for a fixed star before Galle, following the direction of Le Verrier, discovered it as a planet on the 23rd September, 1846, in the PROXIMITY of the place indicated by that astronomer, only that, as we have seen, it was 500,000,000 miles lower in the heavens, nearer home, than expected.

Thus, the CALCULATED, or THEORETICAL, planet NOT having been found as yet, and Neptune neither answering in distance, orbit, mass, or motion, the body that was looked for, though he is claimed as the captive stranger and paraded in triumph as the climax of astronomical achievement, it follows therefrom and the reasons advanced before : that the discovery of Neptune, instead of another planet described and predicted, was a pure accident, lying in wait for the king and catching a tartar, shooting at an eagle and hitting a sparrow, as MOST CLEARLY AND STRIKINGLY exposed and developed in *Baby Worlds* of Von Gumpach ; that its very discovery, more damaging than the prediction of weather, reveals the hollowness of certain astronomical pretensions, however great the talent displayed ; and that consequently all present calculations respecting the masses of the heavenly bodies, whether known or unknown, are as yet nothing but—love's labour lost.

Astronomers, therefore, will do well to strike out from their works EVERY CHAPTER AND VERSE relating to the weight, masses and densities of the celestial bodies, particularly when they consider, that the solar atmosphere will bear aloft the lightest as well as the most solid and gigantic nucleus, provided its being clothed by God in the corresponding envelope of an elastic, light and heat-con- centrating and reflecting atmosphere, in the transparent lens of hydrogen, throwing blessings upon the body it encloses.

We know that in the building of an edifice the material and masses become lighter as it rises from the foundation, from the most solid part of the whole.

It is the same with the earth from its centre to the end of its sphere, and so it is the same with the dome of the solar system, rising from the inner foundation of our magnificent luminary to its vanishing limits, bearing aloft in rising scale and proportion the heralds of its wonderful structure, and, mildly shining among them, the abode still of the man-made Creator Himself.   Hence it is but reasonable, and conformable to analogy, to suppose, without requiring further proof, that the body of the sun cannot be less dense than the body of the earth, or any other planet, but that the centre of our brilliant system should not only in size, but also in weight, in solidity of construction, in intrinsic power, and, as it were, animation, surpass the smaller bodies which he feeds and warms, and be fit to bear upon his shoulders, and carry round with ease, a stupendous solar world, the evidence of Divine power and wisdom, in the conception and simplicity of the creation.

XIII.

COMMON CENTRE OF GRAVITY OF TWO HEAVENLY BODIES; SIR JOHN HERSCHEL'S MARVELLOUS ELUCIDATION.—REVOLUTION OF TWO STARS ROUND EACH OTHER; THE BINARY SYSTEM.—SIRIUS, AND OTHER SUNS, MOVING ROUND A DARK BODY AS THEIR CENTRE OF GRAVITY.—FIXED STARS IN LIFE, ASTRONOMY AND GEOLOGY.—RETROGRADING MINDS AND THEORIES IN SCIENCE.—REVELATION BORNE OUT BY TRUE SCIENCE.—DR. MÆDLER AND THE CENTRE OF GRAVITY OF THE UNIVERSE.—IMPORTANCE OF ACCURATE OBSERVATIONS.—THE THEORY OF NEWTON A LABYRINTH OF ERRORS.

CONNECTED with the masses of bodies is the theory of their motion round a common centre of gravity.

That the calculation of this common centre of gravity must be wrong if that of the masses of bodies be incorrect, and *vice versa*, is evident; but if even these masses were correctly calculated to the ounce, the theory of a common centre of gravity between the heavenly bodies would be a fiction nevertheless, as will be seen very clearly.

"If we connect two solid masses by a rod, and fling them aloft, we see them circulate about a point between them, which is their common centre of gravity; but if one of them be greatly more ponderous than the other, this common centre will be proportionally nearer to that one, and even within its surface; so that the smaller one will circulate, in fact, about the larger, which will be comparatively but little disturbed from its place." (§ 359, Sir J. H.'s *Outlines*).

Applied to ". . . . . . the earth and moon, while in the act of revolving, monthly, in their mutual orbits about their common centre of gravity, ... We may conceive this motion by connecting two unequal balls by a stick, which at their centre of gravity, is tied by a long string, and whirled round. Their joint SYSTEM will circulate as one body about the common centre to which the string is attached, while yet they may go on circulating round each other in subordinate gyrations, as if the stick were quite free from any such tie, and merely hurled through the air," (§ 452.)

"Admitting, then, in conformity with the laws of dynamics, that two bodies connected with and revolving about each other in free space do, in fact, revolve about their common centre of gravity, which remains immovable by their mutual action, it becomes a matter of further enquiry, WHEREABOUTS between them this centre is situated. Mechanics teach us that its place will divide their mutual distance in the inverse ratio of their WEIGHTS or MASSES ; and calculations grounded on phenomena, of which an account will be given further on, inform us that this ratio, in the case of the sun and earth, is actually that of 354,936 to 1,—the sun being in that proportion, more ponderous than the earth. From this it will follow that the common point about which they both circulate is only 267 miles from the sun's centre, or about 1-3000th part of its own diameter." (§ 340).

Error in science, as well as in religion, being as infinite in its variety as truth is infinite in its simplicity, it would require volumes to expose the contradictions of the old theory of the solar system, and pages after pages to dissect all the fallacies in the foregoing extracts from Sir John Herschel's *Outlines*. The necessity of only

noticing the most flagrant errors and anomalies on a limited scale
is already a great drawback to the simplicity of the "Simplicity
of the Creation," and therefore I have tried, and will do so still,
to be as brief as possible in my refutations.

That, as yet, we have no test of calculating masses, I have shown
before, and we have seen also how Le Verrier and other astronomers
themselves have lately put in their own personal evidence to that
effect, and thereby, without intention, no doubt, nullified all
their previous labours.

When we come to calculate to a nicety the mass of a balloon
floating above us, we may perhaps find out a means of calculating
also the mass of the moon; and if we can ascertain and calculate
the density of the solar atmosphere, its pressure upon the sun,
and—knowing distance, dimension, and envelope,—we may find
out the weight of the heavenly bodies revolving within the solar
system; for, the atmospheres themselves of sun and planets are
balances, in the scale of which inflated or expanded bodies rise or
sink according to their weight and volume.

Supposing the masses of the heavenly bodies to be truly found,
we know, that they are not, two-and-two, tied together, or united
by a stick; that there is no one, no kindly force, to hold the
stick or the balance; that there is nothing imaginable to form, or
fix upon, a fulcrum, to establish a real tangible centre or point of
gravity, as is done by the human hand in the case of the balls
united by, or by means of the string tied to, a stick.   Therefore, the point, or centre of gravity, between the two heavenly
bodies is simply simulated or fictitious.

When we see the minute-hand of a clock counterbalanced by a
weight, the minute-hand, for all that, forms but one body with its
balance, whatever may be fixed or attached to either of the
extremities.   This one body turns round an axis, round a centre,
as do the spokes of a wheel; this axis is something tangible,

something intelligible, held and moved by a tangible means; and is in fact the axis, the centre, that turns the hand, whether that axis, common centre, or fulcrum, be central or ex-central, nearer to one end of the hand or nearer to the other; "the joint SYSTEM will circulate (revolve) as one body about the common centre," but the two ends will NOT, as it is said they do, "circulate round each other in subordinate gyrations," nor will "the smaller end (or mass) circulate about the larger," but both will REMAIN IN THEIR RELA-TIVE POSITIONS. If the pointing end of the minute-hand were to circulate about the balance-end, it must needs get to the opposite side of it, which, I think, would be rather strange if it did; or, if the earth and moon were "circulating round each other," the earth, too, should some time or other get to the other side of the moon, which, not being the case, there can, of course, be no question either of their "circulating round each other."

If the circulation round each other, and their motion about a common centre of gravity were true, Jupiter, Saturn, Uranus, and most likely Neptune, would be placed in most awkward positions.

These planets having a plurality of moons, there would be also a plurality of "common centres of gravity," a different common centre between the planet and each one of its moons, and different centres between moon and moon, and moons and moons; but how the planet would manage to move round each one of these several centres of gravity, besides turning on its own axis; and how moons about planet and moons about moons would contrive to "circulate round each other in subordinate gyrations;" or how the JOINT SYSTEM of moons and planet could and would circulate round a mutual ONE common centre of gravity, I must leave to more learned men than myself to unravel; it is an astronomical article of faith beyond anything ever proposed to our belief in revelation.

The same difficulty would apply to the sun.

Between the sun and each one of the planets, according to the old theory, there is a different "common centre of gravity;" each centre "IMMOVABLE by their mutual action," and THEREFORE SHIFTING as each planet progresses in its course round the sun. Now if "both the SUN AND EARTH circulate about the common point, which is only 267 miles from the sun's centre," how is the circulation to go on, "about the other immovable yet shifting common points of gravity" between the sun and each one of all the other planets?

And if "every particle (or body) of matter in the universe attracts every other particle (or body) etc.," there must be then a common centre of gravity between planet and planet in every fathomable direction; and what a wonderful "circulation about common centres of gravity, AND ABOUT EACH OTHER in subordinate—and I suppose insubordinate—gyrations," should we not have?

If, moreover, "the earth and planets be retained in their orbits about the sun by SOLAR ATTRACTION" emanating from the SOLAR CENTRE; and "if gravity, or the attraction of the earth and planets" emanating from their CENTRES, "be the force which retains the moons in their orbits;" and if this gravity, this attraction of the primaries, is counteracted by the centrifugal force said to emanate at the same time from these very primaries, in order to prevent the moons and planets falling down upon them : what then, is the use, and what becomes, of the EX-CENTRAL COMMON CENTRES of gravity alleged to exist somewhere between these primaries and their secondaries, and about which all of them and all about each other are said to move and circulate? Was there ever a more eccentric theory?

Besides this, in the words of Sir John Herschel, *Outlines*, § 841, there still "exist sideral systems, composed of two stars revolving ABOUT EACH OTHER in REGULAR ORBITS, and constituting what may be termed BINARY STARS, to distinguish them from double stars generally so-called, ... ;" they, § 840, " describe orbits round each other, and around their common centre of gravity."—" But it is not," § 847, " with the revolutions of bodies of a planetary or cometary nature round a solar centre that we are concerned; it is with that of sun round sun—each, perhaps, at least in some binary systems where the individuals are very remote, and their period of revolution very long, accompanied with its train of planets and THEIR satellites, closely shrouded from our view by the splendour of their respective suns, and crowded into a space bearing hardly a greater proportion to the enormous interval which separates them, than the distances of the satellites of our planets from their primaries bear to their distances from the sun itself."—" No doubt can therefore, remain as to the prevalence in this remote (binary) system of the Newtonian law of gravitation." § 864. " We have the same evidence, indeed, of their rotation ABOUT EACH OTHER, that we have of those of Uranus and Neptune about the sun," § 846.

Now after what we have seen before of eccentric centres of gravitation, very few, if any, of my readers, will be inclined to accept the preceding allegations of Sir John Herschel: that the Newtonian law of gravitation, which altogether fails in our own solar system, should prevail in systems so remote, that their distance cannot be measured, not even fairly guessed; and that two suns should revolve round EACH OTHER, round a common centre of gravity, and that this common centre of gravity, according to Mädler, who has " signalized himself in this line of enquiry," should be IMMATERIAL, and in IMMATERIAL VACANT SPACE ; that, in the words of Humboldt, *Cosmos* p, 81, " two or more self-luminous bodies do, indeed, revolve around a common centre of gravity ; " that " this centre of gravity falls in a space, occupied, possibly, only by unagglomerated matter, *i. e.* cosmical vapour, whilst in our system, the centre of gravity is included WITHIN THE SURFACE of a visible central body ; " and that, p. 134, " our sun REVOLVES AROUND THE COMMON CENTRE OF GRAVITY OF THE WHOLE SYSTEM, which centre, in consequence of the varying position of the planets, falls sometimes WITHIN and sometimes WITHOUT the body of the sun itself." How the sun gets round this imaginary centre Humboldt does not tell us.

The law of " universal gravitation " would keep systems, suns, planets, satellites, and comets, immovably fixed in their respective positions like one universal still-standing machine, if this " universal gravitation " were not, happily, counteracted and equally matched by centrifugal force, and heavenly systems and bodies, fortunately, not set in motion by an ever-acting, everlasting, primitive impulse, and sundry other forces, acting upon and counteracting each other.   Hence all the confusion and uncertainty of the theory of universal gravitation.

The unfathomable ball of creation, by God gently wafted through eternity, holding within its vast expanse worlds and systems in numbers beyond the reach even of the utmost stretch of our imagination, yet all are harmoniously distributed within the sphere of the universe, all have assigned to them their proper place of permanency and motion ; and the intermediate space between the larger systems is taken up by minor constellations innumerable, yet all so beautifully arranged, that by a divine provision of universal systemal repulsion, each one pursues its destined course and follows the laws laid down by its Creator, uninterfered with by one another, the greater systems neither pressing upon nor absorbing the lesser ones, nor the lesser ones stepping beyond their limits to obstruct the path of the greater systems, yet all held by one common electrical sphere in which they are poised, and, however grand or diminutive, minutely balanced.

It is possible, perhaps probable, as I have said before, that all

the systems of the universe gravitate to one common centre ; and
if this be so, that this common centre of gravity is our own solar
system.   But it is, perhaps, still more likely, that our own solar
system is simply the centre of created worlds, without being their
centre of gravitation, in as much as with an universal systemal
repulsion, no universal gravitation can co-exist, no gravitation to
a common centre, no gravitation of system towards system is
required, though their polarization may effect each other.   This
polarization and mutual repulsion, this keeping apart, does not,
however, affect the interior order and arrangements of the individual systems, whether revolving round a central sun simply as
their centre, or as their centre of gravitation; none encroaches
upon, nor interferes with, the other.

Therefore, as our solar system unchangingly performs its functions, in its INTERNAL ORGANIZATION uninfluenced by any other
system, so every other system similarly and independent of every
other, carries out the work and object of its own constitution.

Without a tangible centre no system can exist, as witnessed even
upon earth, whether religious, political, or social; and two systems,
or more, can still less act in harmony for one common purpose,
without a real tangible centre keeping them linked together in the
bond of unity.

It is the same in the binary system of the fixed stars, where sun
is said to revolve about sun, both revolving round each other.
For, as we have seen before, they have, or at least are said to have,
no common tangible centre, and their revolving about each other
is a physical impossibility.

That this should be so, at first gave me a little trouble when I
began the experiment of turning my forefingers round each other;
it seemed so beautifully to illustrate " two mighty stars " revolving
about each other.   On reflection, however, I found this turning
about each other to be a simple delusion, something like the moon

turning round her own axis, that it existed only in appearance, chiefly caused by both fingers preserving their parallelism whilst moving.

Let us illustrate it in the case of a binary system, confined within a certain space of the heavens, and completely filling the same when moving in a circle. Let the following figure be the space occupied by the two suns a and b with their systems; between N. and s. they are side by side; pursuing their proper course with equal velocity, a will arrive at B., whilst b arrives at w.; and so on it will be found that both systems preserve their relative position, that both move in the same path and circle and FOLLOW EACH OTHER, and that consequently a never gets round b, and b never round a; and if their velocities were unequal, the one would run into the other, but they would never course ABOUT EACH OTHER.

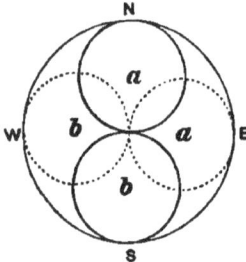

Illustrate it by separating a and b to whatever distance from the centre of their sphere, and the result will precisely be the same.

Let our next figure be the space assigned to the two systems, a and b placed opppsite each other and within the skirt of their sphere; follow the same test and reasoning as before, and you will again find the physical impossibility of their revolving ABOUT EACH OTHER. They revolve round a common centre, to be sure; but this common centre is arbitrary, not real, not tangible, therefore, only an imaginary point nowhere to be fixed, and consequently neither fulcrum nor centre

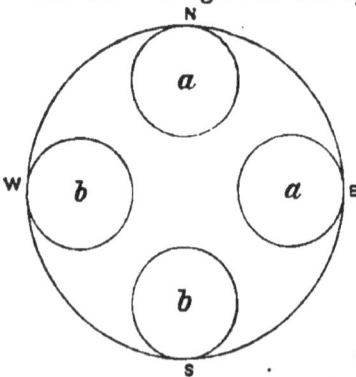

of gravity ; the confines of their spheres, or envelopes, simply touch each other as in the previous figure.

Moreover, being SUNS, each one having its own photosphere they MUST repel each other, they CANNOT gravitate towards each other,' though both MAY gravitate to some real centre of' the universe.

If *b* were to describe a smaller circle within the circle of *a*, whatever their velocity of motion, it would be proof of *a* circu- lating round *b*, not only as its centre, but also as its centre of gravitation.   And this,—always supposing observations to be correct, and taking " circulation round each other " to be but a misapplied expression—no doubt is the case in what are called binary systems, and what takes place also with the so-called double stars, in systems where three or four SUNS SEEM to move round a common centre of gravity, and which centre is considered to be an opaque, a dark, non-shining, central body, towards which the shining stars are gravitating.   This last assumption, however, is in the highest degree improbable, though not impossible.

A shining body, like our own luminary, by its radiating heat, causes the dilatation and expansion of its own atmosphere, and bòdies like our earth and the other planets, gravitating to the centre of the sun, are thus repelled, as it were, to that distance from the sun himself, where in the tension of the solar atmo- sphere they find their level and therein course around the sun by a truly natural process as further back has been described.

If our central body were dark, non-luminous, and particularly without radiating heat : all resistance to the gravitation of the planets would be removed, the atmosphere of the sun would collapse, and they would inevitably fall upon him. If, on the other hand, in their present respective proportions—and it has been surmised that Sirius, and other suns, are moving round dark bodies by which they are far surpassed in bulk and mass—the

U

planets were shining in their own radiating light and heat, they would recede, push back, from their dark central body, towards which they COULD NOT be attracted, towards which, at least, they COULD NOT GRAVITATE, the condition of gravitation being removed. The planets, self-luminous, would, in fact, be suns, and as such maintain their respective positions in the universe, independent of their present attracting central body; and instead of gravitating to him, the sun and planets would MUTUALLY repel each other, and the latter would no longer follow in the track of the former, they would cease to belong to the solar sphere or system.

Nevertheless, as I have said before, it is possible still, that in other systems of the heavens God may have reversed the organization which exists in our own, and that in these other systems the central body, the centre of gravitation should be dark and yet carry within its atmosphere, not planets like our own, but planets shining in their own light and even surpassing our own sun in brightness, splendour and size, and throwing light, though by us unseen, upon the body round which they circulate, and from which, being suns, they can have no light or heat in return, unless it be electricity to feed their photospheres.

Whether, in this case, the dark central body would be suspended and move within an electric helix like self-luminous orbs and centres of systems? Whether it would revolve on its own axis by its own interior and atmospheric exterior organization, without a photosphere? or whether the planetary suns coursing around it would influence its interior life and cause a rotation on its own axis? Whether the presumed central body can gravitate to its solar satellites? Whether between these solar satellites and their dark common centre of gravity, there exist other common centres of gravity—the same as between Jupiter and each one of his moons are said to exist—round which mutually they are moving? And whether these common centres of gravity are eccentric like the one round which the earth and the moon are calculated to revolve ?   Whether these planetary suns to a dark central body, by the very nature of their constitution, are not independent systems ?   And if these splendid luminaries do throw their light upon their presumed common centre : should not its darkness disappear and give us,—the same as with the unassisted eye we see the distinct outline of the body of the moon a little before and a little after NEW moon,—in outline at least, a glimpse of the colossal globe that is said to bear two, three, four, if not more, suns with their systems on his shoulders ?

These and other difficulties connected with the theory of suns circulating round a dark central body, like most other facts assumed in astronomy, cannot be reconciled with, nor explained by, the theory of universal gravitation.   However, the assumption MAY be true, though I do not think it likely, scarcely more so than that suns and systems should circulate round an imaginary immaterial common centre, whether in vacant space or lying within unagglomerated matter, i. e., cosmical vapour.

As to the masses of these distant bodies it would be waste of time to speak of them in the present state of astronomical science, when, as we have seen, the masses of our own sun, earth, and moon, are still unknown, whatever some astronomers may maintain to the contrary, and when on more important questions and calculations leading authorities are by no means agreed, though they may hold general views and some special articles in common.

Before gaslight illuminated the habitations of man, and oil and tallow in their humble capacities still everywhere had undisputed sway over darkness, a private in the army one day asked one of his betters the difference between a fixed star and a planet. The superior replied, that the difference consisted in the fixed star shining in his own light, but the planet in the light he got from the star.   Then I am sure to be a planet, said the soldier, for I always get my light from the master of the barrack.

As thus there are thousands still in the military world receiving their light from the heads of barracks,—that great landmark of TRUE CIVILISATION, "gas," not having penetrated everywhere—so there are thousands of civilian planets receiving their light from self-shining masters of more elevated masonry; and as a few only of the fixed stars of the heavens outshine, as it were, those of second and more subordinate magnitudes: so but a few of the fixed stars of the astronomical sphere here below shine over the lesser lights who crowd around them; but all of them oscillate and scintillate and flicker, and differ in their spectra; yet, with all that, there is an apparent unity, with its imposing mathematical mystery covering the incongruity and moth-holes of their creed, imperceptible to the multitude of planets having no light of their own.

Hence, all this seeming unanimity about the theory of Newton has no reality in fact, the borrowed light of astronomical planets reverts to but a few glittering luminaries, to a certain number of self-shining dominating lights, all differing except in name; there is no unmistakeable sun in which they centre, and thus all is uncertain, their elements ever varying, their calculations always at fault; there are long sights and short sights, there are broad views and narrow views, high ones and low ones, according to the stand point and lens of the observer.

Therefore, no sensible man would pin his faith to such fallible guides, put himself into the relationship of the planet to the star of the barrack, though he admire their light and their talents, and pay deference to their learning and observatory eminence.

In geology it is the same as in astronomy, only that in this branch of science there is still more diversity, uncertainty and speculation.   The different schools hold different theories, but amongst these, almost every independent enquirer, with thought and a mind of his own, entertains different views and deductions ; a good many are agreed only when there is anything seemingly to tell against revelation and the account of the creation given by Moses.   Nevertheless, these ultras of the science follow but a few prominent leaders, wanting no God, no virtue, no sin, no redemption, looking upon beasts as their kindred and venerable companions ; and in this ignoble fraternity having lost the dignity of man, they also lost his prerogative, common sense.

Geological Don Quixotes, they hope with a fossil bone to break up creation, with a piece of old broken earthenware to scratch out the history of mankind and tear up in bits the sacred writings ; with a pointed boulder of flint to put aside Moses and the Prophets and the Martyr-testimony of the Church; with the spear of a savage to sink the ark of Noah and overthrow the tower of Sion ; with a handful of sand taken from the pit of their cranium to blind all the nations of the earth, their gorilla brothers included; and yet these pretended men " call upon our credulity to give them credit for learning ; and upon our faith in science to prefer their geological crotchets to universally-established historical facts."   •

Is there nothing simple to recall the remembrance of these philosophers to future ages ?   Dr. Scheuchzer, a Swiss naturalist, took a fossil skeleton found near the lake of Constanz to be the remains of a man, calling it " homo diluvii testis."   It turned out, however, that this " homo diluvii testis " was nothing but the skeleton of a salamander, of which a living specimen may be seen at the Zoological gardens, Regent's Park, London, and which ever since, in commemoration, bears the name of " Andrias Scheuchzeri."

Upon foundations, such as these, the whole history of man, with all that is dear to him, with his hope in eternal life, is to be thrown overboard! Where then is the presumptuous folly, and where is common sense? Where is the illusion, where the naked fact? Where the retrogradation to animalism, where the progress to the life of angels? Where the vulgar and narrow mind, and where the glowing soul expanding beyond the limits of nature into the regions of a happy eternity? Oh, for the difference! to lean on the ape, and to lean on a Heavenly Father!

To dwell any longer on such DEBASING ABSURDITIES as taught by Lyell, Huxley, Darwin, & Co. would be an insult to the natural perception and understanding of the reader. However, LEARNED men are not wanting either who, on SCIENTIFIC grounds, inch by inch cut away the plank of infidelity, vindicate the high position of man in creation, the sacred character and truth of Holy Writ, and palpably reconcile with it the discoveries made in every science. The most eminent work on the subject I have met with is that of *Bibel und Natur* by Dr. Reusch, Professor of Theology, in Bonn, published by Herder, in Freiburg, Baden.

Revelation and true faith can never be affected by true science, and no science can be held as true if it bear not the test of revealed religion and the sacred writings truly interpreted.

As then, without regard to this test even, there is no end of differences in matters of science, save the science of the saints, so there can as yet be no science as an ESTABLISHED TRUTH except in minor points; and this will be found to be the case the more every branch be sifted. In geology man is lost at a few fathoms deep in the earth, scarcely knowing the surface he treads on; in astronomy he is erring about in the distance, not knowing as yet the bright sun and moon shining upon him, launching out into the deep without catching any fry near the shore.

Thus Dr. Mädler loses himself in space without knowing the weight or interior of the body that bears him; calculates the path and direction of the sun, not knowing the path of comet or meteor, or the correct orbit of earth and moon; calculates the centre of gravity of our solar system and of the WHOLE UNIVERSE to lie somewhere in vacant space in the group of the Pleiades, if not in its principal star Alcyone, yet not knowing the centre of gravity of our own globe and satellite.   In this instance the Doctor alone is the pointer, all are to follow; but the differences of scientific men before spoken of are here manifest also, for, Sir John Herschel affirms, that the Pleiades seem to him rather an eccentric point of gravity for our solar system and the universe.

How then can we have faith in present astronomical propositions and dicta, when doctors differ, and when in the highest of the secular sciences, eccentric common centres of gravity, as between the earth and the moon, and the sun and the earth and the planets, and between each other, are taught and believed in? and when in the calculation of distances, seemingly the most simple of all astronomical problems, they are so seriously at fault, and when, upon their own showing, their observations even cannot be relied upon, as seen already in treating of comets?

Hind, in his *Introduction to Astronomy*, p. 34, says: "ACCURATE OBSERVATIONS . . . . . have PROVED that the mean distance between the Earth and the Sun exceeds 95,370,000 miles."

"It is of the HIGHEST IMPORTANCE in astronomy that this number should be exactly known, because it is used in calculations AS A SCALE by which to estimate planetary distances; and if WE HAVE NOT THE ACCURATE DISTANCE OF OUR OWN GLOBE FROM THE SUN, we cannot find that of any other planet with precision."

Pray, then, remember the lucky finding of Neptune, and also consider what becomes of these ACCURATE OBSERVATIONS, and of Kepler's third law applied to them, and what of the distances of ALL OUR PLANETS, if, in the same year in which the above of the "HIGHEST IMPORTANCE" was written, it is found, that the distance between the earth and the sun is only 8,000,000 of miles less than hitherto PROVED! what of the fact (*Introduction*, p. 212) that "it is from the observations of the transits of Venus which happened in the years 1761 and 1769, that astronomers have fixed the received distance of the Sun from the Earth!" what of the result of Sir John Herschel's proposition that "the first step towards ascertaining the real size of the sun is to determine its distance?" What becomes of universal gravitation if in the "Celestial Measurings and Weighings" of Sir John, in *Good Words*, of June, '64, we read that "there is reason to believe too that the distance of the moon" THOUGH (our knowledge of which does not assume that of the sun as known) "has been somewhat MIS-ESTIMATED," not MIS-CALCULATED, "and that an alteration" (though not nearly to so great a proportional extent as the error in the distance of the sun), "bringing our NEAREST CELESTIAL NEIGHBOUR into somewhat closer proximity than heretofore SUPPOSED, is required."

This, then, is the result of universal gravitation, that as yet WE DO NOT KNOW THE MASS AND DISTANCE OF OUR NEAREST NEIGHBOUR! The calculation of the tides then also wrong? The moon draws ten feet and the sun two feet!

In a fair notice of my first edition in the *Athenæum* I have been told that I did not seem to understand the theory which I assailed, and have confessed already in my *Concise View* that I do not do so, because it CANNOT BE UNDERSTOOD.

In this respect I stand much in the same position to universal gravitation as the two Papal gendarmes to the Dean of Canterbury. They did not understand him; in his letter "From Rome" he is wroth at the poor fellows for their ignorance; but, good innocent man, it never struck him that they took his Italian for broken English!*

---

* They, perhaps, took him for the Dean of Canterbury? However, he does not seem to know LATIN either, or ever to have read the Litany of the

Such is to ME the SINGULAR MODE OF ACTION OF UNIVERSAL GRAVITATION.

XIV.

UNIVERSAL GRAVITATION A VICIOUS CIRCLE.—ELLLIPTIC ORBITS.—GREATER AND LESSER DISTANCE OF THE SUN AND MOON FROM THE EARTH.—PATH AND PASSAGE OF THE MOON.—BACKWARD MOTION OF THE MOONS OF URANUS AND NEPTUNE.—COMETS, PLANETS AND MOONS INFLUENCED BY THE POLES OF THEIR PRIMARIES.—OLD THEORY OF THE TIDES; ITS ANOMALIES AND REFUTATION.—NEW AND CORRECT THEORY.—NON-ROTATION OF THE MOON ROUND HER OWN AXIS; PROOFS.—ECCENTRIC THEORY OF AN ECCENTRIC COMMON CENTRE OF GRAVITY BETWEEN EARTH AND MOON.—TIDE OF THE ATMOSPHERE.—NO TIDES IN THE OPEN SEA; NONE ROUND THE EQUATOR; TABLE OF TIDES.

WE have seen that neither observation, parallax, measurements, the velocity of light, comets, nor the transit of Venus, have given us the TRUE DISTANCE, SIZE and MASS of the sun; and as these

---

Blessed Virgin to which he refers in his classical epistle "From Rome."

In that beautiful composition of titles of the IMMACULATE MOTHER OF OUR LORD, the "MISERERE NOBIS," "HAVE MERCY on us," occurs four times at the beginning and ONCE at the end, as a prayer addressed to God Himself. Intermediate, in our invocation of the "Blessed among women" to intercede for us, there is, once after every title, FORTY-FIVE TIMES RUNNING, the petition "ORA PRO NOBIS," "PRAY for us."

"The priest was kneeling at the altar, singing the Litany of the Virgin in which she is addressed in direct prayer, 'Mother of mercy, have mercy on us:' 'Mother of grace, have Mercy upon us,' etc.: the CONTADINI repeating the 'MISERERE NOBIS' AFTER EACH TITLE of invocation had been given out by the priest." So the Dean.

Now his ear never caught or understood the forty-five times repeated "ORA pro nobis;" but, ro doubt, the last "MISERERE nobis" must have come upon him like the discovery of Neptune, for he actually quotes it in its native Latin, whilst the former he neither gives in Latin nor in English.

Is not this very odd, very, indeed, if he DO know Latin, and if he HAVE read the Litany of that Blessed Virgin? Surely, he cannot have meant to deceive? And yet that REASONABLE, LOVING, anxiety, lest his flock should read CATHOLIC works! What a strange Dean of Canterbury! at once so rich and learned, and yet so poor and ignorant!

"Atrocious translation" is a specimen also of the DEAN'S ENGLISH!

are used as a SCALE in the science of astronomy, it follows: that the error in distance, size and mass of the sun NULLIFIES the ACCURACY of all the observations hitherto made, every parallax and measurement, the velocity of light, the distances, orbits, sizes, masses and velocities of all the planets and moons and comets, their mutual attraction, perturbation, common centres, the tidal theory, etc.; that in fact, universal gravitation itself is nothing but a VICIOUS CIRCLE, that NOT ONE calculation based upon it can be accepted as true even for our own solar system, much less for celestial bodies and systems beyond our own.

My new system, I trust, will give a sound and firmer basis to astronomical science, and that the talent and intellectual labours of so many gifted men hitherto thrown away upon a thankless theory, may be applied to mine and lead to happier results, both as to astronomy itself as well as to the higher questions involved in it, and in the latter of which so many of the learned have failed far more than in the former.

As my theory so simply explains and proves everything connected with our solar system, so also the elliptical orbit of the earth, as well as of all the planets, if not disproved, is at least accounted for by it.

When the moon is near the horizon, either rising or setting, she appears much larger, than when she is higher up. The sun likewise looks much larger in the morning and evening than at mid-day, when he is nearly above our heads. The reason of it is this: that when these bodies are (to us) high in the vault of heaven, their rays fall nearly in a perpendicular line upon the spot we live, without being refracted by the atmosphere; if, however, they are near the horizon, and consequently FURTHER AWAY FROM us, though equally near to the earth at large, their rays are refracted by the atmosphere which they penetrate in a slanting manner, and therefore also to a greater depth or thickness, and hence they appear to us through a magnifying medium, although with diminished brightness.

The same must be the case when the earth is oscillating above the equator of the sun.

When in our summer the northern part of it is attracted, and the southern one repelled, the luminary of our system appears to us small, though NEARER than in winter, because his rays are almost perpendicular to us, we are nearly in the immediate centre of the focus of his concentrated light; and in our winter he appears to us large, though we are repelled and the northern part of the globe retiring from him, because he has apparently receded into the tropic of Capricorn, his rays come to us obliquely through a greater depth or thickness of the refracting atmosphere, the hydrogen lens above has shifted from a perpendicular to an oblique position, from the zenith, as it were, towards the southern horizon, showing to us a larger picture of his disc. But throughout this oscillation of the earth, the sun, probably, is always equally near to it in its orbit in general, and his supposed perigee and apogee most likely an optical deception.

Upon the very opposite assumption the theory of the ellipticity of the earth's orbit is founded. According to this the sun is said to be further off from us in summer than in winter, because he seems so much smaller, and in winter he is said to be so much nearer than in summer, because he looks so much larger.

This reasoning would be true, if the intervening atmosphere and still more the shifting focus of our hydrogen lens, did not counteract, or rather reverse, the law, that, the more distant the object, the smaller its apparent diameter, and the nearer the object, the larger its diameter, as is the case with the moon, who appears greater when she is in perigee than when she is in apogee, not being, like the sun, affected by the hydrogen lens in, or upon, which she floats.

Now if the calculation of the greater or lesser distance of the earth from the sun, and the elliptical orbit of the earth round the sun, have no other foundation than the fact, that the sun appears to us smaller in summer than in winter ; that the time between the vernal and autumnal equinox is about eight days longer than that between the autumnal and vernal equinox, and that therefore the sun is much further away from the earth in our summer than he is winter, we may well doubt it after what has been said before, and according to which the earth is more likely to move always at the same, or nearly the same, relative distance from the sun.

As to the longer journey, or slower motion, of the earth between the vernal and autumnal equinox, than between the autumnal and vernal equinox, it does not follow that therefore the earth must be at a greater distance from the sun, that it has to sweep through a wider curve during the former period. I think it more probable, that the earth this side the equator, having more land, and therefore being more intensely electric than the south, and consequently more strongly attracted by the sun, it travels slower, that is, it makes less progress in its orbital motion, though the axial rotation remains the same when the sun attracts the northern, as when he does the southern hemisphere. It is like the paddle-wheel of a steamer ; the rotation of the wheel remains the same, whether against or with the current, though the progressive motion is different. Or, like the wheels of the locomotive, which for want of sufficient friction with the rail do not make much progress, though the revolutions remain the same. It may be also that in our summer, for the reason before stated, the envelope of our earth becomes more expanded than when the rays of the sun are upon, and absorbed by, the waters of the south of our globe, and that therefore the earth may ascend and float a little higher in the atmosphere of the sun than generally. We should thus have an expansion and contraction of the atmosphere of the earth which would bring it further from the sun in our summer and nearer to him in winter, as the old theory has it, though the expansion and contraction, unlike that of the hair of the comet, would not, perhaps, influence the distance from the sun in any perceptible degree. At all events, my new theory, in either case, seems to me to solve the problem of the ellipticity of the orbit of the earth round the sun, not proved, at least, by the difference of the apparent diameter of the disc of the sun in summer and winter according to the old theory.

It is natural that during the oscillations of the earth, the alternate attraction and repulsion of its poles, and the expansion and contraction of its envelope, perhaps influenced also by the oscillations of the sun, or by the passing planets of our system, it should not move in a straight, undeviating line round the sun, but at times pass a little more to the right, and at others more to the left; and hence, no doubt, what are called PERIODIC AND SECULAR INEQUALITIES.

By my theory the mutual attraction and repulsion of the planets, though not on the law of universal gravitation, is clearly demonstrated if at their respective distances it exist at all; they may affect each other when approximating in their passage round the sun, but they do not disturb, or keep each other, in their respective positions and orbits. My theory, moreover, seems most beautifully to illustrate the passage of the moon round the earth, and assigns a probable reason for her course.

Having, like a nascent meteor, risen from the earth to be converted into a glowing lamp and suspended high above us to shine into the darkness of our nights, but as a body without the electric inflation or constitution of the planets, she is deprived of rotary motion, and only floats upon our sphere like a vessel on the ocean. God placed her on our ethereal atmosphere, THAT AS A LIGHT TO THE EARTH, in turn attracted and repelled, she might pass from pole to pole, and shed her benignant rays on those portions of our globe in particular, which are least in the enjoyment of the light of the sun. And for this purpose the Creator of the world had constructed her in such a manner, that, arriving within the equatorial regions, where there is neither pointing nor dipping of the magnetic needle, her craters and mountains should prevent her turning over, and that they also should act as divinely-arranged rudders, to keep her always in the path that had been assigned to her.

Drifting across her ethereal ocean from south to north she passes slowly on to the ecliptic, to the equatorial region of the earth, gently carried eastward by the revolving motion of our globe, though still making for the north. On nearing the ecliptic, the middle of the ethereal current she has to cross, her speed increases, and the rapid revolution of the earth, in its greatest circumference, would carry her out of her course and whirl her round the equator, if the rudders of the divine Constructor did not keep her in her path; she hastens, as it were, to take her light to the north, and drifts slower again on passing out of the current of the ecliptic.

In this manner the moon passes round the earth from pole to pole, from one side of the south to the other side of the north, crossing her ocean like a steady ferry the river, without ever altering her course or position. Every succeeding passage she fails, like the earth in its orbit round the sun, to reach the exact point from which she started, until after about nineteen years she arrives again at the starting-point; and this is called the retrogression of the line of nodes.

The motion of the moons of the other planets, as well as the rings and moons of Saturn,—whether his rings be solid like the moon, or an electric product like the fire-veil of the sun, or the aurora borealis—rests upon the same foundation; as the planet revolves, it carries them round on its ethereal ocean, on the same principle on which a ship moves slower than the current which bears her; or, the motion of the planet eastward exercises a more powerful influence on the moon than the westerly current of the planet's atmosphere.

The planets Uranus and Neptune, however, as we have seen already page 269, seem to form an exception; for it has been found, that all the satellites of Uranus, and according to the discovery of Mr. Hind, that of Neptune, move in the opposite direction, that is, from east to west; and hence it follows, that the upper atmospheres of these planets must have a westerly current so strong as to counteract, to overcome, the rotary influence from west to east of the planets themselves.*

The same happens with the sun and the comets, of which many move from west to east and many from east to west in the solar atmosphere; those moving from east to west, or having a retrograde motion, belonging almost exclusively to those comets which travel to the furthest limits of our system, into regions beyond Neptune, the most distant of all our hitherto discovered planets. The westerly current, or flow, of the more remote parts or oceans of atmosphere of the sun becomes stronger the nearer we approach the confines of the system; and the westerly motion of these infinitely-attenuated and elastic oceans of fluids or gasses, overcomes the eastward attracting influence of the eastward revolving sun; or the comets, or moons, in question, have no ballast, they are too light, and therefore, as it were, carried back by the superior stream

---

* As the water is left behind by the rotation of the earth, so is its atmosphere of air more and more left behind as its distance from the earth increases; hence also the increased motion or flow of the air, of wind, the higher we ascend the mountain. It is the same with the atmosphere of the sun; it is more and more left behind in proportion to its distance from the solid body. Hence also the planets are left behind in proportion to their size as well as distance; and for this reason they move slower the more they are elevated above the sun.

or streams of the solar or planetary atmosphere.   These moons
and comets may also want a proper or necessary amount of electricity, so that in their passage from one part of the planet to the
other, from one pole of the solar system to the other in almost
every direction across the ecliptic, the attraction of the eastward
rotating planet and orb is insufficient to draw them in their respective train, and thus, by some other law, are firmly guided in their
missionary courses by the legitimate centre that governs the whole.

Comets, in their diversified, spiral, and cyclical course round the
sun, are no doubt influenced, if not positively directed, in their
motions, by the poles of the solar system and the sun himself, as
they are also more numerous about the former than in any other
region of the solar sphere, the same as the auroras are more
abundant in the polar regions than in the temperate zones of the
earth, whilst none have been observed at the equator; the same
also as meteors, mostly from the polar hemispheres, take their
course in the direction of the meridian: but the extraordinary
eccentricity of the orbits of the comets, and the scarcity of their
visits, as already noticed before, seems to be due to the polar
attraction and repulsion of the solar system, and their consequent
passage from one solar ocean into the other; and as thus they
plunge into a denser medium the more they approach the sun,
their hair is driven back and elongated into tails of enormous size,
and often parted into two or more;* sometimes even, according
to the resistance of the matter or current they move in
or come across, these elastic bodies themselves divide, if not
in reality, at least apparently, and for a considerable time,

---

* The tails are often of stupendous magnitude, varying from 5 to 200
millions of miles in length.   But the brief period, in which they are often
formed is no less amazing.   The tail of the comet of 1843, which was long
enough to reach from the sun to the planetoids, was formed in less than
twenty days.   The tails of comets are not always single; some have apeared
with several tails.   The comet of 1744 had six tails.—*Worlds Beyond the
Earth*, p. 219.

into two seemingly-distinct comets, as was the case with Biela's in 1846.

It has hitherto been supposed that the moons of Saturn, (as well as the sun and celestial objects) are carried parallel to the edges of the rings; but it is shown now, that, by the apparent motion of the heavens, produced by the diurnal rotation of Saturn, they are moved so as to pass alternately from side to side of each of these edges, the same, and for the same reason, no doubt, as our moon alternately passes, attracted and repelled from pole to pole.

The tides are caused by the ship-like passage and PRESSURE of the moon upon one of our oceans above; but the present, or Newtonian, theory of the tides, ascribes them to the ATTRACTION of the moon. In one of the most current works upon this subject we read, that "the tides are occasioned by the attraction of the sun and moon upon the waters of the earth.

"Let A B C D be supposed to be the earth, and E its centre: let the dotted circle represent a mass of water covering the earth;

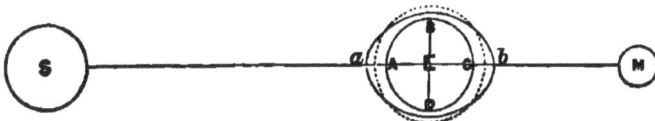

let M be the moon in its orbit, and s the sun.

"Since the force of gravity, or attraction, diminishes as the squares of the distances increase, the waters on the side of C, are more attracted by the moon M, than the central parts at E, and the central parts are more attracted than the waters at A; consequently, the waters at a will recede from the centre; therefore while the moon is in the situation M, the waters will rise toward a and b on the opposite sides of the earth; that is, they will rise at a by the immediate attraction of the moon M, and will rise at b by the centre E receding, and leaving them more elevated there.

"The moon goes round the earth in an elliptical orbit, and

V

therefore she approaches nearer to the earth in some parts of her orbit than in others. When she is nearest, the attraction is the strongest, and consequently it raises the tides most; and when she is farthest from the earth, her attraction is the least and the tides the lowest."

In another work on *Natural Philosophy* we read that "the tides proceed from the attractive force of the sun and moon, which diminish the gravity of the waters of the ocean, or, which is the same thing, draw or lift up the waters towards themselves.

"The force of attraction acts in straight lines; and, therefore, if we draw two straight lines from the moon's centre, M B, M D, to represent this force, acting on the parts B and D, it is obvious that

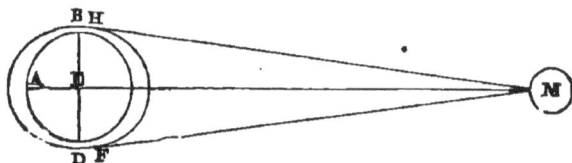

the water at B and D will not be raised, but depressed, by being drawn away from B to H, and from D to F, and so of every part of the circle B E D. But the waters will, at the same time, rise on the side of the earth away from the moon, because the EARTH'S CENTRE being more strongly drawn TOWARDS THE MOON than the point A, recedes from A, which is the same in effect, as if the waters at A receded, or rose up from the earth's centre."

By the earth's rotation on its axis, the two tide-waves will be continually making their round over the ocean.

"The force of attraction acting in straight lines, it should be high water at any place in the OPEN SEA when the moon is upon the meridian of that place; but, in fact, the greatest (and least) heights of the water at such a place do not occur till about three hours after the periods fixed in this supposition. The delay is thus explained: the elevated parts of the sea have received such an impulse towards ascent, that they continue to rise after the earth's rotation has carried them from under the line of the direct attraction of the moon; this impulse being also aided for a time by the moon continuing to attract the water upwards, though in a less degree.

"The sun, though his attractive influence is three times less than that of the moon, acts upon the ocean in the same manner, though in a less degree. When these two bodies unite their influence, which they do at the seasons of new and full moon, the tides naturally rise the highest: they assist each other in raising the ocean as before described, and depressing it in a corresponding manner.

"Of the irregularities of the tides, caused by the continually-changing position of the sun and moon, those are the greatest which are occasioned by the obstacles offered by the land to the ebb and flow of the waters. The impediments created by shallows in the ocean, and by shores, bays, gulfs, and promontories of islands and continents are such, that the tides are greatly delayed, altered both in degree and direction, and in many places so accumulated, that they rise to heights far exceeding what is witnessed in the open ocean."

If, as Humboldt says, the tide-waves of the waters of the ocean in the OPEN SEA, rise hardly to a FEW FEET, they mount, in consequence of the configuration of the coasts, which oppose the coming tide, to 50 feet at St. Malo, and from 65 to 70 feet in the Bay of Fundy.

This is, as far as I have been able briefly to represent it, the present theory of the tides. Let us now consider its

ANOMALIES.

And here I cannot do better than have recourse to a work on the *Anomalies of the Tides*, by Thomas Kerigan, R.N., F.R.S. In his Address to the Reader he says:—

v 2

"I have shown that when the moon is vertical to St. Helena, she is 2,362 miles nearer to it than to the point called the 'Land's End,' in Cornwall; and that as her REPUTED power of gravitation is NOT OF SUFFICIENT FORCE to raise a tide at the former place, it CANNOT be the cause of the high tide which flows at the Land's End, particularly since she is in the MERIDIAN of both places at the same instant, which adds another link to the chain of proofs already given respecting the NEGATIVE INFLUENCE OF THE MOON over the tides of the ocean." •

Besides this palpable instance of the moon's exercising no ATTRACTIVE influence whatever upon the ocean, I continue from his pamphlet, page 8 :—"The THEORY OF THE TIDES expressly states, that 'the action of the moon over the waters of the ocean is the most direct and intense when she is moving in the equinoctial; and, should the times of conjunction or opposition take place when she is so circumstanced, and at her LEAST DISTANCE from the earth, the tides will be the greatest possible. But, in all cases, the highest tides must take place WITHIN THE TROPICS ; and the nearer any port or harbour is to the equator, the greater will be the force of the tide in such harbour.'

"Now we know from actual experience that there ARE NO TIDES at any of the islands within the torrid zone that are 200 or 300 miles from the nearest mainland. At St. Thomas' Island, lying very nearly under the equator, in the middle of the ocean, and free from interference on every side, we find from actual observation that instead of a high tide in that harbour, there is a total absence of anything like a rise and fall of the water, except the FEW INCHES flux and reflux which keep pace with the alternate returns of the sea and land breezes, the same as at most other islands adjacent to the equator, and that are at a proper distance from the mainland, as already noticed."

Though the moon of necessity must and does exercise some

little indirect influence upon the magnetic needle wherever she passes. on account of the electric link which, as it attaches planets and comets to the sun, attaches her to the earth, thereby disturbing,* to a slight degree, the normal electric state of those parts she is passing over, perceptible also in the epileptic, the somnambulist, and other objects in nature immersed into, or overshadowed by, this link: these FACTS, with the preceding and many other valuable remarks of the author, must for ever nullify the theory of her attraction—NOT EVEN EXTENDING TO THE MAGNETIC NEEDLE—of the waters of the ocean. And yet, contrary to these undisputed FACTS, it has been computed by Sir Isaac Newton, and believed to this day, that the attractive force of the moon raises the water in the great ocean ten feet, whereas that of the sun raises it only two feet; that, when both the attraction of the sun and moon act in the same direction, that is, at new and full moon, the combined forces of both raise the tide twelve feet. But when the moon is in her quarters, the attraction of one of these bodies raises the water, while that of the other depresses it; and, therefore, the smaller force of the sun must be subtracted from that of the moon; consequently the tides in THE MIDST of the ocean will be no more than eight feet.

Here we have a theory, universally and tenaciously believed, in positive opposition to a well-ascertained fact; and though the fact for ever nullifies the theory, it is well, that, besides the preceding anomalies, others should still be noticed.

According to Sir Isaac Newton, it is the attraction of the earth,

---

* This disturbing influence must necessarily vary with the phases of the moon, and be different when the side turned towards us is illuminated or darkened, when our satellite is in opposition or in conjunction, when we receive the fulness of the sun's rays, with their reflection, more or less, by the moon, or when, more or less, she interrupts that fulness of light and warmth, preventing its falling upon the earth, and thereby diminishing his electric influence. Hence also a lunar monthly electric tide, and the different effect on plants and seed put into the ground at different times, as we have seen before.

revolving by centrifugal force, which keeps the moon in her orbit. Captain Kerigan, on the other hand, maintains, that by calculation, based upon the laws of UNIVERSAL GRAVITATION, the moon is NOT retained in her orbit by the mere force of terrestrial attraction; for if left entirely to THAT, she would fly off from the present orbit and be lost upon the body of the sun.

Now if it be true, as no doubt it is, that, under equal conditions, the earth attracts the moon, the larger body the smaller one, how can it be said, as in the present theory is done, that the moon attracts, or " draws to herself the earth's centre?" It is a perfect contradiction, and this the more so, as their mutual point of gravitation has been laid down at rather more than about two-thirds the distance from the centre to the surface of the earth. If, moreover, the earth have more attractive power than the moon, if it attract the distant moon herself: should not then the earth also exercise more attractive power over its own waters than the moon? Or is it at all likely, that more affinity should exist between the surface of the water and the moon, than between the water and the earth itself, both forming but one body? And as the earth is drawing down the moon, keeping her forcibly in her orbit: is the moon, in return, thirstingly, as it were, drawing away, and lifting up towards herself, the waters of the ocean? This again, is a pure contradiction, unless it were asserted, that the earth and moon attract each other like positive and negative electricity, and that the earth throws up the tide-wave to attract the moon. But this would ascribe attraction to the water, which no one ever dreamt of doing. The tide-wave, moreover, is more than three hours IN ADVANCE of the moon, and it has never been asserted that the moon follows in its wake like a cart after the horse. If, on the other hand, the moon attracts the water: do her attractive rays act more powerfully when exercised in the vertical, or more in the oblique direction? According to the positive, unalterable, law of attraction, the action is more powerful in the vertical, than in the oblique line; why then is the tide highest three hours after the moon has left the vertical position, that is : why is it three hours away from the vertical line of the moon, and that, too, in ADVANCE of the moon? This has certainly been explained by the holders of the present theory ; but is the explanation reasonable and natural? I think it to be neither the one nor the other.

If a given power is not able to lift up a certain weight perpendicularly, it will not be able to do so in an oblique line. Suppose the body to be drawn up, or to be attracted, to be elastic, very heavy, of great length, as we may consider the water to be ; and that, as in the case of the moon, the attracting, or lifting power, is not stationary, but constantly moving forward : it must be clear to the most ordinary capacity, that the body drawn, or attracted, must drag along until it gets into the perpendicular line with the lifting, or upwards-attracting power, and that in all cases, whether the attracting power be able to lift the attracted body from the ground, or not, the object attracted must always follow in the wake, at the tail of the onward-moving attracting body. But in the case of the moon, the body said to be attracted, is always in ADVANCE, the cart before the horse.

No case can be made out in which an attracted body is, or can be, swung forward, or forced, in advance of the body attracting, except by an extraordinary power, suddenly applied ; and this extra power, suddenly applied, would have to be greater if the attracting body were moving on without intermission, as is the case with the moon, than if it were stationary. This, therefore, cannot be applied to the tides. It might, however, be said, that, though the moon does not give a rising impulse to the tide-wave, notwithstanding what is actually taught, this impulse is given by the earth, carrying the water away from under the moon more than twenty-five times faster than the moon is progressing in the same direction in which the earth revolves; but this is not so either. If, upon this supposition the moon drew, or attracted, the water at all, she would have to draw it from the western side, and that too, with a still greater rapidity than that with which the earth revolves, as otherwise she could not raise the water; and IN THAT CASE, the tide-wave would flow from west to east, whereas the current of the ocean, as well as that of the tide-waves, comes from the opposite side, flowing from east to west. If it were contended, that the rapid revolution of the earth has nothing to do with the raising of the tide, or with the pretended impulse given to its wave, on the comparatively quiescent ocean: then the carrying away of the tide-wave in advance of the moon, and the attraction of the moon after the water has gone away from under her, along with the revolution of the earth, falls of itself to the ground.

Notwithstanding all this, and in order to reconcile a forced theory with fact, to establish the moon's attraction in advance, contrary to the law of attraction being strongest when exercised in the vertical line, it is actually maintained by some, that the moon really does draw the tide in an oblique direction, and this too, on the most simple principles of mechanics. How this is possible, and how it can be reconciled with tidal phenomena, I do not understand.

The earth E revolves from west to east, the moon M goes in the same direction, and the water of the earth flows in the opposite way, as it were, between both. The tide-wave is always on the east-side of the moon, at the same relative distance from her, without any intermission, or change, and flowing from east to west; it is always, according to the divisions in fig. 8,—high water being at A—about three hours and a half, let us say three hours, in ADVANCE of the moon, though flowing TOWARDS HER.

The tidal wave, moving towards the advancing moon, and yet permanent in its relative position to her, makes the theory of

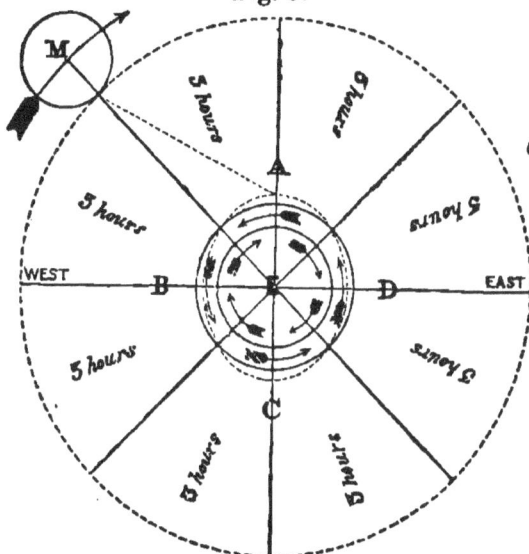

Fig. 3.

oblique attraction very plausible, and would certainly remove the anomaly of high water meeting the spectator on the revolving earth, about three hours after he has passed under the moon's meridian, did it not nullify the positive law of attraction, which acts strongest in the vertical line.   But apart from this, the question will naturally arise : why should the moon attract towards the east, and not towards the west ?   Why before, and not after her ?   Why should she draw water from the east, and not from the west, nor equally from the north and south ?   If the theory of attraction be true at all, why does the moon, a completely round body, not attract on all sides at the same time towards her own centre, and give us high water under her perpendicular line ? The theorists of oblique attraction have yet to prove, that the moon is positive in the east, and negative in the west of her body ;

that she has a pole of attraction for the WATERS of the earth, not for the earth itself, towards the east; one of repulsion, or negative, towards the west; neutral north and south, and no attractive power at all in her perpendicular, and therefore nearest, line to the earth; and yet upon this totally unwarrantable supposition alone, of attraction on one side only, and that in advance too, can the theory of oblique attraction be maintained.

Drawing in an oblique direction, on the most simple principles of mechanics, it is said: that the moon does not draw the tide-wave all at once, but sheet after sheet, globule after globule; and taking the tide to attain a height of thirty feet during the six hours it is rising, would give the one-sixtieth part of an inch for every second, so that, apparently, the operation would be very easy.

Now it must not be forgotten, that the moon goes faster towards the east, than the waters of the earth are flowing from east to west, like a steamer that ascends the river with greater rapidity, than the river itself is descending; whilst, however the waters are flowing westward, they RECEDE AWAY from the face of the moon at the same time, on account of the daily revolution of the earth from west to east round its own axis; the revolution of the earth carries the waters along with it, but imparts to them at the surface in particular, a movement, or flow, in the opposite direction, that is, from east to west. But, as said before: this westerly current of the waters is not as rapid as the passage of the moon from west to east, though the earth itself revolves at the rate of fifteen degrees per hour, whilst the moon in the same direction, progresses but at the rate of thirteen degrees per day.

Therefore, if the moon were drawing forward a sheet of water over the level of the westward moving, yet eastward receding, ocean, and in succession adding another, and another over another: she could not wait drawing the second sheet until she had finished the first, nor wait for the third until she had drawn the second;

for, if the waters were flowing towards the moon at the same rate at which she is moving against the current of the water,—without taking into consideration the revolution of the earth,—the first or preceding sheet would always have moved in advance, away from the next, as the water never stands still ; and in order thus to raise a tide, each successive sheet of water would have to be drawn with an ever-increasing velocity and power, so as to render it possible for every following sheet to overtake and accumulate upon its predecessor ; and this cannot be accounted for if the moon's power of attraction be always steady and equal.   If the attraction of every successive sheet of water were to follow immediately after the preceding sheet had been commenced, then the operation of raising the tide would no longer be little by little, sheet after sheet, as a person would draw blanket over blanket, or globule over globule, over a constantly-rising incline, as shown by figure 4 ; but it would be an uninterrupted simultaneous elevation of the

Fig. 4.

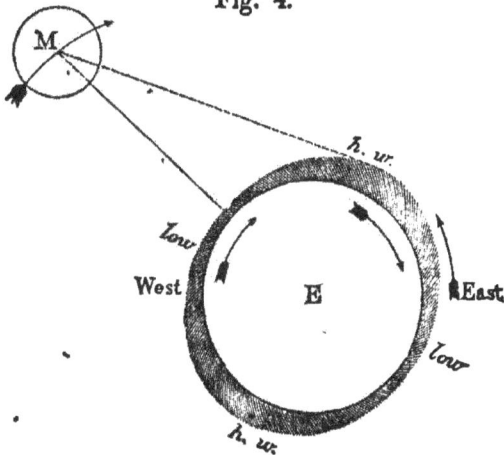

waters, the focus of attraction being exactly at the top of the tide-wave, or high water.

We here have reasoned upon an imaginary case ; but if we take the fact, as it really is, that the waters of the ocean, though flowing towards the advancing moon, are yet much more rapidly receding away from her than just now we have supposed them to advance, the raising of a tide-wave becomes still more difficult and improbable, and not at all easy on the simple principle of mechanics. Indeed, it seems to me more intelligible that the moon should lift and keep up the tide-wave, whether perpendicular or obliquely, with one steady uninterrupted effort of her whole power, than that she should draw it by forming sheet upon sheet. This theory, in fact, would make the attractive power of the moon greatest at the greatest distance, confined to the eastern side only, and completely null everywhere else; the tide-wave once raised, the moon would have to keep it up by supplying it from the most distant eastern point of low water, to the top of high water; and thus, the proposition of oblique attraction cannot for a moment be entertained.

The theory of the tides states, that the water being attracted by the moon, it recedes at the sides of the earth, B and D, figs. 1 and 2, and thus causes the tide-wave on the side of the earth opposite to the moon; but what becomes of this tide-wave, if, on the oblique theory, the moon attracts the water on one side of the earth only? This consideration seems entirely to have been overlooked by the theorists of oblique attraction, and is in itself enough for ever to discard this assumption.

We must, then, pass again to the theory of vertical attraction, and once more notice the constant following of the tide-wave three hours after the moon has receded from our meridian, or, which is the same, after we have been carried forward away from his meridian.

I have shown already, and must notice it once more, that the revolution of the earth cannot carry the tide-wave in advance of the moon, as, in that case, it would have to flow from west to east, whilst it flows, with the waters of the ocean, in the contrary direction. Or, if the earth does carry away the tide-wave, in advance of the moon : where was the wave first formed ? Underneath the moon, or on her western side ? It is curious, if, as we are taught, the moon continues to attract the water after the rotation of the earth is said to have carried the tide-wave from under her to the east, because in that case she MUST ALSO HAVE BEGUN TO ATTRACT the water before it came to her from the west, and thus have drawn it BACKWARD from its westerly flow, QUICKER STILL than by the rotation of the earth it is carried eastward; and HENCE, BY REASON OF THIS NECESSARILY BACKWARD DRAFT FROM ITS WESTERLY CURRENT,—if it be true that the earth carries away the tide-wave in advance of the moon,—the tide-wave OUGHT to flow from west to east; and yet its course and flow is westward. Attractive power itself is not of a sudden, impulse-imparting nature, and for this reason the ATTRACTION OF THE MOON cannot be of that kind, as to swing the tide-wave, if there were any, from her western to her eastern side. Attraction, moreover, is not a thing that accumulates, like the power of a steam engine, which will be followed by the train, though it should be separated, or have passed away, from it. And if the engine were to be placed at the end of the train, it would no longer draw, or attract, but propel or push it.

The rays of attraction, supposing the moon to possess them like the rays of the sun, are strongest in the vertical line; those of the moon, however, do not accumulate on the earth, and least of all, in advance, but they are like ropes, like a plumb-line, by which the moon, in a manner, lifts, as it is said, the water. On the other hand, the rays of the sun do accumulate by refraction and reflection, though they are never so strong as at mid-day; and this accumulation from the early part of the morning gives us the height of heat in the afternoon, two or three hours after he passed the meridian. Thus, the delay of the tide-wave, following the moon three hours after she has passed from her vertical position, cannot be compared with the delay in the warmer state of the air in the afternoon; it is like a room, which will remain warm for some time after the fire of the day has gone out. But, to compare to this the rising of the tide after the moon has passed, would be expecting the water to rise after you have ceased to pump.

It must, besides this, be borne in mind, how enormous the attraction of the moon ought to be, in order to overcome the pressure of the atmosphere upon the water, of 15 lb. to the square inch, particularly if on the vertical theory, she causes the wave still to rise for three hours after she had ceased to influence it; for we cannot raise an inch of water ourselves unless we pump, or suck, the air away from over it first.

The earth attracts the moon, and yet, according to the current theory of the tides, the moon would draw up the entire sea, if the air were not between.

According to the exposition of the Newtonian theory of the tides, as explained before, " the water will rise on the side of the earth away from the moon, because the earth's centre is more strongly drawn towards the moon than the point A, and recedes from A, which is the same in effect, as if the water at A receded, or rose up from the earth's centre."

Now if this were the case, that the side of the earth away from the moon leaves the water behind; would it not be more natural still, that, the earth's centre being so strongly attracted by the moon, the water facing the moon should give way on all sides to the pressure of the atmosphere, and cause the solid earth to peep out like a ball drawn out of a basin of water, rather than allow the water to accumulate upon it ? A momentary attraction might cause a momentary swell, but permanent attraction is permanent rest.

If the moon is able to "draw towards herself the earth's centre," I beg to ask, on the strength of this law : at what time does she do so? At what time does she relax her hold? For, if the CONSTANT production of the two tide-waves depended upon this attraction of the earth's centre, the moon would have to attract the earth powerfully and WITHOUT INTERMISSION, and this would be drawing it out of its sphere, and ultimately bringing them both together ; and if this result were not to follow, it would completely neutralize the attraction, and there would be no attraction at all ; and for the moon at any time relaxing her hold, would be equally nullifying the theory. Nor can the proposition of the moon's attracting the earth's centre be true, if, as it is said also, the moon exercises more power over the water than over the earth.

But this last proposition, illustrated in the following manner, cannot be true either. A, B, C, at first are equally distant from each other ; they are travelling towards a certain point ; but C walks faster than B, and B again faster than A ; the consequence is, that when C arrives at the point, B is left much behind C, and A more behind B ; and this is applied to the tides. In the following diagram (fig. 5) the moon M attracts the water at C more strongly than at B, and at B more than at A ;

Fig. 5.

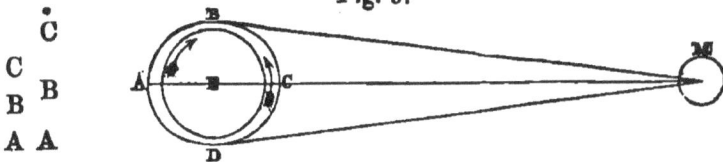

the water at C is consequently raised with velocity, depressed at B by being drawn away towards C, and raised at A by the very WANT of velocity ; or, as the illustration has it : C goes faster than B, and B goes faster than A.

Now this certainly seems very plausible, but would, if true, stand good merely for the time when the moon was created ; this MIGHT have been the first operation for producing the tides; but, when once set in motion, C would no longer go faster than B, nor B faster than A; thenceforward they would travel with EQUAL VELOCITY, or if the proposition were true, C, B, A moving in a circle, C would soon overtake A and gain upon B, and B would overtake A also, and the three ultimately be lost one in the other.

If the moon attracts B, D, equally at the same time, she anticipates, or hastens, the flow of water coming from the east, and retards, or holds back, that of the water going to the west ; and this operation would give us high water at C, immediately under the moon, WHICH IS CONTRARY TO FACT ; and if, on the principle of oblique attraction, the moon draws only at C, D, anticipating, or quickening the flow of water on her eastern side, the illustration becomes still more untenable, apart from the certainty, that want of velocity will not make water rise, nor raise the tide, though it had not to contend against the even pressure of the atmosphere. Thus, in whatever light we view the theory, we meet with nothing but anomalies.

What has been said of the attraction of the moon, is equally applicable to the attraction of the sun. If, by this means, he causes the tide-wave to rise, his attraction must then not only be more powerful on the water than upon the earth, but so strong as to raise the water in opposition to the whole atmosphere pressing upon it; or, the sun must attract the air and water at the same time in a peculiar manner. If the sun attracts the earth itself with a kind of violent effort, then the fluid water, and the still more fluid and elastic air, would give way to the forward impulse of the solid earth, and cause, as with the moon on the same supposition, low, instead of high water. If the sun and moon attract our globe, including its water and atmosphere, in an equal manner, as one whole, then they cannot cause the raising of the tide; the effect of the attraction upon the water would be the same as if there were no attraction at all.

If, in theory, the attraction of the moon is sufficient to raise a high tide on the open ocean, why can she not raise a tide in the Mediterranean, or Baltic, and other smaller inland lakes? "In small collections of water," it is argued, " the moon acts with the same line of attraction, or nearly so, upon every portion of the surface at once, and therefore, the whole of the waters being equally elevated at the same period, no part of them is ever higher than the other."

But then, the same takes place in the open sea; " there the moon acts with the same line of attraction upon a similar portion of surface," beyond which the waters, by degrees, cease more and more to be under her influence; and yet it is said, " that in this wide expanse of waters she raises up the parts immediately under her," which is equal to saying, that a loadstone attracts a large piece of iron better than a small one. " The moon raising up the parts of water immediately under her," that is, in her perpendicular (which is against fact) the depression at the sides naturally FOLLOWS; but this depression does not, as we are told it does, assist the raising, it cannot be cause and effect at the same time; and this raising, moreover, can only be supposed to be done, by the moon's rays of attraction being concentrated in one focus immediately underneath, or, by their being more powerful there than everywhere else; and in either of these cases, she would as strongly attract when above the Mediterranean, or Baltic, as when above any portion of the open sea. And this, moreover, would be less striking and wonderful to us, than " that the whole of the waters should be equally elevated," particularly when we consider, that this EQUAL ELEVATION AT THE SAME PERIOD ALL OVER is not observable at the islands, or the shore,

w

and that the water is not elastic, and consequently incapable of such an equal, imperceptible, and incomprehensible elevation.

If, besides this, the theory of the attraction of the moon, as applied to the absence of a tide in the Mediterranean, were correct, we should have no tide on any of the shores of either the Atlantic, Indian, or Pacific Oceans.

It has, to my knowledge, never been calculated, how many Mediterraneans it would take to enable the moon to raise a tide. If this sea had a tide of but one foot, we might make out, that a sea ten times the size, or less, would give a tide of ten feet, or more ; but, the moon having NO TIDAL INFLUENCE WHATEVER on ONE Mediterranean : how many Mediterraneans are to be added before such influence begins ?   How many more, before such influence becomes perceptible ?   And how many more still before such influence will draw a tide of ten feet only ?   Suppose that it would take ten or twenty times the space of the Mediterranean to render the attraction of the moon upon the water perceptible, or to enable her to draw a tide of a few feet ; she would then, on passing from one shore of the Atlantic to the other, not begin to exercise any perceptible influence on the water, or draw up any tide, unless she had passed over a space more than ten or twenty times that of the Mediterranean ; and on arriving at the other side, the tide would pass away, and her influence again become null, as soon as she should have got within the same space from the land.   The tide-wave, upon this theory, would begin and increase with the distance of the moon from the shore, and, AGAINST FACT, be highest in the middle of the ocean, where she would have abundance of surface to raise a tide.   If, indeed, in the middle of the ocean she were to raise a tide of 100 feet, she would draw the water from such an immense extent of surface, as not to require one drop from the shore ; but, apart from this, the highest tide in the middle of the ocean, if there were one at all, would precisely be as imperceptible at the shores of the Atlantic, as at the shores of the Mediterranean, inasmuch as the influence of the moon would as imperceptibly cease, and that too at the same distance from the shore on the one side, as it began at the other. Like a hill of sand in the middle of a desert; it may be raised and scattered without any sensible effect on the level, and without the slightest encroachment on its borders.

It is almost impossible to notice all the arguments which have been had recourse to to prop up the Newtonian theory of the tides; everyone has tried to explain a simple fact, but no explanation is found fitting. In the meanwhile, since my first edition, the old theory of the absence of a tide in the Baltic and the Mediterranean seems to have been abandoned, for, in his *Introduction*, p. 54, Mr. Hind says: "In the Mediterranean and Baltic, both inland seas, the tides are scarcely appreciable, for the inlets to these seas are so narrow, that the ocean has not time to fill them with the tidal wave, before the ATTRACTION of the moon is REVERSED, or tends to make the waters ebb."

That the attraction of the moon also TENDS to make the waters EBB, is certainly something new to me; but as it is more troublesome to prevent an old house from falling than to build up a new one, to refute fallacies than to establish a truth, I will now, without entering further into the contradictions of the present, proceed to state my

NEW AND CORRECT THEORY OF THE TIDES.

God's will, by fixed laws, governs the whole universe. These laws are arbitrary, but cease to be so to our understanding, when we are able to account for their operations; by them He guides the heavenly bodies, as the will of my soul guides my pen, moves my limbs, and rules my whole frame. So then it is the will of God, that this earth should be girt by the ocean, that its waters should neither displace the air and fly off to the moon, nor be absorbed by the bowels of our globe. So it is the will of God, that this earth should be enveloped in a fluid, called the air ; that this mixture of the air should have no tendency towards the centre of the earth, but withdraw, and keep away from its pits and subterranean cavities, nor a tendency to escape beyond our sphere, as proved by its rarity in the upper regions. It was His will, that this ocean of air should constantly flow round, and cover our globe ; that above this ocean, called the atmosphere, as a matter of extreme probability, if not necessity, and as I have shown already, there should be a still clearer and purer element, commonly called ether, but which, as I have demonstrated before, is almost beyond doubt, an ocean of oxygen ; and that above this ocean of oxgyen there should be other oceans of still more subtle and elastic fluids ; that somewhere above one of these oceans,

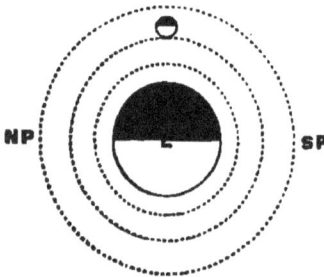

destined like these, and like the water and the air, to remain in the space assigned to her by God, and like a ship on the sea, there should appear the bright moon, AS A LIGHT TO THE EARTH, as the beneficial agitator of the oceans of matter beneath, her upper part enveloped, perhaps, in a substance quite different from that in and upon which her lower part is immersed and floating.

Be this, however, as it may ; whether two, three, or four oceans, or rings, of different matter, separate us from the moon : for the peculiar purpose for which she was created, God placed her, gravitating to the' centre of the earth, on one of the oceans above us, not like a revolving planet, like a revolving wheel, but like a well-laden ship on the sea which she has to traverse ; on this ocean she performs her journey round our globe, according to the laws which I have treated of before, always presenting to us part, or the whole, of her hull, without our ever, like a fish in the water, beholding the deck or the masts.

The theory that the moon rotates round an axis of her own is a totally false assumption, based upon the idea that a body moving in a circle, and a body forming the centre of a sphere or a circle and moving on its pivot, is one and the same thing. If this be proved not to be true, the controversy must naturally be at an end. It is only to be regretted, that in so simple a matter, so many words should have been, and have to be, wasted.

The earth rotates 365 times round its own axis, whilst performing one revolution round the sun. If the earth, in its revolution round the sun, did not rotate round an axis of its own, one side of it only would ever be presented to the sun; if it rotated but once a-year round its own axis, every part of the earth, during that year, would once be presented to the sun.

The moon and the earth occupy the same relative position. If the moon turned but once round her own axis during the month, every part of her would once be presented to the earth during that month; but as the moon does not rotate upon an axis of her own, so but one side of her alone is for ever presented to the earth.

A body moving in a circle, and keeping with the same side, or the same face, the axis, or centre of such circle, constantly in view, would be SUBJECT TO THE CONDITION OF NON-ROTATION. If without intermission you will keep in sight a fixed, or central, object, whatever the line you move in, your non-rotation becomes a condition. The same face of the moon, the same lower part of a balloon, the same bottom of a ship, keeping the earth, the centre of their circular orbit or motion, constantly in view, their non-rotation round an axis of their own, round a centre within themselves, becomes a condition.

Thus, to the inhabitants of the earth, the moon does not seem to, nor does she, rotate round an axis of her own, because they always see the same face; but to the sun, it is argued, she does rotate round an axis of her own, because in turn she presents to him every part of her surface.

A little consideration only, will show, that to the sun even, the rotation of the moon round an axis of her own, is but apparent, supposing the inhabitants of that fiery globe to hold the contrary doctrine.

Moving in a circle, without the earth round which to revolve, the inhabitants of the sun might easily be misled respecting the rotation of the moon round an axis of her own. The space described by the orbit of the moon might, to their eyes, be quite inappreciable; they might consider her rotating on her own axis, subject to certain changes of positions in the heavens. But when they see her move round another dark body, they will at once conclude that she does not move on her own pivot, but in a circle, round the centre of that circle. They would see that there is no rotation, but a simple floating motion round the earth. Let the moon gradually sink down towards the earth, towards the centre of her gravitation, and by degrees the interior part of the circle, the FACE WHICH WE NOW SEE, will be hidden from the sun; let the moon finally float close to the earth, or ride upon the mountain tide she will then have raised by her attraction, (! !) and nothing but the part away from us, the ONLY PART SEEN BY THE SUN NOW WHEN IN CONJUNCTION, will ever be beheld by his inhabitants, and that also but once a month only. The moon to them will then, like the ship, be nothing but part and parcel of the earth which she really is, with this characteristic, that she would present to him her deck, her sun-turned face but once a month, whilst the earth presents to him the same face once every day, and that for this reason: that the moon is too heavy and the fluid upon which she floats too light, to enable the earth to carry her round with it, on her present ocean, within the time of its own rotation.

The axis of the earth, then, is the axis of the moon's rotation, though the old theory does not make it clear where that axis lies.

"Strictly speaking, the centre of the earth is not the point round which the moon revolves; but both bodies have a revolution round their common centre of gravity. This point is situated at an average distance of 2690 miles from the centre of our globe, or about 1270 miles beneath the surface, and subject to a variation of rather more than 165 miles, one way or the other, in consequence of the variable distance between the earth and the moon. The perturbation in the place of the earth, or rather its re-action on the place of the sun, owing to this motion round the centre of gravity, may affect the sun's longitude more than five seconds of arc, and his latitude about 0·7."—*Hind's Solar System.*

If the earth and moon revolve round a COMMON CENTRE of gravity, this common central point involves an axis, and as that point is said to be about two-thirds removed from the centre of the earth towards its surface, this axis or common point must be eccentric.

Now how the earth can revolve round its own CENTRAL axis passing from pole to pole, and how at the same time it can revolve round an ECCENTRIC point or axis, I cannot understand, nor do I understand this proposition when applied to the moon.

Suppose an uneducated person to turn on his own axis, round himself, and at the same time, or any other, trying to turn or circulate round his own nose; would this not be very ridiculous in the poor fellow? Suppose, however, that a LEARNED man, the Astronomer Royal, for instance, were trying the same thing; what should we think of him? Yet this is the theory which scientific men teach and apply to every heavenly body.

Though so evidently absurd, as in the comic instance indicated, I would still ask, when applied to the earth and the moon: is the common centre of gravity in question a point or object fixed, or is it moving? an axis stationary or shifting? a point material or immaterial? Does the earth, in its diurnal rotation, carry this point away with it from under the moon, take it back under her again, and away from her again and again? Or does that common centre remain stationary in a direct vertical line under the moon whilst the earth is going round? How, in any case, can the earth revolve round this common centre of gravity which is either an IMMOVEABLE eccentric somewhere, or a point carried about everywhere? Does this common centre of gravity round which BOTH BODIES are said to revolve, being 2690 miles outside the centre of the earth, ever come in opposition as well as conjunction with the moon in regard to to the CENTRE of our globe? And if this be so, does the moon rise and fall in her course as every 24 hours this fixed centre of gravity, as in fig. 1 and 2, comes in conjunction

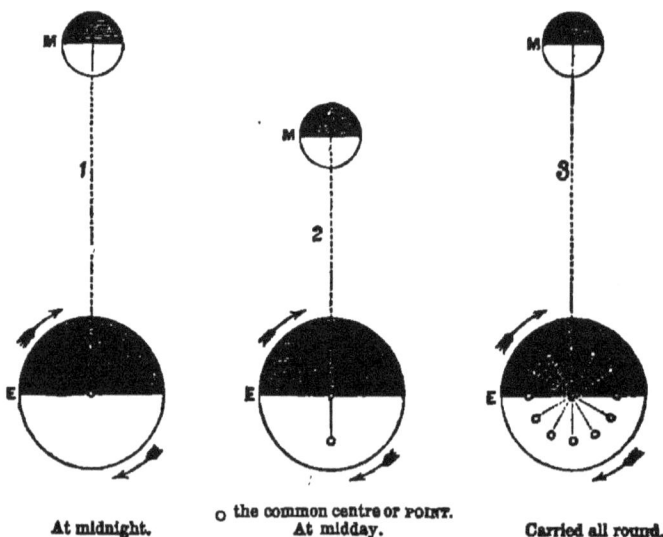

At midnight.   o the common centre or POINT.   Carried all round.
At midday.

or opposition? And if this common centre of gravity exists all round within the earth, forming a ring, a kind of tyre, the POINT of gravity carried all round as in fig. 3, at 1270 miles beneath the surface : why not then at once draw a line from the moon direct to the centre of the earth, the centre of this supposed ring of gravity? what does otherwise become of the POINT? And how can the earth revolve round this interior ring or line of gravity? how at all can it revolve or circulate round an excentric point or spot whether IMMOVEABLE, FIXED, OR SHIFTING, when and whilst it performs its central axial rotation?

To solve this riddle, or even properly to propose and dissect it, is more than I can do ; but the reader will not fail to see, with all the drawbacks of my indistinct exposition, that, though calculated to a mile, the whole affair is a pure fallacy, as already noticed page 4 and 237, and that the central axis of the earth, the centre of the earth itself, IS AND MUST BE the axis of the moon's rotation, her ONLY centre of gravitation, the fulcrum of her motion ; and this truth no distance of the moon from the earth can change.

The moon, therefore, does neither revolve round a centre of gravity which she has in common with the earth, nor has she a rotation on her own axis ; then only could she be said to rotate round an axis of her own, if she were substituted in the place of the earth, or if but once within a given time she were to show every part of her surface to the earth.

A ship sailing round the earth is in exactly the same condition as the moon ; like her she always presents the same face to the centre of the earth, and consequently does not rotate round an axis of her own, but round that of the earth to which she belongs ; remove her to the distance of the moon, and nothing will change her law of motion ; she only is further removed from the centre of her orbit, from the solid body round which she floats. Imagine the ball in the diagram page 229 to be a ship going round the earth, and you will perceive that she CANNOT turn round her own axis, no more than with a constant wind, a primitive impulse, at her back, and attracted by the earth's centre, she can preserve her parallelism, always point to the top of the page.

Ship, moon, or balloon, whether floating on the sea of water, of air, or of ether of our earth, behold every quarter of the globe the same as the earth itself; but it does not follow that for this reason, as it is argued, they should also move round an axis within themselves, round an axis separate from that of the earth; to suppose this, would be a manifest absurdity, because, in fact, they are nothing but part and parcel of the earth round whose axis alone they revolve.

A different thing it would be with the wheels of a carriage drawn round the earth. Whilst no one would say that the carriage, presenting always the same inferior part to the earth, to the centre of its gravitation, was rotating on an axis of its own in going round the earth, everybody would be obliged to acknowledge, that the wheels by themselves, during the same journey, would make many revolutions round their own respective axis; but, put drags to the wheels, and in that instant rotation round their own axis would cease, because, like the moon, the ship, and the carriage, they would always present the same part, the same face, to the centre of the earth. The test of the rotation of a body must, in fact, be taken from the point or centre of gravitation round which such a body moves, by which it is attracted. If, for instance, *vis-à-vis* the sun, the moon were to present to him always the same face during her orbit round the earth, it would amount to non-rotation; hence, astronomers say, the moon DOES ROTATE on an axis of her own because the above does not take place. And yet, the case stands quite different. We do not, in the first place, know of any motion of a body like the one supposed, moving in a circle or orbit, without this circle or orbit having an axis, its point or centre of gravitation; the motion, in fact, according to the laws of nature, is simply impossible, except to an intelligent or living creature.

A person may, for instance, move round a centre, round a chair, so as to present front, and at the opposite side, back to the chair, and a spectator outside see nothing but the person's back ; but this is simply going FORWARD, sideward and BACKWARD, though the motion is in a circle round the chair, the same as the earth is made to go round the sun, preserving its parallelism, page 230, fig. 2 ;—it is not WALKING FORWARD in a circle as shown by fig. 3 same page. To reason, therefore, upon this so HIGHLY INTEL-LECTUAL performance and illustration, would be an absurdity.

The moon, in the second place, would, on the above supposition, present every part of her surface once a month to the earth, and this really would be one rotation round her own interior axis during one of her revolutions ; but, this not being the case, non-rotation is established as a matter of course,

To prove this, stretch out your arms and hold a hat between your hands ; then turn round yourself, round your own axis, as often as you like, and see whether you can see the opposite side of the hat UNLESS YOU TURN IT ROUND ITS OWN AXIS BETWEEN YOUR HANDS. This not being the case UNLESS YOU DO SO, non-rotation, as with the moon, is absolute. Body and axis are equally attracted by the centre of gravitation ; body and axis never change their relative position ; both, like the clogged wheels, present always the same part to the centre of the earth ; and wherever a body and its axis present always the same face to the centre of their gravitation, though that centre be only the axis of a silken thread round which they move, THERE IS NON-ROTATION ; and to a spectator, or body, outside the circle of such motion, the rotation would be in appearance only.

Before the drags were applied to the carriage-wheels, they had a double motion like the earth ; the one motion in a circle round the earth, and the other round their own respective central axis. When clogged, the wheels made a simple revolution only round the earth, they simply moved in a circle, and rotated once round the axis of that circle.

A circle, then, MUST have an axis ; a body simply moving in a circle must naturally move round such axis as its centre, the same as in drawing a circle one leg of the compass moves round the other as its rotating centre, though, like the earth round the sun, like the unclogged wheels round the earth, such a body may at the same time rotate round an axis of its own. In the latter case, the rotating body does not with the same face, with one particular part of its surface, keep the axis of its circular motion constantly in view; and this is the test of its rotation. The moon in her circular orbit, the ship, the carriage and the clogged wheels, in their motion round the earth, do with the same face always look to the centre of the earth, and hence the proof of their non-rotation.

Take another test, and instead of a vulgar hat, let a person hold a mirror between his outstretched hands, and, if fond of it, steadfastly look at his own beloved image whilst turning round his own axis; and if during one complete rotation the image does not turn its back upon him, he may be sure that neither his likeness, nor the mirror, have turned round their respective axis, though he himself has done so round his own. If this be not convincing, nothing will.

A central body, in order to keep another body moving in a circle round it, constantly in view, whether with a particular or any or every other part of its surface : rotation round its own axis becomes a condition. Hence, moving simply in a circle, and moving in, or upon, a centre, forming a moving centre, are two different motions, though both may be combined. That they are not so combined in the moon, and that she does not rotate round an axis of her own, I think I have sufficiently proved ; and no experiments will establish the contrary, unless they grapple with the special arguments, and disprove the illustrations, I here have used.

From the coincidence of the exact time of the tides, with the exact time of the progress of the moon round the earth, in connection with the rotation of the earth round its own axis, there can be no question but that they are occasioned by the moon, contrary to the idea of Capt. Kerigan, who believes her influence to be entirely negative.   Now, if we substitute " pressure " of the moon for " attraction " we shall be able easily to point it out as the cause of high and low water.

It seems natural, that, like a ship on the sea, or a boat on a canal, the passage of the moon over the ocean of matter in which she swims, should produce a pressure, and a corresponding expansion at the sides.   This pressure upon the said ocean, we will call it ether, acts directly, or indirectly, upon the air, the same as the pressure of the air acts upon the water; and the nearer the moon, the stronger the pressure, and the further away, the less her influence; it is like the pressure of a ship according to the width of the river, and her distance from, or nearness to, the shore.   The pressure will also be influenced by the greater or lesser rapidity with which the ship is progressing.

We all know the corresponding effect of the wind, or of an agitated atmosphere, upon the water, in causing the rise of waves. The agitation of the atmosphere itself is caused in a variety of ways, partly known, partly unknown, into which, however, I here cannot enter ; but I have shown already, on treating of thunderstorms, with what force the lowering cloud presses upon, and displaces, the air beneath it.   Yet the passage of the moon is more than the lowering cloud, though her pressure seems to be unfelt, because so even and so far extended.

That, apart from winds and the various currents of the air, there is a regular movement belonging to the whole of the atmosphere, is proved by Humboldt.   In his *Cosmos* he says : " One of the main features of the atmosphere is the variation of the pressure of the air between the tropics, a regular, so easily discernable, hourly oscillation, a kind of ebb and flow of the atmosphere, which is not to be ascribed to the attraction of the moon, and which very much varies according to the geographical latitude, the seasons, and the height of the point of observation above the level of the sea." In another place he speaks of "the hourly variations of the atmospheric pressure, which successively take place from east to west, and which in the tropics are so regular."

Whence, then, comes this regular ebb and flow of the tide of the atmosphere? It does not result from the ATTRACTION of the moon, as Humboldt truly observes, nor from any cause operating on the surface of the earth, no more than the tide of the ocean is caused by some operation at the bottom of the sea. If the surface of our atmosphere were as tangible to observation as the surface of the sea, I have no manner of doubt, but that the tide of the air, in all its features, would resemble the tide of the sea, particularly after what is said by Humboldt of its variation according to place and season.

This, and that the tide, both of the sea and the air, successively takes place from east to west, renders it still more probable, that the tide of the sea is but the effect of the tide of the air, and that both are caused by the PRESSURE of the moon.

My new theory of the tides seems to me, in the highest degree, to be the true one, inasmuch as it requires no forced application, and is simple, without any contradictory points between law, theory, and fact; and because it solves all the anomalies which cannot be reconciled with the theory of Newton. I do not think, for instance, that it is overstraining my reasoning, if I account for the two opposite points of high water, and the two intervening opposite points of low water, in the following manner :—

It is a well-known fact, that, apart from particular currents, produced by particular causes, the general flow, or motion, of the air, as well as of the waters of the sea, is from east to west, the same as their tides ; and this flow from east to west, with Galileo and Kepler, I ascribe to the rotation of the earth in the opposite direction, from west to east.   The moon, sailing, as it were, in the same direction from west to east, though across the earth, goes against the currents of water and air that flow on the earth from east to west.   By her pressure, direct or indirect, upon the atmosphere : the air, an elastic body, is depressed and displaced underneath, and raised in front by her onward course against the contrary current.   The first tide-wave is thus created immediately in front, or to the east of the moon ; and the air, escaping behind, will rise as much, or nearly so, beyond its ordinary level, as it had been depresssed below it, and thus cause the formation of the second tide-wave, on the western side, and at a greater distance from the moon than in front.   These atmospheric tide-waves will press upon the water, and present to us the tide-waves of the ocean in a corresponding manner ;   the pressure of the moon, extending in its effect over at least one-fourth part of the earth, each of the tide-waves will cover a similar space, leaving between them, immediately under the moon, and, by the very want of velocity, also on the other side of the earth, opposite to the moon, one-fourth part each of low water.

Or, the moon may also be considered a floating bridge ; the ethereal current strikes against her bulk, is arrested and accumulates, then forces its way underneath and over, like the air through the arches of, and over, a bulky bridge, and rises and accumulates again on the other side in proportion to the disturbance of its level. And this disturbance of the ethereal level acts upon the air, and the air causes the tides of the sea.

As the height of the tides of the sea is influenced by shallows in the sea, by the configuration of the coasts of the mainland : so the variation of the ebb and flow of the air depends much on the chains of mountains it has to pass by, or over, and on the nearness, or distance, of the moon from such coasts or mountains.

All this appears simple and beautiful, and yet,—excepting the preceding paragraph—without having overstrained my reasoning upon old assumptions, I have nevertheless been labouring in vain to reconcile my own new theory with the received facts and explanations of the tides; in trying to refute the fallacies of the old theory, I have been striving against a phantom of the imagination, because the FACTS of the old theory of the tides, STATED AS SUCH, are nothing but THINGS CALCULATED, and as UNTRUE IN REALITY, as the theory itself is inconsistent and forced, and its result, the child of so much labour, a grand mistake.

It looks very well on paper to see the sun pulling at the earth one way and the moon pulling it another, and thus to make them raise a tremendous tide' on either side of the earth, of course directly under the sun, and under the moon's path and motion ; but, as Capt. Kerigan states, there is not only no tide at St. Helena : there is none either, or scarcely any, under the whole of the ecliptic, under the direct passage and line of attraction of the moon.   And yet it has been assumed—and I have before been misled by the assumption—that, the sun kindly helping her at times, there is a tide raised by the moon, before, behind, or under our satellite; it has been calculated to a fraction how many per cent. of the height of the tidal wave are due to the attraction of the sun, and how many due to that of the moon ;* that the tides OUGHT TO BE of a certain height, and that there OUGHT TO BE the highest tide in the middle of the ocean, particularly under the ecliptic ;† yet notwithstanding all these OUGHT TO BE's and calculations, there is NO TIDE AT ALL in the open sea, neither in front, nor after, nor

---

* See Anomalies, page 142.   † See Anomalies, page 141.

under the moon, and consequently much less a tide at the opposite side of the earth, directly or indirectly caused by the moon.

The old theory, therefore, is IN EVERY RESPECT SIMPLY UNTRUE.

On looking over the authorized tide-tables of the Admiralty for 1861, for places all over the earth, I find the following islands and ports under the ecliptic, within the tropics, where there is a tide,— well observed—a SPRING TIDE* ONLY of from 2 to about 6 feet, but chiefly inclined towards the lower figure, viz. :

St. Helena 3'; Ascension 2'; St. Thomas 4½'; Princes Island 4½'; the ports from the River Niger upwards to Sierra Leone from 2' to 5', and a few only from 6' to 7'; Cape Verde Islands 5'; the Azores, though beyond the tropics, 4' to 6'; all the islands and ports in the Caribbean sea and the Bahamas, including Bermudas ¾' to 4'; Tobago, Trinidad, Carthagena 1½' to 4'; West Coast of Mexico 1¼' to 7'; Galapagos Islands 5' to 6½'; Coast of Peru 2' to 6'; ports of Bolivia and Chili 3' to 6' ; the ISLANDS in the North Pacific 2' to 4½'; the ISLANDS in the South Pacific Ocean 1' to 6'; the Phillipines 2½' to 6'; Java 2' to 6'; Coast of Sumatra 3' to 6'; Crimon Islands, Celebes, Moluccas 3' to 6'; the ISLANDS in the Indian ocean 1½' to 7'; Ceylon 1½' to 3'; Mauritius 3'; and at all these islands and ports, without almost an exception, there are no "neap" or low-tides recorded, simply because there are none.

Some of the coasts and islands under the ecliptic do, however, form exceptions; for, in several places the SPRING-TIDES rise from 6' to 20' and even higher, where the shores of continents, islands and prominent headlands, are exposed to the direct current of the ocean, like Madagascar, the east coast of Africa, of South America from Paraiba to Maranhan and Para, or where the oceanic

---

* Spring tides are those produced at new and full moon, twice a month, when both sun and moon are said to combine for a good pull, a long pull, and a strong pull altogether, at mother earth.

x

current has to pass through narrow straits as at Malacca, the straits of Banca, etc., where the rise is evidently not due to the attraction, and in part only to the pressure, of the moon.

Going from the tropics towards the poles, the tides rise higher and higher at the shore of the mainland, the "neap-tides" even from 9' to 30', and yet the islands do not follow in proportion; for, the Falkland Islands have a "spring-tide" of only from 5' to 8'; the Bermudas only 4', and Newfoundland only from 2' to 8½'; New Zealand 7' to 8'; Japan 3' to 7', and Kamchatka 6¼': proof all this, that the moon draws no tide ANYWHERE IN THE OPEN SEA as she OUGHT TO DO according to the old theory, and that calculation goes for nothing against simplicity of theory, and still more cannot stand against FACT.

But all this makes it clear, that the moon passing along the ecliptic, by her mediate pressure upon the water, drives the tide down towards the poles, and that the second tide within 24 hours is but a reaction of the first one actually caused by the moon; that the first tide occuring regularly, the re-action must be regular. And if, further, there are any tides under the ecliptic, they are not so much to be ascribed to the direct pressure of the moon there, as to the re-action of the tides she produces upon the shores of the islands and continents north and south of the ecliptic, and to the confinement of certain portions of the ocean and its consequent course and accumulation of the water through narrow straits, or along, or upon, an opposing coast.

XV,

I WILL now try to reconcile a few anomalies of the theory of attraction, with my new theory of pressure.

If the attractive power of the moon is not sufficient to raise a
tide at St. Helena, whilst it is said to be the cause of the high
tide at the "Land's End," which is not east of the moon, but on
the same meridian with St. Helena : then this anomaly disappears
by substituting the PRESSURE of the moon for her ATTRACTION ;
for, the water is running away from the moon instead of running
up to her; and this pressure is perhaps the only means of assisting
a mere difference of temperature to drive the warm water of
the tropics over the poles of the earth, and there to make them
more freely circulate.

A ship on the water will displace as much water as her bulk is
immersed in, and the pressure of the ship on the water underneath
will not be greater than that of the water which she displaced,
inasmuch as the water displaced is equal in weight to the body that
has taken its place. But the pressure of the ship will force the
water displaced to the sides, and manifest itself in a rise at the
shore to which it is driven, supposing the ship to be large, and the
channel narrow, enough for the purpose.

The moon is nothing else but a ship, a boat, to which the ocean
is as a large river, and gulfs, bays, and rivers, are as canals; the

elastic air readily gives way to the pressure of the moon, to the pressure of her weight, and therefore exercises but little, if any, influence on the water directly underneath; but it accumulates in a wave, increased by the onward pressure of the moon, and weighing upon the water at a distance, makes it rise at the shore   Hence the imperceptible change in the ebb and flow of the water at St. Helena, though the moon be there in the zenith, and the high tide at the Land's End so far away.   Hence also, as we have seen, NO TIDES at any of the ISLANDS within the torrid zone.   Islands, at the same time, are to the flow of water what single hills, or solitary mountains, are to the motion of the air; they are not sufficiently large, unless the flood be very rapid and strong, to offer any perceptible resistance, so as to be an obstacle to its course; they simply divide the approaching current, which has time and space enough to pass and expand on either side, without rolling up the shore to any tidal degree.

By my theory, the anomaly of the tide-wave continuing to rise, by impulse, or by continued influence, for three hours after the water has been carried away, by the rotation of the earth, from under the vertical line of attraction of the moon, as well as the theory of oblique attraction, is solved without difficulty; for, the moon, by her pressure, DOES give a propelling impulse to the water, which will continue to flow, and to rise at the shore, THOUGH she has long passed away; and the greater the pressure extends into distance, the greater the accumulation of the wave; and the shorter the distance, the lower the tide.   Hence the higher tide when the moon is right above the equator, driving the water on each side all the way to the poles, and the low tide when she is in her quarters nearer the north or south pole.

"It has also been observed, that the height of the barometer is, one time with another, less in the middle of the second quarter than in that of either of the others; and that it is somewhat greater at new and full moon than at the quarter." *Knight's Penny Cyclopædia*, 372.

According to my theory of pressure the rising tide has no superincumbent weight to overcome,—though it is said (*Cyclopædia*, 372) that "the ATMOSPHERE is continually undergoing a slight alteration from the EFFECTS of THE TIDES," and (*Manual of Scientific Enquiry*, 1859, p. 181) "We have already alluded to the influence of its (Mediterranean) WATERS IN ORIGINATING small ATMOSPHERIC WAVES," which are pure fallacies ;—for the water is like the scales of a balance, of which the one will not rise, unless the other is PRESSED DOWN BY WEIGHT. And with the tides it is still more necessary, that one scale should SINK FIRST by accumulated pressure, and thus cause the other to rise, because the weight of the atmosphere is precisely equal all over the ocean, which being entirely covered and pressed upon by the air everywhere, the weight of the ATMOSPHERE would prevent the rise of the tide, unless, on the old theory, the moon, in ATTRACTING the water, were powerful enough to attract, or draw up also, to displace, or overcome the resistance of, the air. The work performed by the ocean in rising against the atmospheric pressure, is precisely equal to the force exercised by the accumulating influence of the moon, the pressure of the air-tide, on the portion which it causes to sink. Therefore, the phenomenon of the tides is the result of weight, of water THROWN, not DRAWN, into motion, by an external force ever varying, both in direction and amount; and this external force is beyond all doubt the pressure, and not the attraction, of the moon ; it is, as it were, the rocking to and fro of our atmosphere upon the ocean. The loftier ocean of our sphere, in the course of the earth pressing against the atmosphere of the sun, is not free from agitation either, and as its mighty tide is oscillating from pole to pole, the moon likewise oscillates, so that at one time we see more of one or the other of her sides than at another, as we do also see more or less of her edge according to her distance and position from us, north or south, east or west; and this is called the libration of the moon.

"It has been stated that the Moon always presents the same face towards the Earth . . . . . . strictly speaking, however, it is not precisely so . . . . . . we occasionally see a little further round each limb than at other times . . . . . we sometimes are able to see a little more than usual of her north and south polar regions." *Hind's Introduction to Astronomy*, p. 91.

Thus, if a learned man, endowed with whiskers, at one time shows a little more of the right, and at another a little more of the left one to the person he may be speaking to, "he then does not always present to him precisely the same face!"

Respecting the absence of tides in the Baltic and Mediterranean, it is easy to conceive that the continents of Europe and Africa, their mountain-chains and configuration, may prevent the tide-waves of the atmosphere having effect upon these seas, or, that they are already BEYOND the influence of the moon. This influence, however, would exist, if the Mediterranean, for instance, were connected with the southern ocean by means of the Red Sea; it would then most certainly have tides at the place of junction, if not all over. But the chief cause, no doubt, is this : that these seas are too small to allow the motion given to the water by the gentle pressure of the moon, to accumulate into a very perceptible tidal wave by the time it reaches the shore; for, it is natural, that, the smaller the sheet of water, the less time and material there is to raise a wave, or tide, at the shore, as we have seen already.

When the moon is in conjunction with, or opposition to, the sun; when she either receives the full reflected light of the earth, or the earth receives that of the moon, her pressure is greatest, because in her passage round the earth, she happens, at that time, to be not only nearest, but also travelling with greatest velocity. For, according to observation, the true velocity of the moon becomes always greater, the nearer she herself comes to New and Full Moon, or, within the ecliptic; and at these points of her orbit she obtains her maximum, as in the two quadratures she moves with her minimum of speed, being then also at her greatest distance from the earth.

The difference of distance, apogee and perigee, and consequently, difference of pressure, of the moon, must, of course, influence the height of the tides, apart from the longer or shorter course of the wave.

The distances of the moon from the earth, from observation, do not agree with the theory of a purely elliptical motion, for these distances are always less at New and Full Moon, and greater in the quadratures, than they should be. Whether my accounting for the apogee and perigee of the moon will clear up this point, or help to do so, I must leave to the reader to judge.

The earth being a magnet, electricity is accumulated at the poles in the form of a brush, and depressed, or neutral, at the equator, as shown in the diagram; if, therefore, the moon

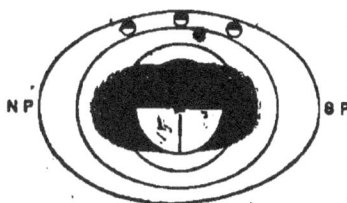

approaches either of the poles, she will be pushed further off from the solid part of the earth, from the normal position or distance when right above the equator, until compelled to return on her course. Hence her path will be elliptical.

Or, the solid part of the earth being surrounded by various envelopes, upon one of which the moon is swimming: these envelopes, especially the outer ones, will become depressed round the equator, and consequently elongated at the poles as the earth is rolling forward on its belly, on its equator, in the solar atmosphere against which it goes and upon which it presses, the same as an egg rolling on its soft inner skin; the envelope, in fact, will assume an elliptical shape, the same as the nucleus of Donati's comet, see page 29, or the diagram page 32.

Thus, by this means also, the moon,—floating upon one of the upper oceans or envelopes of the earth,—repelled or attracted from pole to pole without crossing either, and apart from the special electric repulsion noticed before, would be carried away further from the solid part of the earth after having crossed the line.

The depression of the envelopes of the earth on the equator would change, or shift, become inclined, according to the oscillation of the earth, and thereby influence the PERIODICAL apogee and perigee of the moon.

When at each equinox, the axis of the earth being perfectly horizontal to the sun, this rotary depression on the equator coincides with the full, or new, moon, that the moon is right above the equator of, and, owing to this depression, also nearer to, the earth; the pressure of the earth against the solar atmosphere, and the pressure of the moon, MAY and no doubt DO COMBINE to produce the equinoctial storms and tides which, at that time, on both sides of the equator, have their widest range to blow and roll towards the poles the greatest volumes of air and water, and accumulate them into the most gigantic waves and currents of the atmosphere and the ocean; and the equinoctial gales would be constant if it were not for the oscillation of the earth.

Whether the depression, particularly the daily one, of the envelope of the earth around its equator, does actually, and how, help to produce or influence the tides, I have not made clear to myself as yet; but the daily and regular variation of the barometer,— apart from the dilatation and contraction of the air from heat and cold and the variation noticed page 340,—I think we may safely ascribe to the daily depression of our higher exterior envelopes, and its action upon our atmosphere of air, according to season and the oscillation of the earth.

" Without doubt, the most interesting phenomenon, and one that lies at the root of all the great atmospheric movements, especially those proceeding northward in the northern hemisphere, and southward in the southern, is the equatorial depression of the barometer, with its characteristic belt of calms, known by the name of Doldrums. The barometer, at the level of the sea, does not indicate a mean atmospheric pressure of equal amount in all parts of the earth; but, on the contrary, the equatorial pressure is uniformly less in its mean amount than at and beyond the tropics."

The preceding, taken from the *Manual of Scientific Enquiry* referred to before, is also readily explained by the equatorial depression of the envelopes of the earth.

The position and pressure of the moon will no doubt be influenced by the expansion and contraction of our atmosphere above, according to the seasons, and therefore also affect our atmospheric and oceanic currents and tides and barometric indications.

It has also been observed, that the moon is gradually accelerating in motion, and, like Encke's comet to the sun, approaching nearer to us; this approach has been calculated at less than one inch per annum, and accompanies what is called: the secular acceleration of the moon's motion.

La Place's theory of this acceleration is thus BRIEFLY stated in the Monthly Notices of the Royal Astronomical Society of June, 1859.

" ' The time of revolution of the moon round the earth depends upon the amount of her gravitation towards the earth ; and if her orbit were not disturbed by the sun and PLANETS*, it would be

---

* § 642, Sir John Herschel says: "THEIR (the planet's) action upon the earth are out of the question," and in the present instance the planets are said to affect the moon, the SATELLITE of earth.

absolutely constant. The sun's disturbing force, however, on the whole, during a revolution, diminishes the central or radial force, and her orbit is, therefore, larger and her period greater than it would be if she were undisturbed. The period will, therefore, depend upon the sun's distance from the earth, and will be different for the different months of the year, but will have a mean value which will be pretty nearly constant for each year. On account, however, of the decrease of the eccentricity of the earth's orbit, the sun's average distance in any year is slightly greater than it was in the preceding, and his disturbing power diminishing the moon's gravity towards the earth is less, and, therefore, the moon's gravitation towards the earth is greater, and her motion becomes accelerated.' "

" La Place, with his usual sagacity and power of analysis, deduced the effect of this secular disturbance in the mean motion; but he concluded, Mr. Adams thinks erroneously, that the RADIAL FORCE only is concerned in producing this effect, and that the equal description of areas would not be affected by it, the TANGEN-TIAL FORCE producing no effect. It is a curious fact, that, in the infancy of the lunar theory a similar mistake was made with respect to the motion of the moon's apogee. It was assumed correctly, that only the radial disturbing force was concerned DIRECTLY in producing the motion of the apse, but it was not taken into account that the tangential force, by diminishing or increasing the velocity, caused the position of the moon to be at a given time, different from what it otherwise would have been, and, therefore, INDIRECTLY altered the radial force, and tended to produce a motion of the apse : and it is remarkable that this indirect effect of the tangential disturbing force was in this case of the motion of the apogee equal to the direct effect of the radial disturbing force; and that Clairaut by this discovery relieved men's minds from one of those periodical panics, which, ever since the publication of the *Principia*, have at intervals seized the minds of astronomers with regard to the absolute truth of the theory of gravitation."

There are still great differences among astronomers respecting the true causes of this acceleration. As to my own part, without pretending to understand the operation of the disturbing sun* and the planets, the radial force, the tangential force, their action direct and indirect, and without being insensible to the important relief of the minds of astronomers, to their ease of heart when so enormous a stone fell from it by the discovery, that the "tangential disturbing force" is equal to, neutralized by, the "radial disturbing force," I nevertheless think, that the acceleration of the moon's motion and her gradual but slow approach to us—if true—may arise from one of three simple natural causes, applicable also to the planets and comets, if they should be found to approach the sun, and their progressive motion to accelerate.

1. The earth is constantly increasing in bulk by feeding upon its atmosphere ; this atmosphere, then,—though God, at any time, can re-extend it by a liberation of gases from the earth—must in the meanwhile diminish, contract, or loose in density ; and in consequence of either, the moon would sink lower to find her level.

2. The moon may still be feeding upon the ocean on which she moves, thereby become heavier, and by degrees go deeper towards us. Or :

3. The moon may still give up gases with which she may originally have been inflated; and by giving up an expansive

---

* Of the sun, in connexion with the tides, I have not spoken, as, according to my theory, he has nothing whatever to do with them. He attracts all the heavenly bodies of our system in their ENSEMBLE, whilst upon individual parts—as upon air, water, gases, electricity—he exercises an expanding, a dispersing influence ; hence also the contradiction once before noticed: of his being said to attract the watery envelope of the earth, and consequently contracting it, and to repel, not the nucleus, but the ENVELOPE, of the comet, and dispersing it,

element, she loses more and more of her floating capacity and approaches nearer and nearer.

Every approach of the moon to the earth will bring her more and more into the attracting grasp of the latter by narrowing her orbit and thereby accelerating her motion; but which of these three reasons is actually operating to bring about this gradual approximation and consequent acceleration of the moon's motion, it would be difficult to decide. The contraction, diminution, or rarefaction, of our atmosphere, as already noticed, page 148, seems to me the most natural. As to the others: our satellite appears already—and has appeared so for some time since, most likely from the time of Adam—like a huge piece of coke, all her vitality exploded or burnt out, and consequently unable to attract matter, or to emit any gases, yet so constituted as to be kept afloat within certain limits, and gravitating to the earth's centre.

The earth would be turned into a body like the moon, if the electric power within were to explode and break through the bonds which hold it confined; it would, as it were, become a floating, lifeless body, if by the burning and wasting influence of the liberated and enkindled elements it were to be deprived of the momentum, of the spring, that gave it diurnal rotation, that imparted warmth and fertility to the whole frame, the same as blood does to that of living beings.

If the veil of the earth were removed, or to collapse, the moon would come down upon it. Or, the moon would burn and be blown into a thousand fragments, if, like the earth, she were inflated with the electric fluid.

It is the same with the earth itself, as well as with all the planets. If my theory be true, that, when the sun was placed in the heavens by the Eternal Word, the planets began to rise and to ride on the expanding matter of his atmosphere, or were wafted into it by the same omnipotent power: so, by the contraction or removal of this attenuated atmosphere, they also would be affected. Thus, when at the last day, with the breath of Omnipotence, God will blow out the light of the world, a universal chill and contraction will take the place of warmth, expansion, and life ; and the earth having in the meanwhile outgrown in bulk, its crust become thicker and augmented in density, and thus the electric life-blood within more and more shut up by impeded perspiration, there will come to pass, according to the laws He laid down, what the Creator of heaven and earth Himself predicted : " The sun shall be darkened, and the moon shall not give her light, and the stars shall fall from heaven, and the powers of heaven shall be moved." And from the darkness and universal chill, and the closing of the earth's pores, the evaporation of electricity of the earth will be stopped, and the confined element,—throwing the bowels of the earth into convulsions, often of a sudden rolling mountains of water over devoted cities and places, opening the abyss of the deep and the yawning chasm on land, to engulf the ship and swallow up the habitations of man,—will then, with terrific thunder burst forth in millions of volcanoes on land and water, with the rapidity of thought search the intestines of the sea from pole to pole, in' length and in breadth, set on fire the coalbeds of the earth and its interior reservoirs of gas and lakes of petroleum, and from its unfathomable depth throw up heaven-high millions of surging fountains ; fearfully lighting up darkness, it will pour out, from its newly-opened craters and outlets, electric fire, burning oil, and molten lava, in streams so incessant and enormous, that the ocean itself will evaporate like a drop of water on a heated plate of iron ; countless meteors, as if in anger, will fall from heaven upon the earth to aid in its destruction ; terrible flashes of lightning in fitful embrace will fill and dart about in the air, and threatening auroras and clouds pregnant with wrath girt our sphere ; and above all this the sign of redemption will be blazing before the Son of Man in His glory, a welcome to the blessed who loved, kissed, and dropped tears upon it here, and a horror to the wicked who could not bear its sight; and the light of the sun and moon and the planets being extinguished, the whole universe is invited to witness the judgment of the justice of God; and then indeed, in the words of St. Peter, " the earth will be consumed by fire," and at the approach of all this " there

truly will be distress of nations by reason of the confusion of the roaring of the sea and the waves."

But, we need not go the end of the World for the confirmation of the operation of these laws. A slight, but awful, instance of a mere, as it were, momentary, obscuration of the sun, was witnessed already at the crucifixion of our Lord. When, at that momentous sacrifice of the atonement, the Heavenly Father threw a veil of mourning for His only-begotten Son over the orb of light; when horror-struck at the awful crime which was being committed, the sun was paralysed, its life blood chilled and the evaporation of electricity stopped : all became dark, cold and dismal, the earth expanded and trembled, its bosom heaved from sorrow, rocks were rent, and the waters of the earth threatened to overflow the land. And though then no electric fire issued from the rent on Calvary, yet later, about the same spot, at the right time and place, it broke out from the quaking earth, in fiery balls and flashes, when there was question of destroying the works for the third temple which was being built at Jerusalem under Julian the Apostate, with the avowed object of nullifying the prophecy of Christ, and taking away from His Church the foundation which on Calvary He had laid and sealed with His blood.

There are many subjects connected with my new theory of the Solar System, which would require particular notice, and many which especially and more favourably would elucidate it; but, as an amateur in Natural Philosophy, I have neither the facilities, nor leisure, to prosecute this agreeable study with that continuity and perseverance which it demands.

The work, as it is, labours under the disadvantage of a fortuitous division into chapters, and the absence of proper classification; but this must be excused from want of time, having been commenced without any previous knowledge of astronomy, therefore without any pre-conceived plan, written by degrees, mostly at long intervals, adding here and there, and branching off into other subjects closely connected and intimately linked together.

As, moreover, no new theory can be expected to be received as true without proving the old one to be false, and as it is not sufficient to refute an old theory without having a new and better one to substitute in its place: so the pure treatment of my New Solar System was impossible, and of necessity became interspersed and broken into by references to the fallacious system of Newton and his followers.

Therefore, having gone through the stately building, the work of love, beheld its rising columns, taken notice of its light and shade, its polished and decorated as well as of its cruder parts, but, too often interrupted and impeded, as it were, by falling forces, conical motions and eccentric bodies, by the ruins and rubbish of a crumbling and empty universal-gravitation-rookery, it will be well to recapitulate the chief points of my New Theory, to re-call to the reader the leading traits of the astronomical monument, in order to impress upon his mind the sound foundation and clear and striking outlines of a new construction, in its imposing grandeur and majestic simplicity an edifice worthy of the

Queen of the Sciences to possess, gathering round her throne hosts of wise men of the West, as once BUT THREE came from the East, led by the stars of heaven to do homage to the Star of the Sea, bright in the glory of the SUN AND THE SON OF GOD HIMSELF.

God, the self-existing, self-acting, eternal Being, created matter to which He imparted the property of motion.

Motion is synonymous with force; it is an effect of, and dependent upon, matter. Therefore, motion, or force, cannot come, cannot act, of itself, cannot be self-acting, being dependent upon matter.

All changes in nature are due to the motion of matter, and motion, in all cases, is due to expansion and contraction, to attraction and repulsion, the result of light and darkness, of heat and cold, the product of positive and negative electricity, in their mutual relation the cause of force and motion.

Light and darkness are latent in every created substance, i.e. every substance is electric, and as a matter of fact and theory, has its poles of attraction and repulsion, the cause of motion.

Chaos itself was endowed with motion, both as a whole and in every individual atom.

The seed of the heavenly hosts, visible and invisible, having been cast by God and the law given, worlds began to grow and to enlarge by feeding upon, by precipitation from, chaos,—as solid bodies by degrees to separate from imponderable matter, and system to separate from system, as grain separates from grain, and cell from cell.

The central orb of every system, at first dark, but later bursting out into light, is clothed in a vast envelope of imponderable matter. Within this envelope the planets and their satellites are stationed in rising gradation above the equator of their luminous centre; and this centre, carrying the whole weight of the system, revolves round its own central axis in the universe of worlds and systems, all wonderfully poised and balanced against each other in the electric globe of creation.

The earth,—and the other celestial bodies of our solar system, —nourished by and grown forth from chaos as we see it now, also surrounded with a vast envelope of air and gases, floats and rotates forward upon its equator,—the balance-keeper between its poles,—in the atmosphere of the sun, in the ethereal and lofty ocean of our system, in its proper level according to size and weight or density, bearing in mind, that the envelope of the sun becomes more and more dilated as it reaches the end of the system, and that the envelopes of the planets also become more and more expanded or rarefied as they recede from the solid nucleus until they terminate, the moons at the same time floating in their re-spective levels in the atmospheres of the planets, the same as the planets do in the atmosphere of the sun.

The sun and planets, like all that grows and moves, must of necessity have an interior organization to have grown, to be possessed of, and to evolve, heat; they MUST AND DO POSSESS an electric constitution, the cause of attraction, repulsion, motion, and rotation.

As by polarization system is balanced by system, and system acts upon system, with mutual breaks of contact: so, without disturbing each other, all the millions of systems move onward in inconceivable grandeur and harmony within the sphere of creation; and as in time each system revolved, so its planets were gently borne along and carried forward by their common centre and bond of union and attraction.

Before suns were enkindled this motion was slow, though pro-gressive with the progressing development and maturity of the heavenly bodies; when, however, they had broken out into radiant light, all became more active with life; the light of the sun acted upon the electric element within his own solid body, the

Y

electric blood coursed more quickly, and the contact of radiation from the solar photosphere upon the body below being broken by the spots of the sun, the body itself, the centre of the system, with all that it carried, revolved with greater rapidity; and the sun shining upon the earth and expanding its electric element within, disturbing its balance and causing a constant current more strongly to flow from midnight to mid-day, turned it also more rapidly round its own axis, whilst the moon partook of the accelerated motion of her primary.

The revolution of the heavenly bodies round their own axis is but the revolution of the heated vessel suspended before the fire.

Moons, being without a living interior organization, simply float in the atmospheres of the planets without revolving round their own axis.

Our moon floats round the earth above the equator, under the ecliptic, where there are only low spring tides, or, in fact, no tides at all. In her progress round the earth she displaces the atmosphere on which she swims, it accumulates on each side of the moon, on each side of the equator, weighs and mediately presses upon the ocean below, and drives the water towards the poles, thus causing the tides by making the water run up the shore wherever it has no room to expand, the second flood during the twenty-four hours of the day being but a reaction of the first.

The exterior envelopes of the planets perform the office of reflector and burning-lens, throwing the light and heat of the sun in a focus upon the body beneath.

The seasons are caused by the oscillation of the earth, at one time the south-pole inclining towards the sun, and at another the north-pole; and the burning-lens of our planet being thus shifted, as it were, we have it cold in winter and see the sun larger, and warm in summer whilst seeing him smaller.

All other phenomena in nature are due to electric attraction and repulsion, contraction and expansion, to motion, caused by electricity in its two-fold nature.

Thus the beautiful arrangement of the universe resolves itself into Systemal Repulsion, and Solar and Planetary Gravitation, the clothing of the heavenly bodies within various spheres of air and gases, the inflation of their interior with negative, and the investiture of their exterior with positive electricity.

Of this simplicity of the creation, I trust, to have convinced the unbiassed reader; if, however, I should still have left some doubt upon his mind, he will nevertheless agree with me, that my conception of the solar system is more simple, more consistent, and consequently more natural and true than the theory hitherto received.

Whether, as balls of gas, or globes of fire, God projected the heavenly bodies in straight lines and curved their paths by the reputed law of universal gravitation; whether at once they were created as they are and dropped by His Omnipotent Power and Will into their respective centres and orbits, with the unquenchable electric fire within, surrounded and kept enclosed by a vault, by a crust entwined of gold and silver, copper and iron, coal and layers of stone and clay of every description, harmoniously bound and blended together, but subject to superficial changes, according to the designs of God in regard to man; whether, by electricity moulded out of and separated from chaos, they first formed little specks of matter in the blank of space, and, themselves conceived by the thought of the Omnipotent, suns gave birth to planets and planets to moons; or whether, as I believe, from the seed of worlds cast into chaos by God, suns and planets with their systems grew into existence, comets rose up from the suns and moons from the planets, as meteors still rise from the earth:—no mind can fathom the First Cause, how from the depth of His Godhead the Almighty called everything into existence; the Eternal Wisdom spoke, and, in the language of a divine, chaos heard tremblingly the voice of its Creator, and separated the newly-ordained bodies. At His call, not only our sun, but suns without number, enkindled in the night of the spheres of creation, and to millions of stars and planets He assigned their course. HIS WILL became order, beauty and wonder throughout the whole sweep of creation; suns cast their light through the darkness, and adoring worlds coursed around them in their illuminated paths; the whole new creation became a MIRROR OF HIS ETERNAL GODHEAD, AND THE MANIFESTATION OF HIS SELF-EXISTING GLORY.

SEQUEL
TO
DEDICATION AND PREFACE.

THE LAWS OF GRACE AND JUSTICE AND THE LAWS OF NATURE.

MY DEAR WIFE, the favoured object of the miracle spoken of in my dedication, gently breathed her last on Wednesday, the 30th of November, 1859, about a quarter past 10 o'Clock in the morning. On the 26th, the previous Saturday evening, she became very feverish and was attacked with a slight inflammation of the lungs which caused her death; but the infant which she bore for nearly nine months had already ceased to live within her the day before she died.

Under these circumstances, fearing a speedy decomposition, she was put into her coffin in the evening; but the truly angelic expression which the departing spirit, like a last seal, had impressed upon her countenance, did not change; she retained the complexion almost as of one sleeping, and no offensive vapour escaped from her body. An odour, not unpleasant, was certainly perceptible on the second, and on the four or five succeeding days; every object in the room was examined, and we came to the conclusion that this odour proceeded from the wax-lights we kept burning, the deceased being entirely free from smell.

Not being able to separate from so lovely a corpse, and moreover under the apprehension lest she be in a trance, the solemn Requiem-mass and her burial had to be postponed, and at last indefinitely so, after she had lain for six days.

Thus we kept watching and praying by her side, looking upon her more in joy than in sorrow; friends and strangers came to see her, but no one had ever beheld anything like it, no one could easily quit the lovely picture, all felt a kind of attraction which

kept them gazing at her heavenly countenance. And still no
corruption came, notwithstanding all the elements of corruption
within her, as nothing whatever passed from her body. By
degrees her eyes sank back into their sockets, and afforded the
only evidence of positive change; but still, nothing escaped, no
sign of decomposition appeared, though since her death three or
four strong changes of temperature and weather had taken place,
and though during the last four days before her burial the gas-
stove in her room had been lighted, keeping up a temperature of
52 degrees.

On the 16th of December I received the following letter from
our physician, eminent in his profession :

" 48, FINSBURY SQUARE, 16TH DEC., 1859.

" MY DEAR MR. ADOLPH,

    " I am sorry we had not the good fortune to meet you last night. It
was my intention to have sent you word at what time a medical friend would
call along with me. I found however that Dr. Letheby whom I had in view was
absent in Southampton. I fixed upon him on account of his great experience
as chemist, medical jurist, and officer of health, &c. I subsequently found that
opportunity would present of taking with me Mr. Wells, editor of the *Medical
Times and Gazette*, &c., of Upper Grosvenor Street. He agreed with me that
he had not witnessed anything in the way of preservation of the body like that
presented in the cases of your dear wife. From what I saw last night, however,
it is plain that a change has at length occurred. I am not sure, however, but
that the change of colour may not be attributed to the frost rather than to
DECOMPOSITION. It seems to me in fact that the process going on is one of
simple DESICCATION rather than decomposition.

" I am, &c.,   " ARTHUR LEARED."

On reading this, that the body of my dear wife was under
process of drying up, and no smell having been perceptible in the
room since the first four or five days after her death, we could not
but come to the conclusion, that the odour first perceived, at times
very strongly and at times not at all, of a sweetish nature,
differently compared by different visitors and by the inmates of

the house, must have come from the departed whilst the body was fresh, as a signal mark of God's favour.

No further change having taken place than the one indicated in the above letter, we took her to her last resting place on the 20th December, 1859, the twenty-first day after her death.

On the 21st of January, 1860, the following letter appeared in the *Medical Times and Gazette* :—

Qr. No. 404. SIGNS OF DEATH.

"Can any of your readers afford information with regard to the following circumstance?—A lady entering upon the ninth* month of pregnancy died of pneumonia. All the other phenomena of death ensued, except that the colour of the face was unusually life-like. On the fifteenth day from that of death there was not the least cadaverous odour from the corpse, nor had its appearance much altered, and it was only on the sixteenth day that the lips darkened. The temperature of the atmosphere had undergone many changes during the time mentioned, but although there had been frost for a short period, the weather was in general damp and not very cold.

"I am, &c.   " SIGMA."

To this letter there was no reply.

Why thus in the case of my dear wife, still bearing the infant, with a change of cold, warm, wet and frosty weather against her, with the mild temperature of her room in the end—the law of nature should have been reversed from decomposition into desiccation, was and remains a mystery to the medical men who saw her ; and it will continue so to be until they look for higher laws than those of nature.

To me, her state in death is but a consequence of her life, which was moderate in her meals, fervent in her devotion, generous, noble, and ardent to a fault in her feelings, with a most exalted PURITY OF MIND, and an EXTRAORDINARY PATIENCE IN SUFFERING, all the more pleasing to God, as she was also of a warm temper ; and as once before in life he favoured her with a special mark of

---

* She was very near her confinement, most likely but a few days.

His love, so, no doubt, as a reward for her patient suffering during
seven years in the prime of life, He also bestowed upon her the
privilege of her body being so lovely in death, and, thus far,
preserved from corruption.

In respect to the above it is laid down beautifully by Görres, in
his *Christian Mystic*, how the laws of grace and justice not only
go hand-in-hand with, but also supersede, the laws of nature.

A person living only for this world, satisfying every appetite,
giving himself up only to eating and drinking, and indulging in
vice of one kind or another, more and more darkens, and thereby
animalizes his soul; because his body, being of earth, and living
only by and for the earth and all its lower elements, the soul in
darkness is more and more carried away by the inclinations of the
body, and with it drawn into and incorporated with the earth to
which the body belongs.

A stage further, all spiritual life is drowned, smothered, extinguished, the body indulges in every animal propensity; and,
clothed in mortal sins, and directed by the spirit of evil, the soul
with its body is immersed in every degrading passion; this will
bring disgusting disease, contagion and rottenness, of which the
diseased and abandoned soul will be the governing agency; and like
water meeting water, oil meeting oil, like other kindred substances
meeting and amalgamating, like the wicked holding to the wicked:
so infection of the body will induce infection in the kindred body,
corruption of the soul will induce corruption in the kindred
soul; and when sin and the spirit of evil have arrived at culmination, then the bridge is thrown by which the worst evils pass
into the human frame; resistance is removed, the conductor is
formed by which the infernal demon himself, the origin of evil,
the liar from the beginning, strikes into, cohabits with, and
possesses the soul, possesses the RUINED image of God, and thus
rules the body; and owning both, by guilt or consent, he invents lies

without limit, and often performs false signs and wonders to deceive
the good and retain the wicked, as amply recorded in Holy Writ,
in sacred and profane history, and as witnessed in our own days.

Corrupt and hideous as are such souls, so in proportion will be
their bodies; repulsive in life they become loathsome in death,
and quickly decay and dissolve in the foul flame and vapours of
the infernal regions, as it were, in which they lived.

A person living an ascetic, an abstemious pious life, by degrees
restraining every passion which the human heart is prone to,
subjecting himself to abnegation of every kind, doing violence to
himself and not only rubbing off the film and rust of his soul and
driving off darkness, but by degrees practising every virtue and
letting in more and more light, for, every good deed, and more
particularly every heroic virtue, is an additional lens, a magnifying glass, around the soul, bringing nearer and nearer to it the
vision of God, the Light Eternal: such a person, keeping the
body under the rigourous control of the soul, drawing his chief
nourishment from bread and wine and kindred substances, and being
ruled and inspired therein by true religion and its divine author,
spiritualizes the body, the same as the other animalizes and
degrades the soul.

A stage further, a spiritual instead of a spirituous, a divine fire
instead of an earthly, animal, and hellish fire, pervading the body,
that body, refined, clarified and spiritualized, will be under the
lofty control of the soul and be governed by it; and as the soul
progresses in light, life, and power, derived from a pure spiritual
source, by degrees clothing itself with almost every cardinal virtue :
so the body and all its senses and organs will become partakers of
the interior light and life, as evidenced in the lives of the saints,
in whom even the birds of the air, the beasts of the field and the
forest,* and the fishes in the water, have recognised again the

* As particularly evidenced in the Roman Amphitheatre.

RESTORED image of God, joyfully hovering about the likeness of their Creator, listening and doing homage to it, the same as in paradise they did to Adam and Eve. The wicked also recognise the image of God, but, children of the serpent, like their adopted parent they lie in wait to destroy it.

Hence, whatever there is sublime in morality, virtue, and godliness, will take up its abode in such a soul and body ; kindred will pass over to kindred, and in the end the laws of nature— having at first gone hand-in-hand with those of grace—will be subject to such souls and bodies and apparently reversed.

As has been said before, every good work and act of divine love is a magnet which attracts and draws down the grace of God; and when thus the measure is full, when the spirit of sanctity in body and soul has reached the point of culmination, then every obstacle has been removed, the bridge thrown, the conductor raised, the channel completed by which God Himself enters and performs the most wonderful acts in His servants, until He strikes into, unites Himself with, and possesses the soul itself.

"Blessed are the clean of heart for they shall see God;" but here the clean soul does not only see God, He actually removes it in ecstacy from the consciousness of this world, as He did St. Paul, and holds it in embrace according to His own measure. He also possesses the body, and with the soul often draws it to Himself so as to raise it above the earth, temporarily suspending the law of terrestrial gravitation, as He suspended it when He —and many faithful followers after Him—walked over the water; and when our bodies shall have been transfigured and glorified, gravitation will be suspended for good, the same as He set it aside when He ascended into heaven.

Rooting and reposing in God, the branch truly attached to the Vine, the soul and body, when in that state, are directly

nourished by God, and earthly food is no longer a necessary of
life; it is then that He verifies what He says: "My BODY is
meat indeed and my BLOOD is drink indeed." For, it is no
longer the THOUGHT of God that is embodied in every created
object, not the BLESSING incorporated in the manna which rained
down in the desert: but it is God, the GOD-MAN HIMSELF,
who, by His own act and will, clothed HIMSELF into the forms
of bread and wine; and the bodies of souls thus nourished by
the LIVING BREAD that came down from heaven, though often
diseased and heavily afflicted by illness, having been purified by
the spiritual, the heavenly fire that pervaded their frame, by
the heavenly atmosphere in which they breathed, often not only
do not decay, but emit the odour of sanctity, as it is called, to
rise up and commingle with the incense of nature and the incense
and the prayers of the Church, to be carried by angels to the
throne of the Lamb.

Such, in a few words, is the picture drawn by the soul and
body, by free will and consent covered all over with the disgusting ulcers of vice, and—like the falling stone to the earth—
gravitating to the origin of evil, to the horrible embrace of the
serpent; and of that soul and body, with free-willed self-denial,
violence, and aspiration, radiant in the halo of virtue, detaching
itself from the earth, and like the balloon rising higher and
higher in the scale of perfection, and persevering in the ascent,
at last is attracted by the centre of All Light, and rests in the
bosom of the Eternal Father.

To each of these two conditions the laws of grace and justice are
attached and go hand-in-hand, and finally supersede—already
on earth—the laws of nature; and as the followers and the seed
of the serpent, often like splendid meteors, turned into destructive masses, are gravitating to the earth but to be for ever
buried in oblivion in its bowels, or bursting and leaving nothing

behind but an odious smell; or falling from their height, vanishing like a shooting star, blotted out from the catalogue of pre-
supposed orbs, leaving no trace but the memory and regret
of an extinguished luminary: so the seed of the woman, in
numberless hosts, rise up from the garden. of the Church to
adorn the firmament of the Kingdom of God and cluster round
the Star of the Sea; and as the earth turns round its axis every
day and courses around the sun every year: so each day and
every year brings round to our contemplation the splendid array
of the zodiac and all the other constellations of the Church;
the nebulæ of time disperse, and, never weary, with the telescope
and the microscope alike sweeping the world and searching and
seeking the seed of heaven, the Watchman, triple-crowned, standing
on the elevated Observatory of the Universal Lighthouse and
Beacon on the Rock of Ages, every day still and every year
reveals to us new stars to follow, stars that long ago shed their
light upon the Church, but were hid from our vision; and ever
and anon to the world unexpected, new comets swing aloft in
splendour, coming and going, re-invigorating the life of the
system by the mysterious power derived from the centre, and
transmitting the record of their lustre from generation to generation. Opposed, however, in their progress towards the fountain
of light, they only more lay bare their bosom to the influence
of its heavenly glow, and with renewed and persevering vigour
pursue and hasten their impetuous course to the attracting furnace
of Eternal Love which set their hearts on fire; and the more
they are resisted the more gloriously they shine.

And thus both these heavenly systems go on in infallible order
and harmony, each one undergoing a process of purification and
elevation in its members, until both shall have been transubstan-
tiated into superior existence by their common Author and God.

INDEX.